The Four-Way Test
of the things we think, say or do

1. Is it the **TRUTH**?
2. Is it **FAIR** to all concerned?
3. Will it build **GOODWILL** and better **FRIENDSHIPS**?
4. Will it be **BENEFICIAL** to all concerned?

This dictionary belongs to:

Donated by the
**Westbury-Carle Place
Rotary Club**

We want to thank the individuals who made contributions to this dictionary by sending in their suggestions and comments. They made this edition better than the last one.

M. Ansari, Dr. Richard Barshinger, John Beiter, Gael E. Bohannon, Caroline Brost, Dahmen Brown, David Carr, Nick Cowie, Laurie David, Stanley Dixon, Evan Eckersley, Phil Ferrara, Ryan Fiorillo, Mary Gallagher, Ali Gelineau, Stella Guenther, Chris Hope, Elizabeth Jewell, Arman Kandarian, Timothy Knight, Ruth Kent, Mary Lavarnway, Ann Macaluso, Kaitlin MacAulay, Phyllis Makinster, Richard McArthur, Larry Margolis, Claude and Isaac Mertzenich, Colista Moore, Joan Morgan, Constance Mote, Mark Nabb, Wanda Parker, Judith Pendleton, Lianna Prill, Sasha Sagan, Storrs Warinner, Don White, Dorothy Willard and Betty Williford.

We are grateful for your ideas and contributions. Please write to us, to let us know what you would like to see in the next edition of the dictionary.

www.TheDictionaryProject.com

Cover designed by Darren Ouellette

A STUDENT'S DICTIONARY
& Gazetteer, 9th edition

Printed and made in the U.S.A.
ISBN 0-9745292-8-1

ABCDEFGHIJKLMNOPQRSTUVWXYZ

Table of Contents

abcdefghijklmnopqrstuvwxyz

ABCDEFGHIJKLMNOPQRSTUVWXYZ

It is truly amazing that out of twenty-six letters there can be thousands of combinations of letters to make up the words in our vocabulary.

This little dictionary touches on between 32,000 and 33,000 of the most commonly used words.

Words have power and can govern our lives. Whether talking to a friend or a family member, making a speech, or in the world of business, it is to your advantage to have a storehouse of words to convey your thoughts to whoever is listening. It builds strong self-confidence.

As you carry this book with you through your school years and use it, your vocabulary will grow along with you. New technical words and phrases will be abreast of the times, but the basics will live on to be uttered by you and upcoming generations.

This dictionary is given to you with best wishes for a word-endowed future.

Colista Moore
Rogue River, OR

Aa Bb Cc Dd Ee Ff Gg Hh Ii Jj Kk
Ll Mm Nn Oo Pp Qq Rr Ss Tt Uu
Vv Ww Xx Yy Zz

2

NINE PARTS OF SPEECH

Three little words you often see,
Are Articles – *a, an,* and *the.*
A Noun's the name of anything,
As *school*, or *garden*, *hoop*, or *swing.*
Adjectives tell the kind of Noun,
As *great, small, pretty, white,* or *brown.*
Instead of Nouns the Pronouns stand-
Her head, *his* face, *your* arm, *my* hand.
Verbs tell of something to be done –
To *read, count, sing, laugh, jump,* or *run.*
How things are done, the Adverbs tell,
As *slowly, quickly, ill,* or *well.*
Conjunctions join the words together
As man *and* woman, wind *or* weather.
The Preposition stands before
A noun, as *in*, or *through* a door.
The Interjection shows surprise,
As *oh!* How pretty – *ah!* How wise.
The whole are called Nine Parts of Speech,
Which reading, writing, speaking teach.

Dr. Chase , 1867

3

Punctuation

Punctuation marks and pauses are used to separate written or printed information into sentences and parts of sentences to assist the reader in understanding the meaning of the writer.

A **Period** (.) is placed at the end of a sentence; as,
> *I admit I made a mistake.*
> Or is used after an abbreviation; as
> *Dr. Johnson*   *Jan. 1, 2004*

A **question mark** (?) indicates a question; as
> *What is your opinion?*

An **Exclamation Point** (!) denotes humor or strong feeling; as,
> *Facing it—always facing it – that's the way to get through it. Face it!* Joseph Conrad, 1924

The **Comma** (,), **Semicolon** (;), and **Colon** (:) mark grammatical division in a sentence; as,
> *There are three genders: feminine, masculine, and neuter; in language they are referred to as she, he, and it, respectively.*

A **comma** is used to list items in a series. The **colon** may separate two clauses when the second explains the first, and is also used before an explanatory list. The **semicolon** is used to separate two independent clauses when there is no conjunction. Sometimes it is called a weak period.

A **Hyphen** (-) is used between syllables in a word divided at the end of a line; as,
> *knowl-edge*   *pow-er*
> And between the parts of a compound word, as
> *self-improvement*   *absent-minded*

The **Dash (--)** is used where there is a sudden break or pause in a sentence; as,

Individual commitment to a group effort-- that is what makes a team work, a company work, a society work, a civilization work.
    Vince Lombardi, 1970

Quotation marks (" ") denote the words of another; as

"My name is Margalo," said the little bird softly in a musical voice. "I come from fields once tall with wheat, from pastures deep in fern and thistle; I come from vales of meadow-sweet, and I love to whistle." Stuart Little, E.B. White, 1985

An **apostrophe (')** denotes that a letter or letters are left out; as,

I'd be better off if I had gone to school longer, and so will you.
    Joe Louis

And is also used to show ownership; as

There are spaces between our fingers so that another person's fingers can fill them in. Anonymous

Pronunciation Key

a̧,	for	ŏ,	as in	want	y̆,	for ĭ, as in	chlorophyll
ê,	"	â,	"	tear	c,	" k, "	character
ȩ,	"	ā,	"	eight	ç,	" s, "	city
ï,	"	ē,	"	police	çh,	" sh, "	chivalry
ĭ,	"	ĕ,	"	sir	eh,	" k, "	chorus
ô,	"	ŭ,	"	ton	ġ,	" j, "	general
o̧,	"	o͞o,	"	do	n̦,	" ng, "	sang
o̧,	"	o͝o,	"	book	ș,	" z, "	hands
ô,	".	a̧,	"	pork	s,	" sh, "	sure
õ,	"	û,	"	worm	x,	" ḡz, "	example
u̧,	"	o͝o,	"	pull	gh,	" f, "	laugh
u̧,	"	o͞o,	"	rule	ph,	" f, "	phone
y̆,	"	ĭ,	"	try	qu,	" k. "	plaque
			qu, for kw, as in	quiet			

5

Guide to the Dictionary

entry word **re-spect** [rĭ-spĕkt′] *noun*, 1) an aspect
2) consideration for someone, *verb*, 3) to have a good opinion of someone, to look up to, to hold in esteem

spelling **cour-te-ous** [kûr′tē-əs] *adjective*, showing good manners, polite

meaning **com-pas-sion** [kəm-păsh′ən] *noun*, kindness, tenderness, pity, sympathy for suffering

more than one meaning **char-ac-ter** [kăr′ək-tər] *noun*
1) the moral makeup of a person
2) a person in a book or movie or play
3) a symbol, sign, or written letter used in a language

part of speech **hon-or** [ŏn′ər] *noun*, 1) fame or a rank of high esteem, *verb*, 2) to hold in high respect 3) to keep an agreement

no plural **cour-age** [kûr′ĭj] *noun, no plural*, the strength to face danger when overwhelmed by fear

guide words

persevere   phrase

per-se-vere [pûr′sə-vîr′] *verb*, to work to accomplish a goal despite difficulty

per-sist [pər-sĭst′] *verb*, to keep doing something, to persevere

phrase [frāz] *noun*, 1) a part of a sentence 2) an expression, a saying, *verb*, 3) to express in words or a particular way

syllabic pronunciation **in-teg-ri-ty** [ĭn-tĕg′rĭ-tē] *noun*, honesty, uprighteousness

phonetic spelling **in-spire** [ĭn-spīr′] *verb*, to stimulate to think and to act, to influence

6

a [ā] *article*, used to indicate one before a noun

a-ba-cus [ăb′ə-kəs] *noun*, a frame holding parallel rods with beads, used for doing arithmetic

ab-a-lo-ne [ăb′ə-lō′nē] *noun*, a shell sea mollusk animal whose shell is lined with mother-of-pearl

a-ban-don [ə-băn′dən] *verb*, to leave or give up completely, to walk away from

a-bate [ə-bāt′] *verb*, to gradually become less in amount or degree

ab-bre-vi-ate [ə-brē′vē-āt′] *verb*, to shorten, to cut down, to contract

ab-di-cation [ăb′dĭ-kā-shŭn] *noun*, a surrender, abandonment

ab-do-men [ăb′də-mən] *noun*, the digestive system in the body, belly

ab-duct [ăb-dŭkt′] *verb*, to kidnap

ab-er-ra-tion [ăb′ə-rā′shən] *noun*, a departure from what is right, correct or natural

a-bet [ə-bĕt′] *verb*, to encourage or assist, especially in wrongdoing

ab-hor [ăb-hôr′] *verb*, to withdraw in disgust or horror

a-bide [ə-bīd′] *verb*, 1) to put up with, tolerate 2) to wait patiently for 3) to withstand

a-bil-i-ty [ə-bĭl′ĭ-tē] *noun*, 1) capability, power or knowledge 2) skill, expertise

a-ble [ā′bəl] *adjective*, having the strength or knowledge to do something

ab-nor-mal [ăb-nôr′məl] *adjective*, irregular, unusual, peculiar

a-board [ə-bôrd′] *preposition*, on, in, upon (a ship)

a-bode [ə-bōd′] *noun*, home; a place of residence; a habitation

a-bol-ish [ə-bŏl′ĭsh] *verb*, to do away with entirely

ab-o-li-tion [ăb′ə-lĭsh′ən] *noun*, complete destruction

a-bom-i-na-ble [ə-bŏm′ə-nə-bəl] *adjective*, detestable, odious

ab-o-rig-i-ne [ăb′ə-rĭj′ə-nē] *noun*, a primitive native

a-bor-tive [ə-bôr′tĭv] *adjective*, unsuccessful, fruitless, unfinished

a-bound [ə-bound′] *verb*, to overflow, to prevail, to exist in large numbers

a-bout [ə-bout′] *adverb*, 1) concerning 2) approximately 3) here or there, everywhere

a-bove [ə-bŭv′] *adverb*, 1) at a higher level, *preposition*, 2) over, in a higher place

a-bra-sion [ə-brā′zhən] *noun*, a wearing away or rubbing that causes scratches

a-bridge [ə-brĭj′] *verb*, to shorten by using fewer words

a-bridg-ment [ə-brĭj′mənt] *noun*, abbreviation, shortening

a-broad [ə-brôd′] *adverb*, in a foreign country, overseas

ab-rupt [ə-brŭpt′] *adjective*, happening without warning, suddenly

ab-scess [ăb′sĕs′] *noun*, an accumulation of pus due to an infection from bacteria

ab-sence [ăb′səns] *noun*, 1)the state of being away from someplace or someone 2) lack

ab-sent [ăb′sənt] *adjective*, missing, not present, unavailable

ab-sent-mind-ed [ăb′səntmīn′dĭd] *adjective*, forgetful, not paying attention

ab-so-lute [ăb′sə-lōot′] *adjective*, complete, perfect, total

ab-solve [əb-zŏlv′, -sŏlv′] *verb*, to pardon, to forgive

ab-sorb [əb-sôrb′, -zôrb′] *verb*, 1) to soak up a fluid or moisture 2) to hold someone's attention 3) to dissolve a debt completely and accept the loss 4) to assimilate information

ab-sorb-ent [əb-sôr′bənt] *adjective*, absorbing moisture, fluid or light rays

ab-stain [ăb-stān′] *verb*, to deliberately avoid doing something

ab-sti-nence [ăb′stə-nəns] *noun*, the act of depriving oneself of some indulgence, voluntary denial

ab-stract [ăb-străkt′] *adjective*, difficult to describe or qualify

ab-surd [əb-sûrd′, -zûrd′] *adjective*, outrageous, silly, untrue

ab-surd-i-ty [əb-sûrd′ĭt ē] *noun*, that which is ridiculous or unreasonable

a-bun-dance [ə-bŭn′dəns] *noun*, large quantity, plenty

a-bun-dant [ə-bŭn′dənt] *adjective*, plentiful, more than enough

a-buse [ə-byōo′s] *noun*, 1) hurtful treatment, *verb*[ə-byōoz′], 2) to hurt someone or something intentionally

a-bu-sive [ə-byōo′sĭv, -zĭv] *adjective*, using insulting words or bad treatment

a-but [ə-bŭt′] *verb*, to border upon, to be next to or touching

a-bys-mal [ə-bĭz′məl] *adjective*, 1) immeasurably bad 2) bottomless

a-byss [ə-bĭs′] *noun*, a very deep crack in the earth, a bottomless gulf, a deep chasm

ac-a-dem-ic [ăk′ə-dĕm′ĭk] *adjective*, 1) of or relating to an education in the arts 2) having no practical importance

a-cad-e-my [ə-kăd′ə-mē] *noun*, a school offering specialized training and education

ac-cel-er-a-tion [ăk-sĕl′ə-rā′shən] *noun*, an increase in the speed of something

ac-cent [ăk′sĕnt′] *noun*, 1) the sound and tone of a foreign language, *verb*, 2) to emphasize a sound or letter when pronouncing a word 3) to emphasize something

ac-cept [ăk-sĕpt′] *verb*, 1) to agree with 2) to receive

ac-cep-tance [ăk-sĕp′təns] *noun*, the act of receiving something graciously

ac-cess [ăk′sĕs] *noun*, a way to get to a person, place, or thing

ac-ces-si-ble [ăk-sĕs′ə-bəl] *adjective*, easy to reach or use, approachable

ac-ces-so-ry [ăk-sĕs′ə-rē] *noun*, 1) an accomplice, connected as a subordinate 2) an article or decoration worn to complete an outfit, such as a belt, scarf, hat, etc.

ac-ci-dent [ăk′sĭ-dənt] *noun*, 1) an incident involving bodily injury or death 2) a mistake

ac-claim [ə-klām′] *verb*, to applaud, to announce with great approval

ac-cli-mate [ə-klī′mĭt] *verb*, to change or adjust to a different temperature or environment

ac-com-mo-date [ə-kŏm′ə-dāt′] *verb* 1) to adapt or make fit 2) to do a favor 3) to provide lodging

ac-com-mo-da-tion [ə-kŏm′ə-dā′shən] *noun*, an adaptation, an adjustment made to assist someone

ac-com-pa-ny [ə-kŭm′pə-nē] *verb*, to go with, to join

ac-com-plice [ə-kŏm′plĭs] *noun*, someone who goes along with something, usually something wrong

ac-com-plish [ə-kŏm′plĭsh] *verb*, to finish, to complete a task or reach a goal

ac-com-plish-ment [ə-kŏm′plĭsh-mənt] *noun*, something done, a goal that has been reached, a finished task

ac-cord [ə-kôrd′] *noun*, 1) mutual agreement, *verb*, 2) to be in harmony, to conform

ac-cor-di-on [ə-kôr′dē-ən] *noun*, a musical instrument that is played by fingering a keyboard on one side and squeezing bellows in and out on the other side

ac-cost [ə-kôst′] *verb*, to approach abruptly and speak to first

ac-count [ə-kount′] *noun*, 1) an explanation, a statement of money or inventory, *verb*, 2) to give a reason, to explain

ac-crue [ə-krōō′] *verb*, to be added on to, to increase

ac-cu-mu-late [ə-kyōōm′yə-lāt′] *verb*, to collect, to amass

ac-cu-mu-la-tion [ə-kyōōm′yə-lā′shən] *noun*, gathering a quantity of something

ac-cu-ra-cy [ăk′yər-ə-sē] *noun*, exactness, correctness, precision

ac-cu-rate [ăk′yər-ĭt] *adjective*, having no mistakes

ac-cu-sa-tion [ăk′yōō-zā′shən] *noun*, a claim that someone has done something wrong

ac-cuse [ə-kyōōz′] *verb*, to say someone has committed an offense or crime

ac-cus-tom [ə-kŭs′təm] *verb*, to become familiar with something through use or habit

ace [ās] *noun*, 1) a playing card 2) a fighter pilot who was shot down

ache [āk] *noun*, 1) a painful or sore spot, *verb*, 2) to hurt, to feel pain

a-chieve [ə-chēv′] *verb*, to do what one sets out to do

a-chieve-ment [ə-chēv′mənt] *noun*, something that a person has done to reach a goal

ac-id [ăs′ĭd] *noun*, a chemical substance soluble in water, sour in taste, that reddens litmus paper

ac-knowl-edge [ăk-nŏl′ĭj] *verb*, 1) to look at someone and speak to him or her 2) to admit the truth

ac-knowl-edg-ment [ăk-nŏl′ĭj-mənt] *noun*, 1) admission, recognition of an achievement 2) an expression of thanks

a-corn [ā′kôrn′, ā′kərn] *noun*, the seed of an oak tree

acoustic

a-cous-tic [ə-kōō′stĭk] *adjective*, of
or relating to the ability to carry a
sound without distortion

ac-quaint [ə-kwānt′] *verb*, to cause
to know or be aware of someone
or something

ac-quaint-ance [ə-kwān′təns] *noun*,
a person with whom one is
familiar

ac-qui-esce [ăk′wē-ĕs′] *verb*, to
submit, to consent, to agree

ac-quire [ə-kwīr′] *verb*, to get or to
buy, to gain possession of

ac-qui-si-tion [ăk′wĭ-zĭsh′ən] *noun*,
something gained or acquired

ac-quit [ə-kwĭt′] *verb*, to render a
verdict that an accused person is
not guilty of a crime, to exonerate

ac-quit-tal [ə-kwĭt′l] *noun*, a setting
free from a charge, a finding of
"not guilty"

a-cre [ā′kər] *noun*, a plot of land
measuring 43,560 square feet

ac-ri-mo-ny [ăk′rə-mō′nē] *noun*,
bitterness or caustic temper or
speech, bad feelings

ac-ro-bat [ăk′rə-băt′] *noun*, a person
skilled in tumbling or in using
gymnastics equipment

ac-ro-nym [ăk′rə-nĭm′] *noun*, a
word formed from the initial
letters of a series of words, e.g.
AIDS: *Acquired Immune
Deficiency Syndrome*

a-cross [ə-krôs′, ə-krŏs′] *adverb*,
1) from one side of a place to the
other, *preposition*, 2) on the
opposite side from something

act [ăkt] *noun*, 1) something that has
been done, a deed 2) a part of a
play containing different scenes,

verb, 3) to pretend to be something
or someone 4) to behave 5) to carry
out, to perform

ac-tion [ăk′shən] *noun*, 1) something
done 2) the process of doing

ac-tive [ăk′tĭv] *adjective*, 1) always
busy doing something 2) taking
part

ac-tiv-i-ty [ăk-tĭv′ĭ-tē] *noun*, state of
action, energy, a task

ac-tor [ăk′tər] *noun*, a male or
female performer in a play, a film
or television program

ac-tu-al [ăk′chōō-əl] *adjective*, real,
existing in fact

ac-tu-al-ly [ăk′chōō-əl-ē] *adverb*,
really, truly, in reality

a-cu-i-ty [ə-kyōō′ĭ-tē] *noun*,
keenness, as of thought and vision

a-cute [ə-kyōōt′] *adjective*, 1) sharp,
pointed, keen, sensitive 2)
describing an angle formed by
intersecting lines that is less than
90 degrees

A.D. abbreviation for the Latin
words **Anno Domini**, the time
after the birth of Christ

ad-age [ăd′ĭj] *noun*, a proverb or a
saying with a moral message

ad-a-mant [ăd′ə-mənt] *adjective*,
inflexible, unyielding

a-dapt [ə-dăpt′] *verb*, to change to
fit the environment

ad-ap-ta-tion [ăd′ăp-tā′shən] *noun*,
a change to improve the way
things work

add [ăd] *verb*, 1) to put together with
something else 2) to count

ad-der [ăd′ər] *noun*, a snake with a
poisonous bite, a viper

ad-dict [ə-dĭkt′] *noun*, someone who has a strong habit or a physical despendency, as for drugs

ad-dic-tion [ə-dĭk′shən] *noun*, a dependency, such as a physical craving for a drug

ad-di-tion [ə-dĭsh′ən] *noun*, 1) the act of putting or joining things together 2) an annexation

ad-di-tion-al [ə-dĭsh′ən-əl] *adjective*, added on, extra

ad-dress [ə-drĕs′] *noun*, 1) the location of something, usually the street number, city, and state, *verb*, 2) to speak or write to someone 3) to direct in writing, as in a letter 4) to speak to an audience, as in a speech

a-dept [ə-dĕpt′] *adjective*, proficient, expert, highly skilled

ad-e-quate [ăd′ĭ-kwĭt] *adjective*, just enough, sufficient, all right

ad-here [ăd-hîr′] *verb*, to hold, to be attached, to stick to

ad-her-ent [ăd-hîr′ənt] *adjective*, 1) sticking securely, *noun*, 2) a supporter or follower

ad-he-sive [ăd-hē′sĭv, -zĭv] *noun*, a sticky substance that holds things in place

a-dieu [ə-dyōō′, ə-dōō′] *noun and interjection*, farewell, goodbye

ad in-fi-ni-tum [ăd ĭn′fə-nī′təm] *adverb*, to infinity, endlessly

ad-ja-cent [ə-jā′sənt] *adjective*, next to, lying near or touching

ad-jec-tive [ăj′ĭk-tĭv] *noun*, a word that describes a person, place or thing

ad-join [ə-join′] *verb*, to be next to and connected to something

ad-journ [ə-jûrn′] *verb*, to put off a meeting until later, to postpone

ad-just [ə-jŭst′] *verb*, to reposition or reset something

ad-just-a-ble [ə-jŭst′ə-bəl] *adjective*, capable of being arranged or altered

ad-just-ment [ə-jŭst′mənt] *noun*, a minor modification or change

ad lib [ăd lĭb′] *verb*, to make up words or actions at the moment, not before

ad-min-is-ter [ăd-mĭn′ĭ-stər] *verb*, 1) to look after or tend 2) to distribute, to disperse 3) to manage, to run

ad-min-is-tra-tion [ăd-mĭn′ĭ-strā′shən] *noun*, those who are empowered to make decisions, management

ad-min-is-tra-tor [ăd-mĭn′ĭ-strā′tər] *noun*, a person authorized to carry out certain duties

ad-mi-ra-ble [ăd′mər-ə-bəl] *adjective*, excellent, deserving the highest praise

ad-mi-ral [ăd′mər-əl] *noun*, the highest ranking officer in the navy

ad-mi-ra-tion [ăd′mə-rā′shən] *noun*, appreciation, esteem

ad-mire [ăd-mīr′] *verb*, 1) to look up to someone, to revere 2) to look at with pleasure

ad-mis-si-ble [ăd-mĭs′ə-bəl] *adjective*, that may be allowed

ad-mis-sion [ăd-mĭsh′ən] *noun*, 1) access, admittance 2) acknowledgment, confession

ad-mit [ăd-mĭt′] *verb*, 1) to allow someone to enter 2) to acknowledge as true

a-do [ədōō] *noun*, fuss, excitment

a-do-be [ə-dō′bē] *noun*, 1) a type of clay or the sun-dried bricks made from it, 2) a home made with baked mud, usually in the Southwest

ad-o-les-cent [ăd′l-ĕs′ənt] *noun*, a teenager

a-dopt [ə-dŏpt′] *verb*, 1) to use someone else's idea 2) to make someone a member of your family, legally

a-dore [ə-dôr′, ə-dōr′] *verb*, to like or love very much

a-dorn [ə-dôrn′] *verb*, to decorate, to add beauty, to embellish

a-drift [ə-drĭft′] *adverb*, out in water or space without means of control, drifting

a-dult [ə-dŭlt′] *noun*, a grown person or animal

ad-vance [ăd-văns′] *adjective*, 1) ahead of time, done beforehand, *noun*, 2) a positive change, progress *verb*, 3) to move forward

ad-van-tage [ăd-văn′tĭj] *noun*, something that helps a person

ad-van-ta-geous [ăd′văn-tā′jəs,] *adjective*, profitable, useful, beneficial, favorable

ad-vent [ăd′vĕnt′] *noun*, the arrival of someone or something

ad-ven-ture [ăd-vĕn′chər] *noun*, 1) a bold or hazardous undertaking, 2) an exciting or extraordinary experience

ad-verb [ăd′vûrb] *noun*, a word that tells how, when, or where something is done

ad-ver-sar-y [ăd′vər-sĕr′ē] *noun*, a rival, foe, or opponent

ad-verse [ăd-vûrs′] *adjective*, opposed, hostile

ad-ver-si-ty [ăd-vûr′sĭ-tē] *noun*, hard times, trouble

ad-ver-tise [ăd′vər-tīz′] *verb*, 1) to announce publicly 2) to try to sell something by promoting it

ad-ver-tise-ment [ăd′vər-tīz′mənt] *noun*, a notice or short film about something that is for sale

ad-vice [ăd-vīs′] *noun*, counsel, opinion given for acceptable conduct, guidance

ad-vis-a-ble [ăd-vī′zə-bəl] *adjective*, proper, recommended, prudent

ad-vise [ăd-vīz′] *verb*, to suggest to someone what to do, to inform, to recommend, to counsel

ad-vis-er [ăd-vī′zər] *noun*, one who advises, one who gives counsel or information

ad-vo-cate [ăd′və-k t′] *noun*, 1) one who pleads the cause of another or himself, [ăd′və-kāt′] *verb*, 2) to defend, to urge, to support, to endorse

aer-i-al [âr′ē-əl] *adjective*, pertaining to or inhabiting the air

aer-o-bic [â-rō′bĭk] *adjective*, requiring the presence of free oxygen to live

aer-o-dy-nam-ics [âr′ō-dī-năm′ĭks] *noun*, the science of the movement of flying bodies in air

aer-o-sol [âr′ə-sôl′] *noun*, a gaseous suspension of fine particles, a mist

aer-o-space [âr′ō-spās′] *adjective*, having to do with the atmosphere and space

aes-thet-ic [ĕs-thĕt′ĭk] *adjective*, relating to beauty or the arts

af-fair [ə-fâr′] *noun*, matter, concern, business of any kind

af-fect [ə-fĕkt′] *verb*, 1) to act upon to bring about a change 2) to move the feelings of someone 3) to pretend to do something

af-fec-ta-tion [ăf′ĕk-tā′shən] *noun*, insincerity, pretense

af-fec-tion [ə-fĕk′shən] *noun*, tenderness, warmth, love, devotion, kindness

af-fec-tion-ate [ə-fĕk′shə-nĭt] *adjective*, feeling or showing love

af-fec-tion-ate-ly [ə-fĕk′shə-nĭt-lē] *adverb*, showing love and care

af-fi-da-vit [ăf′ĭ-dā′vĭt] *noun*, a written declaration, an oath

af-fil-i-ate [ə-fĭl′ē-āt′] *verb*, to closely connect or associate with someone or groups of people

af-fil-i-a-tion [ə-fĭl-ē-ā′shən] *noun*, an association, a connection

af-fin-i-ty [ə-fĭn′ĭ-tē] *noun*, a feeling of attraction to a person or thing

af-firm [ə-fûrm′] *verb*, to approve, to confirm, to endorse

af-fir-ma-tion [ăf′ər-mā′shən] *noun*, a solemn pledge

af-firm-a-tive [ə-fûr′mə-tĭv] *adjective*, 1) being positive and helpful, *noun*, 2) a word or phrase expressing a positive reply

af-flict [ə-flĭkt′] *verb*, to hurt, to torment, to cause pain

af-flic-tion [ə-flĭk′shən] *noun*, state or cause of distress, pain or misery

af-flu-ence [ăf′lōō-əns] *noun*, an abundant supply, wealth

af-ford [ə-fôrd′] *verb*, to have enough money to buy something

a-float [ə-flōt′] *adjective*, on or above the water, buoyant

a-fraid [ə-frād′] *adjective*, scared

Af-ri-ca [ăf′rĭ-kə] *noun*, one of the seven continents, surrounded by the Atlantic and Indian Oceans, containing 12.7% of the world's population

af-ter [ăf′tər] *adverb*, 1) following, next, *preposition*, 2) behind

af-ter-math [ăf′tər-măth′] *noun*, the end result, outcome, consequence

af-ter-noon [ăf′tər-nōōn′] *noun*, time between noon and evening

af-ter-ward [ăf′tər-wərd] *adverb*, later

a-gain [ə-gĕn′] *adverb*, once more

a-gainst [ə-gĕnst′] *preposition*, 1) in opposition to 2) in contact with, touching 3) in the other direction, opposite

age [āj] *noun*, 1) the amount of time something has lived 2) a particular period of time, *verb*, 3) to grow or cause to become old

aged [ā′jĭd] *adjective*, old

a-gen-cy [ā′jən-sē] *noun*, an organization designed to distribute funds and services to people

a-gen-da [ə-jĕn′də] *noun*, a list of items that need attention

a-gent [ā′jənt] *noun*, 1) a person who represents a group 2) a force of change

ag-gran-dize [ə-grăn′dīz′] *verb*, to make greater in power, rank or riches

ag-gra-vate [ăg′rə-vāt′] *verb*, to make worse, to annoy

ag-gre-gate [ăg′rĭ-gĭt] *noun*, the sum total of a group of things

ag-gres-sion [ə-grĕsh′ən] *noun*, anger in an attack

ag-gres-sive [ə-grĕs′ĭv] *adjective*, forceful, likely to attack

ag-ile [ăj′əl, -īl′] *adjective*, swift, nimble, able to move easily

ag-i-tate [ăj′ĭ-tāt′] *verb*, to shake up

ag-i-ta-tion [ăj′ĭ-tā′shən] *noun*, confusion, frenzy, excitement

ag-nos-tic [ăg-nŏs′tĭk] *noun*, a person who believes that the human mind cannot comprehend the beginning of the universe or the existence of God

a-go [ə-gō′] *adverb*, in the past

ag-o-ny [ăg′ə-nē] *noun*, very bad physical or emotional pain

a-gree-a-ble [ə-grē′ə-bəl] *adjective*, easy to get along with

a-gree-ment [ə-grē′mənt] *noun*, a mutual understanding

ag-ri-cul-ture [ăg′rĭ-kŭl′chər] *noun*, the science and art of cultivating soil and growing crops

a-head [ə-hĕd′] *adverb*, 1) in front, preceding 2) into the future

aid [ād] *noun*, 1) a form of help 2) assistance, support, *verb*, 3) to help or assist

aide [ād] *noun*, an assistant, a helper

ail-ment [āl′mənt] *noun*, a sickness,

aim [ām] *noun*, 1) target, direction 2) intention, purpose, goal, *verb*, 3) to point toward a target when ready to throw something, fire something, or take a picture

air [âr] *noun* 1) the mixture of gases surrounding the earth 2) an appearance or atmosphere, *verb*, 3) to refresh or ventilate

air-borne [âr′bôrn′] *adjective*, carried by or through the air

air-craft [âr′krăft′] *noun*, a machine capable of flying

air force [âr′fôrs′] *noun*, a group of soldiers trained to fight and destroy using aircraft

air-plane [âr′plān′] *noun*, a flying machine used to transport people or cargo moving through the air by means of motor driven propellers or engines and kept aloft by the upward thrust of the air on its fixed wings

air pol-lu-tion [âr - pə-lōō′shən] *noun*, particles that contaminate the air and make it unsafe to breathe

air-port [âr′pôrt′] *noun*, a place where aircraft land and take off

aisle [īl] *noun*, the passageway between seats in a bus, auditorium, church, etc., or between shelves in a store

a-jar [ə-jär′] *adjective and adverb*, slightly open

Al-a-bama [ăl′ə-băm′ə] *noun*, one of the southern United States the capital is Montgomery. The state flower for **Alabama** is the camellia and the motto is "We dare to defend our rights."

a-lac-ri-ty [ə-lăk′rĭ-tē] *noun*, 1) cheerful willingness or readiness 2) speed

a-larm [ə-lärm′] *noun*, 1) a feeling of fear or danger 2) a programmed sound to awaken persons from sleep or warn of danger, *verb*, 3) to worry or frighten

a-las [ə-lăs′] *interjection*, an expression of pity or concern

Alas-ka [ə-lăs′kə] *noun*, one of the 50 United States located in the North, the capital city is Juneau. The state flower of **Alaska** is the forget me-not, and the motto is "North to the Future."

al-ba-tross [ăl′bə-trôs′] *noun*, a large bird with long narrow wings found near the ocean, especially in the South Seas

al-bi-no [ăl-bī′nō] *noun*, a person or animal lacking pigmentation, and therefore extraordinarily pale

al-bum [ăl′bəm] *noun*, 1)a book that holds a collection of pictures, stamps or autographs 2) a collection of recordings

al-che-my [ăl′kə-mē] *noun*, medieval chemistry that attempted to turn ordinary things into gold

al-co-hol [ăl′kə-hôl′] *noun*, 1) an intoxicating beverage, such as wine, beer or liquor, the result of fermentation 2) an organic compound used as an antiseptic

al-co-hol-ic [ăl′kə-hô′lĭk] *adjective*, 1) containing alcohol as an ingredient, *noun*, 2) a person with the disease of alcoholism

al-der-man [ôl′dər-mən] *noun*, a member of a city council

a-lert [ə-lûrt′] *adjective*, 1) awake and ready to act, *noun*, 2) a warning of danger

al-fal-fa [ăl-făl′fə] *noun*, a plant in the pea family

al-ga [ăl′gə] *noun*, *plural*, **algae**, an organism, often one-celled, that continuously divides itself, with no root or stem that lives in water

al-ge-bra [ăl′jə-brə] *noun*, a branch of mathematics in which the expressions of quantity are in symbols that express general relationships

a-li-as [ā′lē-əs] *noun*, an assumed name, a fake name

al-i-bi [ăl′ə-bī′] *noun*, the plea of having been elsewhere at the time a crime was committed

al-ien [ā′lē-ən] *adjective*, 1) strange, wholly different in nature, *noun*, 2) a person from another country

al-ien-ate [āl′yə-nāt′] *verb*, to separate, to make hostile

a-light [ə-līt′] *verb*, to come from, to get off

a-lign-ment [ə-līn′mənt] *noun*, 1) arrangement in a line 2) the condition of cooperation

a-like [ə-līk′] *adjective*, 1) similar, *adverb*, 2) in the same manner

al-i-men-ta-ry [ăl′ə-měn′tə-rē] *adjective*, supplying nourishment

al-i-mo-ny [ăl′ə-mō′nē] *noun*, an allowance given to a spouse for his or her support after a divorce

a-live [ə-līv′] *adjective*, 1) living, not dead 2) active, moving

al-ka-li [ăl′kə-lī′] *noun*, any soluble substance having marked basic qualities as a mineral salt or mixture of salts that can neutralize acids and turns litmus blue

all [ôl] *adjective*, 1) the whole of, any, *adverb*, 2) completely, totally, *pronoun*, 3) everyone, the whole thing

al-lay [ə-lā′] to ease, to calm, to appease

al-le-ga-tion [ăl′ĭ-gā′shən] *noun*, the statement of something that is to be proved, an accusation

al-lege [ə-lĕj′] *verb*, to urge as a reason, to assert, to suggest

al-le-giance [ə-lē′jəns] *noun*, strong support of something, an obligation to a government or ruler, loyal support

al-le-go-ry [ăl′ĭ-gôr′ē] *noun*, a story with a moral in which characters are used as symbols, a fable

al-ler-gy [ăl′ər-jē] *noun*, an abnormal reaction to substances in the environment such as pollen, dust or insect bites

al-le-vi-ate [ə-lē′vē-āt′] *verb*, to lighten or lessen pressure

al-ley [ăl′ē] *noun*, a very narrow road between buildings

al-li-ance [ə-lī′əns] *noun*, a partnership based on trust, common objectives, etc.

al-li-ga-tor [ăl′ĭ-gā′tər] *noun*, a large reptile that lives near water and is a carnivore

al-lit-er-a-tion [ə-lĭt′ə-rā′shən] *noun*, repetition of the beginning sound, e.g., *Peter Piper picked a peck of pickled peppers*

al-lo-cate [ăl′ə-kāt′] *verb*, to set something aside for a special use

al-lot [ə-lŏt′] *verb*, to divide something and then to give someone his or her portion

al-lot-ment [ə-lŏt′mənt] *noun*, the portion given to each

al-low [ə-lou′] *verb*, to permit someone to do something

al-low-ance [ə-lou′əns] *noun*, a definite sum granted, an abatement or deduction

al-loy [ăl′oi′] *noun*, a substance composed of two or more metals mixed together

al-lude [ə-lood′] *verb*, to hint or suggest indirectly

al-lu-sion [ə-loo′zhən] *noun*, an indirect reference, knowledge of something not actually mentioned

al-ly [ə-lī′] *noun*, 1) someone who helps you fight against someone else, a person or group who has the same purpose or cause, *verb*, 2) to unite, to join

al-ma ma-ter [ăl′mə mä′tər] *noun*, a school from which a person has graduated

al-ma-nac [ôl′mə-năk′] *noun*, a book containing a calendar and certain statistics

al-might-y [ôl-mī′tē] *adjective*, having very great power

al-mond [ä′mənd] *noun*, a nut-like kernel of the fruit of the almond tree

al-most [ôl′mōst′] *adverb*, nearly, not quite

alms [ämz] *noun*, *plural*, money or items given to charity

a-lo-ha [ə-lō-ha′] *noun and interjection*, Hawaiian word used for greeting or parting

a-lone [ə-lōn′] *adjective and adverb* without others, singular

a-long [ə-lông′] *adverb*, 1) following the length of something from end to end 2) moving forward 3) as a companion, *preposition*, 4) over, through, by the course of

a-loud [ə-loud'] *adverb*, in a clear voice that can be easily heard

al-pha-bet [ăl'fə-bět'] *noun*, the letters of a language arranged in customary order

al-pha-bet-i-cal [ăl'fə-bět'ĭ-kəl] *adjective*, arranged in the same order as the letters of the alphabet

al-read-y [ôl-rĕd'ē] *adverb*, even now, before a certain time

al-so [ôl'sō] *adverb*, in addition, too

al-tar [ôl'tər] *noun*, a table in a holy place where religious rites are performed

al-ter [ôl'tər] *verb*, to change

al-ter-a-tion [ôl'tə-rā'shən] *noun*, a change or an adjustment

al-ter-nate [ôl'tər-nāt'] *verb*, to take turns, to switch off

al-ter-na-tive [ôl-tûr'nə-tĭv] *adjective*, 1) other, *noun*, 2) something you can choose

al-though [ôl-thō'] *conjunction*, supposing that, notwithstanding

al-ti-tude [ăl'tĭ-tōōd'] *noun*, extent upward, the height of something measured in relation to a reference level, such as sea level

al-to [ăl'tō] *noun*, a singing voice between tenor and soprano

al-to-geth-er [ôl'tə-gĕth'ər] *adverb*, wholly, completely, including everyone, united

al-tru-is-tic [ăl'trōō-ĭs'tĭk] *adjective*, unselfish, concerned about others

al-u-mi-num [ə-lōō'mə-nəm] *noun*, a very soft silver metal made chiefly of bauxite ore

al-ways [ôl'wāz] *adverb*, at all times, forever

am [ăm] *verb*, the part of the verb **be** used with **I**

a.m. in the morning, the hours between midnight and noon, from the Latin ante meridiem "before midday"

a-mass [ə-măs'] *verb*, to collect, to accumulate

am-a-teur [ăm'ə-tûr'] *adjective*, 1) for enjoyment rather than money, not professional, *noun*, 2) someone who is not paid for doing something but does it for pleasure

a-maze [ə-māz'] *verb*, to surprise or bewilder

a-maze-ment [ə-māz'mənt] *noun*, a state of surprise, awe, wonder

am-a-zon [ăm'ə-zŏn'] *noun*, a mythical female warrior

am-bas-sa-dor [ăm-băs'ə-dər] *noun*, an appointed official who represents his or her country in another country

am-ber [ăm'bər] *noun*, the fossilized sap from a pine tree

am-bi-dex-trous [ăm'bĭ-dĕk'strəs] *adjective*, capable of using either hand with equal ease

am-bi-ence [ăm'bē-əns] *noun*, the environment, the atmosphere in a room or building

am-big-u-ous [ăm-bĭg'yōō-əs] *adjective*, capable of more than one interpretation, unclear

am-bi-tion [ăm-bĭsh'ən] *noun*, a strong desire to achieve

am-bi-tious [ăm-bĭsh'əs] *adjective*, wanting to have power, driven to achieve

am-biv-a-lent [ăm-bĭv′ə-lənt]
adjective, having conflicting
emotional attitudes

am-bul-ance [ăm′byə-ləns] *noun,* a
vehicle equipped to carry the
injured or sick

am-bush [ăm′bŏŏsh] *noun,* a
surprise attack

a-mel-io-rate [ə-mēl′yə-rāt′] *verb,* to
improve a deplorable condition

A-men [ā-mĕn′, ä-mĕn′] *interjection,*
an expression of approval, to say I
agree

a-mend [ə-mĕnd′] *verb,* to correct,
to change generally for the better

a-mend-ment [ə-mĕnd′mənt] *noun,*
a change made in a bill or motion

a-men-i-ty [ə-mĕn′ĭ-tē] *noun,* a
courtesy, a thoughtful action of
consideration, a convenience

A-mer-i-ca [ə-mĕr′ĭ-kə]*noun,* a
country in North America with a
democratic form of government

A-mer-i-can In-di-an [ə-mĕr′ĭ-kən
ĭn′dē-ən] *noun,* the people whose
ancestors lived in North America
for thousands of years

am-e-thyst [ăm′ə-thĭst] *noun,* a
purple or bluish-violet stone

a-mi-a-ble [ā′mē-ə-bəl] *adjective,*
friendly, easy to get along with

a-mid [ə-mĭd′] *preposition,* among,
surrounded by

am-mo-nia [ə-mōn′yə] *noun,* a
colorless compound of hydrogen
and nitrogen

am-mu-ni-tion [ăm′yə-nĭsh′ən]
noun, rockets, bombs and bullets
used in loading or charging guns
and rifles

am-ne-sia [ăm-nē′zhə] *noun,* loss of
memory

am-nes-ty [ăm′nĭ-stē] *noun,* a pardon
that absolves or discounts past offenses

a-mong [ə-mŭng′] *preposition,*
between, surrounded by

a-mount [ə-mount′] *noun,* 1) a sum,
the total number 2) quantity, mass,
verb, 3) to add up

am-phib-i-an [ăm-fĭb′ē-ən] *noun,*
1) a cold-blooded animal living
both on land and in the water
2) an airplane designed to take off
on the water or on land

am-phib-i-ous [ăm-fĭb′ē-əs]
adjective, able to live or operate
on land or in water

am-phi-the-a-ter [ăm′fə-thē′ə-tər]
noun, a theater with seats all
around the stage or arena

am-ple [ăm′pəl] *adjective,* abundant,
plentiful, more than enough

am-pli-fy [ăm′plə-fī′] *verb,* to
increase in size, to make louder

am-pu-tate [ăm′pyŏŏ-tāt′] *verb,*
to cut off

am-u-let [ăm′yə-lĭt] *noun,* a good
luck charm, a talisman

a-muse [ə-myŏŏz′] *verb,* to entertain
or delight, to make someone smile

a-muse-ment [ə-myŏŏz′mənt] *noun,*
entertainment, a distraction

an [ən; ăn when stressed] *article,* used
instead of a in front of a vowel

an-aer-o-bic [ăn′ə-rō′bĭk] *adjective,*
able to live only in the absence of
free oxygen

a-nal-o-gous [ə-năl′ə-gəs] *adjective,*
similar, comparable, resembling

a-nal-o-gy [ə-năl′ə-jē] *noun*, an explanation based on resemblances between different objects or cases

a-nal-y-sis [ə-năl′ĭ-sĭs] *noun*, a thorough examination

an-a-lyze [ăn′ə-līz′] *verb*, to examine critically elements or constituent parts

an-ar-chist [ăn′ər-kĭst] *noun*, a person who rebels against the established order or laws

an-ar-chy [ăn′ər-kē] *noun*, a lawless condition of society, chaos

a-nat-o-my [ə-năt′ə-mē] *noun*, the study of the structure of people, animals or plants

an-ces-tor [ăn′sĕs′tər] *noun*, a person from whom one is descended

an-ces-try [ăn′sĕs′trē] *noun*, family members or ethnic descent

an-chor [ăng′kər] *noun*, 1) a heavy piece of metal on the end of a rope or chain lowered into the water to keep a boat from drifting, *verb*, 2) to secure or fasten firmly, to fix in place

an-cient [ān′shənt] *adjective*, belonging to a remote period of time, very old

and [ənd, ən; ănd when stressed] *conjunction*, a word that adds or joins other words

an-ec-dote [ăn′ĭk-dōt′] *noun*, a short story or fact of an interesting nature

a-ne-mi-a [ə-nē′mē-ə] *noun*, a condition marked by a low red blood cell count

an-e-mom-e-ter [ăn′ə-mŏm′ĭ-tər] *noun*, an instrument to measure the force or speed of wind

an-es-the-sia [ăn′ĭs-thē′zhə] *noun*, the loss of sensation or consciousness, especially when caused by anesthetic

an-es-thet-ic [ăn′ĭs-thĕt′ĭk] *noun*, a substance that causes loss of sensation or consciousness

an-gel [ān′jəl] *noun*, 1) a messenger of God 2) a kind, loveable person

an-ger [ăng′gər] *noun*, feelings of hostility and hurt, a desire to fight or argue

an-gle [ăng′gəl] *noun*, 1) the shape made when two lines or planes intersect 2) a point of view, *verb*, 3) to fish with a hook and line

an-gry [ăng′grē] *adjective*, feeling extreme emotion due to hurt feelings, hostile

an-guish [ăng′gwĭsh] *noun*, extreme pain either of body or mind, acute suffering, deep sadness

an-i-mal [ăn′ə-məl] *noun*, a living creature that is not a plant

an-i-mate [ăn′ə-māt′] *verb*, to inspire with energy, to enliven

an-i-mos-i-ty [ăn′ə-mŏs′ĭ-tē] *noun*, a feeling of hatred leading to opposition, hostility

an-kle [ăng′kəl] *noun*, the joint connecting the foot to the leg

an-nals [ăn′əlz] *noun*, records, history, chronicles, accounts

an-neal [ə-nēl′] *verb*, to subject to high heat, with subsequent cooling, for the purpose of softening thoroughly and making less brittle

an-nex [ə-nĕks′] *noun*, 1) a building added onto a larger building, *verb*, 2) to add on, to attach a smaller thing to something larger

an-nex-a-tion [ăn′ĭk-sā′shən] *noun*, the territory or thing that has been added

an-ni-hi-late [ə-nī′ə-lāt′] *verb*, to reduce to nothing, to destroy entirely, to obliterate

an-ni-ver-sa-ry [ăn′ə-vûr′sə-rē] *noun*, a day celebrated each year to remember something important from the past

an-no-tate [ăn′ō-tāt′] *verb*, to write comments or explanatory notes

an-nounce [ə-*nouns*′] *verb*, to give notice of the arrival of something, to proclaim, to declare

an-nounce-ment [ə-*nouns*′mənt] *noun*, 1) act of making something publicly known 2) a printed notice

an-noy [ə-noi′] *verb*, to disturb or irritate, to bother

an-noy-ance [ə-noi′əns] *noun*, anything that irritates

an-nu-al [ăn′yōo-əl] *adjective*, 1) occurring once a year, *noun*, 2) a plant that lives for one year

a-noint [ə-noint′] *verb*, to baptize, to purify, to sprinkle with oil or holy water

a-nom-a-ly [ə-nŏm′ə-lē] *noun*, an irregularity, a deviation from the normal, one of a kind

a-non-y-mous [ə-nŏn′ə-məs] *adjective*, of unknown name

an-o-rex-i-a [ăn′ə-rĕk′sē-ə] *noun*, a lack of appetite and inability or refusal to consume food

an-oth-er [ə-nŭth′ər] *adjective, and pronoun*, 1) an additional, one more 2) someone else

an-swer [ăn′sər] *noun*, 1) a reply to a question, *verb*, 2) to make a response, to reply

ant [ănt] *noun*, a small insect that lives in colonies, known for its social order and industry

an-tag-o-nist [ăn-tăg′ə-nĭst] *noun*, one who opposes with active resistance, an opponent

Ant-arc-ti-ca [ănt-ärk′tĭ-kə] *noun*, one of the seven continents in the world, the area containing the South Pole

an-te-ced-ent [ăn′tĭ-sēd′nt] *adjective*, going before, preceding

an-te-lope [ăn′tl-ōp′] *noun*, a wild animal, similar to a deer, which runs fast and usually has antlers

an-ten-na [ăn-tĕn′ə] *noun, plural*, **antennae**, 1) the sensory organ of an insect or crustacean 2) wire supported in the air for directly transmitting and receiving electromagnetic waves

an-them [ăn′thəm] *noun*, a song or hymn of praise

an-thro-po-log-ist [ăn′thrə-pŏl′ə-jĭst] *noun*, a student of the history and science of mankind

an-thro-po-mor-phic [ăn′thrə-pə-môr′-fĭk] *adjective*, having human form or characteristics

an-ti-bi-ot-ic [ăn′tĭ-bī-ŏt′ĭk] *noun*, a substance such as penicillin, effective in fighting destructive microorganisms and used for treatment of diseases

an-tic-i-pate [ăn-tĭs′ə-pāt′] *verb*, to look forward to

an-ti-dote [ăn′tĭ-dōt′] *noun*, 1) a remedy to counteract poison 2) anything that counteracts evil

an-tip-a-thy [ăn-tĭp′ə-thē] *noun*, a strong dislike, a natural aversion

an-ti-quat-ed [ăn′tĭ-kwā′tĭd] *adjective*, old-fashioned, obsolete

an-tique [ăn-tēk′] *adjective*, 1) old-fashioned, from another time, *noun*, 2) something made long ago and valued for its age

an-ti-sep-tic [ăn′tĭ-sĕp′tĭk] *adjective*, 1) clean, sterile, *noun*, 2) a substance that tends to prevent harmful effects of bacteria growth

an-tith-e-sis [ăn-tĭth′ĭ-sĭs] *noun*, the exact opposite

ant-ler [ănt′lər] *noun*, one of a pair of branched horns on the head of a deer

an-to-nym [ăn′tə-nĭm′] *noun*, a word that has the opposite meaning to another

anx-i-e-ty [ăng-zī′ĭ-tē] *noun*, uneasiness of mind, concern about future events

anx-ious [ăngk′shəs] *adjective*, worried, concerned, nervous

any [ĕn′ē] *adjective*, 1) some, an uncertain number, quantity, volume or degree 2) every 3) one

an-y-bod-y [ĕn′ē-bŏd′ē] *pronoun*, any person

an-y-one [ĕn′ē-wŭn′] *pronoun*, anybody

an-y-thing [ĕn′ē-thĭng′] *noun*, something

an-y-way [ĕn′ē-wā′] *adverb*, 1) no matter what happens, in any case, 2) haphazardly, in any fashion

an-y-where [ĕn′ē-hwâr′] *adverb*, in, at, or to any place

a-part [ə-pärt′] *adverb*, separately in place or time

a-part-ment [ə-pärt′mənt] *noun*, rooms or a part of a building where someone lives

ap-a-thy [ăp′ə-thē] *noun*, lack of interest or concern

ape [āp] *noun*, 1) a large mammal similar to a monkey without a tail, for example, a gorilla, chimpanzee, etc., *verb*, 2) to copy, to imitate

aph-o-rism [ăf′ə-rĭz′əm] *noun*, a statement of truth, an adage

a-poc-a-lyp-tic [ə-pŏk′ə-lĭp′tĭk] *adjective*, prophetic, pertaining to revelations about the end of the world

a-pol-ogize [ə-pŏl′ə-jīz′] *verb*, to say you're sorry for what you have done, to express regret

a-pol-o-gy [ə-pŏl′ə-jē] *noun*, an explanation by way of asking for forgiveness

a-pos-tle [ə-pŏs′əl] *noun*, any person zealously advocating a cause

a-pos-tro-phe [ə-pŏs′trə-fē] *noun*, a mark (') used to indicate an omission of a letter, i.e. *I'll*, or letters, or to denote the possessive case i.e. *Kara's*

ap-o-thegm [ăp′ə-thĕm′] *noun*, a pithy, compact saying, a proverb

a-poth-e-o-sis [ə-pŏth′ē-ō′sĭs] *noun*, deification, glorification, making into a god

ap-pall [ə-pôl′] *verb*, to fill with horror or dismay, to shock

ap-pa-rat-us [ăp′ə-rā′təs] *noun*, equipment, a collection or set of implements for a specific purpose

ap-par-el [ə-păr′əl] *noun*, clothing, garments

ap-par-ent [ə-păr′ənt] *adjective*, capable of seeming or appearing to be evident, easily understood

ap-par-ent-ly [ə-păr′ənt] *adverb*, clearly, evidently, obviously

ap-pa-ri-tion [ăp′ə-rĭsh′ən] *noun*, a ghost, a phantom

ap-peal [ə-pēl′] *noun*, 1) the proceeding by which a case is brought to a superior court for reexamination or review 2) a request, *verb*, 3) to plead, to ask urgently 4) to seem attractive

ap-pear [ə-pîr′] *verb*, 1) to come into view unexpectedly 2) to seem to be

ap-pear-ance [ə-pîr′əns] *noun*, 1) the act of appearing or becoming visible 2) the way something or someone looks

ap-pease [ə-pēz′] *verb*, to satisfy or often by granting demands

ap-pel-lant [ə-pĕl′ənt] *noun*, one who asks for a judicial decree

ap-pel-la-tion [ăp′ə-lā′shən] *noun*, a name or title

ap-pen-di-ci-tis [ə-pĕn′dĭ-sī′tĭs] *noun*, inflammation of the appendix

ap-pen-dix [ə-pĕn′dĭks] *noun*, 1) information added to a section at the end of a book 2) a small tube-like part growing out from the intestine

ap-pe-tite [ăp′ĭ-tīt′] *noun*, craving, the desire for food

ap-plaud [ə-plôd′] *verb*, to show approval especially by clapping hands together in praise

ap-plause [ə-plôz′] *noun*, praise for someone, usually expressed by clapping hands

ap-ple [ăp′əl] *noun*, the firm-fleshed, round, juicy fruit of an apple tree

ap-pli-ance [ə-plī′əns] *noun*, an apparatus or device, a machine

ap-pli-ca-ble [ăp′lĭ-kə-bəl] *adjective*, relevant, appropriate

ap-plic-ant [ăp′lĭ-kənt] *noun*, one who makes a request, a petitioner

ap-pli-ca-tion [ăp′lĭ-kā′shən] *noun*, 1) work or effort 2) a use 3) a form used to make a request

ap-pli-que [ăp′lĭ-kā′] *noun*, ornamental patterns put on textiles

ap-ply [ə-plī′] *verb*, 1) to seek or ask 2) to put on 3) to pertain to

ap-point [ə-point′] *verb*, to choose for a job

ap-point-ment [ə-point′mənt] *noun*, 1) a meeting, a date, a rendezvous 2) a selection, a nomination

ap-prais-al [ə-prā′zəl] *noun*, value or worth, assessment

ap-praise [ə-prāz′] *verb*, to evaluate something to establish its worth

ap-pre-ci-ate [ə-prē′shē-āt′] *verb*, 1) to recognize the value of something, 2) to increase in value

ap-pre-ci-a-tion [ə-prē′shē-ā′shən] *noun*, grateful recognition

ap-pre-hend [ăp′rĭ-hĕnd′] *verb*, 1) to arrest a criminal 2) to anticipate

with anxiety, dread, or, fear
3) to understand

ap-pre-hen-sion [ăp′rĭ-hĕn′shən] *noun*, 1) the ability to understand 2) fear, anxiety, dread

ap-pren-tice [ə-prĕn′tĭs] *noun*, a novice or learner who works with a master to learn a skill

ap-prise [ə-prīz′] *verb*, to give notice to, to inform

ap-proach [ə-prōch′] *noun*, 1) a way of dealing with something, 2) the way that leads to a place, *verb*, 3) to move toward

ap-proach-able [ə-prō′chə-bəl] *adjective*, friendly, nice, easy to talk to

ap-pro-pri-ate [ə-prō′prē-ĭt] *adjective*, 1) suitable, proper, [ə-prō′prē-āt]*verb*, 2) to allocate, to set aside for a purpose

ap-pro-pri-a-tion [ə-prō′prē-ā′shən] *noun*, a setting apart for a particular purpose

ap-prov-al [ə-prōō′vəl] *noun*, 1) praise 2) the go-ahead or okay to proceed, permission

ap-prove [ə-prōōv′] *verb*, to say that something is favorable

ap-prox-i-mate [ə-prŏk′sə-mĭt] *adjective*, 1) almost accurate or exact, near, [ə-prŏk′sə-māt]*verb*, 2) to estimate or guess

ap-prox-i-mate-ly [ə-prŏk′sə-mĭt-lē] *adverb*, closely, nearly

ap-pur-te-nance [ə-pûr′tn-əns] *noun*, something connected but not as important, accessory

A-pril [ā′prəl] *noun*, the fourth month of the year, having 30 days

a-pron [ā′prən] *noun*, a cover worn over the front of one's clothing to protect it

ap-ro-pos [ăp′rə-pō′] *adjective*, appropriate, pertinent

apt [ăpt] *adjective*, 1) suited to, fitting, 2) likely or almost certain

ap-ti-tude [ăp′tĭ-tōōd′] *noun*, natural ability, fitness, talent

a-quar-i-um [ə-kwâr′ē-əm] *noun*, a container, usually with glass sides, to keep animals or plants in water

a-quat-ic [ə-kwăt′ĭk] *adjective*, occurring on or in the water

aq-ue-duct [ăk′wĭ-dŭkt′] *noun*, a passage used to bring water from a distant source

aq-ui-fer [ăk′wə-fər] *noun*, an underground geologic formation in which the cracks in rock, sand, soil, or gravel are filled with water

Ar-ab [ăr′ə-b] a native or inhabitant of Arabia, 2) any of a Semitic people originating in Arabia

Ar-a-bic [ăr′ə-bĭk] *adjective*, 1) relating to the Arab language and culture, *noun*, 2) the Semitic language of the Arabs

ar-bi-ter [är′bĭ-tər] *noun*, a judge, an umpire or referee, a mediator

ar-bi-trar-y [är′bĭ-trĕr′ē] *adjective*, chosen without any qualifications

ar-bi-tra-tion [är′bĭ-trā′shən] *noun*, the hearing and determination of a matter in dispute by a third party

ar-bor [är′bər] *noun*, a place shaded by trees or shrubs

arc [ärk] *noun*, a segment of a circle, a curve in an arch

arch [ärch] *noun*, a curved part of a structure sometimes supporting.

weight above it, as in bridges, doorways, etc

ar-chae-ol-o-gist [är′kē-ŏl′ə-jĭst] *noun*, an expert in the study of past civilizations and ancient cultures, mainly through digging up remains

ar-chae-ol-o-gy [är′kē-ŏl′ə-jē] *noun*, the study of ancient things made by man from an earlier civilization

ar-cha-ic [är-kā′ĭk] *adjective*, antiquated, ancient, very old

arch-bish-op [ärch-bĭsh′əp] *noun*, the chief bishop in the Christian church

ar-cher-y [är′chə-rē] *noun*, the practice or art of shooting with a bow and arrow

ar-chi-pel-a-go [är′kə-pĕl′ə-gō′] *noun*, a group or chain of islands

ar-chi-tect [är′kĭ-tĕkt′] *noun*, one who draws plans and specifications for buildings

ar-chi-tec-ture [är′kĭ-tĕk′chər] *noun*, the art and study of building design and structure

ar-chive [är′kīv′] *noun*, a place in which public records and historic documents are kept

arc-tic [ärk′tĭk, är′tĭk] *adjective*, relating to the area around the North Pole

ar-dent [är′dnt] *adjective*, passionate, zealous

ar-dor [är′dər] *noun*, passion, zeal, enthusiasm, great excitement

ar-du-ous [är′jōō-əs] *adjective*, difficult, attended with great labor or exertion, strenuous

are [är] *verb*, the part of the *verb* **is** used with we, you and they

ar-e-a [âr′ē-ə] *noun*, 1) a location or vicinity 2) a surface measured by multiplying the width times the length

a-re-na [ə-rē′nə] *noun*, a playing field or a stadium surrounded by seats used for sports or entertainment

aren't [ärnt, är′ənt] the *contraction,*for the two words **are** and **not**

ar-gon [är′gŏn′] *noun*, one of the noble gases, a colorless, odorless, inert gas

ar-gue [är′gyōō] *verb*, to fight by using words, to disagree, to debate, to dispute

ar-gu-ment [är′gyə-mənt] *noun*, a disagreement, a dispute

ar-id [ăr′ĭd] *adjective*, dry, parched, dusty

a-rise [ə-rīz′] *verb*, to come up, to issue, to spring forth, to get up

a-ris-to-crat [ə-rĭs′tə-krăt′] *noun*, a man or woman of fashion, a member of the nobility

a-rith-me-tic [ə-rĭth′mĭ-tĭk] *noun*, the science of numbers

Ar-i-zo-na [ăr′ĭ-zō′nə] *noun*, one of the 50 United States, located in the Southwest, the capital is Phoenix. The state flower of **Arizona** is the blossom of the saguaro cactus and the motto is "God enriches."

ark [ärk] *noun*, a large flat-bottomed ship

Ar-kan-sas [är′kən-sô′] *noun*, a state in the southern part of the United States, the capital is Little Rock. The **Arkansas** state flower is the apple blossom and the motto is "The people rule."

arm [ärm] *noun*, 1) the part of the body between the shoulder and the hand, *verb*, 2) to provide with weapons, to prepare for combat

ar-ma-ments [är′mə-mənts] *noun, plural*, military forces and weapons

ar-mi-stice [är′mĭ-stĭs] *noun*, a treaty of peace, a truce

ar-mor [är′mər] *noun*, a covering of metal worn by soldiers for protection

ar-mor-y [är′mə-rē] *noun*, a military storehouse

arms [ärmz] *noun, plural*, weapons, such as bombs and guns

ar-my [är′mē] *noun*, a large number of soldiers training for war on land, ground troops

a-ro-ma [ə-rō′mə] *noun*, smell, scent, fragrance

a-round [ə-round′] *adverb*, 1) in a circle, surrounding all sides, 2) nearby 3) approximately, *preposition*, 4) on every side

a-rouse [ə-rouz′] *verb*, 1) to awaken, to animate, to coax 2) to excite, to provoke, to stimulate

ar-raign [ə-rān′] *verb*, to call to account, to bring into court to answer charges, to accuse publicly

ar-range [ə-rānj′] *verb*, to put in proper order or sequence

ar-range-ment [ə-rānj′mənt] *noun*, 1) the act of putting in an orderly condition 2) plan

ar-ray [ə-rā′] *noun*, 1) a display, an exhibition, *verb*, 2) to arrange in a display, to place, to order

ar-rest [ə-rĕst′] *noun*, 1) the act of taking someone into custody,

verb, 2) to stop the movement or progress of 3) to seize by legal authority and charge with breaking the law

ar-ri-val [ə-rī′vəl] *noun*, the moment when a place or object is reached

ar-rive [ə-rīv′] *verb*, to reach a destination or a goal

ar-ro-gance [ăr′ə-gəns] *noun*, an overbearing sense of self-importance. excessive pride

ar-ro-gant [ăr′ə-gənt] *adjective*, having a high opinion of oneself

ar-row [ăr′ō] *noun*, 1) a pointed stick that is shot from a bow 2) a mark shaped like an arrow that shows direction

ar-se-nal [är′sə-nəl] *noun*, a collection of weapons, a storehouse of ammunition

ar-son [är′sən] *noun*, the crime of purposely setting property on fire

art [ärt] *noun*, the creation of beautiful or significant things through music, drawing, painting, sculpture, cooking, dancing, etc.

ar-ter-y [är′tə-rē] *noun*, the thick blood vessels that carry blood from the heart

ar-te-sian well [är-tē′zhən] *noun*, a well that taps ground water that is under pressure and therefore does not require a pump, so called from Artois, France

ar-thri-tis [är-thrī′tĭs] *noun*, inflammation of the joints

ar-ti-cle [är′tĭ-kəl] *noun*, 1) a particular thing 2) a section, provision, item 3) a brief written composition, as in a newspaper or magazine 4) one of the limiting articles, *a, an*, or *the*

ar-tic-u-late [är-tĭk′yə-lĭt]
adjective, 1) effective, distinct,
able to express ideas clearly, [är-
tĭk′yə-lāt] *verb*, 2) to speak
clearly to express an idea

ar-ti-fact [är′tə-făkt′] *noun*,
something made by humans
from an earlier civilization

ar-ti-fice [är′tə-fĭs] *noun*, skill,
scheme, ability, trick

ar-ti-fi-cial [är′tə-fĭsh′əl] *adjective*,
1) made of or contrived by art
2) not natural, not made by
nature, made by humans 3)
assumed, affected

ar-t-il-ler-y [är-tĭl′ə-rē] *noun*, the
mounted guns in an army, the
soldiers in charge of such guns

art-ist [är′tĭst] *noun*, a person
skilled in the techniques of the
fine arts, such as painting,
dancing, drama, etc.

as [ăz] *conjunction*, 1) when, while
2) because, *preposition*, 3) in
such a way, like

ascend [ə-sĕnd′] *verb*, to climb, to
go upward, to rise

as-cen-sion [ə-sĕn′shən] *noun*, the
act of going up, a rising

as-cent [ə-sĕnt′] *noun*, the act of
climbing up an incline, a slope

as-cer-tain [ăs′ər-tān′] *verb*, to
determine, to establish

as-cet-ic [ə-sĕt′ĭk] *adjective*, self-
denying, austere, disciplined

ash [ăsh] *noun*, 1) the gray powder
residue left after something has
burned 2) a tree of the olive
family, with hard wood

a-shamed [ə-shāmd′] *adjective*,
upset because you have done

something that you think is silly,
not good enough or wrong,
embarrassed, feeling inadequate

a-shore [ə-shôr′] *adverb*, on to land

A-sia [ā′zhə] *noun*, one of the
seven continents in the world,
holding 57% of the population

a-side [ə-sīd′] *adverb*, 1) to or
toward one side, away, *noun*,
2) departure from the main topic,
3) a whisper

ask [ăsk] *verb*, 1) to pose or speak
a question, to inquire 2) to
request something

a-skew [ə-skyōō′] *adjective*, out of
line, crooked

a-sleep [ə-slēp′] *adjective*,
sleeping, not awake

as-pect [ăs′pĕkt] *noun*,
1) appearance, look 2) part, or
point of view

as-phalt [ăs′fôlt′] *noun*, a tar like
material mixed with sand or
gravel for highway surfaces or
pavement,

as-phyx-i-a-tion [ăs-fĭk′sē-ā′-shən]
noun, death by lack of oxygen

as-pi-ra-tion [ăs′pə-rā′shən] *noun*,
an ambition, a strong desire

as-pire [ə-spīr′] *verb*, to desire, to
strive, to hope for

as-pi-rin [ăs′pər-ĭn, -prĭn] *noun*, a
medicine used to relieve pain,
acetylsalicylic acid

as-sas-si-nate [ə-săs′ə-nāt′] *verb*,
to murder a public official or
other important person

as-sault [ə-sôlt′] *noun*, 1) an attack,
verb, 2) to strike, to hit

as-sem-ble [ə-sĕm′bəl] *verb*, to
gather or put together

astrology

as-sem-bly [ə-sĕm′blē] *noun*, 1) a
gathering, for educational,
religious, or entertainment
programs of students and
teachers 2) a legal body such as
Congress or a legislature

as-sert [ə-sûrt′] *verb*, to declare,
to state forcefully

as-sess [ə-sĕs′] *verb*, to evaluate

as-sess-ment [ə-sĕs′mənt] *noun*, an
appraisal, an estimation

as-set [ăs′ĕt′] *noun*, something of
value or worth

as-sign [ə-sīn′] *verb*, 1) to give out
or allot 2) to delegate a task

as-sign-ment [ə-sīn′mənt] *noun*,
homework, an obligation, work
given to someone to accomplish

as-sim-i-late [ə-sĭm′ə-lāt′] *verb*, to
take in food or knowledge, to
make part of oneself

as-sist [ə-sĭst′] *verb*, to aid or help

as-sis-tance [ə-sĭs′təns] *noun*, help
or aid given

as-sis-tant [ə-sĭs′tənt] *noun*, a
helper, an aide, a subordinate,

as-so-ci-ate *noun*, [ə-sō′shē-ĭt]
1) a colleague or co-worker,
[ə-sō′shē-āt′] *verb*, 2) to think of
something in relation with
something else 3) to keep
company, to socialize with
someone

as-so-ci-a-tion [ə-sō′sē-ā′shən,]
noun, a group of people joined
together for one purpose

as-sort-ment [ə-sôrt′mənt] *noun*, a
variety of things

as-suage [ə-swāj′] *verb*, to lessen,
to ease, to comfort

as-sume [ə-sōōm′] *verb*, to think
something is true without proof

as-sump-tion [ə-sŭmp′shən] *noun*,
something taken for granted

as-sur-ance [ə-shōōr′əns] *noun*, a
promise, a pledge

as-sure [ə-shōōr′] *verb*, to tell
someone firmly, to promise

as-ter-isk [ăs′tə-rĭsk′] *noun*, the
sign *

as-ter-oid [ăs′tə-roid′] *noun*, small
planets, most of which are located
between Mars and Jupiter, that
revolve around the sun

asth-ma [ăz′mə] *noun*, a
respiratory disorder that causes
difficulty in breathing
accompanied by wheezing

a-stig-ma-tism [ə-stĭg′mə-tĭz′əm]
noun, an eye defect that prevents
proper focus

as-ton-ish [ə-stŏn′ĭsh] *verb*, to
give a feeling of surprise, amaze

as-ton-ish-ment [ə-stŏn′ĭsh-mənt]
noun, complete surprise,
amazement

as-tound [ə-stound′] *verb*, to
amaze, to shock, to surprise

as-tral [ăs′trəl] *adjective*, relating
to the stars

a-stray [ə-strā′] *adverb*, off course,
in error

as-trin-gent [ə-strĭn′jənt] *adjective*,
capable of drawing tissue together

as-tro-naut [ăs′trə-nôt′] *noun*, a
person who travels in space

as-trol-o-gy [ə-strŏl′ə-jē] *noun*, the
study of the assumed influence
of planets and stars on the course
of human affairs

27

as-tron-o-mer [ə-strŏn′ə-mər] *noun*, a person who studies the stars and the heavens

as-tro-nom-i-cal [ăs′trə-nŏm′ĭ-kəl] *adjective*, enormously large or extensive, huge

as-tron-o-my [ə-strŏn′ə-mē] *noun*, the study of the sun, moon, stars, and the heavens

as-tute [ə-stōōt′] *adjective*, having a clever, shrewd, or cunning mind, perceptive, insightful

a-sy-lum [ə-sī′ləm] *noun*, protection from persecution

at [ăt] *preposition*, 1) showing where 2) showing when 3) showing how much

ate [āt] *verb*, the *past tense* of **eat**

a-the-ism [ā′thē-ĭz′əm] *noun*, the belief that there is no God

ath-lete [ăth′lēt′] *noun*, someone who is good at sports

ath-let-ic [ăth-lĕt′ĭk] *adjective*, strong, vigorous, having to do with sports

at-las [ăt′ləs] *noun*, a book or section of a book that has maps

at-mos-phere [ăt′mə-sfîr′] *noun*, 1) the mass of air surrounding the earth 2) the mood or feeling of a place, a tone conveyed

at-om [ăt′əm] *noun*, the smallest particle of an element

a-tom-ic [ə-tŏm′ĭk] *adjective*, tiny, infinitesimal, having to do with atoms

at-om-iz-er [ăt′ə-mī′zər] *noun*, an instrument for reducing a liquid to a fine spray

a-tri-um [ā′trē-əm] *noun*, **atria** or **atriums**, *plural*, 1) one of the

chambers of the heart 2) the central court in a Roman building

a-tro-cious [ə-trō′shəs] *adjective*, very bad, outrageously cruel

a-troc-i-ty [ə-trŏs′ĭ-tē] *noun*, a brutal deed, a cruel act

a-tro-phy [ăt′rə-fē] *verb*, to stop the development of an organ, to wither from lack of use

at-tach [ə-tăch′] *verb*, 1) to connect something with something else 2) to feel affection toward someone 3) to ascribe

at-tach-ment [ə-tăch′mənt] *noun*, 1) an affection, fondness 2) an additional part for a machine

at-tack [ə-tăk′] *noun*, 1) fight, an attempt to harm someone, *verb*, 2) to assail, to try to injure someone, to assault

at-tain [ə-tān′] *verb*, to succeed in getting something done, to reach a goal

at-tempt [ə-tĕmpt′] *noun*, 1) a try, an endeavor, *verb*, 2) to try, to endeavor, to make an effort

at-tend [ə-tĕnd′] *verb*, 1) to be present at 2) to take care of 3) to pay attention to

at-tend-ance [ə-tĕn′dəns] *noun*, the act of showing up or putting in an appearance, presence

at-tend-ant [ə-tĕn′dənt] *noun*, one who accompanies or attends, as a companion, a servant

at-ten-tion [ə-tĕn′shən] *noun*, application of the mind to something, concentration

at-ten-tive [ə-tĕn′tĭv] *adjective*, careful to notice little things

at-tic [ăt′ĭk] *noun*, the top floor of a house, right under the roof

at-tire [ə-tīr′] *noun*, anything that dresses or adorns, clothing

at-ti-tude [ăt′ĭ-tood′] *noun*, a mood or feeling, a way of thinking

at-tor-ney [ə-tûr′nē] *noun*, one legally authorized to act for another, a lawyer

at-tract [ə-trăkt′] *verb*, 1) to draw the attention of someone or something 2) to cause someone or something to come near

at-trac-tion [ə-trăk′shən] *noun*, 1) the power to bring things together 2) a fascination that holds someone's attention

at-trac-tive [ə-trăk′tĭv] *adjective*, pleasing to look at, appealing

at-trib-ute *noun*, [ə-trăb′yoot] 1) a distinguishing feature that belongs to someone, [ətrĭb′yoot] *verb*, 2) to explain how one thing is associated with another

at-tri-tion [ə-trĭsh′ən] *noun*, a gradual wearing down

auc-tion [ôk′shən] *noun*, a public sale at which buyers bid on items to be sold

auc-tion-eer [ôk′shə-nîr′] *noun*, one licensed to hold a public sale at which items go to the highest bidder

au-dac-i-ty [ô-dăs′ĭ-tē] *noun*, impertinence, boldness, self-assurance, daring

au-di-ble [ô′də-bəl] *adjective*, capable of being heard

au-di-ence [ô′dē-əns] *noun*, 1) a group gathered to watch or listen to a performance 2) a formal hearing

au-di-o [ô′dē-ō′] *adjective*, relating to sound or sound reproduction

au-dit [ô′dĭt] *noun*, 1) a special examination of accounting records, *verb*, 2) to examine or investigate, to evaluate

au-di-tion [ô-dĭsh′ən] *noun*, a tryout for a performance

au-di-to-ri-um [ô′dĭ-tôr′ē-əm] *noun*, an assembly hall with a stage and seats

au-ger [ô′gər] *noun*, a tool for making holes, a drill

aug-ment [ôg-mĕnt′] *verb*, to make larger, to increase, to add on to

au-gur [ô′gər] *verb*, to foretell, to see into the future, to portend, to be an omen

Au-gust [ô′gəst] *noun*, the eighth month of the year, having 31 days

aunt [ănt, änt] *noun*, the sister of one of your parents or the wife of the brother of one parent

aus-pic-es [ô′spĭ-sēz′] *noun*, *plural*, 1) patronage, protection 2) omens

aus-pi-cious [ô-spĭsh′əs] *adjective*, fortunate, showing signs that promise success

aus-tere [ô-stîr′] *adjective*, 1) severe or strict in judging, harsh, rough or bitter acting, 2) very plain, simple, bare, ascetic

Aus-tra-lia [ô-strāl′yə] *noun*, one of the seven continents in the world, the only continent that is also a country containing .5% of the world's population

au·then·tic [ô-thĕn′tĭk] *adjective*, genuine, known to be true

au·then·ti·cate [ô-thĕn′tĭ-kāt′] *verb*, to prove genuine

au·thor [ô′thər] *noun*, a person who writes books, articles, etc.

au·thor·i·ta·tive [ə-thôr′ĭ-tā′tĭv] *adjective*, reliably accurate

au·thor·i·ty [ə-thôr′ĭ-tē] *noun*, 1) a person who runs something or controls it 2) command, prestige, influence, control 3) an expert

au·thor·ize [ô′thə-rīz′] *verb*, 1) to empower 2) to make legal 3) to give permission to act

au·to·bi·og·ra·phy [ô′tō-bī-ŏg′rə-fē] *noun*, a story written about the author's life

au·to·crat [ô′tə-krăt′] *noun*, a ruler with supreme power

au·to·graph [ô′tə-grăf′] *noun*, a person's signature

au·to·mat·ic [ô′tə-măt′ĭk] *adjective*, acting mechanically without human direction

au·to·mat·i·cal·ly [ô′tə-măt′ĭk-lē] *adverb*, 1) by a machine 2) without trying or thinking first

au·tom·a·tion [ô′tə-mā′shən] *noun*, a machine that imitates the actions of humans, a robot

au·to·mo·bile [ô′tə-mō-bēl′] *noun*, a vehicle on wheels, powered by an engine to transport people on roads, a car

au·ton·o·mous [ô-tŏn′ə-məs] *adjective*, self-governing, acting independently from government

au·top·sy [ô′tŏp′sē] *noun*, the dissection of a dead body to determine the cause of death

au·tumn [ô′təm] *noun*, the third season of each year, between summer and winter, September 22-December 21, known as fall

aux·il·ia·ry [ôg-zĭl′yə-rē] *adjective*, additional, serving alongside another, supporting

a·vail·a·ble [ə-vā′lə-bəl] *adjective*, capable of being used, accessible

av·a·lanche [ăv′ə-lănch′] *noun*, a large overwhelming quantity of snow, rocks, or mud that falls from a mountain suddenly, gaining size as it falls

av·a·rice [ăv′ə-rĭs] *noun*, greed for money or power

a·venge [ə-vĕnj′] *verb*, to punish in kind, to retaliate, to get even

av·e·nue [ăv′ə-nōō′] *noun*, a wide road, often with many roads intersecting at right angles

av·er·age [ăv′ər-ĭj] *adjective*, 1) ordinary, usual, typical, *noun*, 2) the mean value of a set of numbers, *verb*, 3) to find the mean

a·ver·sion [ə-vûr′zhən] *noun*, a dislike, antipathy, opposition

a·vert [ə-vûrt′] *verb*, to avoid, to prevent or turn away

a·vi·a·tion [ā′vē-ā′shən, ăv′ē-] *noun*, air transportation

a·vi·a·tor [ā′vē-ā′tər] *noun*, an airplane or helicopter pilot

av·id [ăv′ĭd] *adjective*, eager, enthusiastic, excited

av·o·ca·tion [ăv′ō-kā′shən] *noun*, a hobby

a·void [ə-void′] *verb*, to stay away from, to evade, to shun

a-wait [ə-wāt'] *verb*, to expect, to wait, to anticipate

a-wake [ə-wāk'] *adjective*, not sleeping, alert

a-ward [ə-wôrd'] *noun*, 1) a prize or an honor to recognize exceptional service, achievement, etc., *verb*, 2) to present a certificate or plaque or trophy in recognition of an honor

a-ware [ə-wâr'] *adjective*, knowing, conscious, informed

a-way [ə-wā'] *adverb*, 1) off in a different direction 2) from one's possession

awe [ô] *noun*, a solemn wonder

aw-ful [ô'fəl] *adjective*, very bad, appalling, frightful, horrible

awk-ward [ôk'wərd] *adjective*, 1) clumsy, ungraceful 2) embarrassing

awl [ôl] *noun*, a pointed tool used to make holes in wood or leather

awn-ing [ô'nĭng] *noun*, a sunshade, a canvas covering for a porch, patio, deck, etc.

a-wry [ə-rī'] *adjective*, 1) amiss, crooked, wrong, *adverb*, 2) distorted, twisted to one side

ax or axe [ăks] *noun*, a tool with a metal blade fixed onto a handle, used for chopping

ax-i-om [ăk'sē-əm] *noun*, a self-evident truth or principle

ax-is [ăk'sĭs] *noun*, a real or imaginary line around which something rotates

ax-le [ăk'səl] *noun*, the rod on which a wheel revolves

B

ba-by [bā'bē] *noun*, an infant, a very young child or animal

bach-e-lor [băch'ə-lər] *noun*, a man who has never been married

back [băk] *adverb*, 1) to or in a place where something or someone was before 2) toward the back part, away from the front, *noun*, 3) the part of the body opposite the face, extending from the neck to the hips, *verb*, 4) to move in the opposite direction from 5) to give support to

back-ground [băk'ground'] *noun*, 1) the surroundings behind something 2) past experience, culture

back-ward [băk'wərd] *adjective*, 1) having the part designed for the back in the front 2) following tradition and resisting progress, *adverb*, 3) in reverse 4) beginning at the end

ba-con [bā'kən] *noun*, meat from the back and sides of a pig, salted and smoked

bac-te-ri-um [băk-tîr'ē-əm] *noun*, **bacteria**, *plural,* a single-celled living thing that cannot be seen without a microscope, reproducing rapidly and sometimes causing diseases

bad [băd] *adjective*, 1) not good, rotten 2) severe, unpleasant

badge [băj] *noun*, a pin or emblem worn to show authority or accomplishment

baf-fle [băf'əl] *verb*, to mix up, to confuse, to puzzle

bag [băg] *noun*, a container made of paper, cloth, plastic, or leather that has an opening at the top or in the side

bag-gage [băg′ĭj] *noun*, *no plural*, suitcases, bags, or trunks one takes along on a journey

bail [bāl] *noun*, 1) a bond or pledge backed with money or property, *verb*, 2) to ladle or empty out

bail-iff [bā′lĭf] *noun*, an official who keeps order in the court

bait [bāt] *noun*, 1) food used to lure or entice fish or animals, *verb*, 2) to annoy or taunt

bake [bāk] *verb*, to cook by dry heat, usually in an oven

bak-er [bā′kər] *noun*, someone who makes pastries, cakes and breads for a living

bak-er-y [bā′kə-rē] *noun*, a place where pastries, cakes, and breads are made and sold

bal-ance [băl′əns] *noun*, 1) steadiness 2) an instrument used to weigh something 3) the difference between the debits and credits of an account, *verb*, 4) to keep oneself steady 5) to make equal and stable, to compensate for

bal-co-ny [băl′kə-nē] *noun*, a porch outside of a building and above the first floor

bald [bôld] *adjective*, having no hair

bale [bāl] *noun*, 1) a large amount of materials bundled tightly together, *verb*, 2) to bundle into a package

balk [bôk] *verb*, 1) to hesitate, to refuse to act 2) to hinder

ball [bôl] *noun*, 1) a round object used in games or as a toy 2) a large party for dancing

bal-lad [băl′əd] *noun*, a poem or song that tells a story

bal-last [băl′əst] *noun*, 1) heavy material carried by a ship to balance it 2) crushed stones

bal-let [bă-lā′] *noun*, a play without speech in which the story is told through dance

bal-loon [bə-lōōn′] *noun*, a rubber bag that expands as it is filled with air or gas

bal-lot [băl′ət] *noun*, 1) a printed ticket used in voting to make a choice 2) the whole vote cast

balm [bäm] *noun*, 1) something that relieves pain 2) a soothing ointment

balm-y [bä′mē] *adjective*, mild, refreshing, soft and warm

bam-boo [băm-bōō′] *noun*, a woody grass with hollow stems, abundant in the tropics

ban [băn] *noun*, 1) an order that does not allow something, *verb*, 2) to not allow something or to prevent it from taking place

ba-nal [bə-năl′, bā′nəl] *adjective*, ordinary, commonplace, trite

ba-nan-a [bə-năn′ə] *noun*, a long yellow tropical fruit having a tasty soft pulp

band [bănd] *noun*, 1) a group of musicians who get together to play their instruments, usually brass, percussion, and woodwinds 2) a strip of something 3) a group 4) a piece that wraps around something, *verb*, 5) to join together

band-age [băn′dĭj] *noun*, 1) a piece of cloth or adhesive strip put over a wound, *verb*, 2) to cover a wound with a cloth or adhesive strip

ban-dan-na [băn-dăn′ə] *noun*, a cotton scarf or handkerchief

ban-dit [băn′dĭt] *noun*, a robber

ban-dy [băn′dē] *verb*, to exchange, to pass to and fro

bane [bān] *noun*, the cause of ruin

bane-ful [bān′fəl] *adjective*, ruinous, having poisonous qualities, causing misery

bang [băng] *noun*, 1) a startling loud sound, *verb*, 2) to produce a loud sound by striking one thing against another

ban-ish [băn′ĭsh] *verb*, 1) to send away, to dismiss 2)2 to exile as a form of punishment

ban-is-ter [băn′ĭ-stər] *noun*, a railing that guards the edges of stairs to prevent someone from falling

ban-jo [băn′jō] *noun*, a stringed musical instrument, like a ukulele or guitar, but with a circular sound box

bank [băngk] *noun*, 1) the land along a body of water, the side of a hill 2) a commercial institution where money is kept and loaned

bank-er [băng′kər] *noun*, one who conducts the business of a bank

bank-rupt [băngk′rŭpt] *adjective*, broke, out of business

ban-ner [băn′ər] *noun*, a flag, a cloth bearing an emblem

ban-quet [băng′kwĭt] *noun*, a feast or ceremonious dinner

ban-ter [băn′tər] *noun*, 1) good-natured ridiculing or teasing 2) to speak or to address in a witty or teasing manner

bap-tism [băp′tĭz′əm] *noun*, the ceremony of initiation into Christianity

bap-tize [băp-tīz′] *verb*, to sprinkle someone with holy water and bestow a name

bar [bär] *noun*, 1) a counter where drinks are sold 2) a long solid object made of wood or metal 3) the legal profession, *verb*, 4) to exclude or block

barb [bärb] *noun*, the sharp projections from a fishhook, arrow, etc.

bar-ber [bär′bər] *noun*, a person who cuts men's hair

bare [bâr] *adjective*, 1) uncovered, naked 2) unadorned, empty

bare-ly [bâr′lē] *adverb*, almost, not, hardly, only, scarcely

bar-gain [bär′gĭn] *noun*, 1) something bought at a lower price than expected 2) an agreement or compact between two parties, *verb*, 3) to haggle for a lower price

barge [bärj] *noun*, a large boat with a flat bottom, used to carry things on a river or a canal

bar-i-um [bâr′ē-əm] *noun*, one of the elements of the alkaline earth group

bark [bärk] *noun*, 1) the outer covering of a tree trunk and branches 2) a sound made by a

dog or a brusque voice, *verb*,
3) to speak like a dog

bar-ley [bär′lē] *noun*, one of the
cereal grains

barn [bärn] *noun*, a farm building
where animals and harvested
crops are kept

bar-na-cle [bär′nə-kəl] *noun*, a
marine crustacean that attaches
itself to rocks and boats

ba-rom-e-ter [bə-rŏm′ĭ-tər] *noun*,
a device that measures air
pressure in the atmosphere

bar-on [băr′ən] *noun*, a nobleman,
an aristocrat, a tycoon

ba-roque [bə-rōk′] *adjective*, an
ornate style of the 17th and 18th
centuries

bar-racks [băr′əks] *noun*, *plural*,
buildings for lodging soldiers

bar-rage [bär′ĭj] *noun*, an artillary
attack covering a wide area

bar-rel [băr′əl] *noun*, 1) a large
container shaped like a cylinder
2) the long metal tube forming
part of a gun

bar-ren [băr′ən] *adjective*,
1) bleak, lacking in good things
2) unable to bear children,
unable to support growth

bar-ri-cade [băr′ĭ-kād] *noun*,
1) an obstruction or means of
defense, a barrier, *verb*, 2) to
block, to fortify, to bar

bar-ri-er [băr′ē-ər] *noun*, 1) a
stumbling block, a deterrent, an
obstacle, something blocking the
way 2) a fence or wall to mark
the limits of a place

bar-ris-ter [băr′ĭ-stər] *noun*, in
England, a counselor at law, a
lawyer

bar-ter [bär′tər] *verb*, to trade

base [bās] *adjective*, 1) low,
menial, inferior, *noun*, 2) the
bottom on which something
stands 3) an area designated
safe, in baseball one of four
sandbags

base-ball [bās′bôl′] *noun*, a game
played with a bat and a ball by
two teams with nine players on a
side, on a field with four bases at
the corners of a diamond

base-ment [bās′mənt] *noun*,
storage or living space in the
foundation of a building

ba-sic [bā′sĭk] *adjective*,
1) forming a starting point,
fundamental 2) plain, simple

ba-sin [bā′sĭn] *noun*, 1) a low,
bowl-shaped area, a depression in
the land 2) a shallow sink

ba-sis [bā′sĭs] *noun*, the starting
point or central idea of
something, the core

bask [băsk] *verb*, to lay in the sun

bas-ket [băs′kĭt] *noun*, a container
woven from reeds or twigs and
used to carry things

bas-ket-ball [băs′kĭt-bôl′] *noun*, a
game in which two teams of five
players each compete to throw a
ball through a hoop

bass [bās] *adjective*, 1) the lowest
range of the male voice, as in
sound, [băs] *noun*, 2) a fresh-
water fish

bas-tion [băs′chən] *noun*, a
fortress, a defense

bat [băt] *noun*, 1) a piece of wood
used to hit a ball in a game 2) a
flying mammal, *verb*, 3) to hit a
ball with a piece of wood

batch [băch] *noun*, a number of things in a group

bath [băth, bäth] *noun*, 1) a container large enough to hold water to wash a body 2) the act of washing one's body

bathe [bāth] *verb*, to wash with water, usually a part of a body

bath-room [băth/rōōm] *noun*, a room with a toilet and a sink where people wash

ba-ton [bə-tŏn] *noun*, a wand, a night-stick, a rod, a stick used for conducting an orchestra

bat-tal-ion [bə-tăl/yən] *noun*, a large group of army soldiers ready for battle

bat-ter [băt/ər] *noun*, 1) the person holding the bat in baseball 2) a mixture of flour, eggs, and milk, *verb*, 3) to hit repeatedly

bat-ter-y [băt/ə-rē] *noun*, an apparatus for storing chemical energy and converting it to electrical power

bat-tle [băt/l] *noun*, 1) a fight or skirmish between people or different groups, *verb*, 2) to fight

bay [bā] *adjective*, 1) reddish-brown, *noun*, 2) part of a sea that stretches into land 3) a recess in a room 4) a tree or shrub of the laurel family, *verb*, 5) to howl, to bark at

bay-o-net [bā/ə-nĭt] *noun*, a blade on the end of a rifle

ba-zaar [bə-zär/] *noun*, a marketplace where miscellaneous articles are sold

B.C., signifying a time before the birth of Christ

be [bē] *verb*, 1) used to give information about people or things and to join words for people or things to the qualities or position they have 2) the present indicative of verb **be**

Singular	Plural
I **am**	We **are**

I am proud of you.
We are happy to see you.

You **are**   You **are**

You are what you eat.
You are the students in the class.

He/she/it **is**   They **are**

He is a gentleman, he is kind.
They are all our friends.

Past Tense

I **was**   We **were**

I was waiting for you.
We were hoping you could stay.

You **were**   You **were**

You were a true friend.
You were good children.

He/she it **was**   They **were**

It was time to go.
They were our heroes.

beach [bēch] *noun*, a shore of sand or pebbles along a lake or ocean

bea-con [bē/kən] *noun*, a lantern, a beam of light, a signal

bead [bēd] *noun*, a small ball of glass or other material, with a small hole for string or wire to pass through

beak [bēk] *noun*, the hard, pointed mouth of a bird

beak-er [bē/kər] *noun*, 1) a deep, open-mouthed, thin, glass vessel with a lip 2) a tall wide-mouthed cup

beam [bēm] *noun*, 1) a large piece of timber used to brace or support a ceiling 2) a ray, a shaft of light, *verb*, 3) to smile happily 4) to send or transmit, to shine

bean [bēn] *noun*, the seed of any bean plant, often used for food

bear [bâr] *adjective*, 1) referring to unfavorable market conditions. *noun*, 2) a large, furry mammal with four feet, *verb*, 3) to carry 4) to suffer or endure 5) to produce, to bring forth, to yield 6) to exhibit, to show, to relate

beard [bîrd] *noun*, hair on the face below the mouth

beast [bēst] *noun*, an animal, especially a savage animal

beat [bēt] *noun*, 1) a stroke or movement that is part of a rhythm, *verb*, 2) to pound, to tap 3) to conquer, to defeat

beau-ti-ful [byōō′tə-fəl] *adjective*, charming and delightful to the senses, very pretty, lovely

beau-ty [byōō′tē] *noun*, qualities pleasing to the eye or ear, charm

bea-ver [bē′vər] *noun*, an amphibious rodent with brown fur that builds dams as homes

be-cause [bǐ-kôz] *conjunction*, since, for the reason that

beck-on [bĕk′ən] *verb*, to signal, call or summon to come over by motioning silently

be-come [bǐ-kŭm′] *verb*, to change or to grow to be, to suit

bed [bĕd] *noun*, 1) a piece of furniture on which someone sleeps 2) the bottom of something 3) a garden plot

bed-lam [bĕd′ləm] *noun*, confusion, noise, uproar

bed-room [bĕd′rōōm] *noun*, the room for sleeping in

bee [bē] *noun*, an insect with a stinger and wings that makes honey and carries pollen from flower to flower

beech [bēch] *noun*, a family of hardwood trees that includes oak and elm

beef [bēf] *noun*, the meat from cattle

bee-hive [bē′hīv′] *noun*, a house made for bees to live in

been [bǐn] the *past tense* of the *verb* **to be**

beer [bîr] *noun*, an alcoholic beverage made of grain and hops

beet [bēt] *noun*, a purplish root vegetable, used for dye and eaten like a turnip

bee-tle [bēt′l] *noun*, an insect whose outside wings make a hard, protective cover for its body

be-fore [bǐ-fôr] *adverb*, 1) happening in advance of, earlier than, *preposition*, 2) in front of, ahead of

be-fore-hand [bǐ-fôr′hănd] *adverb*, ahead of, in advance

beg [bĕg] *verb*, to ask as a favor

be-gan [bǐ-găn′] *verb*, *past tense* of **begin**

beg-gar [bĕg′ər] *noun*, one who asks for favors or money

be-gin [bǐ-gǐn′] *verb*, to start, to commence

be-gin-ning [bĭ-gĭn′ĭng] *noun*, the start of something, the origin

be-guile [bĭ-gīl′] *verb*, to divert or to entertain, to trick, to charm

be-gun [bĭ-gŭn′] *verb, past tense* of **begin**

be-half [bĭ-hăf] *noun*, interest, side

be-have [bĭ-hāv′] *verb*, to act, to conduct yourself

be-hav-ior [bĭ-hāv′yər] *noun*, manner of acting or bearing oneself, conduct

be-he-moth [bĭ-hē′məth] *noun*, a huge creature

be-hind [bĭ-hīnd′] *adverb*, 1) further away from, late, *preposition*, 2) following, at the back of, remaining after

be-hoove [bĭ-hoov′] *verb*, to be fitting or proper, to require

be-ing [bē′ĭng] *noun*, 1) a living thing, *verb*, 2) *present participle of the verb* to **be**

be-la-bor [bĭ-lā′bər] *verb*, to repeat over and over

be-lat-ed [bĭ-lā′tĭd] *adjective*, too late, delayed, tardy

be-lea-guered [bĭ-lē′gərd] *adjective*, in a difficult position, besieged, attacked

bel-fry [bĕl′frē] *noun*, a tower where bells hang above a building

be-lie [bĭ-lī′] *verb*, to contradict, to give a false impression

be-lief [bĭ-lēf′] *noun*, faith, acceptance, conviction

be-lieve [bĭ-lēv′] *verb*, 1) to have faith or confidence 2) to think

be-lit-tle [bĭ-lĭt′l] *verb*, to tease, to make fun of

bell [bĕl] *noun*, a round, hollow object of metal or glass that sounds when it is struck

bel-li-cose [bĕl′ĭ-kōs′] *adjective*, warlike, ready to fight

bel-lig-er-ent [bə-lĭj′ər-ənt] *adjective*, hostile, pugnacious, preparing to fight, aggressive

bel-low [bĕl′ō] *verb*, to make a loud deep sound like a bull

bel-lows [bĕl′ōz] *noun, plural*, a machine that draws air through a valve and expels it through a tube by alternate expansion and contraction, used in an organ or to blow air on a fire

bel-ly [bĕl′ē] *noun*, stomach, gut,

be-long [bĭ-lông] *verb*, 1) to be a part of, to be a member 2) to own or have possession of

be-long-ings [bĭ-lông′ĭngs] *noun, plural*, one's personal possessions

be-lov-ed [bĭ-lŭv′ĭd] *adjective*, cherished, highly valued, adored

be-low [bĭ-lō′] *adverb*, 1) at a lower level, *preposition*, 2) at a lower place or rank

belt [bĕlt] *noun*, 1) a band worn around the waist,, often made of cloth or leather 2) a band or strip indicating a region or zone, *verb*, 3) to hit or smack, to clobber 4) to strap, to surround, to attach by wrapping a cord around something

bench [bĕnch] *noun*, 1) a long seat often wooden 2) the table at which someone works 3) a judge's seat in court

bend [bĕnd] *noun*, 1) a curve, *verb*, 2) to shape into a curve or make crooked 3) to stoop

be-neath [bĭ-nēth′] *preposition*, below, under, at a lower level than

ben-e-dic-tion [bĕn′ĭ-dĭk′shən] *noun*, a blessing

ben-e-fac-tor [bĕn′ə-făk′tər] *noun*, a gift-giver, a patron

be-nef-i-cent [bə-nĕf′ĭ-sənt] *adjective*, performing acts of kindness and charity, generous

ben-e-fi-cial [bĕn′ə-fĭsh′əl] *adjectiven*, useful, profitable

ben-e-fi-ci-ar-y [bĕn′ə-fĭsh′ē-ĕr′ē] *noun*, one who is going to receive something valuable

ben-e-fit [bĕn′ə-fĭt] *noun*, 1) an advantage, 2) a fundraiser *verb*, 3) to be helpful, to aid

be-nev-o-lence [bə-nĕv′ə-ləns] *noun*, goodwill or kindness, charity given, generosity

be-nev-o-lent [bə-nĕv′ə-lənt] *adjective*, generous, kind-hearted

be-nign [bĭ-nīn′] *adjective*, harmless, not cancerous, kind

bent [bĕnt] *past tense* of **bend**

be-quest [bĭ-kwĕst′] *noun*, the act of leaving by a written will, a legacy for the next generation

be-rate [bĭ-rāt′] *verb*, to criticize or scold vigorously

be-reave-ment [bĭ-rēv′-mĕnt] *noun*, the state of being deprived of something, sorrow, mourning

be-reft [bĭ-rĕft′] *adjective*, lacking, deprived of, doing without

ber-ry [bĕ-rĕ′] *noun*, a small, round fruit from a bush

ber-serk [bər-sûrk] *adjective*, frenzied with rage, crazy

be-seech [bĭ-sēch′] *verb*, to ask or entreat, to implore, to beg

be-side [bĭ-sīd′] *preposition*, next to, at the side of, adjacent

be-sides [bĭ-sīdz′] *adverb*, also, in addition to, as an extra

be-siege [bĭ-sēj′] *verb*, to surround with armed forces, to harass, to encircle, to crowd around

best [bĕst] *adjective*, finest, exceeding all others in quality

be-stow [bĭ-stō′] *verb*, to give something, to confer

bet [bĕt] *noun*, 1) an agreement to risk money on the odds of winning, *verb*, 2) to risk money on a gamble, to wager

be-tray [bĭ-trā′] *verb*, to break a trust, to reveal a secret

bet-ter [bĕt′ər] *adjective*, 1) superior, *adverb*, 2) more, larger, *verb*, 3) to improve

bet-ween [bĭ-twēn′] *adverb*, 1) showing where in relation to things or places on either side 2) showing when in relation to events before and after, *preposition*, 3) showing how things are joined 4) showing how things are divided

bev-er-age [bĕv′ər-ij] *noun*, a drink, a liquid refreshment

bev-y [bĕv′ē] *noun*, 1) a large group, 2) a group of birds

be-ware [bĭ-wâr′] *verb*, to be wary, to be careful, to be on guard of

be-wild-er-ed [bĭ-wĭl′dərd] *adjective*, confused, puzzled, perplexed

be-witch [bǐ-wǐch′] *verb* to charm, to put under a magic spell

be-yond [bē-ǒnd] *preposition*, at a distance farther away, on the other side of, past

bi-as [bī′əs] *adjective*, 1) in a slanting manner, obliquely, diagonally, *noun*, 2) inclination, tendency, *verb*, 3) to influence, to show prejudice, to sway

Bi-ble [bī′bəl] *noun*, the religious book of the Christian and Jewish faiths

bib-li-og-ra-phy [bǐb′lē-ǒg′rə-fē] *noun*, a list of books on any subject usually found at the end of a manuscript

bick-er [bǐk′ər] *verb*, to quarrel about petty things

bi-cy-cle [bī′sǐk′əl] *noun*, a two wheeled vehicle propelled by pedals

bid [bǐd] *noun*, 1) an offer to buy something, *verb*, 2) to offer a price for something 3) to ask, plea 4) to utter a greeting

bi-en-ni-al [bī-ěn′ē-əl] *adjective*, happening every two years

big [bǐg] *adjective*, large in size, weight, importance, etc.

big-ot [bǐg′ət] *noun*, a person filled with stubborn hatred

big-ot-ry [bǐg′ə-trē] *noun*, prejudice, intolerance

bi-lin-gual [bī-lǐng′gwəl] *adjective*, capable of speaking two languages

bill [bǐl] *noun*, 1) a piece of paper stating what you owe 2) a plan for a new law 3) the beak of a bird, *verb*, 4) to send a list of charges to someone

bil-liards [bǐl′yərdz] *noun, plural*, a game similar to pool

bil-lion [bǐl′yən] *noun, adjective*, a thousand million, written 1,000,000,000

bil-low [bǐl′ō] *noun*, 1) a wave, a surge of water or smoke, *verb*, 2) to surge, to swell

bin [bǐn] *noun*, a large container with a lid, for flour, grain, etc.

bind [bīnd] *verb*, 1) to tie with rope or string 2) to constrict 3) to put under an obligation

bind-er [bīn′dər] *noun*, that which fastens or binds together, such as a notebook with large rings to hold paper

binge [bǐnj] *noun*, a spree, uncontrolled eating

bin-oc-u-lars [bə-nǒk′yə-lərz] *noun, plural*, a pair of special glasses that make things in the distance look closer

bi-o-de-grad-a-ble [bī′ō-dǐ-grā′də-bəl] *adjective*, capable of being broken down by microorganisms into simple, stable compounds such as carbon dioxide and water

bi-og-ra-phy [bī-ǒg′rə-fē] *noun*, a written history of a person's life

bi-ol-o-gy [bī-ǒl′ə-jē] *noun*, the branch of knowledge that studies all living things

birch [bûrch] *noun*, a hardwood tree known for its thin, peeling bark and yellow leaves in the fall

bird [bûrd] *noun*, any of a class of warm-blooded, two-legged, egg-laying vertebrates with feathers and wings

birth [bûrth] *noun*, the act of being born, the beginning of existence

birth-day [bûrth′dā′] *noun*, the day someone is born

bis-cuit [bĭs′kĭt] *noun*, a small soft bread

bish-op [bĭsh′əp] *noun*, a senior member of the Christian clergy

bi-son [bī′sən, -zən] *noun*, a large mammal native to North America and Europe, including the buffalo

bit [bĭt] *noun*, 1) a very small amount 2) the metal mouthpiece on a bridle used to control a horse 3) the cutting edge of a tool, *verb*, 4) *past tense* of **bite**

bite [bīt] *noun*, 1) a wound made by biting or stinging, *verb*, 2) to seize or cut with teeth

bit-ter [bĭt′ər] *adjective*, 1) having a sharp, sour taste 2) grievous, painful 3) piercingly cold

bi-zarre [bĭ-zär′] *adjective*, very strange or unusual, weird

black [blăk] *adjective*, 1) lacking color or light, the opposite of white, *noun*, 2) a Negro or person of African descent

black-mail [blăk′māl′] *noun*, 1) compliance with another person's terms, hush money, *verb*, 2) to threaten to reveal a secret unless money is paid or favors given

blad-der [blăd′ər] *noun*, a sack attached to the kidneys to hold urine

blade [blād] *noun*, 1) the flat cutting part of anything sharp 2) a long flat leaf of grass, or anything with such a shape

blame [blām] *noun*, 1)responsibility for something bad, *verb*, 2) to say that someone is the cause of something unfortunate

blanch [blănch] *verb*, 1) to bleach, to whiten, to fade 2) to cook briefly in boiling water

bland [blănd] *adjective*, mild, almost tasteless, neutral

blank [blăngk] *adjective*, 1) without writing or other marks 2) without expression

blan-ket [blăng′kĭt] *noun*, a warm cloth cover used on a bed

blas-phe-my [blăs′fə-mē] *noun*, speech or actions that are disrespectful to God or sacred things

blast [blăst] *noun*, 1) a strong gust of wind 2) an explosion, *verb*, 3) to launch into space 4) to use explosions to break something apart 4) to make a loud sound

bla-tant [blāt′nt] *adjective*, loudly offensive, very obvious

blaze [blāz] *noun*, 1) a strong fire, *verb*, 2) to burn, to burst into flame 3) to shine brightly

blaz-er [blā′zər] *noun*, a light sports jacket of wool or silk

bleach [blēch] *noun*, 1) a substance used to whiten or remove color, *verb*, 2) to make white, to remove color

bleach-ers [blē′chərs] *noun*, *plural*, seats for spectators

bleak [blēk] *adjective*, cold and unpleasant, empty

bleed [blēd] *verb*, 1) to lose blood 2) to become mixed or run (e.g., dyes in wet cloth)

blem-ish [blĕm/ĭsh] *noun*, a flaw, an imperfection, a pimple

blend [blĕnd] *noun*, 1) a thorough mixture, *verb*, 2) to mix together into one 3) to go together

bless [blĕs] *verb*, to make holy, to ask God's favor, to praise

bless-ing [blĕs/ĭng] *noun*, 1) words that thank God for His gifts 2) good fortune

blew [blo͞o] *verb*, *past tense* of **blow**

blight [blīt] *noun*, 1) a disease on plants that makes the leaves dry up, *verb*, 2) to injure, to damage, to ruin a crop

blind [blīnd] *adjective*, 1) unable to see, without the sense of sight 2) concealed, hard to see 3) oblivious, unaware, ignorant, *noun*, 4) a window covering to keep out light, *verb*, 5) to make unable to see

blink [blĭngk] *noun*, 1) the quick opening and closing of an eyelid, *verb*, 2) to open and shut the eyes quickly

bliss [blĭs] *noun*, complete joy, happiness, or delight

blis-ter [blĭs/tər] *noun*, a sack under the skin filled with water or blood, from a burn or rubbing

bliz-zard [blĭz/ərd] *noun*, a snow storm with strong winds and cold temperatures

block [blŏk] *noun*, 1) a cube, a chunk 2) the area bounded by four city streets 3) something that stands in the way and prevents movement, *verb*, 4) to hinder, clog or jam, to impede, prevent, stop, or obstruct 5) to form, shape, or press

block-ade [blŏ/kād] *noun,* the closure of an area , port, etc.

blond or blonde [blŏnd] *adjective*, a light hair color the shade of straw or wheat

blood [blŭd] *noun*, a red fluid in the veins of people and animals

blood ves-sel [blŭd vĕs/əl] *noun*, any of the tubes in the body that carry blood, veins, arteries, capillaries

bloom [blo͞om] *noun*, 1) a flower, *verb*, 2) to have a flower

blos-som [blŏs/əm] *noun*, 1) a flower, *verb*, 2) to bloom or open from a bud to come in to a flower

blot [blŏt] *noun*, 1) a spot or stain, *verb*, 2) to absorb a liquid with cloth or a sponge to remove it

blouse [blous] *noun*, a garment similar to a shirt covering the top of a woman's body, from the neck to the waist

blow [blō] *noun*, 1) a sudden strike with a hand or weapon, *verb*, 2) to force air into a current, to expel air from the mouth

blue [blo͞o] *adjective*, 1) having the color of a cloudless sky, 2) unhappy, depressed, *noun*, 3) one of the three primary colors: red, blue, yellow 3) a feeling of unhappiness, sadness

bluff [blŭf] *noun*, 1) a high, steep bank, *verb,* 2) to try to trick or fool someone 3) to pretend

blun-der [blŭn/dər] *noun*, a stupid mistake, an error

blunt [blŭnt] *adjective*, 1) not sharp 2) frank, candid, abrupt, *verb*, 3) to deaden, insensitize, dull

blur [blûr] *verb*, to make unreadable or indistinct, to obscure

blush [blŭsh] *verb*, to become red in the face from embarrassment

board [bôrd] *noun*, 1) a long flat piece of wood 2) a group of directors who act as a body 3) a piece of cardboard or wood on which a game is played, *verb*, 4) to cover with wood 5) to climb onto a vessel or vehicle

board-er [bôr′dər] *noun*, a paying guest who receives meals

boast [bōst] *verb*, to brag, to praise oneself, to show off

boat [bōt] *noun*, a small ship or vessel that floats on water

bode [bōd] *verb*, to predict, to portend, to foretell

bod-y [bŏd′ē] *noun*, 1) the form of a person or animal or object 2) a group of persons who act as a unit 3) a corpse

bog [bôg] *noun*, a swamp, a marsh

boil [boil] *noun*, 1) a painful sore under the skin, *verb*, 2) to heat to 212 degrees Fahrenheit 3) to be angered, to be enraged

bold [bōld] *adjective*, brave, assertive, forward, fearless

bolt [bōlt] *noun*, 1) a metal screw used with a nut to hold something in place, a metal bar 2) a stroke of lightning, a thunderbolt, *verb*, 3) to lock with a bar across a doorway 4) to run quickly from the scene

bomb [bŏm] *noun*, 1) a device of explosive materials, *verb*, 2) to bombard with explosives

bom-bard [bŏm-bärd] *verb*, to attack repeatedly

bond [bŏnd] *noun*, 1) a mutual feeling of trust that joins people together 2) a written promise to pay money, *verb*, 3) to join together, to unite

bond-age [bŏn′dĭj] *noun*, confinement, servitude

bone [bōn] *noun*, a hard white substance forming the skeleton of an animal

bon-fire [bŏn′fīr′] *noun*, a large controlled fire

bo-nus [bō′nəs] *noun*, a sum paid in addition to a salary

book [bŏŏk] *noun*, 1) sheets of paper bound together for reading or writing, a long work of literature, etc., in electronic form, *verb*, 2) to make reservations for a future date

book-keep-ing [bŏŏk′kē′pĭng] *noun*, a systematic record of business transactions

boo-mer-ang [bōō′mə-răng′] *noun*, an angular shaped object that returns after it is thrown

boor [bŏŏr] *noun*, a rude person, a jerk, a brash individual

boost [bōōst] *noun*, 1) a lift, a helping hand, *verb*, 2) to support, to encourage

boot [bōōt] *noun*, 1) a shoe that covers above the ankle, *verb*, 2) to kick 3) to load the operating system on a computer

booth [bōōth] *noun*, an enclosure, a cubicle, a compartment

bor-der [bôr′dər] *noun*, 1) the dividing line between two countries or territories 2) an edge or trim, *verb*, 3) to lie along the edge of a geographic boundary

bore [bôr] *noun*, 1) a tiresome person, a nuisance, *verb*, 2) to drill a hole, to penetrate 3) to make someone feel disinterested

bore-dom [bôr′dəm] *noun*, dullness, lack of interest

boring [bôr′ĭng] *adjective*, not interesting, dull, tedious

born [bôrn] *adjective*, having been given life, existing as a result of birth

bor-ough [bûr′ō] *noun*, a self-governing incorporated town

bor-row [bŏr′ō] *verb*, 1) to use a thing that does not belong to you, then return it 2) to adopt or use as one's own

bos-om [bŏŏz′əm] *adjective*, 1)re-lating to the closeness between friends or family, *noun*, 2) chest area 3) a woman's breast

boss [bôs] *noun*, 1) the person in charge, someone in control, the employer, *verb*, 2) to give orders in a domineering way

bot-a-nist [bŏt′n-ĭst] *noun*, one who studies plants

bot-a-ny [bŏt′n-ē] *noun*, the science of plants

both [bōth] *adjective*, *pronoun*, the two together in conjunction

both-er [bŏth′ər] *noun*, *no plural*, 1) trouble or difficulty, a nuisance, *verb*, 2) to trouble or annoy, to harass or pester

bot-tle [bŏt′l] *noun*, 1) a container with a narrow neck, usually made of glass or plastic, used to hold liquids *verb*, 2) to store in a glass container

bot-tom [bŏt′əm] *noun*, 1) the lowest part or end of something 2) the buttocks, seat

bough [bou] *noun*, a tree branch

bouil-lon [bŏŏl′yŏn′, -yən] *noun*, a broth, consommé

boul-der [bōl′dər] *noun*, a very large stone or rock

bounce [bouns] *verb*, to spring back from a surface, to jump

bound [bound] *adjective*, 1) constrained, restricted, confined 2) going to, destined for, *noun*, 3) a jump, a leap

bound-a-ry [boun′də-rē] *noun*, a real or imaginary limit, especially between two properties

boun-ti-ful [boun′tə-fəl] *adjective*, generous, plentiful, provided freely, abundant

bow [bō] *noun*, 1) a ribbon worn as a hair ornament 2) a piece of wood bent with a string, used to shoot arrows 3) a piece of wood with hairs fastened, to create vibrations when pulled across the strings of an instrument, such as a violin, [bou] 4) the front of a boat, *verb*, 5) to bend the body forward to show respect 6) to submit to another, to consent

bowl [bōl] *noun*, a round deep dish

box [bŏks] *noun*, 1) a container like a cube, constructed from wood, cardboard, metal, etc. *verb*, 2) to fight with clenched fists

boy [boi] *noun*, a male child

boy-cott [boi′kŏt′] *noun*, 1) the refusal to purchase the products of an individual, corporation, or nation as a way to bring economic pressure for social or political change, *verb*, 2) to refuse to deal with or trade with, to desert, to abandon

brace [brās] *noun*, 1) a support, a vise, 2) two like items, a pair, *verb*, 3) to fortify

brack-et [brăk'ĭt] *noun*, 1) a shelf supported against a wall 2) one of the signs [ ] in a sentence that encloses additional information

brack-ish [brăk'ĭsh] *adjective*, a mixture of fresh and salt water

brag [brăg] *verb*, to praise oneself, to sound conceited, to boast

braid [brād] *noun*, 1) a rope, twine, *verb*, 2) to weave hair or rope

Braille [brāl] *noun*, a system of writing for people who are blind using raised dots as letters and numbers to be read with the fingertips

brain [brān] *noun*, 1) the mass of nerve tissue in the heads of people and animals, the center of the nervous system 2) the intellect, where we get our thoughts and ideas

brake [brāk] *noun*, 1) a mechanical device used to stop a bicycle, car, train, etc., *verb*, 2) to stop with the help of a mechanical device

branch [brănch] *noun*, 1) a part of a tree that extends from the trunk 2) an office that is located some distance from the headquarters 3) a stream or river that flows into a larger one, *verb*, 4) to move away or divide from the main body into branches

brand [brănd] *noun*, 1) a label indicating the name of a product and the manufacturer, *verb*, 2) to make a permanent mark on an animal to identify it

brass [brăs] *noun*, 1) a yellow metal that is a copper and zinc alloy 2) musical instruments, such as a trumpet, that are made out of metal and that give a loud sound when blown

bra-va-do [brə-vä'dō] *noun*, an air of defiance, fake bravery

brave [brāv] *adjective*, willing to make a personal sacrifice

brav-er-y [brā'və-rē] *noun*, courage, daring, boldness

brawl [brôl] *noun*, 1) a fight, an altercation, a dispute, *verb*, 2) to argue and wrestle or fight

breach [brēch] *noun*, the breaking of a contract, a gap in a fence

bread [brĕd] *noun*, a food made of flour, water, and yeast that is baked

breadth [brĕdth] *noun*, the distance from one side to the other, width

break [brāk] *noun*, 1) an opening made when something is torn or broken, 2) a pause, a rest, a time away from school or work, *verb*, 3) to cause to fall apart 4) to violate, to disobey

break-fast [brĕk'fəst] *noun*, the first meal of the day, eaten in the morning

breast [brĕst] *noun*, 1) one of the two parts on the front of a woman's body that can give milk 2) the top part of the front of the body

breath [brĕth] *noun*, air inhaled and exhaled in respiration

breathe [brĕth] *verb*, to inhale air into the lung and then exhale

bred [brĕd] *verb*, *past tense and past participle* of **breed**

breed [brĕd] *noun*, 1) a type or group of animals descended from a common ancestor, *verb*, 2) to reproduce, to multiply 3) to train and raise animals

breed-ing [brē′dĭng] *noun*, values a person learns, upbringing

breeze [brēz] *noun*, a gentle wind

brev-i-ty [brĕv′ĭ-tē] *noun*, shortness, ability to say something briefly, succinctly

brew [brōō] *verb*, to make drinks such as tea, coffee, or beer

bribe [brīb] *noun*, 1) money given to buy cooperation, *verb*, 2) to get someone's cooperation by paying them, to pay off

brick [brĭk] *noun*, a block of baked clay used in construction

brid-al [brīd′əl] adjective, referring to the bride in a wedding

bride [brīd] *noun*, a woman about to be married, is in the process of being married, or who has just been married

bride-groom [brīd′grōōm] *noun*, a man who is going to be married, is in the process of being married, or has just been married

bridge [brĭj] *noun*, 1) an elevated structure such as a viaduct or an overpass that crosses over water or a highway 2) a card game 3) a link, a connection, *verb*, 4) to join, connect, bring together

bri-dle [brīd′l] *noun*, 1) a leather band on the head of a horse used

for control, *verb*, 2) to show anger or resentment

brief [brēf] *adjective*, 1) for a short period of time, *noun*, 2) a summary of a case for the court, *verb*, 3) to inform

bri-er [brī′ər] *noun* prickly vegetation

bri-gade [brĭ-gād′] *noun*, a large group of soldiers

bright [brīt] *adjective*, 1) giving light 2) shining, not dull 3) smart, intelligent 4) cheerful

bril-liant [brĭl′yənt] *adjective*, 1) very bright, glittering, splendid 2) very clever

brim [brĭm] *noun*, 1) the edge of a cup, glass, or bowl 2) the part of the hat that is the edge

bring [brĭng]*verb*, to carry something or to lead someone

brink [brĭngk] *noun*, the edge of a steep slope, the verge

brisk [brĭsk] *adjective*, 1) quick and active, energetic 2) curt

bris-tle [brĭs′əl] *noun*, 1) a coarse brush, *verb*, 2) to feel irritated or annoyed, to show indignation

brit-tle [brĭt′l] *adjective*, hard but easily broken or snapped

broach [brōch] *verb*, to bring up, to begin discussion of

broad [brôd] *adjective*, 1) wide, extensive 2) tolerant

broad-cast [brôd′kăst′] *noun*, 1) a program aired on television or radio, *verb*, 2) to convey a message through radio or television or other electronic media

broad-en [brôd′n] *verb*, to expand

bro-chure [brō-shōōr'] *noun*, a pamphlet dealing with a subject of passing interest, a leaflet

broc-co-li [brŏk'ə-lē] *noun*, a plant with dense green flowers eaten as a vegetable

broil [broil] *verb*, to cook by exposing to direct, intense heat

broke [brōk] *adjective*, 1) without money, bankrupt, *verb,* 2) *past tense* of **break**

bro-ker [brō'kər] *noun*, an agent who buys and sells for others

bron-chi-tis [brŏn-kī'tĭs] *noun*, an inflammation of the lungs that causes chronic coughing

bronze [brŏnz] *noun*, a hard metal made by mixing copper and tin

brooch [brōch] *noun*, an ornamental dress-clasp, a decorative pin

brood [brōōd] *noun*, 1) a group of animals born together, or the children of a family, *verb,* 2) to think deeply and sadly about something

brook [brōōk] *noun*, 1) a small river or stream, *verb,* 2) to allow

broom [brōōm] *noun*, a brush with a long handle

broth [brôth] *noun*, watery soup, bouillon

broth-er [brŭth'ər] *noun*, a boy or man with the same parents as another person

brought [brôt] *verb*, *past tense* of **bring**

brow [brou] *noun*, the part of the face between the eyes and the hair, the forehead

brown [broun] *adjective*, a dark color like coffee or earth

browse [brouz] *verb*, to leaf through, to scan, to pass over

bruise [brōōz] *noun*, 1) a red or blue mark left on damaged skin caused by blood vessels breaking beneath the skin, *verb,* 2) to injure in such a way that the blood vessels beneath the skin are broken

brunch [brŭnch]*noun*, a meal between breakfast and lunch

bru-nette [brōō-nĕt'] *noun*, a woman with brown hair

brush [brŭsh] *noun*, 1) a device with wires or plastic bristles spread from a handle for combing, or cleaning, or painting, *verb,* 2) to clean, straighten, or paint with a brush

brusque [brŭsk] *adjective*, rough and short in manner, blunt,

bru-tal [brōōt'l] *adjective*, fierce, harsh, destructive

brute [brōōt] *noun*, a cruel person who acts like a wild animal

bub-ble [bŭb'əl] *noun*, 1) a sphere containing liquid or gas, *verb,* 2) to boil, to froth

buck [bŭk] *noun*, a male deer, goat, or rabbit

buck-et [bŭk'ĭt] *noun*, a container, usually with a curved handle to carry something, a pail

buck-le [bŭk'əl] *noun*, a device for uniting two loose ends of a belt or to fasten shoes, etc.

bu-col-ic [byōō-kŏl'ĭk] *adjective*, rustic, pastoral, in the country

bud [bŭd] *noun*, a sprout on a stem that becomes a flower or branch

Bud-dhism [bōō′dĭz′əm] *noun*, one of the world's major religions, practiced primarily in Asia and taught by monks

budge [bŭj] *verb*, to make something heavy move a little

budg-et [bŭj′ĭt] *noun*, a plan that indicates how money will be spent

buf-fa-lo [bŭf′ə-lō′] *noun*, a large, shaggy, wild ox

buff-er [bŭf′ər] *noun*, 1) a solution that neutralizes different substances 2) a person or thing between two opposing forces that softens the impact of their collision, *verb*, 3) to lessen

buf-fet [bə-fā′] *noun*, 1) a counter or table for refreshments, [bə-fĭt] *verb*, 2) to batter, to flap, to hit

buf-foon [bə-fōōn′] *noun*, one who clowns around

bug [bŭg] *noun*, 1) an insect that drinks juices from plants or animals 2) an error in data, *verb*, 3) to eavesdrop electronically

bug-gy [bŭg′ē] *noun*, 1) a hand-pushed carriage for babies 2) a one-horse carriage for two people

bu-gle [byōō′gəl] *noun*, a musical instrument, a horn used mainly in the military

build [bĭld] *verb*, 1) to create or construct 2) to increase

build-ing [bĭl′dĭng] *noun*, a structure with a foundation, roof, and walls

built [bĭlt] *verb*, *past tense* of **build**

bulb [bŭlb] *noun*, 1) the underground part of a plant formed by thick leaves, which separates and becomes a new plant 2) any object of this shape, such as a light bulb

bulge [bŭlj] *noun*, 1) a swelling shape, *verb*, 2) to swell outwards

bulk-head [bŭlk′hĕd′] *noun*, a partition separating the compartments of a vessel

bull [bōōl] *adjective*, 1) describing the stock market when prices are going up and the economy is strong, *noun*, 2) male animal of the ox, cow, whale, or other animals

bull-doz-er [bōōl′dō′zər] *noun*, a heavy machine that moves earth

bul-let [bōōl′ĭt] *noun*, a piece of metal fired from a gun

bul-le-tin [bōōl′ĭ-tn] *noun*, an announcement made by an organization, a newsletter

bul-lion [bōōl′yən] *noun*, a block of metal such as gold or silver

bul-ly [bōōl′ē] *noun*, 1) a person who likes to hurt or tease weaker people, *verb*, 2) to intimidate someone, to make someone feel afraid or overpowered

bul-wark [bōōl′wərk] *noun*, protection, a defensive wall

bum-ble-bee [bŭm′bəl-bē′] *noun*, a large yellow and black hairy social bee

bump [bŭmp] *noun*, 1) a sudden blow 2) a raised round lump due to swelling 3) an elevated place in the road or on a surface, *verb*, 4) to knock against, to strike

bun [bŭn] *noun*, 1) a small sweet roll or dinner roll 2) a hairstyle in which the hair is lifted off the neck and wrapped into a circle

bunch [bŭnch] *noun*, 1) a cluster of similar things grouped together, *verb*, 2) to gather together in a mass

bun-dle [bŭn′dl] *noun*, 1) a group of things tied together to make a package, *verb*, 2) to make into a bunch by tying things together

bu-oy [boo′ē] *noun*, an anchored floating device used as a navigation marker

buoy-ant [boi′ənt] *adjective*, 1) having the quality of rising or floating in fluid 2) light-hearted

bur-den [bûr′dn] *noun*, 1) a heavy thought or object that is carried, *verb*, 2) to load

bu-reau [byoor′ō] *noun*, 1) a chest of drawers 2) a satellite office for a news agency

bu-reauc-ra-cy [byoo-rŏk′rə-sē] *noun*, a system of government through agencies to which tasks and authority are delegated

bur-geon [bûr′jən] *verb*, to grow and develop rapidly

bur-glar [bûr′glər] *noun*, one who breaks into other people's property to rob them

bur-i-al [bĕr′ē-əl] *noun*, the act of putting a dead body in the ground, internment

burn [bûrn] *noun*, 1) a wound caused by fire or heat, or friction, *verb*, 2) to be destroyed by fire, heat or acid

burnt [bûrnt] *adjective*, showing damage by fire

burr [bûr] *noun*, a round seed covered with spurs or barbs

bur-ro [bûr′ō] *noun*, a donkey

bur-row [bûr′ō] *noun*, 1) a tunnel or hole dug by an animal as a den, *verb*, 2) to dig or hide in a hole

burst [bûrst] *noun*, 1) a sudden action, *verb*, 2) to break due to force from inside, to combust

bury [bĕr′ē] *verb*, to dig a hole to conceal something underground, to cover, to conceal

bus [bŭs] *noun*, a large vehicle that carries many people to a destination by a road

bush [boosh] *noun*, a shrub

bush-el [boosh′əl] *noun*, dry measure equal to 64 pints or 32 quarts

busi-ness [bĭz′nĭs] *noun*, 1) regular occupation, work, profession 2) something to be transacted 3) affairs involving money

bust [bŭst] *noun*, a sculpture of a person from the chest up

bus-y [bĭz′ē] *adjective*, 1) having a lot to do, *verb*, to work actively

but [bŭt; bət when unstressed] *adverb*, 1) only, merely, *conjunction*, 2) yet, however, *preposition*, 3) except, save 4) other than

butch-er [booch′ər] *noun*, a person who cuts meat for sale

but-ter [bŭt′ər] *noun*, a substance made from churning cream

but-ter-fly [bŭt′ər-flī′] *noun*, an insect with four colorful wings and a slender body

but-ton [bŭt′n] *noun*, 1) a variety of forms or materials used to fasten a dress, pants, shirt, etc. 2) a badge, a form of identification, 3) a knob pressed to operate a device, *verb*, 4) to secure with a button or snap

but-ton-hole [bŭt′n-hōl′] *noun*, the hole in a piece of clothing that a button goes through in order to fasten two sides together

but-tress [bŭt′rĭs] *noun*, a support projecting outwards to strengthen the walls of a building

buy [bī] *verb*, to acquire something by giving an accepted price, to purchase with money

buy-er [bī′ər] *noun*, a customer, one who makes a purchase

by [bī] *adverb*, 1) near, aside, past *preposition*, 2) near, beside 3) through, by way of, via 4) before 5) past

by-pro-duct [bī′prŏd′əkt] *noun*, anything produced in the process of making something else

byte [bīt] *noun*, 1) a small unit of information 2) a fixed number of bits (usually eight)

C

cab [kăb] *noun*, 1) a vehicle for hire, a taxi 2) the part of a truck or train where the driver sits

cab-bage [kăb′ĭj] *noun*, a vegetable with a short stem and thick green or purple overlapping leaves that form a dense head

cab-in [kăb′ĭn] *noun*, 1) a small wooden house 2) a room for living quarters on a ship 3) the passenger area on an airplane

cab-i-net [kăb′ə-nĭt] *noun*, 1) a cupboard with drawers or shelves for keeping items 2) a group of people who advise the head of state

ca-ble [kā′bəl] *noun*, 1) a strong, large diameter wire or rope 2) a bundle of insulated wires that carry electrical current 3) a system delivering television signals by cable

ca-boose [kə-boōs′] *noun*, the last car used on a freight train, containing facilities for the crew

cache [kăsh] *noun*, 1) a hiding place for food or treasures, *verb*, 2) to hide or store something

cack-le [kăk′əl] *noun*, 1) a loud laugh, *verb*, 2) to laugh loudly to sound like a hen

ca-coph-o-ny [kə-kŏf′ə-nē] *noun*, harsh or dissonant noise, a discordant mixture of sounds

cac-tus [kăk′təs] *noun*, *plural*, **cacti or cactuses**, a prickly, succulent plant with thick stems that grows in hot, dry climates

ca-dav-er [kə-dăv′ər] *noun*, a dead body

ca-det [kə-dĕt′] *noun*, a student at a military school

caf-e-te-ri-a [kăf′ĭ-tîr′ē-ə] *noun*, a place to eat where customers select food from a serving line

caf-feine [kă-fēn] *noun*, a stimulant found especially in coffee, tea, or some sodas

cage [kāj] *noun*, a box-like enclosure with bars in which birds or animals are kept, a pen

cake [kāk] *noun*, 1) a sweet, baked food made of flour, fat, and eggs, *verb*, 2) to harden from a wet substance into a solid crust

ca-lam-i-ty [kə-lăm′ĭ-tē] *noun*, an unfortunate event, a disaster

cal-ci-um [kăl′sē-əm] *noun*, a silver-white, rather soft metal that occurs in bones, shells, etc.

cal-cu-late [kăl′kyə-lāt′] *verb*, to find the answer using arithmetic, to plan using reasoning

cal-cu-la-tion [kăl′kyə-lā′shən] *noun*, the result of numbers added together, subtracted, etc.

cal-cu-la-tor [kăl′kyə-lā′tər] *noun*, a small computer used to perform math problems

cal-cu-lus [kăl′kyə-ləs] *noun*,
1) a branch of mathematics that provides a method for describing change 2) a buildup of materials within the body, such as a gallstone

cal-en-dar [kăl′ən-dər] *noun*, a table of the days and the months of the year

calf [kăf] *noun*, 1) the newborn or young of any large mammal, such as a cow, whale, or elephant 2) part of the leg between the knee and the ankle, the back of the shin

cal-i-ber [kăl′ə-bər] *noun*, 1) the diameter of a cylinder or tube 2) a person's degree of ability

Cal-i-for-nia [kăl′ĭ-fôr′nyə] *noun*, a state on the Pacific coast of the United States, the capital is Sacramento. The state flower of **California** is the golden poppy and the motto is "Eureka!"

cal-lis-then-ics [kăl′ĭs-thĕn′ĭks] *noun*, *plural*, exercise to promote strength and grace

call [kôl] *noun*, 1) the act of reaching someone on the telephone or paying a visit 2) a shout or plea, *verb*, 3) to name 4) to shout 5) to ask someone to come 6) to telephone 7) to visit

cal-lig-ra-phy [kə-lĭg′rə-fē] *noun*, beautiful writing, the art of hand lettering

cal-lous [kăl′əs] *adjective*,
1) lacking feeling, hard-hearted, feeling no pity 2) thick skinned

cal-lus [kăl′əs] noun, a hard thickened spot (of skin)

calm [käm] *adjective*, quiet, not excited

cal-o-rie [kăl′ə-rē] *noun*, a unit used to measure the amount of energy produced

cam-el [kăm′əl] *noun*, a large animal with one or two humps on its back, used for transportation

cam-e-o [kăm′ē-ō′] *noun*, a piece of jewelry with a stone or shell on which figures are carved

cam-er-a [kăm′ər-ə] *noun*, an instrument used to take photographs or film pictures

cam-ou-flage [kăm′ə-fläzh] *noun*,
1) special clothing for people or colors in an animal that help them blend into their surroundings, *verb*, 2) an organism's ability to blend in with the environment

camp [kămp] *noun*, 1) a place with tents or cabins for temporary shelter, *verb*, 2) to live outdoors and sleep in a tent under the stars, etc.

cam-paign [kăm-pān′] *noun*,
1) activities organized to gain or win something, *verb*, 2) to create a connected series of operations to achieve a desired result

cam-pus [kăm′pəs] *noun*, the grounds surrounding a college, school, or organization

can [kăn] *noun*, 1) a metal container, *verb*, 2) able to, to know how to 3) to preserve food in a can

ca-nal [kə-năl′] *noun*, 1) an artificial waterway designed for navigation 2) a duct in the body

ca-nar-y [kə-nâr′ē] *noun*, a songbird, usually yellow

can-cel [kăn′səl] *verb*, to annul, to revoke, to end or stop

can-cel-la-tion [kăn′sə-lā′shən] *noun*, a stop or end to something, withdrawal

can-cer [kăn′sər] *noun*, a serious illness in which abnormal cells multiply rapidly

can-di-date [kăn′dĭ-dāt′] *noun*, one who is being considered or who applies for a certain job

can-dle [kăn′dl] *noun*, a cylinder of wax with a wick embedded in it designed to burn to give off light

can-dor [kăn′dər] *noun*, the quality of being fair, unprejudiced, and honest

cane [kān] *noun*, 1) a hollow stick made from some plants 2) a stick with a curved handle used for support when walking

ca-nine [kā′nīn] *adjective*, relating to dogs

can-is-ter [kăn′ĭ-stər] *noun*, a small box or container, often of metal, for holding tea, coffee, etc.

can-ni-bal [kăn′ə-bəl] *noun*, a person who eats people or an animal who eats one of its own kind

can-non [kăn′ən] *noun*, a mounted gun or piece of artillery

can-not [kăn′ŏt] unable to

ca-noe [kə-nōō′] *noun*, a narrow light boat tapered at the ends, moved with paddles

can-o-py [kăn′ə-pē] *noun*, a covering from the sun, an awning or umbrella

cant [kănt] *noun*, 1) empty talk, deceit, 2) jargon, slang, *verb*, 3) to speak insincerely

can't [kănt] the *contraction* of the words **can** and **not**

can-ta-loupe [kăn′tl-ōp′] *noun*, a variety of muskmelon with an orange-red color fruit

can-tan-ker-ous [kăn-tăng′kər-əs] *adjective*, irritable, cross

can-vas [kăn′vəs] *noun*, 1) a strong, coarse, water-repellant cloth used to make tents, sails, etc. 2) a piece of heavy cloth, stretched over a frame, on which pictures are painted

can-vass [kăn′vəs] *verb*, 1) to examine thoroughly, scrutinize 2) to solicit votes or opinions

can-yon [kăn′yən] *noun*, a valley with high, steep sides between hills or mountains

cap [kăp] *noun*, 1) a soft hat 2) the covering for a bottle or tube

ca-pa-ble [kā′pə-bəl] *noun*, able to do things well, having ability, competent or efficient

ca-pac-i-ty [kə-păs′ĭ-tē] *noun*, 1) the maximum amount of liquid a container can hold, 2) the ability or aptitude for a specified task

cape [kāp] *noun*, 1) a piece of land that goes out into the sea 2) a coat without sleeves that covers shoulders, arms, and back

cap-il-lar-y [kăp′ə-lĕr′ē] *noun*, one of the tiniest blood vessels

cap-i-tal [kăp′ĭ-tl] *adjective*, 1) first, chief, 2 punishable by death, *noun*, 3) resources, means or money 4) the city where the rules for a state or county or country are made 5) a letter that begins a sentence or a proper noun

cap-i-tal-ism [kăp′ĭ-tl-ĭz′əm] *noun*, an economic system based on profit and private ownership

cap-i-tal-ize [kăp′ĭ-tl-īz′] *verb*, 1) to finance or invest in for profit 2) to use a capital letter

Cap-i-tol [kăp′ĭ-tl] *noun*, 1) the building in which the Congress of the United States meets to create laws, **capitol** 2) the building in which a state legislature meets to create laws

ca-price [kə-prēs′] *noun*, a silly thought, a notion, whimsy

cap-size [kăp′sīz′] *verb*, to upset or overturn as of a boat

cap-sule [kăp′səl] *noun*, 1) a container with a matching top and bottom that holds medicine 2) a small spacecraft

cap-tain [kăp′tən] *noun*, the leader of a team or a company of soldiers, or the commander of a ship or aircraft

cap-tion [kăp′shən] *noun*, 1) a short title heading 2) text under an illustration

cap-tive [kăp′tĭv] *adjective*, 1) unable to escape, *noun*, 2) a prisoner

cap-ture [kăp′chər] *verb*, 1) to seize and hold as a prisoner 2) to take hold of something

car [kär] *noun*, an automobile, a form of transportation with four wheels and an engine, usually fueled by gasoline

ca-rafe [kə-răf′] *noun*, a glass bottle, a decanter

car-at [kăr′ət] *noun*, a unit of weight for precious stones

car-a-van [kăr′ə-văn′] *noun*, 1) a group of people traveling together with vehicles or animals 2) a mobile home

car-bo-hy-drate [kär′bō-hī′drāt′] *noun*, a class of foods, that supply energy to the body

car-bon [kär′bən] *noun*, a nonmetallic element, symbolized as C, the required common substance in organic chemistry

car-bo-nate [kär′bə-nāt′] *noun*, a salt of carbonic acid

car-bon di-ox-ide [kär′bən dī-ŏk′sīd] *noun*, a colorless, odorless gas that is absorbed by plants

car-bon mon-ox-ide [kär′bən mə-nŏk′sīd′] *noun*, a colorless, odorless poisonous gas

car-cass [kär′kəs] *noun*, the dead body of an animal or bird

car-cin-o-gen [kär-sĭn′ə-jən] *noun*, a substance that causes cancer

card [kärd] *noun*, a small, flat piece of thick, stiff paper (e.g., a playing card, a postcard, a greeting or business card)

car-di-ac [kär′dē-ăk′] *adjective*, pertaining to the heart, stimulating the heart action

car-di-gan [kär′dĭ-gən] *noun*, a sweater that buttons or opens down the front

car-di-nal [kär′dn-əl] *adjective*, 1) chief, most important, *noun*, 2) an important leader of the Roman Catholic church, appointed by the pope as one of his advisors 3) a bright red bird common in North America

care [kâr] *noun*, 1) thought or interest 2) worry, anxiety, *verb*, 3) to show concern for others 4) to look after someone and tend to his or her needs 5) to show affection or love

ca-reen [kə-rēn′] *verb*, to sway or lean over to one side

ca-reer [kə-rîr′] *noun*, a profession, a chosen pursuit

care-ful [kâr′fəl] *adjective*, cautious, watchful

care-ful-ly [kâr′fə-lē] *adverb*, cautiously, with much care

care-less [kâr′lĭs] *adjective*, neglectful, reckless

ca-ress [kə-rĕs′] *noun*, 1) a tender or loving touch, *verb*, 2) to stroke with affection, to touch in a loving way

car-et [kär′ĭt] *noun*, a mark used by writers to indicate something – like a word – to be inserted

car-go [kär′gō] *noun*, the goods or freight carried by a ship, train, plane, or any vehicle

car-i-bou [kär′ə-bōō′] *noun*, a type of reindeer

car-i-ca-ture [kär′ĭ-kə-chŏŏr′] *noun*, a representation of a person or thing in which the peculiarities are exaggerated to produce a ridiculous effect

car-nage [kär′nĭj] *noun*, the killing of many people, a massacre

car-na-tion [kär-nā′shən] *noun*, a garden flower of the pink family

car-ni-val [kär′nə-vəl] *noun*, a festival with parades and dancing, a fair

car-niv-o-rous [kär-nĭv′ər-əs] *adjective*, meat-eating

car-ol [kär′əl] *noun*, a song of joy

ca-rouse [kə-rouz′] *verb*, to drink and be merry, especially with a group

carp [kärp] *noun*, 1) a kind of fish, *verb*, 2) to find fault, to criticize

car-pen-ter [kär′pən-tər] *noun*, a builder or repairer of wooden structures or objects

car-pen-try [kär′pən-trē] *noun*, the art of making things out of wood

car-pet [kär′pĭt] *noun*, a woven or felted floor covering of wool, cotton, or a synthetic, a rug

car-riage [kär′ĭj] *noun*, 1) a wheeled vehicle 2) a manner of bearing or posture

car-ri-er [kär′ē-ər] *noun*, 1) a messenger 2) one engaged in the business of carrying goods for others for hire

car-rot [kăr′ət] *noun*, a plant with an edible root of orange color used as a vegetable

car-ry [kăr′ē] *verb*, 1) to transport something, to convey, to move, to haul 2) to support a weight

cart [kärt] *noun*, a small wagon used to carry goods

car-ti-lage [kär′tl-ĭj] *noun*, elastic animal tissue attached to bones near the joints

car-tog-ra-pher [kär-tŏg′rə-fûr] *noun*, a maker of maps or charts

car-ton [kär′tn] *noun*, a box or container made of cardboard

car-toon [kär-tōōn′] *noun*, a comical drawing of a person, event, or place

car-tridge [kär′trĭj] *noun*, a case holding something that is dispensed mechanically

cart-wheel [kärt′hwēl′] *noun*, a sideways flip that is done by springing the body onto one hand and then the other, followed by the feet

carve [kärv] *verb*, 1) to reshape in an artistic or decorative manner by cutting 2) to cut meat for serving

cas-cade [kă-skād′] *noun*, a small waterfall, a profusion

case [kās] *noun*, 1) a container to hold things 2) a condition, an example of an illness 3) a question of law decided in court 4) something that has happened, an instance

cash [kăsh] *noun*, 1) money consisting of dollars or loose

change, *verb*, 2) to receive bills and coins in exchange for a check

cash-ier [kă-shîr′] *noun*, one who has charge of payments, etc., as at a store

cas-ket [kăs′kĭt] *noun*, a coffin

cas-se-role [kăs′ə-rōl′] *noun*, a dish consisting of several different foods combined together

cas-sette [kə-sĕt′, kă-] *noun*, a magnetic tape used for electronic recording enclosed in a plastic case that can fit directly into a recorder

cast [kăst] *noun*, 1) the performers in a play 2) a plaster dressing used to hold surgery in place, *verb*, 3) to direct a glance at someone or something 4) to assign a role in a play 5) to form a shape from clay on a potter's wheel 6) to throw or toss 7) the act of throwing a fishing line 8) to overshadow 9) to register a vote 10) to discard or shed

caste [kăst] *noun*, 1) one of the hereditary classes in Hindu society 2) any exclusive social class

cas-tle [kăs′əl] *noun*, 1) a fortress or palace usually the residence of nobility 2) a chess piece also known as the rook

cas-u-al [kăzh′ōō-əl] *adjective*, 1) not planned or prearranged 2) informal 3) careless

cas-u-al-ty [kăzh′ōō-əl-tē] *noun*, one injured or killed, as in a battle

cat [kăt] *noun*, a furry animal kept in the house as a pet or to catch mice

cat-a-combs [kăt′ə-kōmz′] *noun*, *plural*, a burial ground of underground rooms

cat-a-log [kăt′l-ôg′] *noun*, 1) a list of names, titles, or articles arranged in order, *verb*, 2) to list names and objects with a description of each one

cat-a-lyst [kăt′l-ĭst] *noun*, an agitator or force of change

cat-a-pult [kăt′ə-pŭlt′] noun, 1) a device that uses tension to throw objects, a weapon, *verb*, 2) to hurl or make an object go as far as possible

cat-a-ract [kăt′ə-răkt′] *noun*, 1) a great waterfall 2) an eye abnormality, cloudiness on the lens of the eye resulting in blurry vision

ca-tas-tro-phe [kə-tăs′trə-fē] *noun*, a sudden calamity, a disaster

catch [kăch, kĕch] *noun* 1) something that has been captured 2) the metal clasp that connects a door to a door frame, *verb*, 3) to get hold of something and not let go, to grasp 4) to get someone's attention 5) to trap

cat-e-chism [kăt′ĭ-kĭz′əm] *noun*, a set of questions and answers, especially for teaching about religion

cat-e-go-ry [kăt′ĭ-gôr′ē] *noun*, division or class to which something belongs

ca-ter [kā′tər] *verb*, to provide food or personal attention, to serve up

ca-ter-pil-lar [kăt′ər-pĭl′ər] *noun*, the larva of a butterfly

ca-thar-sis [kə-thär′sĭs] *noun*, 1) purging or cleansing of any part of the body 2) a release of tension or strong feelings

ca-the-dral [kə-thē′drəl] *noun*, a very large church, the main church of a diocese

Cath-o-lic [kăth′ə-lĭk] *noun*, a member of the church that recognizes the pope as its leader

cat-tle [kăt′l] *noun*, livestock such as cows, bulls, or steer kept for meat, milk, or skins

cau-cus [kô′kəs] *noun*, a group of politicians who gather to decide a policy or select a candidate

caught [kôt] *verb*, *past tense* of **catch**

caulk [kôk] *verb*, to make water-tight, to seal cracks

cause [kôz] *noun*, 1) the reason, the thing that made something happen 2) something you believe in strongly enough to take action, *verb*, 3) to bring about, to make something happen

caus-tic [kô′stĭk] *adjective*, 1) burning 2) sarcastically biting

cau-tion [kô′shən] *noun*, care and watchfulness, a warning

cau-tious [kô′shəs] *adjective*, wary, prudent, careful, watchful

cave [kāv] *noun*, a hollow place under the ground or in the side of a mountain or rock, sheltered from the elements

ca-ve-at [kā′vē-ăt′] *noun*, an indication of limitations

cav-ern [kăv′ərn] *noun*, a hollow place in a rock, a cave

cav-i-ar [kăv′ē-är′] *noun*, the prepared and salted eggs of the sturgeon or other large fish

cav-i-ty [kăv′ĭ-tē] *noun*, 1) a hollow place or a hole 2) a hole in the tooth caused by decay

cease [sēs] *verb*, to end, to stop

cease-less [sēs′lĭs] *adjective,* without end or interruption

ce-dar [sē′dər] *noun*, a large evergreen tree with reddish, fragrant wood

ceil-ing [sē′lĭng] *noun*, 1) the covering of a room 2) the top of a space 3) upper limit

cel-e-brate [sĕl′ə-brāt′] *verb*, to do certain things because of a special occasion, to rejoice

cel-e-bra-tion [sĕl′ə-brā′-shən] *noun*, a festive occasion

ce-leb-ri-ty [sə-lĕb′rĭ-tē] *noun*, fame, renown, a famous person

ce-ler-i-ty [sə-lĕr′ĭ-tē] *noun*, speed,

cel-er-y [sĕl′ə-rē] *noun*, a plant of the parsley family, whose leaf - stalks are eaten raw or cooked

ce-les-tial [sə-lĕs′chəl] *adjective*, in the stars, heavenly

cel-i-ba-cy [sĕl′ə-bə-sē] *noun*, the state of being unmarried

cell [sĕl] *noun*, 1) a small room where a prisoner is kept or where a nun or monk lives 2) a small, microscopic piece of a living organism

cel-lar [sĕl′ər] *noun*, a room under a house, partly or completely underground, a basement

cel-lu-lar [sĕl′yə-lər] *adjective,* 1) formed of cells 2) relating to a mobile telephone system

Cel-si-us [sĕl′sē-əs] *noun*, a scale on which the freezing point is 0

ce-ment [sĭ-mĕnt′] *noun*, a substance that consists of burned lime and clay, which is mixed with water and sand to make concrete

cem-e-ter-y [sĕm′ĭ-tĕr′ē] *noun*, a burial place, a graveyard

cen-sor [sĕn′sər] *noun*, 1) one who filters inappropriate material from books, movies, etc., before they are published or released *verb*, 2) to filter out information considered unsuitable

cen-sus [sĕn′səs] *noun*, an official count of a population

cent [sĕnt] *noun*, a metal coin equal to one penny or 1/100 of a dollar

cen-ten-ni-al [sĕn-tĕn′ē-əl] *adjective,* of or pertaining to a hundredth anniversary

cen-ter [sĕn′tər] *noun*, 1) the middle, a point equidistant from all sides 2) a place where people gather 3) a hub of activity

cen-ti-grade [sĕn′tĭ-grād′] *noun*, the temperature scale in which 0 degrees is the freezing point and 100 degrees is the boiling point of water

cen-ti-me-ter [sĕn′tə-mē′tər] *noun*, a measure of length somewhat less than ½ an inch, a standard unit in the metric system

cen-tral [sĕn′trəl] *adjective*, located in the middle

cen-trif-u-gal [sĕn-trĭf′yə-gəl] *adjective*, relating to forces caused by rotation directed outward from the center

cen-trip-e-tal [sĕn-trĭp′ĭ-tl] *adjective*, directed toward the center

cen-tu-ry [sĕn′chə-rē] *noun*, a period of one hundred years

ce-re-al [sîr′ē-əl] *noun*, any edible grain, including oats, rye, rice

cer-e-bel-lum [sĕr′ə-bĕl′əm] *noun*, the part of the brain related to muscular movements and coordination

ce-re-bral [sĕr′ə-brəl] *adjective*, appealing to the intellect and not the emotions

cer-e-mo-ni-al [sĕr′ə-mō′nē-əl] *adjective*, following a tradition or certain rituals, formal

cer-e-mo-ny [sĕr′ə-mō′nē] *noun*, a special occasion that honors traditions or rituals

cer-tain [sûr ′tn] *adjective*, 1) sure 2) unquestionable 3) not named, but thought to be known

cer-tain-ty [sûr′tn-tē] *noun*, a state of being sure of something

cer-tif-i-cate [sər-tĭf′ĭ-kĭt] *noun*, a document verifying qualification or ability or fact

cer-ti-fy [sûr′tə-fī′] *verb*, to authorize the validity of something, to guarantee

chafe [chāf] *verb*, 1) to rub together leaving a sore 2) to annoy

chaff [chăf] *noun*, 1) a worthless or trivial thing 2) the outer coating of wheat

cha-grin [shə-grĭn′] *noun*, a feeling of embarrassment because one has failed, deep disappointment

chain [chān] *noun*, 1) links or rings joined together, *verb*, 2) to hold with a chain, to restrain

chair [châr] *noun*, furniture with four legs and a back used to seat someone

chal-ice [chăl′ĭs] *noun*, a goblet

chalk [chôk] *noun*, a soft white limestone substance used to write on a blackboard

chal-lenge [chăl′ənj] *noun*, 1) an offer to test ability or skill 2) a formal objection, *verb*, 2) to call or invite to a contest of any kind

cham-ber [chām′bər] *noun*, 1) a room 2) a space surrouded by walls 3) a compartment in a firearm that holds the ammunition

cham-pi-on [chăm′pē-ən] *noun*, the winner of a competition

cham-pi-on-ship [chăm′pē-ən-shĭp′] *noun*, a competition among people or teams to decide who is best

chance [chăns] *noun*, 1) luck, fate 2) possibility or probability 3) risk 4) a turn to do something, an opportunity 5) a coincidence

chan-de-lier [shăn′də-lîr′] *noun*, a branched lighting fixture that hangs from the ceiling in a frame

change [chānj] *noun*, 1) a transformation, the process of becoming different 2) loose coins, *verb*, 3) to make or become different 4) to transform

change-a-ble [chān′jə-bəl] *adjective*, variable, adaptable

chan-nel [chăn′əl] *noun*, 1) a narrow stream of flowing water that connects two large bodies of water, *verb*, 2) to make something, such as energy or water, flow in a certain direction

chant [chănt] *noun*, 1) words repeated in rhythm, *verb*, 2) to speak in a singing way

cha-os [kā'ŏs'] *noun*, disorder, confusion, upheaval

chap-el [chăp'əl] *noun*, a small room within a larger building set aside for religious purposes

chap-er-one [shăp'ə-rōn'] (also **chaperon**) *noun*, an older person who accompanies as a protector

chap-lain [chăp'lĭn] *noun*, a clergyman, minister priest or rabbi serving in a hospital, the armed forces, etc.

chap-ter [chăp'tər] *noun*, a division of a book

char [chär] *verb*, to burn, to scorch

char-ac-ter [kăr'ək-tər] *noun*, 1) the moral makeup of a person 2) a person in a book or movie or play 3) a symbol, sign, or written letter used in a language

char-ac-ter-is-tic [kăr'ək-tə-rĭs'tĭk] *noun*, a trait, an attribute

char-ac-ter-ize [kăr'ək-tə-rīz'] *verb*, to describe, to indicate the character of someone or something

char-coal [chär'kōl'] *noun*, a black porous material made by partially burning wood

charge [chärj] *noun*, 1) an accusation 2) care or instruction given 3) the cost, expense, *verb*, 4) to request money 5) to accuse someone of doing something wrong 6) to rush or attack 7) to fill with power 8) to control

char-i-ot [chăr'ē-ət] *noun*, a horse-drawn cart

cha-ris-ma [kə-rĭz'mə] *noun*, attraction, charm, and allure, having a strong effect on others

char-i-ta-ble [chăr'ĭ-tə-bəl] *adjective*, 1) giving freely and generously 2) lenient, kind

char-i-ty [chăr'ĭ-tē] *noun*, 1) goodness, generosity, and kindness 2) an agency that provides a service to those in need, especially the poor

charm [chärm] *noun*, 1) a pleasant, courteous manner 2) a thing like a talisman or a group of words spoken to cause good luck, *verb*, 3) to please someone, to mesmerize, to infatuate

charm-ing [chär'mĭng] *adjective*, pleasing, beautiful, enchanting

chart [chärt] *noun*, 1) a piece of paper or spreadsheet with information on it indicating, patterns, information, or changes, *verb*, 2) to plan a route on a map of the ocean or other large area

char-ter [chär'tər] *noun*, 1) a document establishing the creation of an entity 2) a reservation of a vehicle, ship, etc., *verb*, 3) to hire a ship, bus, plane, etc.

chase [chās] *noun*, 1) the act of running after someone, *verb*, 2) to follow someone while trying to capture them, to pursue

chasm [kăz'əm] *noun*, a deep hole

chas-sis [shăs'ē] *noun*, the under part of a car, consisting of the frame with the wheels and machinery

chaste [chāst] *adjective*, 1) pure, celibate 2) simple in style

chat [chăt] *noun*, 1) a pleasant conversation, *verb*, 2) to make conversation or small talk

chauf-feur [shō′fər] *noun*, a person employed as a driver of a private automobile

chau-vin-ist [shō′və-nĭst] *noun*, a prejudiced belief in the superiority of your own race, sex, etc.

cheap [chēp] *adjective*, low in price, inexpensive, poorly made

cheat [chēt] *verb*, to do something that is not honest, to deceive

check [chĕk] *noun*, 1) an order for money, *verb*, 2) to make sure everything is in order

check-ers [chĕk′ərz] *noun*, *plural*, a board game played by two people, each with twelve pieces

cheek [chēk] *noun*, 1) the fleshy skin covering the bone beneath the eye socket on both sides of the face 2) impudent speech

cheer [chîr] *noun*, 1) a shout of support 2) gladness, a joyful state of mind, *verb*, 3) to make happy 4) to encourage with shouts

cheer-ful [chîr′fəl] *adjective*, smiling and happy

cheese [chēz] *noun*, the consolidated curd of milk, used as food

chef [shĕf] *noun*, a head cook

chem-i-cal [kĕm′ĭ-kəl] *noun*, a substance or mixture of substances used in or produced by changes in atoms or molecules

chem-ist [kĕm′ĭst] *noun*, a person who studies matter such as gases, metal, and liquids to see what they do and what they are made of

chem-is-try [kĕm′ĭ-strē] *noun*, the study of the composition, properties, and reactions of matter, particularly at the level of atoms and molecules

cher-ish [chĕr′ĭsh] *verb*, 1) to love dearly, to adore 2) to cling to an idea or feeling

cher-ry [chĕr′ē] *noun*, an edible, fleshy, red fruit containing a pit

chess [chĕs] *noun*, a game of skill played on a checkered board by two players each with 16 pieces

chest [chĕst] *noun*, 1) the front of the body between the shoulders and the waist 2) a large box made of metal, wood, etc.

chew [cho͞o] *verb*, to grind food between the teeth

chic [shēk] *noun*, great artistic cleverness or skill, style

chick-en [chĭk′ən] *noun*, fowl kept for eggs or meat

chide [chīd] *verb*, to scold

chief [chēf] *adjective*, 1) highest in office or rank, *noun* 2) the leader of a group or organization

chif-fon [shĭ-fŏn′] *noun*, 1) a soft gauzy silk material used for dresses 2) a light airy dessert made with egg whites

child [chīld] *noun*, *plural* **children**, a young person, offspring

child-ish [chīl′dĭsh] *adjective*, to be selfish and inconsiderate

chil-i or chili con carne [chĭl′ē] *noun*, a dish of tomatoes, meat, and sometimes beans, flavored with hot spices

chill [chĭl] *adjective*, 1) cold, frosty, *noun*, 2) coldness,

freezing, *verb*, 3) to make or
become cold

chime [chīm] *verb*, to make a
tinkling sound like a bell

chim-ney [chĭm′nē] *noun*, a
structure by which smoke passes
from a building to the outdoors

chim-pan-zee [chĭm′păn-zē′] *noun*,
an ape native to Africa larger
than a monkey, without a tail

chin [chĭn] *noun*, the part of the
face below the lips

chi-na [chī′nə] *noun*, *no plural*,
dishes, such as plates, cups, and
saucers, made of fine clay or
porcelain, used to serve a meal

chip [chĭp] *noun*, 1) a small piece
of something that has broken off
2) a cracker or a thinly sliced
fried food 3) the small silicon
object that enables the computer
to process information, *verb*,
4) to break off small pieces from
the main part

chip-munk [chĭp′mŭngk′] *noun*, a
tan rodent like a small squirrel
with a black stripe from its head
to the tip of its tail

chirp [chûrp] *noun*, 1) a high
pitched sound, *verb*, 2) to
warble, to tweet like a bird

chis-el [chĭz′əl] *noun*, 1) a metal
tool with a cutting edge at the
end of a blade, *verb*, 2) to carve
a sculpture or use a chisel as a
tool 3) to swindle, to cheat

chiv-al-ry [shĭv′əl-rē] *noun*,
bravery, heroism, courtesy,

chlo-ride [klôr′īd] *noun*, a
compound of chlorine with
another substance

chlo-rine [klôr′ēn′] *noun*, a yellow-
green gas added to water
supplies as a disinfectant

chlo-ro-form [klôr′ə-fôrm′] *noun*,
a colorless, sweetish liquid used
as an anesthetic

chlo-ro-phyll [klôr′ə-fĭl] *noun*,
a green pigment in plants,
essential for photosynthesis

choc-o-late [chô′kə-lĭt] *noun*, a
candy made from cocoa, the
roasted seeds of cacao

choice [chois] *adjective*, 1) first
rate, *noun*, 2) an act of making a
selection or choosing

choir [kwīr] *noun*, a band of
singers, especially in church

choke [chōk] *verb*, 1) to feel
suffocated because the throat is
obstructed 2) to repress

chol-er [kŏl′ər] *noun*, anger

choose [chōōz] *verb*, to pick or
select from a group, to make a
choice

chop [chŏp] *noun*, 1) a piece of
meat near the bone, *verb*, 2) to
cut with a sharp blade

cho-ral [kôr′əl] *adjective*, for or
relating to a chorus

chord [kôrd] *noun*, a combination
of three or more tones sounded
together in harmony

chore [chôr] *noun*, a job or
distasteful task

cho-re-og-ra-phy
[kôr′ē-ŏg′rə-fē, kōr′-] *noun*,
1) the plan of the steps and
movements in dancing 2) the
creation of ballets or dances

chor-tle [chôr′tl] *verb*, to make a
chuckling or snorting noise

cho-rus [kôr′əs] *noun*, 1) a
company of singers singing in
concert 2) the part of a song that
is repeated, the refrain

chow-der [chou′dər] *noun*, a thick
soup or stew usually featuring
fish, corn, etc.

chris-ten [krĭs′ən] *verb*, to give a
name to someone who is
baptized into the Christian faith

Chris-tian [krĭs′chən] *noun*, a
person who believes in Jesus
Christ

chromosome [krō′mə-sōm′] *noun*,
a thread-like structure in the
nucleus of a cell that carries the
genes

chron-ic [krŏn′ĭk] *adjective*,
continuing for a long time,
habitual, prolonged

chron-i-cle [krŏn′ĭ-kəl] *noun*, 1) a
historical register of facts in
order of their occurrence,
verb, 2) to record in a factual
and detailed way

chron-o-log-i-cal [krŏn′ə-lŏj′ĭ-kəl]
adjective, documented according
to the order of time

chuck-le [chŭk′əl] *noun*, 1) a
quick laugh or giggle, *verb*, 2) to
laugh softly, in a gentle manner

chunk [chŭngk] *noun*, a thick piece
of something, a block, a slab

church [chûrch] *noun*, a building
in which Christians meet to
worship and pray

churn [chûrn] *verb*, to move about
with force, to turn violently

ci-der [sī′dər] *noun*, a drink made
from pressed apples

ci-gar [sĭ-gär′] *noun*, tobacco
leaves rolled together in paper
for smoking

cig-a-ret-te [sĭg′ə-rĕt′] *noun*, a
cylinder of finely cut tobacco
rolled in paper

cin-der [sĭn′dərz] *noun*, a hot coal,
an ember

cin-e-ma [sĭn′ə-mə] *noun*, 1) a
building where films are shown
2) the movie business

cin-na-mon [sĭn′ə-mən] *noun*, the
bark of an East Indian tree,
ground into a spice

cir-cle [sûr′kəl] *noun*, 1) a closed
curve in which all points are
equidistant from the center, a
ring 2) a group of people

cir-cuit [sûr′kĭt] *noun*, the
circumference or distance
around an area

cir-cuit-ous [sər-kyōō′ĭ-təs]
adjective, roundabout, indirect,
meandering, winding

cir-cu-lar [sûr′kyə-lər] *adjective*,
1) round, *noun*, 2) a leaflet, a
pamphlet, a newsletter

cir-cu-late [sûr′kyə-lāt′] *verb*, to
move about or around

cir-cu-la-tion [sûr′kyə-lā′shən]
noun, 1) the act of moving
around, especially of blood
through the body 2) the extent or
amount of distribution,
dissemination

cir-cum-fer-ence [sər-kŭm′fər-əns]
noun, the distance around a
circle or circular figure

cir-cum-lo-cu-tion
[sûr′kəm-lō-kyōō′shən] *noun*, an
indirect or roundabout way of
expressing something

cir-cum-spec-tion
[sûr′kəm-spĕk′shən] *noun*,
caution, care, forethought

61

cir-cum-stance [sûr′kəm-stăns′]
noun, a particular incident, a
state of affairs, events

cir-cum-stan-tial
[sûr′kəm-stăn′shəl] *adjective*,
relating to evidence that tends to
prove a fact at issue by proving
other basic events

cir-cum-vent [sûr′kəm-vĕnt′] *verb*,
to find a way around, to evade a
difficulty

cir-cus [sûr′kəs] *noun*, a traveling
show usually performed by
clowns and acrobats and tamed
animals presented in an arena

cit-a-del [sĭt′ə-dəl] *noun*, fortress

ci-ta-tion [sī-tā′shən] *noun*, 1) a
summons to appear in court, a
ticket 2) a quote or reference to a
document 3) honorary mention
to receive an award or degree for
meritorious service

cite [sīt] *verb*, 1) to quote or refer
to a document 2) to recognize
someone for commendable
service 3) to write a ticket for
violating the law

cit-i-zen [sĭt′ĭ-zən] *noun*, a person
whose legal home is a certain
place or nation

cit-i-zen-ship [sĭt′ĭ-zən-shĭp′]
noun, 1) the fact of being a
citizen 2) community
participation

cit-y [sĭt′ē] *noun*, a large important
town, a metropolis

civ-ic [sĭv′ĭk] *adjective*, pertaining
to a community or city

civ-il [sĭv′əl] *adjective*, 1) polite
2) relating to a government
employee who is not in the
armed forces

ci-vil-ian [sĭ-vĭl′yən] *noun*, a
person not in the armed forces

ci-vil-i-ty [sĭ-vĭl′ĭ-tē] *noun*, good
manners, courtesy

civ-i-li-za-tion [sĭv′ə-lĭ-zā′shən]
noun, 1)the condition of being
civilized or having culture
2)people sharing their way of
life and living in one place at a
time, any human society

civ-i-lize [sĭv′ə-līz′] *verb*, to
reclaim from a savage state

civ-il war *noun*, a war fought in a
country between two different
groups of the same nation

claim [klām] *noun*, 1) a statement,
an assertion *verb*, 2) to say and
demand possession of something
by demonstrating ownership

claim-ant [klā′mənt] *noun*, a
person submitting an application
or making a claim

clair-voy-ant [klâr-voi′ənt]
adjective, having foresight

clam [klăm] *noun*, a shellfish
similar to an oyster

clam-or [klăm′ər] *noun*, 1) a loud
steady noise, an outcry, *verb*,
2) to make a lot of noise

clan [klăn] *noun*, a group of people
sharing one ancestor

clan-des-tine [klăn-dĕs′tĭn]
adjective, secret, concealed

clang [klăng] *noun*, the sound of
one piece of metal hitting
another, a ringing sound

clap [klăp] *noun*, 1) a loud noise,
as in thunder, *verb*, 2) to make a
sound by hitting your hands
together, applause

clar-i-fy [klăr′ə-fī′] *verb*, to make or become clear, to explain

clar-i-on [klăr′ē-ən] *adjective*, shrill, loud and clear

clar-i-ty [klăr′ĭ-tē] *adjective*, clearness

clash [klăsh] *noun*, 1) a fight or angry dispute 2) the sound of metal on metal, *verb*, 3) to hit or fight 4)of colors, to look awful together 5) to collide noisily

clasp [klăsp] *noun*, 1) a fastening, *verb*, 2) to hold tightly

class [klăs] *noun*, 1) a group of similar things or people 2) students meeting together for instruction, *verb*, 3) to classify

clas-sic [klăs′ĭk] *adjective*, showing excellence, exemplary

clas-si-cal [klăs′ĭ-kəl] *adjective*, 1)pertaining to ancient Greeks and Romans 2)excellent in a traditional style, esp. of music

clas-si-fi-ca-tion [klăs′ə-fĭ-kā′shən] *noun*, categorizing, grouping

clas-si-fy [klăs′ə-fī′] *verb*, to sort, to form into classes or groupings

clause [klôz] *noun*, 1) a group of words in a sentence that contains a subject and a verb 2) a single part in a contract, law, or treaty

claus-tro-pho-bi-a [klô′strə-fō′bē-ə] *noun*, a fear of being in a close space

claw [klô] *noun*, 1) the sharp, hard nails on the foot of an animal or bird, *verb*, 2) to tear or scratch with the hand or nails

clay [klā] *noun*, a common earth used for making pottery and bricks when baked

clean [klēn] *adjective*, 1) free from dirt, neat, *verb*, 2) to remove dirt and soil or impurities

clean-li-ness [klĕn′lē-nĭs] *noun*, the state of being clean, neatness of person, good hygiene

cleanse [klĕnz] *verb*, 1) to scrub, to wash, to disinfect 2) to absolve

clear [klîr] *adjective*, 1) easily understood 2) transparent 3) distinct or easy to hear, *verb*, 4) to remove items 5) to remove an obstacle, to bring into focus

clear-ance [klîr′əns] *noun*, 1) the act of removing whatever may obstruct 2) permission

clear-ly [klîr′lē] *adverb*, plainly

cleave [klēv] *verb*, 1) to cling, to adhere to 2) to separate or split

cleft [klĕft] *noun*, a space or opening made by or as if by splitting, a crack, a fissure

clem-en-cy [klĕm′ən-sē] *noun*, mildness, leniency, mercy

clench [klĕnch] *verb*, to hold tightly, to grip, to grasp

cler-gy [klûr′jnoun, ministers of the church as a group

clerk [klûrk] *noun*, a person who does office work, a salesperson

cle-ver [klĕv′ər] *adjective*, quick at learning, smart, good at doing things, resourceful, intelligent

cli-ché [klē-shā′] *noun*, a phrase repeated often, e.g. *down and out*

click [klĭk] *verb*, 1) to make a small noise like a tick on a clock 2) to press the mouse button once

cli-ent [klī′ənt] *noun*, a customer who uses the professional help of another, a patron

cliff [klĭf] *noun*, a high, steep wall of rock or bank on a coast

cli-mate [klī′mĭt] *noun*, the typical weather or atmosphere of a place year after year

cli-max [klī′măks′] *noun*, the crucial point when everything comes together, culmination

climb [klīm] *noun*, 1) the distance traveled up a hill or mountain, *verb*, 2) to go up, to ascend

clime [klīm] *noun*, region, climate

cling [klĭng] *verb*, to hold on tightly, to clutch, to grip

cling-ing [klĭng′ĭng] *adjective*, 1) devoted 2) holding fast 3) dependent

clin-ic [klĭn′ĭk] *noun*, a place that provides medical treatment

clin-ic-al [klĭn′ĭ-kəl] *adjective*, 1) related to observation and treatment of patients 2) scientific and observable

clip [klĭp] *noun*, 1) an object used to fasten things 2) a brisk pace, *verb*, 3) to cut or trim

clique [klēk] *noun*, a small group with joint interests, an exclusive set of friends, a gang

cloak [klōk] *noun*, 1) a loose piece of clothing, usually sleeveless, worn as an outer garment, *verb*, 2) to conceal

clock [klŏk] *noun*, an instrument used to measure time

clog [klôg] *noun*, 1) a blockage 2) a thick-soled shoe with an open back, *verb*, 3) to dance in the shoes 4) to hinder or obstruct, to be blocked up

close [klōs] *adjective*, 1) near 2) alike, similar, *verb*, 3) to shut out 4) to bring or come to an end

clos-et [klŏz′ĭt] *noun*, a small room or space where clothes are hung and items are stored

clot [klŏt] *noun*, a thickened mass of blood tissue

cloth [klôth] *noun*, a piece of fabric made of natural or synthetic fibers, such as cotton or rayon

cloth-ing [klō′thĭng] *noun*, *no plural*, garments, apparel, things worn as clothes, articles of dress

cloud [kloud] *noun*, 1) a visible mass of water droplets floating in the sky 2) a mass of dust

cloud-y [klou′dē] *adjective*, filled with clouds, overcast

clown [kloun] *noun*, a person who has the ability to turn himself or herself into a comical character that brings happiness and joy; especially to children, the sick, and sad

club [klŭb] *noun*, 1) a group of people who meet for a common purpose 2) a heavy stick

clue [kloō] *noun*, information or evidence that helps you find the answer to a puzzling question

clum-sy [klŭm′zē] *adjective*, awkward, unhandy, unwieldy

clus-ter [klŭs′tər] *noun*, 1) a clump, a bunch, a small, close group, *verb*, 2) to gather together into a small group

clutch [klŭch] *noun*, 1) the part of a machine that connects and disconnects the power from the rest of the machine, *verb*, 2) to hold something tightly, to grip, to cling to

clut-ter [klŭt′ər] *noun*, things lying about, a mess

coach [kōch] *noun*, 1) one who trains athletes, actors, or singers 2) a part of a train or bus that carries passengers 3) a carriage drawn by horses, *verb*, 4) to assist talented people develop their capability, to train

co-ag-u-late [kō-ăg′yə-lāt′] *verb*, to form into a dense mass, solidify from a liquid, to congeal

coal [kōl] *noun*, a fossil fuel in solid form that gives off heat when burned

co-a-lesce [kō′ə-lĕs′] *verb*, to combine, to fuse, to join together

co-a-li-tion [kō′ə-lĭsh′ən] *noun*, a union of people of differing views in a single body or group

coarse [kôrs, kōrs] *adjective*, 1) rough in surface or texture 2) crude, ill-mannered

coast [kōst] *noun*, 1) the edge of land that touches the ocean, *verb*, 2) to slide or glide

coat [kōt] *noun*, a piece of clothing with sleeves, worn over everything else, a jacket

coax [kōks] *verb*, to gently persuade

cob-bler [kŏb′lər] *noun*, someone who makes and repairs shoes

cob-web [kŏb′wĕb′] *noun*, threads woven by a spider in which it catches insects

co-caine [kō-kān′] *noun*, a bitter, crystalline alkaloid used as a local anesthetic and illegally as a stimulant

cock [kŏk] *noun*, a male bird

cock-pit [kŏk′pĭt′] *noun*, the compartment in a vehicle where the driver of a race car or the pilot of an airplane sits

co-co-nut [kō′kə-nŭt′] *noun*, the fruit of a variety of palm tree

co-coon [kə-kōōn′] *noun*, the silky case insects live in during the pupa stage

cod [kŏd] *noun, no plural*, a family of sea fish used for food

cod-dle [kŏd′l] *verb*, to treat gently, to pamper, to spoil

code [kōd] *noun*, 1) a collection of laws 2) a system of symbols given certain meanings

co-erce [kō-ûrs′] *verb*, to persuade by force

co-er-cion [kō-ûr′zhən] *noun*, persuasion by force or threats, pressuring

cof-fee [kô′fē] *noun*, the hot beverage made from the seeds of a coffee plant, a pale brown color

cof-fin [kô′fĭn] *noun*, a box in which a dead body is placed

cog [kŏg] *noun*, the tooth projecting from a wheel rim or a bar

cog-ni-zant [kŏg′nĭ-zənt] *adjective*, conscious, aware, knowledgeable

co-her-ent [kō-hîr′ənt] *adjective*, 1) able to be understood 2) forming a connection

co-he-sion [kō-hē′zhən] *noun*, the act of sticking together

coil [koil] *noun*, 1) a wire in a continuous circling shape, *verb*, 2) to wrap up around in a ring

coin [koin] *noun*, 1) a piece of money made of metal, legally authorized by a government 2) a quantity of value, *verb*, 3) to make coins 4)to invent

co-in-cide [kō′ĭn-sīd′] *verb*, to occur at the same time, to correspond exactly

co-in-cid-ence [kō-ĭn′sĭ-dəns] *noun*, random events that happen at the same time, chance, luck

col-an-der [kŭl′ən-dər] *noun*, a bowl-shaped container that is perforated for use as a strainer

cold [kōld] *adjective*, 1) feeling no heat, not warm 2) unfriendly, *noun*, 3) a virus with symptoms of a sore throat and runny nose

col-lab-o-rate [kə-lăb′ə-rāt′] *verb*, to work together as a group

col-lab-o-ra-tion [kə-lăb′ə-rā′shən] *noun*, a project that includes ideas from two or more people working together

col-lapse [kə-lăps′] *noun*, 1) any sudden or complete breakdown, *verb*, 2) to break down suddenly 3) to fall down

col-lar [kŏl′ər] *noun*, 1) a band worn around the neck of an animal 2) the part of a shirt around the neck

col-late [kə-lāt′] *verb*, 1) to examine in order to verify a special authenticity 2) to combine and arrange in order

col-lat-er-al [kə-lăt′ər-əl]*adjective*, 1)secondary, *noun*, 2) security in the form of property for repayment of a loan

col-league [kŏl′ēg′] *noun*, an associate from work, a friend or a peer, a co-worker

col-lect [kə-lĕkt′] *verb*, 1) to accumulate, to compile, to scrape together 2) to gather, to meet, to flock, to convene

col-lec-tion [kə-lĕk′shən] *noun*, an accumulation of things of a similar nature

col-lec-tive [kə-lĕk′tĭv] *adjective*, gathered, compiled, joint

col-lege [kŏl′ĭj] *noun*, a place where people study after high school that gives degrees

col-lide [kə-līd′] *verb*, to come together with great force

col-li-sion [kə-lĭzh′ən] *noun*, the act of striking or dashing together, a crash

co-lon [kō′lən] *noun*, 1) the sign [:] which is used before a listing or examples 2) the part of the large intestine extending to the rectum

colo-nel [kûr′nəl] *noun*, the commanding officer of a regiment, below brigadier general

co-lo-ni-al [kə-lō′nē-əl] *adjective*, relating to a period in America before independence

col-o-nist [kŏl′ə-nĭst] *noun*, someone who settles a territory

col-o-ni-za-tion [kŏl′ə-nĭ-zā′shən] *noun*, the act of setting up homes and starting a community in a new place with the purpose of living there a long time

col-o-ny [kŏl′ə-nē] *noun*, a country, area, etc.under the jurisdiction of another country and occupied by settlers

col-or [kŭl′ər] *noun*, 1) any of the hues of the rainbow, including white, *verb*, 2) to paint, dye, tint, to change the hue 3) to affect

Col-o-rado [kŏl′ə-răd′ō] *noun*, one of the United States, the capital city is Denver. The state flower of Colorado is the rocky mountain columbine, and the motto is "Nothing without the divine will."

co-los-sal [kə-lŏs′əl] *adjective*, huge, gigantic, immense

colt [kōlt] *noun*, a young male horse

col-umn [kŏl′əm] *noun*, 1) a supporting pillar 2) a perpendicular set of lines of a text in a book or periodical or newspaper or web page

col-um-nist [kŏl′əm-nĭst] *noun*, a writer who is featured regularly in a newspaper, or magazine, or Web site

co-ma [kō′mə] *noun*, state of being unresponsive, unconsciousness

comb [kōm] *noun*, 1) a piece of plastic, wood, or metal with teeth to smooth and adjust hair, *verb*, 2) to tidy hair with a comb

com-bat [kəm-băt′] *noun*, 1) an armed battle, *verb*, 2) to fight

com-bi-na-tion [kŏm′bə-nā′shən] *noun*, the act or result of combining, a union, a mixture

com-bine [kəm-bīn′] *verb*, to mix together, to blend, to unite

com-bus-ti-ble [kəm-bŭs′tə-bəl] *adjective*, apt to catch fire, flammable

come [kŭm] *verb*, 1) to arrive, to approach 2) to happen

com-e-dy [kŏm′ĭ-dē] *noun*, a funny performance that makes the audience laugh, humor

come-ly [kŭm′lē] *adjective*, attractive, beautiful, or handsome

com-et [kŏm′ĭt] *noun*, a celestial object of a frozen mass of dust that vaporizes as it nears the sun

com-fort [kŭm′fərt] *noun*, 1) consolation in trouble, *verb*, 2) to show kindness to someone, to console

com-fort-a-ble [kŭm′fər-tə-bəl] *adjective*, to feel at ease

com-ic [kŏm′ĭk] *adjective*, funny, humorous

com-ma [kŏm′ə] *noun*, a punctuation mark (,) used to indicate pauses and to separate elements within a sentence

com-mand [kə-mănd′] *noun*, 1) an order 2) mastery, rule, *verb*, 3) to be in charge of, to order someone to do something

com-man-deer [kŏm′ən-dîr′] *verb*, to take, often for military use

com-mem-o-rate [kə-mĕm′ə-rāt′] *verb*, to honor, to memorialize

com-mence-ment [kə-mĕns′mənt] *noun*, 1) the beginning 2) the ceremony of graduation

com-mend [kə-mĕnd′] *verb*, to praise, to compliment, to support

com-men-su-rate [kə-mĕn′sər-ĭt] *adjective*, equal in extent, in proportion, corresponding

com-ment [kŏm′ĕnt] *noun*, 1) a remark or something said, *verb*, 2) to make a remark about something, to express an opinion

com-merce [kŏm′ərs] *noun*, business, the buying and selling of goods, all forms of trade

com-mer-cial [kə-mûr′shəl] *adjective*,
1) for business purposes, having
profit as an aim, *noun*, 2) an
advertisement on television or radio

com-mis-er-ate [kə-mĭz′ə-rāt′]
verb, to feel or express sorrow

com-mis-sion [kə-mĭsh′ən] *noun*,
1) a committee or board that
studies an issue 2) the
percentage paid to an agent

com-mit [kə-mĭt′] *verb*, 1) to
entrust 2) to pledge 3) to act out,
to perform 4) to put into custody

com-mit-ment [kə-mĭt′mənt] *noun*, a
pledge, a promise, a responsibility

com-mit-tee [kə-mĭt′ē] *noun*, a
group appointed or formed to
consider or accomplish some
matter or business

com-mod-i-ty [kə-mŏd′ĭ-tē] *noun*,
an item that can be bought and
sold, merchandise

com-mon [kŏm′ən] *adjective*,
1) ordinary, found everywhere,
general 2) joint, shared, *noun*,
3) a public outdoor space

com-mon-wealth [kŏm′ən-wĕlth′]
noun, 1) a group of independent
states in a country 2) a self-
governed nation or state

com-mo-tion [kə-mō′shən] *noun*,
noise, confusion, disruption,

com-mune [kŏm′yōon′] *noun*,
a living arrangement in which
everyone shares possessions,
work, etc.

com-mu-ni-ca-ble
[kə-myōo′nĭ-kə-bəl] *adjective*,
contagious, infectious

com-mu-ni-cate [kə-myōo′nĭ-kāt′]
verb, 1) to give or exchange
information 2) to transmit

com-mu-ni-ca-tion
[kə-myōo′nĭ-kā′shən] *noun*,
1) exchange of information
between people 2) that which is
made known 3) transmission

com-mun-ion [kə′myōonyən]
noun, 1) the mutual sharing of
feelings and thoughts 2) a
religious fellowship among
members of a church

com-mu-nism [kŏm′yə-nĭz′əm]
noun, a political system in which
property is owned by the state

com-mu-nist [kŏm′yə-nĭst] *noun*, a
person who believes the state
should own all industry and land

com-mu-ni-ty [kə-myōo′nĭ-tē]
noun, a town, city, suburb, or
other place where people live
and work, a society

com-mute [kə-myōot′] *verb*, 1) to
make a sentence of punishment
less severe 2) to travel, to go back
and forth to work

com-pact [kəm-păkt′] *adjective*,
1) taking up a small amount of
space, *noun*, 2) a small case
with makeup and a mirror 3) an
agreement, *verb*, 4) to press
close together

com-pan-ion [kəm-păn′yən] *noun*,
a friend or associate, one who
accompanies another

com-pa-ny [kŭm′pə-nē] *noun*,
1) an association for business or
social purposes 2) a corporation
or firm 3) guests, visitors 4) a
subdivision of a regiment

com-pa-ra-ble [kŏm′pər-ə-bəl]
adjective, similar, equivalent

com-pare [kəm-pâr′] *verb*, to look
at similarities and differences

com·par·i·son [kəm-păr′ĭ-sən] *noun*, an examination of two or more objects with the view to discovering the similarities or differences, contrast

com·pass [kŭm′pəs] *noun*, 1) an instrument using a magnetic needle that always points to the north 2) a tool used to draw a circle 3) a range, scope, or limits

com·pas·sion [kəm-păsh′ən] *noun*, sympathy for suffering, kindness

com·pat·i·ble [kəm-păt′ə-bəl] *adjective*, in harmony with something, well-suited

com·pel [kəm-pĕl′] *verb*, 1) to drive or urge with force 2) to coerce

com·pen·di·um [kəm-pĕn′dē-əm] *noun*, an abridgement, collection, summary of information

com·pen·sate [kŏm′pən-sāt′] *verb*, to make up for something, to make amends, to pay

com·pen·sa·tion [kŏm′pən-sā′shən] *noun*, an amount paid to make up for something, payment

com·pete [kəm-pēt′] *verb*, to try to win a race, contest, etc.

com·pe·tent [kŏm′pĭ-tənt] *adjective*, capable, qualified, adequate, proficient, able

com·pe·ti·tion [kŏm′pĭ-tĭsh′ən] *noun*, 1) rivalry, free market conditions 2) a contest to see who performs better, a game

com·pet·i·tor [kəm-pĕt′ĭ-tər] *noun*, a rival, one who competes

com·pile [kəm-pīl′] *verb*, to gather information

com·pla·cen·cy [kəm-plā′sən-sē] *noun*, self-satisfaction, smugness

com·plac·ent [kəm-plā′sənt] *adjective*, 1) self-satisfied, smug 2) agreeable, lazy

com·plain [kəm-plān′] *verb*, 1) to express unhappiness or dissatisfaction or resentment 2) to make a formal accusation

com·plaint [kəm-plānt′] *noun*, 1) a criticism, reproach 2) an illness

com·ple·ment [kŏm′plə-mənt] *noun*, that which fills up, completes, or balances

com·plete [kəm-plēt′] *adjective*, 1) lacking nothing, full, whole, 2) finished, *verb*, 3) to finish

com·plex [kəm-plĕks′] *adjective*, 1) complicated, puzzling 2) consisting of connected parts, *noun*, 3) a group of buildings

com·plex·ion [kəm-plĕk′shən] *noun*, 1) the color or hue of the skin of the face 2) general appearance or aspect

com·pli·ance [kəm-plī′əns] *noun*, the act of yielding to demand, consent, agreement

com·plic·ate [kŏm′plĭ-kāt′] *verb*, to confuse, to perplex, to make more difficult or complex

com·pli·ment [kŏm′plə-mənt] *noun*, 1) nice words said about someone, praise, *verb*, 2) to express admiration, to laud

com·ply [kəm-plī′] *verb*, to follow, to conform, to go along with an order or a rule

com·po·nent [kəm-pō′nənt] *noun*, an essential part, a fundamental piece, a constituent part

com-port [kəm-pôrt′] *verb*, formal conduct of oneself, behavior

com-pose [kəm-pōz′] *verb*, 1) to write or create music, literature, etc. 2) to blend the ingredients or components of something

com-pos-ite [kəm-pŏz′ĭt] *adjective*, 1) made up of separate distinctive parts 2) relating to a combination of the Corinthian and Ionic styles

com-po-si-tion [kŏm′pə-zĭsh′ən] *noun*, 1) something written, such as a story or music 2) a creation or formation

com-post [kŏm′pōst′] *noun*, the product from the decomposition of organic materials, such as yard waste or dead plants

com-po-sure [kəm-pō′zhər] *noun*, a settled state, calmness

com-pound [kŏm-pound′] *noun*, 1) a composite, a mixture 2) a fenced-in area with one or more buildings in it, *verb*, 3) to make by mixing together 4) to increase

com-pre-hend [kŏm′prĭ-hĕnd′] *verb*, to understand

com-pre-hen-sive [kŏm′prĭ-hĕn′sĭv] *adjective*, thorough, all-inclusive

com-press [kŏm′prĕs′] *noun*, 1) a bandage, a pad, *verb*, [kŭm′prĕs] 2) to pack, to squeeze

com-prise [kəm-prīz′] *verb*, to include, consisting of

com-pro-mise [kŏm′prə-mīz′] *noun*, 1) the result of adjustments made by concessions to make a mutual agreement, *verb* 2) to adjust by making concessions

com-pul-sion [kəm-pŭl′shən] *noun*, an obsession, a fixation

com-pul-so-ry [kəm-pŭl′sə-rē] *adjective*, required, mandatory

com-pute [kəm-pyōōt′] *verb*, to figure mathematically

com-put-er [kəm-pyōōt′ər] an electronic device for storing and processing data, usually in binary form according to instructions given to it in a variable program

com-rade [kŏm′răd′] *noun*, a faithful friend, a colleague

con-cave [kŏn-kāv′] *adjective*, sunken, as the inside of a sphere or a spoon

con-ceal [kən-sēl′] *verb*, 1) to hide, to keep from sight 2) to disguise

con-cede [kən-sēd′] *verb*, to surrender, to grant, to allow

con-ceit [kən-sēt′] *noun*, vanity, an exaggerated opinion of oneself

con-ceive [kən-sēv′] *verb*, 1) to imagine, to take into one's mind 2) to become pregnant

con-cen-trate [kŏn′sən-trāt′] *verb*, 1) to focus all one's thoughts and efforts on 2) to bring or come together in one place to increase the strength of

con-cen-tra-tion [kŏn′sən-trā′shən] *noun*, 1) density, consistency 2) close attention, intense thought

con-cept [kŏn′sĕpt′] *noun*, an abstract idea, a thought

con-cep-tion [kən-sĕp′shən] *noun*, 1) the act of conceiving life 2) the beginning or understanding of an idea

con-cern [kən-sûrn′] *noun*, 1) anxiety 2) something that

involves a person or people,
verb, 3) to be about, to interest
4) to trouble or worry

con-cert [kŏn′sûrt′] *noun*, 1) music
played for an audience,
[kŏn′sûrt] *verb* 2) to plan or
arrange together

con-ces-sion [kən-sĕsh′ən] *noun*,
the act of yielding, a grant

con-cil-i-ate [kən-sĭl′ē-āt]*verb*, to
gain goodwill, to placate

con-cil-i-a-to-ry [kən-sĭl′ē-ə-tôr′ē]
adjective, making less angry or
hostile to overcome animosity

con-cise [kən-sīs′] *adjective*, short,
to the point, brief, succinct

con-clude [kən-klōōd′] *verb*, 1) to
form an opinion after thinking
about something 2) to finish to
bring or come to an end

con-clu-sion [kən-klōō′zhən] *noun*,
1) the end, an outcome 2) a
decision, a judgment

con-coct [kən-kŏkt′] *verb*, 1) to
prepare by combining different
ingredients 2) to devise a plan or
scheme, to invent

con-cor-dant [kən-kôr′dant]
adjective, in agreement,
harmonious

con-crete [kŏn-krēt′] *adjective*,
1) not abstract, particular, *noun*,
2) a compound mass of cement,
sand, gravel, and water used for
building

con-cus-sion [kən-kŭsh′ən] *noun*, a
head injury that is the result of a
blow or a fall, a shock

con-demn [kən-dĕm′] *verb*, 1)to
pronounce sentence against
someone 2) to disapprove of or
censure, to denounce

con-dem-na-tion
[kŏn′dĕm-nā′shən] *noun*,
criticism, blame, an objection

con-den-sa-tion [kŏn′dĕn-sā′shən]
noun, 1) the act, process, or state
of compressing or being
compressed into a smaller
enclosed space 2) the process by
which a substance changes from
its gaseous state to a liquid state
3) the liquid formed by this
process

con-dense [kən-dĕns′] *verb*, to
shorten or compress

con-de-scend [kŏn′dĭ-sĕnd′] *verb*,
to act superior to someone else

con-di-ment [kŏn′də-mənt] *noun*, a
type of relish or seasoning

con-di-tion [kən-dĭsh′ən] *noun*,
1) a stipulation or provision
2) the state in which a person or
thing exists, *verb*, 3) to prepare,
to adapt, to train, to influence

con-di-tion-ing [kən-dĭsh′ən-ing]
noun, preparation, training

con-dol-ence [kən-dō′ləns] *noun*,
an offering of sympathy, an
expression of regret

con-done [kən-dōn′] *verb*, to
pardon, to forgive, to overlook

con-du-cive [kən-dōō′sĭv]
adjective, tending to promote,
helpful, beneficial

con-duct *noun*, [kən-dŭkt′]
1) behavior, [kən-dŭkt′] *verb*, 2)
to lead or guide 3) to direct

con-duc-tor [kən-dŭk′tər] *noun*,
1) someone who sells tickets on
a train 2) something that
transmits energy 3) the leader of
a musical ensemble

con-duit [kŏn′dōō-ĭt] *noun*, a channel or passage through which electrical wires run

cone [kōn] *noun*, 1) a cylinder that tapers to a point at the end 2) the reproductive structures of the conifer trees, a pine cone 3) a cylindrical wafer on which a scoop of ice cream is placed

con-fed-er-ate [kən-fĕd′ər-ĭt] *noun*, 1) an ally, an accomplice **Condederate** 2) someone who fought for the Confederacy during the Civil War

con-fed-er-a-tion [kən-fĕd′ə-rā′shən] *noun*, an alliance, a coalition

con-fer [kən-fûr′] *verb*, 1) to give as a gift 2) to meet and discuss

con-fer-ence [kŏn′fər-əns] *noun*, a meeting for an exchange of ideas, a conversation with someone to share information

con-fess [kən-fĕs′] *verb*, to acknowledge or admit a fault

con-fes-sion [kən-fĕsh′ən] *noun*, an admission of guilt or wrongdoing

con-fet-ti [kən-fĕt′ē] *noun*, small pieces of colorful paper scattered at celebrations

con-fi-dant [kŏn′fĭ-dănt′] *noun*, one to whom secrets are entrusted, a friend

con-fide [kən-fīd′] *verb*, to trust, to disclose a secret, to reveal

con-fi-dence [kŏn′fĭ-dəns] *noun*, complete trust, a secret

con-fi-dent [kŏn′fĭ-dənt] *adjective*, assured beyond doubt, self-reliant, feeling certain, sure

con-fi-den-tial [kŏn′fĭ-dĕn′shəl] *adjective*, secret, private, trusted

con-fine [kən-fīn′] *verb*, to restrain, within a specific place

con-firm [kən-fûrm′] *verb*, to give proof of something, to ratify

con-fir-ma-tion [kŏn′fər-mā′shən] *noun*, 1) proof, collaboration 2) a ceremony recognizing a member of the church

con-fis-cate [kŏn′fĭ-skāt′] *verb*, to seize, to take without consent

con-flict [kŏn′flĭkt′] *noun*, 1) a fight or argument, a disagreement, *verb*, 2) to disagree or clash

con-form [kən-fôrm′] *verb*, to comply, to obey social norms

con-form-i-ty [kən-fôr′mĭ-tē] *noun*, harmony, agreement

con-fuse [kən-fyōōz′] *verb*, to mix up in the mind, to perplex

con-fu-sion [kən-fyōō′zhən] *noun*, a state of disorder, chaos, bewilderment

con-geal [kən-jēl′] *verb*, 1) to change from a fluid state to a solid state 2) to thicken

con-gen-ial [kən-jēn′yəl] *adjective*, naturally adapted, pleasant and sympathetic, good-natured

con-gen-i-tal [kən-jĕn′ĭ-tl] *adjective*, existing at birth

con-ges-tion [kən-jĕs′shən] *noun*, 1) a gathering or accumulation 2) an overcrowded state

con-glom-er-a-tion [kən-glŏm′ə-rā′shən] *noun*, material sticking together

con-grat-u-la-tions [kən-grăch′ə-lā′shənz] *noun,plural*, expressions of good wishes for a happy event or achievement

con-gre-gate [kŏng'grĭ-gāt'] *verb*, to gather together, to assemble

con-gre-ga-tion [kŏng'grĭ-gā'shən] *noun*, a gathering or assembly

con-gress [kŏng'grĭs] *noun*, 1) a meeting of a group of people **Congress** 2)the legislature of the United States consisting of the Senate and the House of Representatives

con-gru-ent [kŏng'grōō-ənt] *adjective*, in agreement, exactly alike, matching, equal

con-jec-ture [kən-jĕk'chər] *noun*, theory, an unfounded remark

con-ju-gate [kŏn'jə-gāt'] *verb*, to change the form of a verb depending on the subject and the tense, e.g. *I am, you are, he is, we are, you are, they are*

con-junc-tion [kən-jŭngk'shən] *noun*, a word that joins two other words or phrases together

con-jure [kŏn'jər] *verb*, 1) to seem to create out of thin air as if by magic 2) to entreat, to summon

con-nect [kə-nĕkt'] *verb*, to join or bring together, to link, to unite

Con-nect-i-cut [kə-nĕt'ĭ-kət] *noun*, one of the 50 United States, the capital is Hartford. The state flower of **Connecticut** is the mountain laurel, and the motto is "He who transplanted still sustains."

con-nec-tion [kə-nĕk'shən] *noun*, the act of joining, a relationship

con-nive [kə-nīv'] *verb*, to cooperate in a crime or fault, to secretly allow, to deceive

con-nois-seur [kŏn'ə-sûr'] *noun*, one who is an expert judge of art and style or taste

con-no-ta-tion [kŏn'ə-tā'shən] *noun*, suggested meaning

con-quer [kŏng'kər] *verb*,
1) to gain or acquire by force, to be victorious, to win, to defeat
2) to master, to overcome

con-quest [kŏn'kwĕst'] *noun*, a takeover, a triumph, a victory

con-science [kŏn'shəns] *noun*, a sense of moral goodness

con-sci-en-tious [kŏn'shē-ĕn'shəs] *adjective*, influenced by conscience, careful, thoughtful

con-scious [kŏn'shəs] *adjective*, aware of, awake, alert

con-scious-ly [kŏn'shəs-lē] *adverb*, after thorough consideration, with full awareness

con-se-crate [kŏn'sĭ-krāt'] *verb*, to dedicate, to declare sacred

con-sec-u-tive [kən-sĕk'yə-tĭv] *adjective*, one after another

con-sen-sus [kən-sĕn'səs] *noun*, agreement in opinion, custom, or function, the popular choice

con-sent [kən-sĕnt'] *noun*, 1) an agreement, permission, *verb*, 2) to agree, to comply

con-se-quence [kŏn'sĭ-kwĕns'] *noun*, 1) that which follows as a result, outcome 2) importance

con-se-quen-tial [kŏn'sĭ-kwĕn'shəl] *adjective*, significant, very important

con-ser-va-tion [kŏn'sûr-vā'shən] *noun*, the planned management of natural resources to prevent loss, destruction, or waste

con-serv-a-tive [kən-sûr′və-tĭv] *adjective*, 1) opposed to change or innovation 2) traditional 3) a low estimated number

con-ser-va-tor-y [kən-sûr′və-tôr′ē] *noun*, 1) a small greenhouse for growing plants 2) a college of music, an academy

con-serve [kən-sûrv′] *verb*, to keep from loss or waste, to preserve

con-sid-er [kən-sĭd′ər] *verb*, 1) to think about carefully, 2) to take into account 3) to believe

con-sid-er-a-ble [kən-sĭd′ər-ə-bəl] *adjective*, important, great, large

con-sid-er-ate [kən-sĭd′ər-ĭt] *adjective*, caring about a person's feelings, courteous

con-sid-er-a-tion [kən-sĭd′ə-rā′shən] *noun*, 1) thought and attention 2) a payment

con-sign [kən-sīn′] *verb*, 1) to ship something to an agent in another place, to deliver 2) to assign

con-sign-ment [kən-sīn′mənt] *noun*, 1)the entrusting of something to another's care 2) a quantity of goods for delivery

con-sist [kən-sĭst′] *verb*, to be made of or composed of

con-sist-ent [kən-sĭs′tənt] *adjective*, sticking to the same principles, the same throughout

con-so-la-tion [kŏn′sə-lā′shən] *noun*, comfort, sympathy

con-sole [kən-sōl′] *verb*, to comfort, to calm, to soothe

con-sol-i-date [kən-sŏl′ĭ-dāt′] *verb*, to unite, to bring together , to strengthen, to make solid

con-so-nant [kŏn′sə-nənt] *adjective*, 1) in agreement, *noun*, 2) a letter in the alphabet other than a vowel

con-spic-u-ous [kən-spĭk′yoo-əs] *adjective*, plainly visible, prominent, attracting attention

con-spir-a-cy [kən-spîr′ə-sē] *noun*, a combination of people often for an evil purpose, a secret plan

con-stant [kŏn′stənt] *adjective*, continuous, unceasing, faithful

con-stel-la-tion [kŏn′stə-lā′shən] *noun*, an arrangement of stars that resembles a figure

con-sti-tute [kŏn′stĭ-toot′] *verb*, to make up, to empower

con-sti-tu-tion [kŏn′stĭ-too′shən] *noun*, the fundamental law of a nation, state, or society

con-sti-tu-tion-al [kŏn′stĭtoo′shənl] *adjective*, according to law

con-strain [kən-strān′] *verb*, to force, to restrain unnaturally

con-struct [kən-strŭkt′] *verb*, to build a structure, to erect

con-struc-tion [kən-strŭk′shən] *noun*, 1) something that is built, a structure 2) the arrangement of words in parts of a sentence

con-struc-tive [kən-strŭk′tĭv] *adjective*, helpful, useful, productive, beneficial

con-strue [kən-stroo′] *verb*, to interpret, to try to understand

con-sult [kən-sŭlt′] *verb*, to ask for help looking for more information, to confer

con-sult-ant [kən-sŭl′tənt] *noun*, an expert hired to study data to resolve a problem or give advice

con-sume [kən-sōōm′] *verb*, 1) to eat
2) to exhaust, to expend 3) to make
use of, to apply 4) to destroy

con-sum-er [kən-sōō′mər] *noun*,
1) a higher level organism that gets
its food from other living things
2) someone who buys products, a
customer

con-sum-mate [kŏn′sə-māt′] *verb*, to
bring to completion, to perfect

con-sump-tion [kən-sŭmp′shən]
noun, 1) the use of goods 2) a
disease of the lungs, such as
tuberculosis 3) the eating of food

con-tact [kŏn′tăkt′] *noun*, 1) a
physical touch 2) a connection
through writing or speaking, *verb*,
3) to talk or write to, to touch

con-ta-gi-ous [kən-tā′jəs] *adjective*,
of a disease passed from one person
to another by touch or proximity

con-tain [kən-tān′] *verb*, 1) to have
inside, to hold 2) to include

con-tain-er [kən-tā′nər] *noun*,
anything that can hold and carry
something, eg., a box, jar, basket

con-tam-i-nant [kən-tăm′ə-nənt]
noun, a compound that pollutes,
making the original substance
impure, a poison, a toxin

con-tam-i-nate [kən-tăm′ə-nāt′]
verb, to soil, to stain or corrupt
by contact, to pollute, to infect

con-tam-i-na-tion
[kən-tăm′ə-nā′shən] *noun*, the act
of making a substance impure
or unusable

con-tem-plate [kŏn′təm-plāt′]
verb, to consider with continued
attention, to think about, to
ponder

con-tem-po-rar-y [kən-tĕm′pə-
rĕr′ē] *adjective*, 1) of the present
day, current, *noun*, 2) a person
of the same age, a peer

con-tempt [kən-tĕmpt′] *noun*, a
feeling that someone or
something is bad

con-tempt-i-ble [kən-tĕmp′tə-bəl]
adjective, despicable, vile

con-tent [kən-tĕnt′] *adjective*,
1) happy, satisfied, [kŏn′-tĕnt]
noun, 2) that which is inside
3) the subject matter of a book

con-ten-tion [kən-tĕn′shən] *noun*, a
struggle, an argument, a dispute

con-ten-tious [kən-tĕn′shəs]
adjective, argumentative, quarrel-
some, causing disagreement

con-test [kŏn′tĕst] *noun*, 1) a game
in which there is a winner
[kən-tĕst′]*verb*, 2) to disagree, to
object to, to challenge

con-text [kŏn′tĕkst′] *noun*, the words
or actions that provide a necessary
link or explanation to the meaning
of a word or event

con-tig-u-ous [kən-tĭg′yōō-əs]
adjective, adjoining, touching

con-ti-nent [kŏn′tə-nənt] *noun*, a
major land area in the world
unbroken by major oceans

con-tin-gent [kən-tĭn′jənt]
adjective, conditional, relying
on chance or something
happening

con-tin-u-al [kən-tĭn′yōō-əl]
adjective, happening repeatedly

con-tin-ue [kən-tĭn′yōō] *verb*, 1) to
go on, to resume 2) to persevere

con-tin-u-ous [kən-tĭn′yōō-əs]
adjective, uninterrupted

con-tor-tion [kən-tôr'shən] *noun*, something bent or twisted out of shape, a distortion

con-tour [kŏn'tŏor'] *noun*, the outline of a figure, body, hill, coastline, etc.

con-tra-band [kŏn'trə-bănd'] *noun*, illegal goods, smuggled materials

con-tract [kŏn'trăkt'] *noun*, 1) a written agreement between two or more persons that has the force of law, [kən'-trăkt'] *verb*, 2) to draw closer together 3) to make a binding agreement

con-trac-tion [kŏn'trăk' shən] *noun*, 1) the shortening of a word or group of words by omitting a letter or letters and replacing them with an apostrophe, e.g. y*ou'll* 2) the act of drawing together like muscles

con-tra-dict [kŏn'trə-dĭkt'] *verb*, to say the opposite, to oppose in words, to conflict with

con-trar-y [kŏn'trĕr'ē] *adjective*, opposite, perverse, wayward

con-trast [kən-trăst'] *noun*, 1) the difference, *verb*, 2) to compare in order to show the differences between two things

con-trib-ute [kən-trĭb'yŏot] *verb*, to give as part of a group effort, to help in causing

con-tri-bu-tion [kŏn'trĭ-byŏo'shən] *noun*, an offering, a donation

con-trite [kən-trīt'] *adjective*, asking forgiveness, remorseful

con-trive [kən-trīv'] *verb*, to invent or make a clever device

con-trol [kən-trōl'] *noun*, 1) power, guidance, *verb*, 2) to manage or oversee a situation or machine

con-tro-ver-sy [kŏn'trə-vûr'sē] *noun*, a disagreement or debate

con-tu-sion [kən-tŏo'zhən] *noun*, a wound or bruise that does not break the skin

con-va-les-cent [kŏn'və-lĕs'ənt] *adjective*, recovering from sickness

con-vene [kən-vēn'] *verb*, to assemble, to call together

con-ven-ience [kən-vēn'yəns] *noun*, the state of being easily used or immediately available

con-ven-ient [kən-vēn'yənt] *adjective*, adapted to one's comfort or ready use, handy

con-vent [kŏn'vənt] *noun*, a home where nuns live

con-ven-tion [kən-vĕn'shən] *noun*, 1) an assembly of members of a delegation 2) a practice or custom 3) a formal agreement

con-ven-tion-al [kən-vĕn'shə-nəl] *adjective*, contemporary

con-verge [kən-vûrj'] *verb*, to come together, to assemble

con-ver-sant [kən-vûr'sənt] *adjective*, familiar with

con-ver-sa-tion [kŏn'vər-sā'shən] *noun*, an informal discourse, a chat or familiar talk

con-verse *noun*, [kən- vûrs'] 1) the contrary or opposite, [kŏn'vûrs'] *verb*, 2) to talk or exchange ideas

con-ver-sion [kən-vûr′zhən] *noun*, a change from one thing, state, or religion to another

con-vert [kən-vûrt′] *verb*, to change into something else

con-vex [kŏn′věks′] *adjective*, curved out like the outside of a sphere or ball or spoon

con-vey [kən-vā′] *verb*, 1) to take or transport, to carry 2) to communicate, to pass on, to impart, to give 3) to transmit

con-vict [kən-vĭkt′] *noun*, 1) a person who has been judged guilty of a crime, and sentenced *verb*, 2) to decide in court that someone has committed a crime

con-vic-tion [kən-vĭk′shən] *noun*, a strong belief, the state of being convinced, complete confidence

con-vince [kən-vĭns′] *verb*, to satisfy by proof

con-vo-lut-ed [kŏn′və-loo′tĭd] *adjective*, coiled around, intricate, complicated

con-vul-sions [kŏn′və-loo′shənz] *noun*, *plural*, strong and uncontrolled involuntary muscle contractions

cook [kook] *noun*, 1) someone who prepares food to eat, *verb*, 2) to prepare food for eating on the stove or in an oven, to bake

cool [kool] *adjective*, 1) moderately cold 2) calm, *verb*, 3) to lower temperature 4) to calm down

coop [koop] *noun*, a pen, a hen-house, a cage for poultry

co-op-er-ate [kō-ŏp′ə-rāt′] *verb*, to work together, to collaborate

co-op-er-a-tion [kō-ŏp′ə-rā′shən] *noun*, the process of working together toward a common goal

co-op-er-a-tive [kō-ŏp′ər-ə-tĭv] *adjective*, willing to join together to help other people

co-or-di-nate [kō-ôr′dn-āt′] *verb*, to make the parts of something work together, to harmonize

co-or-di-na-tion [kō-ôr′dn-ā′shən] *noun*, 1)the organization of separate elements working together 2) dexterity

cope [kōp] *verb*, to deal successfully with something

cop-per [kŏp′ər] *noun*, a common metal of a reddish color, a good conductor of heat and electricity

cop-y [kŏp′ē] *noun*, 1) an imitation, reproduction, *verb*, 2) to make something that is the same as something else

cop-y-right [kŏp′ē-rīt′] *noun*, the exclusive right to publish and sell the matter and form of a literary or artistic work

cor-al [kôr′əl] *noun*, 1) a pink reddish color, salmon 2) a stony substance formed from the skeletons of tiny animals that live on reefs in the ocean

cord [kôrd] *noun*, thin rope

cor-dial [kôr′jəl] *adjective*, sincere, friendly, encouraging, gracious

core [kôr] *noun*, 1) the heart, essence 2) the center of focus

cork [kôrk] *noun*, a light substance that comes from the outer bark of a tree

corn [kôrn] *noun*, 1) a sore that sometimes grows on toes when shoes are tight 2) a grain grown on a stalk as an ear (maize)

cor-ner [kôr′nər] *noun*, 1) a nook, a niche 2) where two lines, edges or streets meet, *verb*, 3) to put someone in a place they cannot escape 4) to monopolize

cor-nu-co-pi-a [kôr′nə-kō′pē-ə] *noun*, 1) a horn containing an abundantly overflowing supply of food 2) abundance, plenty

cor-ol-lary [kôr′ə-lĕr-ē] *noun*, what follows logically, the consequence or effect

cor-o-na-tion [kôr′ə-nā′shən] *noun*, a ceremony to crown a king or queen

cor-po-ra-tion [kôr′pə-rā′shən] *noun*, a business that operates as a person under the law

corps [kôr] *noun*, a troop, a main subdivision of an army

corpse [kôrps] *noun*, a dead body

cor-pu-lent [kôr′pyə-lənt] *adjective*, very fat

cor-pus-cle [kôr′pə-səl] *noun*, a minute particle of matter, a cell, a red or white blood cell

cor-ral [kə-răl′] *noun*, a pen for livestock

cor-rect [kə-rĕkt′] *adjective*, 1) without error, *verb*, 2) to point out an error 3) to make right

cor-rec-tion [kə-rĕk′shən] *noun*, a change or revision, to remove error, to make right

cor-rel-ate [kôr′ə-lāt′] *verb*, to be linked to something

cor-re-la-tion [kôr′ə-lā′shən] *noun*, a mutual relationship, a connection

cor-re-spond [kôr′ĭ-spŏnd′] *verb*, 1) to communicate by writing and receiving letters 2) to fit in with, to match, to agree

cor-re-spond-ence [kôr′ĭ-spŏn′dəns] *noun*, an exchange of letters

cor-ri-dor [kôr′ĭ-dər] *noun*, a hallway connecting rooms or apartments opening onto it

cor-rob-o-rate [kə-rŏb′ə-rāt′] *verb*, to confirm or support with facts,

cor-ro-sive [kə-rō′sĭv] *adjective*, 1) eating away gradually by chemicals 2) sarcastic

cor-ru-gat-ed [kôr′ə-gā′tĭd] *adjective*, bent into a series of alternate ridges and grooves for added stiffness

cor-rupt [kə-rŭpt′] *adjective*, 1) crooked, dishonest, *verb*, 2) to cause to become dishonest

cor-rup-tion [kə-rŭp′shən] *noun*, dishonesty, depravity

cort-ege [kôr-tĕzh′] *noun*, a funeral procession

cos-met-ic [kŏz-mĕt′ĭk] *noun*, a beauty product

cos-mic [kŏz′mĭk] *adjective*, universal, vast

cost [kôst] *noun*, 1) the price you pay when you buy something, *verb*, 2) to have as a price

cos-tume [kŏs′tōōm′] *noun*, clothes worn for a special occasion to represent a particular period in history, traditions of a country, or an assumed character

cot [kŏt] *noun*, a lightweight collapsible bed that usually folds up to a smaller size

cot-tage [kŏt′ĭj] *noun*, a small country house, a bungalow

cot-ton [kŏt′n] *noun*, 1) the white part of the cotton plant used to make cloth 2) the fabric made from this

couch [kouch] *noun*, an upholstered piece of furniture that seats several people

cough [kôf] *noun*, 1) a hacking sound made when air is expelled from the lungs, *verb*, 2) to make a sharp noise when air comes up from the lungs

could [kŏŏd]*verb past tense* of **can**

could-n't [kŏŏd/nt] the *contraction* of the words **could** and **not**

coun-cil [koun/səl] *noun*, an assembly of men and women convened for consultation

coun-sel [koun/səl] *noun*, 1) an attorney, a lawyer 2) an interchange of opinions, advice, *verb*, 3) to advise

coun-se-lor [koun/sə-lər] *noun*, someone who helps you make the best choice

count [kount] *noun*, 1) the total number added together, *verb*, 2) to add up 3) to recite numbers in the correct order 4) to list

coun-te-nance [koun/tə-nəns] *noun*, a facial expression, aspect, appearance, composure

coun-ter [koun/tər] *noun*, 1) the table between customers and workers in a business, *verb*, 2) to make an offer against another

coun-ter-act [koun/tər-ăkt/] *verb*, to act against, to neutralize, to prevent the effects of

coun-ter-clock-wise [koun/tər-klŏk/wīz/] *adjective*, in the opposite direction from the movement of the hands of a clock

coun-ter-feit [koun/tər-fĭt/] *noun*, 1) an imitation intended to deceive, *verb*, 2) to imitate or make a copy of

coun-ter-part [koun/tər-pärt/] *noun*, a thing that completes another, a mate

coun-try [kŭn/trē] *noun*, 1) the land ruled by a government, a nation 2) the land outside of a city, usually farmland or forests

coun-ty [koun/tē] *noun*, an administrative division of a state

coup [kōō] *noun*, a successful sudden attack to seize power

coupe [kōōp] *noun*, a sports car with a sloping back

cou-ple [kŭp/əl] *noun*, 1) two people or things that are usually thought of together, a pair, *verb*, 2) to fasten

cou-pon [kōō/pŏn/] *noun*, a certificate that gives a discount on a purchase, a rebate

cour-age [kûr/ĭj] *noun*, *no plural*, the strength to face danger when overwhelmed by fear

cou-ra-geous [kə-rā/jəs] *adjective*, brave or bold in danger

cour-i-er [kōōr/ē-ər] *noun*, a messenger who carries documents

course [kôrs] *noun*, 1) the path of something, the sequence of events 2) a portion of a meal 3) a series of lessons in a school

court [kôrt] *noun*, 1) a place where judges decide how people who break laws should be punished and civil cases heard 2) an open space next to a house in an enclosed area 3) the members of

a royal family 4) a royal palace
verb, 5) to date

cour-te-ous [kûr′tē-əs] *adjective*,
showing good manners, polite

cour-te-sy [kûr′tĭ-sē] *noun*,
politeness, civility, consideration

court-ship [kôrt′shĭp′] *noun*, going
steady, a period of dating before
marriage

cous-in [kŭz′ĭn] *noun*, the son or
daughter of your uncle or aunt

cove [kōv] *noun*, an inlet

cov-e-nant [kŭv′ə-nənt] *noun*, 1) a
legal undertaking to do or to
refrain from doing some act or
thing 2) a contract under seal

cov-er [kŭv′ər] *noun*, 1) something
that is placed over something to
protect it, *verb*, 2) to protect by
placing a lid, wrapping, etc.,
around or on top of 3) to hide
4) to go over

co-vert [kŭv′ərt] *adjective*,
concealed, hidden, or sheltered

cov-et [kŭv′ĭt] *verb*, to want to
have a thing belonging to
another person

cow [kou] *noun*, a grown female
cattle or certain other large
animals

cow-ard [kou′ərd] *noun*, someone
who is easily frightened, a
person who lacks courage

cow-ard-ice [kou′ər-dĭs]*noun*, lack
of courage to face danger,
timidity, fearfulness

cow-ard-ly [kou′ərd-lē] *adjective*,
timid, afraid, scared, fearful

cow-boy [kou′boi′] *noun*, a man
who rides a horse and looks after
livestock on a ranch

cow-er [kou′ər] *verb*, to shrink
from as if from fear, to tremble

coy-o-te [kī-ō′tē] *noun*, a North
American, wolflike dog

co-zy [kō′zē] *adjective*, snug, warm

crab [krăb] *noun*, a sea animal with
eight legs, a pair of pincers and a
hard shell that lives in a hole

crack [krăk] *noun*, 1) a line
showing where something is
broken, a crevice 2) a sharp
noise, such as a gunshot or
thunder, *verb*, 3) to break while
separating

cra-dle [krād′l] *noun*, 1) a small
bed for an infant 2) the holder
for a telephone, *verb*, 3) to rock
or shelter an infant or comfort a
crying baby

craft [krăft] *noun*, 1) something
made by hand to be used as a
decoration 2) a boat or plane 3)
guile, slyness, skill

cram [krăm] *verb*, to fill or force in

cramp [krămp] *noun*, 1) a painful
muscle spasm, *verb*, 2) to keep
within narrow limits, to confine

crane [krān] *noun*, a machine
designed to lift heavy loads from
one place to another

cra-ni-um [krā′nē-əm] *noun*, the
skull of a vertebrate animal

crank [krăngk] *noun*, a handle that
is attached to a machine and that
turns, causing motion

crash [krăsh] *noun*, 1) a loud noise
caused by a collision or
something falling, *verb*, 2) to
shatter violently, scattering
pieces everywhere

crass [krăs] *adjective*, grossly
insensible, rude coarse

crate [krāt] *noun*, a large usually wooden box or container, a case

cra-ter [krā′tər] *noun*, a bowl-shaped cavity, a hole

crave [krāv] *verb*, to want something badly

crawl [krôl] *verb*, to move on one's hands and knees, to move slowly

cray-on [krā′ŏn′] *noun*, a small pencil of colored wax used for drawing or writing

craze [krāz] *noun*, a fad, a popular idea, a trend

cra-zy [krā′zē] *adjective*, insane, wild, daft, very foolish

creak [krēk] *verb*, to make the sound of a door that has not been oiled, a long squeak

cream [krēm] *adjective*, 1) the color yellowish white, ivory 2) a thick lotion, *noun*, 3) the fatty part of milk

crease [krēs] *noun*, the mark or line left by a fold, a wrinkle, a furrow

cre-ate [krē-āt′] *verb*, to make, to bring into being, to produce for the first time

cre-a-tion [krē-ā′shən] *noun*, the act of causing to exist, that which is produced

cre-a-tive [krē-ā′tĭv] *adjective*, having new or original ideas

cre-a-tiv-i-ty [krē-ā′tĭv-ĭ-tē] *noun*, the ability to make or invent things, originality

crea-ture [krē′chər] *noun*, an animal or insect or person

cre-den-tials [krĭ-dĕn′shəlz] *noun*, a document verifying a person's ability or authority, references

cred-i-ble [krĕd′ə-bəl] *adjective*, 1) believable 2) reliable or trustworthy

cred-it [krĕd′ĭt] *noun*, 1) a system in which merchandise is given with the understanding that it will be paid for later 2) attention and praise received for doing something

cred-o [krē′dō] *noun*, creed, a belief

cred-u-lous [krĕj′ə-ləs] *adjective*, easily imposed upon, gullible

creed [krēd] *noun*, a statement of religious or ethical beliefs, principles

creek [krēk, krĭk] *noun*, a stream smaller than a river and larger than a brook, an inlet of water

creep [krēp] *verb*, to move slowly and quietly, to move furtively

cre-o-sote [krē′ə-sōt′] *noun*, an oily liquid obtained from coal tar and used as a wood preservative

cre-scen-do [krə-shĕn′dō] *noun*, a musical direction used to indicate increasing loudness

cres-cent [krĕs′ənt] *noun*, a curved shape like a new moon

crest [krĕst] *noun*, 1) the feathers that stick up on top of a bird's head 2) the top of something

crev-ice [krĕv′ĭs] *noun*, a narrow opening in a surface resulting from a split, a cleft, a gap

crew [krōō] *noun*, a group of people who work together

crib [krĭb] *noun*, 1) a bed made for an infant 2) a container for grain

crick-et [krĭk′ĭt] *noun*, 1) a ball game played with two teams of eleven players 2) a small brown

insect that makes a shrill
chirping noise at night

cried [krīd] *verb, past tense* of **cry**

crime [krīm] *noun*, something that
is against the law, illegal acts

crim-i-nal [krĭm′ə-nəl] *adjective*,
1) illegal, wrong, *noun*, 2) one
who has broken the law and is
guilty of a crime

cringe [krĭnj] *verb*, to shrink back
in fear, to cower in fear

crip-ple [krĭp′əl] *verb*, to disable or
make lame

cri-sis [krī′sĭs] *noun, plural*, **crises**
[krī′-sēz], a stressful time when
something serious is happening

crisp [krĭsp] *adjective*, 1) dry,
brittle, easily broken 2) firm and
fresh 3) very clean and pressed

criss cross [krĭs′krôs′] *noun* , a
pattern of intersecting lines

cri-te-ri-on [krī-tîr′ē-ən] *noun*,
plural, **criteria**, a standard for
judgment, a benchmark

crit-ic [krĭt′ĭk] *noun*, 1) a person
who is paid to review a
performance or art 2) one who
finds faults

crit-i-cal [krĭt′ĭ-kəl] *adjective*,
1) very important, able or
inclined to pass judgment
2) crucial, pivotal, key

crit-i-cism [krĭt′ĭ-sĭz′əm] *noun*,
faultfinding, review

crit-i-cize [krĭt′ĭ-sīz′] *verb*, to
examine as a critic, to censure

croak [krōk] *noun*, 1) a hoarse
sound like that made by frogs,
verb, 2) to make a low hoarse
sound in the throat

croc-o-dile [krŏk′ə-dīl′] *noun*, a
large reptile, like an alligator,
that lives in tropical climates

crook-ed [krŏŏk′ĭd] *adjective*,
1) bent, not straightened
2) dishonest, deceitful

crop [krŏp] *noun*, food that is
planted and grown in a field

cross [krôs] *adjective*, 1) irritated,
angry, *noun*, 2) a shape with four
arms, *verb*, 3) to go over, to
extend across

crouch [krouch] *verb*, to stoop

crow [krō] *noun*, 1) any of a genus
of large black birds including
the raven, *verb*, 2) to make a
sound like a rooster

crowd [kroud] *noun*, 1) a large
gathering of people or things,
verb, 2) to press or cram

crown [kroun] *noun*, 1) a circular
head covering, often made of
metal with precious stones, worn
by royalty, *verb*, 2) to top, to
culminate, to climax

cru-ci-al [krōō′shəl] *adjective*,
important or essential as decisive
in determining a doubtful crisis

cru-ci-ble [krōō′sə-bəl] *noun*, a
vessel used for melting substances

cru-ci-fix-ion [krōō′sə-fĭk′shən]
noun, the killing of someone by
nailing them to a cross

crude [krōōd] *adjective*, 1) roughly
made 2) rude, ill-mannered

cru-el [krōō′əl] *adjective*, mean,
deliberately causing pain

cru-el-ty [krōō′əl-tē] *noun*,
unkindness, oppression

cruise [krōōz] *noun*, 1) a holiday on
ship, *verb*, 2) to sail or drive around

crum-ble [krŭm′bəl] *verb*, to fall or break apart into small pieces

cru-sade [krōō-sād′] *noun*, 1) a campaign conducted by a band of people devoted to advancing a cause, *verb*, 2) to battle for a cause

crush [krŭsh] *verb*, to hurt or flatten by applying pressure

crust [krŭst] *noun*, a hard surface on the outside of something that often preserves what is inside, such as bread, pie, or the earth

crutch [krŭch] *noun*, a piece of wood or metal extending from under the arm to the ground, to support a person who cannot walk unaided

crux [krŭks] *noun*, the crucial point

cry [krī] *noun*, 1) a shout or call for help, *verb*, 2) to call out loudly 3) tears flowing from the eyes

crypt [krĭpt] *noun*, a vault where bodies are buried, a tomb

cryp-tic [krĭp′tĭk] *adjective*, having a hidden meaning, puzzling

crys-tal [krĭs′təl] *adjective*, 1) clear, transparent, *noun*, 2) a clear mineral 3) a high-quality clear glass

crys-tal-lize [krĭs′tə-līz′] *verb*, 1) to form crystals 2) to become clear

cub [kŭb] *noun*, a young animal of certain mammals, such as a bear, lion, or fox

cube [kyōōb] *noun*, a solid shape with six equal square sides

cu-bic-le [kyōō′bĭ-kəl] *noun*, a small space for work partitioned off in a larger room

cu-cum-ber [kyōō′kŭm′bər] *noun*, a long, green-skinned fruit, pickled or eaten green as a salad

cue [kyōō] *noun*, 1) a signal or hint 2) a rod used to hit a billiard ball

cul-de-sac [kŭl′dĭ-săk′] *noun*, a small dead end off the main road

cu-li-nar-y [kyōō′lə-nĕr′ē] *adjective*, relating to cooking

cull [kŭl] *verb*, to select or gather, to pick out from others

cul-mi-nate [kŭl′mə-nāt′] *verb*, to reach the highest point of altitude, power, climax

cul-pa-ble [kŭl′pə-bəl] *adjective*, deserving blame, liable

cul-prit [kŭl′prĭt] *noun*, a person charged with or found guilty of a crime or misdeed

cul-ti-vate [kŭl′tə-vāt′] *verb*, 1) to improve or develop, to refine 2) to foster and promote healthy growth, as in crops, to farm

cul-tur-al [kŭl′chər-əl] *adjective*, relating to culture, the arts, good taste, etc.

cul-ture [kŭl′chər] *noun*, 1) advancement of knowledge of a civilization 2) a level of education that appreciates art and style 3) a growth of micro-organisms and bacteria

cum-ber-some [kŭm′bər-səm] *adjective*, heavy, hard to manage, unwieldy

cu-mu-la-tive [kyōōm′yə-lā′tĭv] *adjective*, increasing by successive additions

cun-ning [kŭn′ĭng] *noun*, cleverness, craftiness, skill

cup [kŭp] *noun*, 1) a bowl-shaped drinking container with a handle 2) a unit of measure equal to 8 ounces

cup-board [kŭb′ərd] *noun*, a small closet or recessed shelf

cu-ra-tor [kyŏŏ-rā′tər] *noun*, one who has the care of a museum, a custodian, keeper

curb [kŭrb] *noun*, 1) control, restraint 2) a ledge, a boundary, especially that along a street

cur-dle [kûr′dl] *verb*, to sour and coagulate from liquid to a solid, e.g., milk to cheese

cure [kyŏŏr] *noun*, 1) the remedy that made someone who was sick well again, *verb*, 2) to make someone who is ill, healthy, to heal

cur-few [kûr′fyŏŏ] *noun*, a set time enforced by law when people are to be off the streets

cu-ri-os-i-ty [kyŏŏr′ē-ŏs′ĭ-tē] *noun*, 1) the feeling of wanting to know about something 2) something that is strange or unusual

cu-ri-ous [kyŏŏr′ē-əs] *adjective*, 1) eager to know about all things 2) singular, unusual, odd

curl [kûrl] *noun*, 1) a ringlet or spiral shape, *verb*, 2) to make straight hair wavy and curly

cur-mudg-eon [kər-mŭj′ən] *noun*, a mean, bad-tempered person

cur-rant [kûr′ənt] *noun*, a small berry, similar to a raisin

cur-ren-cy [kûr′ən-sē] *noun*, money in circulation

cur-rent [kûr′ənt] *adjective*, 1) up to date, pertaining to the present, *noun*, 2) the powerful flow of water or electricity or air

cur-ric-u-lum [kə-rĭk′yə-ləm] *noun*, the classes that a school requires or offers

cur-ry [kûr′ē, kŭr′ē] *noun*, 1) a yellow powder containing several spices, including turmeric, used to flavor cooking, *verb*, 2) to seek favor by flattery 3) to brush or groom a horse

curse [kûrs] *noun*, 1) something said hoping harm will come to someone, *verb*, 2) to wish bad luck or injury on someone 3) to use profanity

cur-sive [kûr′sĭv] *adjective*, characteristic of handwriting in which the letters flow together

cur-tail [kər-tāl′] *verb*, to cut short, to reduce, to decrease

cur-tain [kûr′tn] *noun*, a piece of cloth hung over a window or doorway as a screen or cover

cur-tsy [kûrt′sē] *noun*, 1) a formal bow made with bent knees by a woman, *verb*, 2) to make a bow with one foot forward, as a gesture of respect

cur-va-ture [kûr′və-chŏŏr′] *noun*, a curve, a bend

curve [kûrv] *noun*, a bend or arch

curv-ed [kûrvd] *adjective*, having a line with no part straight or flat

cush-ion [kŏŏsh′ən] *noun*, a case or bag stuffed with some soft and elastic material

cus-tard [kŭs′tərd] *noun*, a pudding made of milk and eggs

cus-to-di-an [kŭ-stō′dē-ən] *noun*, a caretaker or porter, a janitor

cus-to-dy [kŭs′tə-dē] *noun*, safe-keeping under penalty or under a judge's ruling, imprisonment

cus-tom [kŭs′təm] *noun*, a tradition passed on from one generation to the next

cus-tom-ar-y [kŭs′tə-měr′ē] *adjective*, established by common usage, habitual

cus-tom-er [kŭs′tə-mər] *noun*, one who makes regular purchases from a business, a patron

cut [kŭt] *noun*, 1) a wound or injury that may cause the skin to bleed 2) an insult, a derogatory remark 3) something that has been made shorter, *verb*, 4) to break with a knife or blade

cute [kyōōt] *adjective*, lovable, adorable, delightful

cu-ti-cle [kyōō′tĭ-kəl] *noun*, the skin surrounding the fingernail

cy-cle [sī′kəl] *noun*, 1) events that happen in the same order over and over again 2) a bicycle, *verb*, 3) to move in a circuit controlled by alternating negative and positive currents

cy-clone [sī′klōn′] *noun*, a violent windstorm, sometimes called a typhoon or hurricane, depending on location and intensity

cyl-in-der [sĭl′ən-dər] *noun*, a solid spherical figure with two faces that are circles

cy-lin-dri-cal [sə-lĭn′drĭ-kəl] *adjective*, a circular or spherical shape with straight sides

cym-bal [sĭm′bəl] *noun*, a large round metal plate used as a percussion instrument

cyn-ic [sĭn′ĭk] *noun*, a doubter, a skeptic, an unbeliever

cyn-i-cism [sĭn′ĭ-sĭz′əm] *noun*, an attitude given to sneering at honesty or virtue, a belief that people are self-interested

cy-press [sī′prĭs] *noun*, an evergreen tree with scale-like needles

cyst [sĭst] *noun*, a fluid-filled pouch attached to a body part

czar [zär, tsär] *noun*, the supreme ruler of an empire, especially the ruler of Russia before 1917

D

dad [dăd] *noun*, father

daf-fo-dil [dăf′ə-dĭl] *noun*, a spring flowering bulb, yellow or white, with a trumpet-like center

dag-ger [dăg′ər] *noun*, a weapon used specifically for stabbing, a knife

dai-ly [dā′lē] *adjective*, done or occurring every day

dain-ty [dān′tē] *adjective*, delicate, fragile, small, pretty, refined

dair-y [dâr′ē] *noun*, a place where milk is produced and pasteurized

da-is [dā′ĭs] *noun*, a raised platform for honored guests

dai-sy [dā′zē] *noun*, a flower with a yellow or brown center surrounded by petals

dal-ly [dăl′ē] *verb*, to loiter, to fiddle around, to flirt

dam [dăm] *noun*, 1) a wall built across a river or lake that controls water flow, *verb*, 2) to construct a wall in a lake to retain the water and to regulate its flow

dam-age [dăm′ĭj] *noun*, 1) loss or detriment due to injury or harm, *verb*, 2) to hurt or cause harm or injury to something, to destroy

damp [dămp] *adjective*, humid, somewhat wet, moist

dance [dăns] *noun*, 1) a set of movements designed to accompany music, *verb*, 2) to move in rhythm with music, to skip or leap, to perform

dan-de-li-on [dăn′dl-ī′ən] *noun*, a bitter herb with yellow flowers, often cooked and eaten

dan-ger [dān′jər] *noun*, the possibility of harm

dan-ger-ous [dān′jər-əs] *adjective*, full of risk, perilous, hazardous, likely to harm

dan-gle [dăng′gəl] *verb*, to hang, to suspend loosely

dap-pled [dăp′əld] *adjective*, spotted with color or shade

dare [dâr] *verb*, to be brave, bold

dark [därk] *adjective*, 1) without light, not reflecting light, shadowy, *noun*, 2) a space without light

dark-ness [därk′nĕs] *noun*, *no plural*, a lack of light

darl-ing [där′lĭng] *adjective*, 1) a term of affection for someone who is loved, *noun*, 2) a sweetheart, the favorite

darn [därn] *verb*, to mend socks

dart [därt] *noun*, 1) a small weapon like a tiny arrow that can be thrown by hand, *verb*, 2) to move quickly or suddenly

dash [dăsh] *noun*, 1) a punctuation mark used to indicate a pause 2) a small amount, *verb*, 3) to move quickly 4) to hit, to shatter, to break, to strike

da-ta [dā′tə, dăt′ə, dä′tə] *noun*, information collected on a subject for reference

da-ta-base [dā′tə-bās′, dăt′ə-] *noun*, information arranged for computer access

date [dāt] *noun*, 1) the date, month, and year 2) a small, sweet, brown fruit grown in tropical climates, *verb*, 3) to measure, to mark 4) to court

daub [dôb] *verb*, to dab, to smear, to spread paint or another liquid

daugh-ter [dô′tər] *noun*, a human female considered with reference to her parents

daunt [dônt] *verb*, to intimidate

daunt-less [dônt′lĭs] *adjective*, bold

daw-dle [dôd′l] *verb*, to dilly dally, to waste time

dawn [dôn] *noun*, the break of day when the sun rises

day [dā] *noun*, 1) the period of 24 hours when the earth completes one rotation on its axis 2) the opposite of night, when the sun is in the sky

daze [dāz] *noun*, 1) a stupor, amazement, *verb*, 2) to cause to feel stunned

daz-zle [dăz′əl] *verb*, to awe, to impress with splendor

dead [dĕd] *adjective*, 1) not alive, inactive, *noun*, 2) referring to people who are no longer alive

dead-line [dĕd′līn′] *noun*, a set time when an assignment is to be finished, a time limit

dead-lock [dĕd′lŏk′] *noun*, a stalemate, a state when no progress can be made

dead-ly [dĕd′lē] *adverb*, dangerous, possibly fatal, causing death, mortal

deaf [dĕf] *adjective*, 1) not able to hear 2) not listening

deal [dēl] *noun*, 1) a business agreement 2) an amount, *verb*, 3) to barter, or buy and sell 4) to be concerned, to cope 5) to pass out or to distribute

deal-er [dē′lər] *noun*, 1) a person who buys and sells things 2) in a card game the person who holds the cards and passes them out

dear [dîr] *adjective*, a term of fondness or affection

dear-ly [dîr′lē] *adverb*, a feeling of affection or closeness

death [dĕth] *noun*, the end of life

de-ba-cle [dĭ-bä′kəl] *noun*, a downfall, falling apart

de-bat-a-ble [dĭ-bā′tə-bəl] *adjective*, disputable, open to question

de-bate [dĭ-bāt′] *noun*, 1) a public discussion of a topic in which opposing points of view are presented by panel members, *verb*, 2) to argue different points of view in a public forum

de-bil-i-tate [dĭ-bĭl′ĭ-tāt′] *verb*, to weaken or temporarily disable

deb-it [dĕb′ĭt] *noun*, a debt, an entry on an account for a sum owed

de-bris [də-brē′, dā′brē′] *noun*, rubbish, especially such as results from destruction

debt [dĕt] *noun*, money owed

debt-or [dĕt′ər] *noun*, one who owes something

de-but [dā-byōō′] *noun*, 1) a beginning 2) first public appearance

dec-ade [dĕk′ād′] *noun*, a period of ten years

de-cay [dĭ-kā′] *noun*, 1) the condition of being rotten or degenerating, *verb*, 2) to rot, to decompose 3) to decline

de-ceased [dĭ-sēst′] *adjective*, dead, defunct, without life

de-ceit [dĭ-sēt′] *noun*, fraud, trickery, deception, treachery

de-ceit-ful [dĭ-sēt′fəl] *adjective*, dishonest, deceptive, false

de-ceive [dĭ-sēv′] *verb*, to lead into error, to mislead, to delude

De-cem-ber [dĭ-sĕm′bər] *noun*, the twelfth month of the year, having 31 days

de-cen-cy [dē′sən-sē] *noun*, that which is proper, in good taste

de-cent [dē′sənt] *adjective*, honorable, good, fitting, ethical

de-cep-tion [dĭ-sĕp′shən] *noun*, the act of deceiving, fraud, false representation, dishonesty

de-cide [dĭ-sīd′] *verb*, to determine, to make a choice to do something, to resolve, to choose

de-cid-u-ous [dĭ-sĭj′ōō-əs] *adjective*, the falling out of something at a certain stage of development as some leaves, antlers, insect wings, or milk teeth

dec-i-mal [dĕs′ə-məl] *noun*, pertaining to or founded on the number ten

de-ci-pher [dĭ-sī′fər] *verb*, 1) to translate from secret characters 2) to make out or read

de-ci-sion [dĭ-sĭzh′ən] *noun*, act or result of deciding or settling, a conclusion, a resolution

de-ci-sive [dĭ-sī′sĭv] *adjective*, positive, absolute, unwavering

de-ci-sive-ly [dĭ-sī′sĭv-lē] *adverb*, in a clean manner, precisely

deck [dĕk] *noun*, 1) the floor on each level of a ship 2) a pack of playing cards

dec-la-ra-tion [dĕk′lə-rā′shən] *noun*, an announcement or statement, a proclamation

de-clare [dĭ-klâr′] *verb*, to formally announce what someone thinks or decides, to state firmly

de-cline [dĭ-klīn′] *noun*, 1) a descent or slope, *verb*, 2) to refuse 3) to steadily become less

de-com-pose [dē′kəm-pōz′] *verb*, 1) to rot or decay 2) to separate the constituent parts of something

de-com-po-si-tion [dē-kŏm′pə-zĭsh′ən] *noun*, decay, something that has rotted

dec-o-rate [dĕk′ə-rāt′] *verb*, to beautify, to adorn, to enhance

dec-o-ra-tion [dĕk′ə-rā′shən] *noun*, 1) an ornament 2) an award ribbon

dec-o-rous [dĕk′ər-əs] *adjective*, proper, correct, appropriate

de-coy [dē′koi′] *noun*, lure or bait

de-crease [dĭ-krēs′] *noun*, 1) a measured amount after something has been taken away, *verb*, 2) to diminish and lessen the amount, to reduce

de-cree [dĭ-krē′] *noun*, 1) a pronouncement, a declaration, *verb*, 2) to announce, to proclaim

de-crep-it [dĭ-krĕp′ĭt] *adjective*, broken down with age, feeble, worn out, dilapidated

de-cry [dĭ-krī′] *verb*, to discredit, to denounce publicly

ded-i-cate [dĕd′ĭ-kāt′] *verb*, 1) to recognize in a solemn ceremony 2) to devote oneself to something 3) to inscribe a message in the front of a book

ded-i-ca-tion [dĕd′ĭ-kā′shən] *noun*, 1) devotion to something 2) a person acknowledged by the author

de-duce [dĭ-dōōs′] *verb*, to reach a conclusion by reasoning

de-duct [dĭ-dŭkt′] *verb*, to subtract, to take away

deed [dēd] *noun*, 1) an action taken, something someone has done 2) a legal document proving that someone owns a piece of land

deep [dēp] *adjective*, 1) a long distance beneath the surface of the earth 2) intense, heartfelt, very strong feelings

deer [dîr] *noun*, a beautiful, graceful animal of the woods that runs fast

de-face [dĭ-fās′] *verb*, to mar, to vandalize, to destroy

def·a·ma·tion [dĕf′ə-mā′shən] *noun*, harm to a person's good reputation, slander, libel

de·fault [dĭ-fôlt′] *noun*, failure to fulfill an obligation

de·feat [dĭ-fēt′] *noun*, 1) a loss to be overcome, *verb*, 2) to beat, to conquer, to cause to fail

de·fect [dē′fĕkt′] *noun*, 1) a fault, a failing, a drawback, *verb*, 2) to desert your country or cause

de·fec·tive [dĭ-fĕk′tĭv] *adjective*, faulty, broken, imperfect

de·fend [dĭ-fĕnd′] *verb*, to fight to protect something, to guard

de·fend·ant [dĭ-fĕn′dənt] *noun*, a person required to make an answer in a lawsuit

de·fense [dĭ-fĕns′] *noun*, armor, protection from harm

de·fen·sive [dĭ-fĕn′sĭv] *adjective*, guarding, protective

de·fer [dĭ-fûr′] *verb*, 1) to put off, to postpone 2) to submit respectfully

def·er·ence [dĕf′ər-əns] *noun*, a yielding of judgment with respect to another

de·fi·ance [dĭ-fī′əns] *noun*, a state of open opposition, a challenge

de·fi·ant [dĭ-fī′ənt] *adjective*, opposing or resisting power or authority, disobedient

de·fi·cient [dĭ-fĭsh′ənt] *adjective*, incomplete, imperfect

def·i·cit [dĕf′ĭ-sĭt] *noun*, an amount that falls short of a goal or perfect balance

de·fine [dĭ-fīn′] *verb*, 1) to determine the boundaries of something 2) to give the meaning of

def·i·nite [dĕf′ə-nĭt] *adjective*, fixed, having certain limits, precise, clearly defined

def·i·nite·ly [dĕf′ə-nĭt-lē] *adverb*, precisely, clearly stated

def·i·ni·tion [dĕf′ə-nĭsh′ən] *noun*, a brief explanation of the meaning of a word or term

de·fin·i·tive [dĭ-fĭn′ĭ-tĭv] *adjective*, final, conclusive, decisive

de·flate [dĭ-flāt′] *verb*, 1) to let the air out 2) to dishearten, to discourage

de·flect [dĭ-flĕkt′] *verb*, 1) to turn aside 2) to fend off

de·form·i·ty [dĭ-fôr′mĭ-tē] *adjective*, want of proper form or symmetry, distortion

de·fraud [dĭ-frôd′] *verb*, to deprive someone of a right or property through deceit, to swindle

de·fray [dĭ-frā′] *verb*, to meet the cost of something, to pay

de·frost [dē-frôst′] *verb*, to melt, to thaw, to remove ice by melting

deft [dĕft] *adjective*, able, adroit, skillful, quick, dexterous

de·funct [dĭ-fŭngkt′] *adjective*, ended, dead, vanished

de·fy [dĭ-fī′] *verb*, to challenge a stronger force, to disobey

de·gen·er·ate [dĭ-jĕn′ər-ĭt] *verb*, to become worse physically, mentally or morally, to decay

de·grade [dĭ-grād′] *verb*, to lower someone's reputation or character, to treat disrespectfully

de·gree [dĭ-grē′] *noun*, 1) a unit for measuring temperature 2) a grade or rank conferred by colleges or universities upon graduates or other scholars 3) a measured portion determined on a scale 4) a proportion

de-i-fy [dē'ə-fī'] *verb*, to idolize or worship as a god

de-jec-tion [dĭ-jĕk'shən] *noun*, sadness, depressed, discouraged, disheartened, in low spirits

Del-a-ware [dĕl'ə-wâr'] *noun*, one of the 50 United States located in the Northeast, the capital is Dover. The state flower of **Delaware** is the peach blossom and the motto is "Liberty and Independence."

de-lay [dĭ-lā'] *noun*, 1) a period of waiting, *verb*, 2) to postpone, to make something take longer

de-lec-ta-ble [dĭ-lĕk'tə-bəl] *adjective*, delicious

del-e-gate [dĕl'ĭ-gāt'] *noun*, 1) one sent and empowered to act for another, a deputy, *verb*, 2) to entrust a duty or task to another, to assign

de-lete [dĭ-lēt'] *verb*, to erase, to mark for omission, to remove

de-lib-er-ate [dĭ-lĭb'ər-ĭt] *adjective*, 1) planned or done on purpose, *verb*, 2) to think carefully in or to make up one's mind, to consider reasons for and against

de-lib-er-ate-ly [dĭ-lĭb'ər-ĭt-lē] *adverb*, after mature consideration, consciously, intentionally

del-i-ca-cy [dĕl'ĭ-kə-sē] *noun*, 1) good taste, tact 2) choice food

del-i-cate [dĕl'ĭ-kĭt] *adjective*, 1) fine, fragile, not heavy 2) tender, pleasant, soft and fair

del-i-ca-tes-sen [dĕl'ĭ-kə-tĕs'ən] *noun*, a shop where meat, cheese, and salads are sold

de-li-cious [dĭ-lĭsh'əs] *adjective*, giving pleasure to the senses, especially taste

de-light [dĭ-līt'] *noun*, *no plural*, 1) a joy, *verb*, 2) to please, to cause joy or happiness

de-light-ful [dĭ-līt'fəl] *adjective*, pleasing, causing pleasure and satisfaction

de-lin-quent [dĭ-lĭng'kwənt] *adjective*, 1) failing to do what is required, *noun*, 2) a child who persistently breaks the law

de-lir-i-ous [dĭ-lîr'ē-əs] *adjective*, wandering in mind, light-headed

de-liv-er [dĭ-lĭv'ər] *verb*, 1) to carry or take to a person to whom something is addressed 2) to assist in the birth of a baby 3) to utter, to speak 4) to rescue or set free

de-liv-er-y [dĭ-lĭv'ə-rē] *noun*, goods that have been brought to be handed over

del-ta [dĕl'tə] *noun*, a fan-shaped area of land where a river splits into channels

de-lude [dĭ-lōōd'] *verb*, to deceive, to fool, to defraud, to mislead

de-lu-sion [dĭ-lōō'zhən] *noun*, 1) a mirage, a fantasy 2) a mistake, a misconception, a false belief

delve [dĕlv] *verb*, to examine, to probe, to look through, to search

dem-a-gogue [dĕm'ə-gôg'] *noun*, a person who wins support by appealing to popular feelings and prejudices

de-mand [dǐ-mǎnd′] *noun*,
1) a strong request, to express a
need, *verb*, 2) to ask for
something very strongly, to
insist or require, to claim

de-mean [dǐ-mēn′] *verb*, to
humiliate, to lower

de-mean-or [dǐ-mē′nər] *noun*, a
person's posture or conduct,
behavior, attitude

de-mer-it [dǐ-měr′ǐt] *noun*, a bad
mark, a point against you

de-mise [dǐ-mīz′] *noun*, the death
of any person, the end of
something, the failure

de-moc-ra-cy [dǐ-mŏk′rə-sē] *noun*,
a form of government where the
power is retained by the people

dem-o-crat [děm′ə-krǎt′] *noun*, a
person who advocates
democracy, one who believes in
government in which all people
have a vote

Dem-o-crat-ic Party *noun*, a U.S.
political party that grew out of
the Republican-Democratic
Party led by Thomas Jefferson

de-mol-ish [dǐ-mŏl′ǐsh] *verb*, to
destroy, to ruin, to take down

dem-o-li-tion [děm′ə-lǐsh′ən]
noun, destruction

dem-on-strate [děm′ən-strāt′]
verb, 1) to establish beyond
doubt 2) to show how something
is done

dem-on-stra-tion
[děm′ən-strā′shən] *noun*, 1) a
presentation of evidence, proof
2) an exhibit, a display 3) a
peaceful march 4) a protest

de-mor-al-ize [dǐ-môr′ə-līz′] *verb*,
to corrupt, to discourage

de-mur [dǐ-mûr′] *noun*,
1) hesitation, objection, *verb*,
2) to make an objection, to delay

de-mure [dǐ-myŏŏr′] *adjective*,
quiet modesty, reserved

den [děn] *noun*, a place in which a
wild animal lives, a lair

den-i-grate [děn′ǐ-grāt′] *verb*, to
ridicule, to belittle, to defame

den-im [děn′ǐm] *noun*, a heavy
blue cotton fabric used to make
clothing such as jeans

de-nom-i-na-tion
[dǐ-nŏm′ə-nā′shən] *noun*,
1) classification, identification,
2) a branch of a religion

de-nom-i-na-tor [dǐ-nŏm′ə-nā′tər]
noun, the number base or the
quantity below the line of a
fraction

de-nounce [dǐ-*nouns*′] *verb*, to
attack as deserving of
punishment or censure

dense [děns] *adjective*, compact,
closely packed together

den-si-ty [děn′sǐ-tē] *noun*, the
amount of something in a
specified volume or area

dent [děnt] *noun*, 1) an indentation,
a nick, *verb*, 2) to make a gouge

den-tal [děn′tl] *adjective*, referring
to teeth or dentistry

de-nun-ci-a-tion
[dǐ-nŭn′sē-ā′shən] *noun*, a public
accusation, condemnation

de-ny [dǐ-nī′] *verb*, to declare
something untrue

de-part [dǐ-pärt′] *verb*, to leave, to
go away from 2) deviate, to
withdraw 3) to die

de-part-ment [dĭ-pärt′mənt] *noun*, a separate division of a government or a business

de-par-ture [dĭ-pär′chər] *noun*, when someone leaves a place or the subject

de-pend [dĭ-pěnd′]*verb*, to be based on for support, to rely on

de-pend-a-ble [dĭ-pěnd′ā′bəl] *adjective*, reliable

de-pend-ant [dĭ-pěn′dənt] *noun*, a member of a family supported by another

de-pend-ent [dĭ-pěn′dənt] *adjective*, relying on something else for support

de-pict [dĭ-pĭkt′] *verb*, to portray

de-plete [dĭ-plēt′] *verb*, to reduce, to lessen by overuse

de-plor-able [dĭ-plôr′ə-bəl] *adjective*, shockingly bad, vile, horrible, reprehensible

de-plore [dĭ-plôr′] *verb*, to look down on with disgust

de-ploy [dĭ-ploi′] *verb*, to move troops to the battle line to make it stronger, to position

de-port [dĭ-pôrt′] *verb*, to force someone to leave a country

de-pose [dĭ-pōz′] *verb*, to remove from office or power

de-pos-it [dĭ-pŏz′ĭt] *noun*, 1) money placed in a bank for safe-keeping 2) money to hold something until the full price is paid 3) sediment that has settled on the bottom, *verb*, 4) to put aside for safe keeping

de-po-si-tion [děp′ə-zĭsh′ən] *noun*, testimony under oath

de-pot [dē′pō] *noun*, a building for accommodations of passengers or freight, a warehouse

de-praved [dĭ-prāvd′] *adjective*, morally corrupt

de-pre-ci-ate [dĭ-prē′shē-āt′] *verb*, to diminish in value

de-pre-ci-a-tion [dĭ-prē′shē-ā′shən] *noun*, a falling of value

dep-re-da-tion [děp′rĭ-dā′shən] *noun*, stealing, pillage, plundering, destruction

de-press [dĭ-prěs′] *verb*, 1) to make someone feel sad, melancholy, despondent 2) to hold down

de-pres-sion [dĭ-prěsh′ən] *noun*, 1) a feeling of despair, hopelessness, sadness 2) a low place or hole or basin 3) a slump in the economy, a long recession

de-prive [dĭ-prīv′] *verb*, to deny, to keep from having, to prevent from enjoying something

depth [děpth] *noun*, 1) the distance of how far something goes down *plural*, 2) the innermost or deepest part

de-ranged [dĭ-rānjd′] *adjective*, insane, mad, or crazy

der-by [dûr′bē] *noun*, a contest or race open to anyone

der-e-lict [děr′ə-lĭkt′] *adjective*, 1) abandoned, left to fall into ruin, *noun*, 2) a vagrant

de-ride [dĭ-rīd′] *verb*, to scoff at

de-ri-sion [dĭ-rĭzh′ən] *noun*, ridicule, scorn, mockery

de-riv-a-tive [dĭ-rĭv′ə-tĭv] *noun*, unoriginal, something derived from another source

de-rive [dĭ-rīv′] *verb*, to obtain or issue from a source

de-rog-a-to-ry [dĭ-rŏg′ə-tôr′ē] *adjective*, negative, mean, detracting, disparaging

der-rick [dĕr′ĭk] *noun*, a tall tower that fits over an oil well, that supports drilling machinery

de-scend [dĭ-sĕnd′] *verb*, to go or move down, to fall, to drop

de-scend-ant [dĭ-sĕn′dənt] *noun*, a member of a younger generation of a family, such as a child or a grandchild, an offspring

de-scent [dĭ-sĕnt′] *noun*, a change from higher to lower, a downward movement

de-scribe [dĭ-skrīb′] *verb*, to set forth in words, to mark out, to draw, to tell about, to recount

de-scrip-tion [dĭ-skrĭp′shən] *noun*, act or result of representing by words, a narration

des-e-crate [dĕs′ĭ-krāt′] *verb*, to destroy something sacred or of religious significance

des-ert [dĕz′ərt] *noun*, 1) a dry region that is sandy and without trees, [dĭ-zûrt′] *verb*, 2) to leave, to runaway, to abandon

de-ser-tion [dĭ-zûr′shən] *noun*, 1) abandonment 2) the act of leaving service in the Armed Forces without permission

de-serve [dĭ-zûrv′] *verb*, to merit or be worthy of, to earn by service

de-sign [dĭ-zīn′] *noun*, 1) a plan, a pattern, *verb*, 2) to conceive, to invent 3) to create a plan for something, to scheme

des-ig-nate [dĕz′ĭg-nāt′] *verb*, to point out, to indicate, to give a name, to appoint

de-sign-er [dĭ-zī′nər] *noun*, a person who creates new variations of a product

de-sir-a-ble [dĭ-zīr′′ā′bəl] *adjective*, pleasing, agreeable

de-sire [dĭ-zīr′] *noun*, 1) a want, wish, craving, *verb*, 2) to long for, to set your heart on

desk [dĕsk] *noun*, a work table with drawers for paper, pencils, etc.

des-o-late [dĕs′ə-lĭt] *adjective*, deserted, forsaken, comfortless, destitute, dismal, abandoned

des-o-la-tion [dĕs′ə-lā′shən] *noun*, emptiness, extinction

de-spair [dĭ-spâr′] *noun*, 1) the state of being without hope, feeling discouraged, *verb*, 2) to have no hope or optimism

des-per-ate [dĕs′pər-ĭt] *adjective*, hopeless, downcast, frantic, reckless because of despair

des-per-a-tion [dĕs′pə-rā′shən] *noun*, a great need, an anxious feeling out of hopelessness

des-pi-ca-ble [dĕs′pĭ-kə-bəl] *adjective*, contemptible, wretched, vile

de-spise [dĭ-spīz′] *verb*, to look down upon with contempt, to consider unworthy of respect

de-spite [dĭ-spīt′] *preposition*, in spite of, not withstanding

de-spoil [dĭ-spoil'] *verb*, to plunder

de-spond-ent [dĭ-spŏn'dənt] *adjective*, discouragement, depression of mind

des-pot-ic [dĭ-spŏt'ĭk] *adjective*, possessing and abusing unlimited power

des-pot-ism [dĕs'pə-tĭz'əm] *noun*, tyranny, dictatorship, repression

des-sert [dĭ-zûrt'] *noun*, the last course of a meal which is usually sweet

des-ti-na-tion [dĕs'tə-nā'shən] *noun*, the place where someone is going, a goal

des-ti-ny [dĕs'tə-nē] *noun*, fate, a condition predetermined by Divine will, providence

des-ti-tute [dĕs'tĭ-toot'] *adjective*, without means of existence, extremely poor, penniless

de-stroy [dĭ-stroi'] *verb*, to put an end to, to ruin or annihilate

de-struc-tion [dĭ-strŭk'shən] *noun*, the act of tearing down, causing harm or damage

de-struc-tive [dĭ-strŭk'tĭv] *adjective*, harmful, troublesome, ruinous, to cause to make useless

de-tach [dĭ-tăch']*verb*, to separate, to remove, to disconnect, to undo

de-tail [dĭ-tāl'] *noun*, a small part of a larger whole, an item

de-tain [dĭ-tān'] *verb*, 1) to restrain, to confine, to hold back, 2) to delay, to hinder, to impede

de-tect [dĭ-tĕkt'] *verb*, to discover the presence of, to find

de-tec-tive [dĭ-tĕk'tĭv] *noun*, a special policeman who is trained to investigate crimes

de-ter-gent [dĭ-tûr'jənt] *noun*, a soap used in a machine to wash clothes or dishes

de-te-ri-o-rate [dĭ-tîr'ē-ə-rāt'] *verb*, to grow worse or less valuable, to rot, to decay

de-te-ri-o-ra-tion [dĭ-tîr'ē-ə-rā'shən] *noun*, a condition that is getting worse

de-ter-mi-na-tion [dĭ-tûr'mə-nā'shən] *noun*, a firm purpose, resolve, will power

de-ter-mine [dĭ-tûr'mĭn] *verb*, to come to a decision, to realize

de-test [dĭ-tĕst'] *verb*, to dislike intensely, to hate, to loathe

de-tour [dē'toor'] *noun*, a temporary way around a blockage in the road

de-tract [dĭ-trăkt'] *verb*, to belittle, to humiliate, to take away from

det-ri-men-tal [dĕt'rə-mĕn'tl] *adjective*, harmful, injurious

dev-as-tate [dĕv'ə-stāt'] *verb*, to lay waste, to ravage, to destroy

de-vel-op [dĭ-vĕl'əp] *verb*, 1) to go through a process of growth 2) to apply chemicals to film to bring out the picture

de-vel-op-ment [dĭ-vĕl'əp-mənt] *noun*, expansion, growth, gain

de-vi-ate [dē'vē-āt'] *verb*, to turn away from an accepted course of action or standard, to stray

de-vice [dĭ-vīs'] *noun*, an instrument used for a particular purpose, a scheme, a trick

dev-il [dĕv'əl] *noun*, 1) an evil spirit, thought to cause destruction and bad things 2) a wicked or mischievous person

de-vi-ous [dē′vē-əs] *adjective*, crafty, dishonest, gone astray

de-vise [dĭ-vīz′] *verb*, to contrive, to invent, to plan, to create

de-void [dĭ-void′] *adjective*, empty, entirely without, lacking

de-vote [dĭ-vōt′] *verb*, to dedicate oneself to something, to truly believe, to give entirely

de-vo-tion [dĭ-vō′shən] *noun*, a strong love or attachment

de-vour [dĭ-vour′] *verb*, to consume greedily, to eat

de-vout [dĭ-vout′] *adjective*, sincere, honest, dedicated

dew-drop [dōō′drŏp′] *noun*, drops of water that form on anything exposed to air over night

dex-ter-i-ty [dĕk-stĕr′ĭ-tē] *noun*, the ability to operate skillfully by hand, adroitness

di[dī] *noun*, *plural*, **dice**, a small square cube with a different number of dots on each side ranging from 1-6, often used in games of chance

di-a-be-tes [dī′ə-bē′tĭs] *noun*, a disease that keeps the body from changing food into insulin so that it must be supplied with an injection from a needle

di-a-bol-ic-al [dī′ə-bŏl′ĭ-kəl] *adjective*, evil, devilish

di-ag-no-sis [dī′əg-nō′sĭs] *noun*, a scientific determination of signs or symptoms of a disease

di-ag-o-nal [dī-ăg′ə-nəl] *adjective*, a line extending from one corner of a polygon to another, that is not next to it or adjacent

di-a-gram [dī′ə-grăm′] *noun*, 1) a figure, or plan, or idea drawn to illustrate any statement, *verb*, 2) to represent by a drawing

di-al [dī′əl] *noun*, a round face on a machine often with numbers on it that sets the instrument to perform a specific task

di-a-lect [dī′ə-lĕkt′] *noun*, a form of speech characterized by local peculiarities, jargon

di-a-logue [dī′ə-lôg′] *noun*, 1) a conversation between two or more people, especially in a play or narrative 2) a discussion

di-am-e-ter [dī-ăm′ĭ-tər] *noun*, 1)a straight line that joins two points of a figure or bodyand psses through the center 2) the distance through the center of an object from one side to the other

dia-mond [dī′ə-mənd] *noun*, 1) a hard clear stone that is very precious 2) a plane shape with four equal sides with two acute angles and two obtuse angles 3) a baseball field

di-a-phragm [dī′ə-frăm′] *noun*, the muscle for breathing located between the lungs and the stomach

di-ar-rhe-a [dī′ə-rē′ə] *noun*, a purging or looseness of the bowels

di-a-ry [dī′ə-rē] *noun*, a journal in which a person makes daily entries of observations and personal experiences

di-a-tribe [dī′ə-trīb′] *noun*, a tirade, an harangue, a scolding, a forceful and bitter verbal attack against someone or something

dice

dice [dīs] *verb*, to cut into small cubes

di-chot-o-my [dī-kŏt′ə-mē] *noun*, branching into two parts

dic-tate [dĭk′tāt′] *verb*, 1) to speak a letter or memo for someone else to write 2) to impose, to decree, to instruct, to command

dic-ta-tor [dĭk′tā′tər] *noun*, an absolute ruler, a tyrant

dic-tion [dĭk′shən] *noun*, enunciation of words

dic-tion-ar-y [dĭk′shə-nĕr′ē] *noun*, a word list with definitions

did [dĭd] *past tense* of **do**

di-dac-tic [dī-dăk′tĭk] *adjective*, teaching a moral lesson

did-n't [dĭd′nt] the *contraction* of the words **did** and **not**

die [dī] *verb*, 1) to stop living or existing 2) to lose strength, to fade away

die-sel [dē′zəl, -səl] *noun*, an engine that burns oil instead of gasoline

di-et [dī′ĭt] *noun*, 1) the food someone eats, *verb*, 2) to reduce the intake of food

dif-fer [dĭf′ər] *verb*, to be unlike, to modify, to vary, to disagree

dif-fer-ence [dĭf′ər-əns] *noun*, 1) contrast, unlikeness 2) disagreement, dissension 3) the answer in subtraction

dif-fer-ent [dĭf′ər-ənt] *adjective*, 1) not alike 2) not the same as the rest, strange 3) separate

dif-fer-ent-ial [dĭf′ə-rĕn′shəl] *adjective*, the difference or the amount of difference

dif-fi-cult [dĭf′ĭ-kŭlt′] *adjective*, hard to do, not easy, hard to manage or please, tough

dif-fi-cul-ty [dĭf′ĭ-kŭl′tē] *noun*, a hardship, a problem

dif-fuse [dĭ-fyōōz′] *adjective*, 1) not concentrated, *verb*, 2) to spread or circulate 3) to cause to flow on all sides

dif-fu-sion [dĭ-fyōō′zhən] *noun*, a spreading in all directions like gas

dig [dĭg] *verb*, 1) to unearth dirt with tools or a machine to make a hole 2) to search, to probe

di-gest [dī-jĕst′] *verb*, 1) to convert food into an absorbable form 2) to think over and arrange

di-ges-tion [dī-jĕs′chən] *noun*, the process by which food can be converted into a source of energy that can be easily assimilated by the body

di-ges-tive tract *noun*, the alimentary canal where food is processed and broken down into sugars for energy

dig-it [dĭj′ĭt] *noun*, 1) a finger or toe 2) any number 1 through 9, including 0

dig-ni-fy [dĭg′nə-fī] *verb*, to give honor to

dig-ni-tar-y [dĭg′nĭ-tĕr′ē] *noun*, 1) a person who is well known and respected, e.g., national leaders 2) a person holding high rank or position

dig-ni-ty [dĭg′nĭ-tē] *noun*, 1) an office or rank 2) nobleness of manner, aspect, or style 3) the quality of being respected and respecting oneself

di-gres-sion [dī-grĕsh′ən, dĭ-] *noun*, wandering away from the subject temporarily

dike [dīk] *noun*, a dam or wall that holds back water

di-lap-i-dated [dĭ-lăp′ĭ-dā′tĭd] *adjective*, falling into ruin

di-late [dī-lāt′] *verb*, to expand, to enlarge or become wider

dil-a-to-ry [dĭl′ə-tôr′ē] *adjective*, delaying, avoiding

di-lem-ma [dĭ-lĕm′ə] *noun*, a case in which a person must make a difficult choice, a quandary

dil-i-gence [dĭl′ə-jəns] *noun*, careful attention and application to work, constant in effort to accomplish something

dil-i-gent [dĭl′ə-jənt] *adjective*, careful, tireless effort

di-lute [dī-loot′] *verb*, to make thinner or weaker as by mixing or dispersion

dim [dĭm] *adjective*, 1) poorly lit, 2) become less bright or distinct

dime [dīm] *noun*, a coin worth ten cents in the USA

di-men-sion [dĭ-mĕn′shən] *noun*, proportion, size, scope, extent

di-min-ish [dĭ-mĭn′ĭsh] *verb*, to make or grow less, to reduce

dim-i-nu-tion [dĭm′ə-noo′shən] *noun*, a lessening, a reduction in size, something made smaller

dim-ple [dĭm′pəl] *noun*, a small dent in a surface

dine [dīn] *verb*, to eat supper

din-gy [dĭn′jē] *adjective*, dirty, soiled, dull and drab, shabby

din-ner [dĭn′ər] *noun*, a meal served in the evening

di-no-saur [dī′nə-sôr′] *noun*, any of a group of extinct reptiles of the Mesozoic era that were either herbivores, carnivores or omnivores that inhabited the earth millions of years ago

dip [dĭp] *noun*, 1) a decline 2) a plunge, a soaking, an immersion 3) an appetizer, *verb*, 4) to put something into a liquid and then take it out again 5) to slope

diph-the-ri-a [dĭf-thîr′ē-ə] *noun*, a disease in which the throat is coated with a false membrane obstructing the nose and throat

diph-thong [dĭf′thông′] *noun*, a combination of two vowels forming a compound sound

di-plo-ma [dĭ-plō′mə] *noun*, a document under seal attesting a degree conferred to a person

di-plo-ma-cy [dĭ-plō′mə-sē] *noun*, the business of conducting international negotiations

dip-lo-mat [dĭp′lə-măt′] *noun*, an ambassador, a tactful person

dip-per [dĭp′ər] *noun*, a long handle utensil, a ladle

dire [dīr] *adjective*, disastrous, terrible, dreadful, very bad

di-rect [dĭ-rĕkt′, dī-] *adjective*, 1) straight to something without stopping, *verb*, 2) to aim 3) to guide, to control, to manage

di-rec-tion [dĭ-rĕk′shən] *noun*, 1) the way in which someone is going or pointing 2) an instruction explaining a plan

di-rect-ly [dĭ-rĕkt′lē] *adverb*, at once, immediately

di-rec-to-ry [dĭ-rĕk′tə-rē] *noun*, 1) a list of people and their addresses 2) a computer file listing other files

dirge [dûrj] *noun*, sorrowful music

dir-i-gi-ble [dĭr'ə-jə-bəl] *noun*, an airship, a blimp

dirt [dûrt] *noun*, loose soil, something which causes filth

dirt-y [dûr'tē] *adjective*, 1) not clean, soiled 2) obscene

dis-a-bil-i-ty [dĭs'ə-bĭl'ĭ-tē] *noun*, a physical or mental incapacity, an inability to do something

dis-a-ble [dĭs-ā'bəl] *verb*, to impair the activity of something

dis-ad-van-tage [dĭs'əd-văn'tĭj] *noun*, something that makes things more difficult

dis-a-gree [dĭs'ə-grē'] *verb*, to be of a different opinion, to differ

dis-a-gree-a-ble [dĭs'ə-grē'ə-bəl] *adjective*, unpleasant, irritating, obnoxious, bad-tempered

dis-a-gree-ment [dĭs'ə-grē'mənt] *noun*, an argument, a dispute

dis-ap-pear [dĭs'ə-pîr'] *verb*, to vanish, to go away or become hidden, to fade away

dis-ap-point [dĭs'ə-point'] *verb*, to frustrate the hopes of someone

dis-ap-point-ment [dĭs'ə-point'mənt] *noun*, failure, frustration, feeling let down

dis-ap-prove [dĭs'ə-prōōv'] *verb*, to object, to blame, to dislike

dis-arm [dĭs-ärm'] *verb*, to weaken, to subdue, to win over

dis-as-ter [dĭ-zăs'tər] *noun*, something very bad that happens, a catastrophe

dis-as-tr-ous [dĭ-zăs'trəs] *adjective*, unfortunate, a total failure, devastating

dis-be-lief [dĭs'bĭ-lēf'] *noun*, doubt, skepticism, unable to believe

dis-burse [dĭs-bûrs'] *verb*, 1) to pay out money 2) to scatter

dis-burse-ment [dĭs-bûrs'mənt] *noun*, 1) the act of spreading around 2) paying out money

dis-card [dĭ-skärd'] *verb*, to shed, to reject, to eliminate

dis-cern [dĭ-sûrn'] *verb*, to distinguish, to perceive

dis-charge [dĭs-chärj'] *noun*, 1) something that is sent out from its source, *verb*, 2) to unload, to send away, to fire

dis-ci-ple [dĭ-sī'pəl] *noun*, a follower, an apostle

dis-ci-pline [dĭs'ə-plĭn] *noun*, 1) strict training to teach self-control, *verb*, 2) to instruct, to prepare 3) to punish

dis-close [dĭ-sklōz'] *verb*, to reveal, to make known

dis-com-fort [dĭs-kŭm'fərt] *noun*, annoyance, embarrassment

dis-con-nect [dĭs'kə-někt'] *verb*, to cut off, to detach, to sever

dis-con-so-late [dĭs-kŏn'sə-lĭt] *adjective*, sad, gloomy

dis-con-tent [dĭs'kən-těnt'] *adjective*, dissatisfaction, restlessness, unhappiness

dis-con-tent-ed [dĭs'kən-těn'tĭd] *adjective*, unhappy, disgruntled

dis-con-tin-ue [dĭs'kən-tĭn'yōō] *verb*, to cease, to stop

dis-cord [dĭs'kôrd'] *noun*, a conflict, opposition, differences

dis-count [dĭs'kount'] *verb*, 1) a deduction made from the price of something 2) to discredit or disbelieve someone

dis-cour-age [dĭ-skûr′ĭj] *verb*, to dishearten, to deter, to perplex

dis-course [dĭs′kôrs′] *noun*, a talk, a chat, a conversation

dis-cov-er [dĭ-skŭv′ər] *verb*, 1) to make known the existence of something 2) to find out through study or observation

dis-cov-er-y [dĭ-skŭv′ə-rē] *noun*, something seen or learned for the first time

dis-cred-it [dĭs-krĕd′ĭt] *verb*, to destroy confidence, to jeopardize

dis-creet [dĭ-skrēt′] *adjective*, prudent, careful, not giving away secrets, tactful, unobtrusive

dis-crep-an-cy [dĭ-skrĕp′ən-sē] *noun*, disagreement, inconsistency, failure to match

dis-cre-tion [dĭ-skrĕsh′ən] *noun*, discernment, wise conduct and management, freedom to decide something, being discreet

dis-crim-i-nate [dĭ-skrĭm′ə-nāt′] *verb*, to make a difference or distinction, to distinguish

dis-crim-i-na-tion [dĭ-skrĭm′ə-nā′shən] *noun*, *no plural*, distinction, focusing on differences, treated unfairly

dis-cuss [dĭ-skŭs′] *verb*, to talk about, to exchange ideas

dis-cus-sion [dĭ-skŭsh′ən] *noun*, the exchange of ideas, disputation, a debate

dis-dain [dĭs-dān′] *verb*, to treat with scorn or contempt

dis-ease [dĭ-zēz′] *noun*, a change in a living body that interferes with

its normal functioning

dis-en-gage [dĭs′ĕn-gāj′] *verb*, to release, to detach

dis-fig-ure [dĭs-fĭg′yər] *verb*, to deform, to mangle, to injure

dis-grace [dĭs-grās′] *noun*, 1) something to feel ashamed of *verb*, 2) to shame, to degrade

dis-grace-ful [dĭs-grās′fəl] *adjective*, causing shame

dis-grunt-led [dĭs-grŭn′tld] *adjective*, resentful, annoyed, discontented, unhappy

dis-guise [dĭs-gīz′] *noun*, 1) things worn to change the appearance, *verb*, 2) to change the appearance of someone or something, to cover up

dis-gust [dĭs-gŭst′] *noun*, 1) a strong feeling of dislike, revulsion, *verb*, 2) to give someone a strong feeling of dislike, to offend someone

dis-gust-ing [dĭs-gŭs′tĭng] *adjective*, revolting, horrible

dish [dĭsh] *noun*, 1) a container used at a meal to serve food 2) food served

dis-hon-est [dĭs-ŏn′ĭst] *adjective*, deceitful, lying, corrupt

disk [dĭsk] *noun*, any flat round object, a circular plate

dis-like [dĭs-līk′] *verb*, 1) distaste, hostility 2) to detest, to hate

dis-mal [dĭz′məl] *adjective*, causing feelings of unhappiness, showing sadness or gloom

dis-man-tle [dĭs-măn′tl] *verb*, to take apart, to break down to basic parts to see how something is made

dis-may [dĭs-mā′] *noun*, 1) an
uneasy feeling of alarm, *verb*,
2) to cause feelings of sadness

dis-miss [dĭs-mĭs′] *verb*, to send
away, to cause or permit to go

dis-obed-i-ent [dĭs′ə-bē′dē-ənt]
adjective, refusing to obey

dis-o-bey [dĭs′ə-bā′] *verb*, not to do
what you are told, to resist

dis-or-der [dĭs-ôr′dər] *noun*,
1) sickness, ailment
2) confusion, chaos

dis-or-gan-ized [dĭs-ôr′gə-nīzd′]
adjective, not properly planned

dis-par-age [dĭ-spăr′ĭj] *verb*, to
belittle, to criticize

dis-pa-rate [dĭs′pər-ĭt] *adjective*,
miscellaneous, diverse, very
different in kind, a wide variety

dis-par-i-ty [dĭ-spăr′ĭ-tē] *noun*, a
noticeable difference, inequality

dis-patch [dĭ-spăch′] *noun*, 1) a
message 2) haste, *verb*, 3) to
send off to a destination

dis-pel [dĭ-spĕl′] *verb*, to drive or
clear away

dis-pense [dĭ-spĕns′] *verb*, to allot,
to distribute, to pass around

dis-perse [dĭ-spûrs′] *verb*, to
scatter, to distribute in different
directions, to disseminate

dis-per-sion [dĭ-spûr′zhən] *noun*,
scattering, diffusion, distribution

dis-place [dĭs-plās′] *verb*, to
rearrange, to move around

dis-play [dĭ-splā′] *noun*, 1) a show,
an exhibit, *verb*, 2) to show or
exhibit something for many
people to see, to present

dis-please [dĭs-plēz′] *verb*, to
anger, to annoy, to irritate

dis-pose [dĭ-spōz′] *verb*, 1) to get
rid of 2) to adjust, to settle, to
adapt 3) to place or arrange

dis-po-si-tion [dĭs′pə-zĭsh′ən]
noun, 1) nature, personality,
makeup 2) arrangement

dis-pro-por-tion-ate
[dĭs′prə-pôr′shə-nĭt] *adjective*,
relatively too large or too small

dis-prove [dĭs-prōōv′] *verb*, to
refute, to prove false

dis-pute [dĭ-spyōōt′] *noun*, 1) an
argument, a strong difference of
opinion, *verb*, 2) to question the
truth or value of, to debate

dis-qual-i-fy [dĭs-kwŏl′ə-fī′] *verb*,
to invalidate, to judge ineligible

dis-re-gard [dĭs′rĭ-gärd′] *noun*,
1) indifference, disrespect,
neglect, *verb*, 2) to fail to notice

dis-re-spect-ful [dĭs′rĭ-spĕkt′fəl]
adjective, rudeness, contempt

dis-rupt [dĭs-rŭpt′] *verb*, to
interrupt, to disturb, to confuse

dis-sat-is-fac-tion
[dĭs-săt′ĭs-făk′shən] *adjective*,
discontent, uneasiness

dis-sat-is-fied [dĭs-săt′ĭs-fīd′]
adjective, displeased, fed up,
disgruntled, unhappy

dis-sect [dĭ-sĕkt′] *verb*, to divide
into separate parts

dis-sec-tion [dĭ-sĕk′shən] *noun*,
analysis, cutting apart in order to
examine internal structure

dis-sem-i-nate [dĭ-sĕm′ə-nāt′] *verb*,
to scatter, to spread around

dis-sem-i-na-tion
[dĭ-sĕm′ə-nā′shən] *noun*, a
scattering or spreading around,
diffusion

dis-sen-sion [dĭ-sĕn′shən] *noun*, a disagreement causing strife

dis-sent [dĭ-sĕnt′] *verb*, disagreement in opinion, a quarrel

dis-sim-i-lar [dĭ-sĭm′ə-lər] *adjective*, not alike, different

dis-si-pate [dĭs′ə-pāt′] *verb*, to lessen, to decrease, to diminish

dis-so-ci-ate [dĭ-sō′shē-āt′] *verb*, to disband, to separate from

dis-solve [dĭ-zŏlv′] *verb*, 1) to make a solid disappear into a liquid solution 2) breaking up an assembly, to end an agreement

dis-so-nance [dĭs′ə-nəns] *noun*, discord, noise, harsh sounds

dis-suade [dĭ-swād′] *verb*, to discourage by persuasion, to deter

dis-tance [dĭs′təns] *noun*, 1) the space between two points, an interval 2) a place far away

dis-tant [dĭs′tənt] *adjective*, far

dis-tend [dĭ-stĕnd′] *verb*, to expand, to bloat, to stretch

dis-tinct [dĭ-stĭngkt′] *adjective*, 1) distinguished by nature or station 2) individual, unique 3) well-defined, very clear

dis-tinc-tion [dĭ-stĭngk′shən] *noun*, 1) notable difference as in a feature 2) excellence, an outstanding accomplishment

dis-tin-guish [dĭ-stĭng′gwĭsh] *verb*, 1) to stand out, to be different 2) to discern, to recognize

dis-tort [dĭ-stôrt′] *verb*, to twist, to bend, to misconstrue, to falsify

dis-tract [dĭ-străkt′] *verb*, to take one's attention from something

dis-trac-tion [dĭ-străk′shən] *noun*, confusion, complication

dis-traught [dĭ-strôt′] *adjective*, very worried, disturbed, overcome with anxiety, upset

dis-tress [dĭ-strĕs′] *noun*, *no plural*, 1) a feeling of worry, sadness or difficulty, *verb*, 2) to make someone sad or aggravated

dis-trib-ute [dĭ-strĭb′yŏot] *verb*, 1) to allot or divide among several 2) to spread out over an area

dis-tri-bu-tion [dĭs′trə-byŏo′shən] *noun*, *no plural*, passing out things, circulation

dis-trict [dĭs′trĭkt] *noun*, 1) a division of territory 2) a defined portion of a state, or county

dis-trust [dĭs-trŭst′] *noun*, 1) suspicion, doubt, *verb*, 2) to lack confidence in someone

dis-turb [dĭ-stûrb′] *verb*, to bother or interrupt, to make anxious

ditch [dĭch] *noun*, 1) a deep place where water can run off to avoid flooding, *verb*, 2) to discard

dive [dīv] *noun*, 1) a downward plunge, *verb*, 2) to plunge head first into water 3) to move downward, to drop sharply

div-er [dī′vər] *noun*, a person who works under water and wears a special instrument containing oxygen to help him breathe

di-verge [dĭ-vûrj′] *verb*, 1) to go off in different directions from the same place 2) to differ

di-verse [dĭ-vûrs′] *adjective*, different features, various

di-ver-sion [dĭ-vûr′zhən] *noun*, amusement, play, pastime

di-ver-si-ty [dĭ-vûr′sĭ-tē] *noun*, variety, a wide range

di-vert [dǐ-vûrt′] *verb*, to distract, to draw aside, to entertain

di-vest [dǐ-věst′] *verb*, to strip, to take away from

di-vide [dǐ-vīd′] *verb*, 1) learning how many times one number can go into another 2) to separate 3) to share or distribute

div-i-dend [dǐv′ǐ-děnd′] *noun*, 1) the share of surplus earned by any policy 2) a number to be divided by another number 3) a benefit from an action

di-vine [dǐ-vīn′] *adjective*, of or relating to God, holy, sacred

di-vis-i-ble [dǐ-vǐz′ə-bəl] *adjective*, capable of being divided or separated

di-vi-sion [dǐ-vǐzh′ən] *noun*, 1) the act or result of separating anything into parts 2) difference in opinion or feeling 3) an operation on two numbers that results in a quotient

di-vorce [dǐ-vôrs′] *noun*, 1) a legal document that finalizes the end of a marriage, *verb*, 2) to arrange by law for a husband and wife to separate and end their marriage

di-vulge [dǐ-vŭlj′] *verb*, to reveal

do [dōō] *verb*, 1) to cause to happen, carry out, to perform 2) used with another verb to ask a question 3) used with not to say not to do something

doc-ile [dŏs′əl] *adjective*, easy to control, gentle, submissive

dock [dŏk] *noun*, 1) a pier where ships load and unload cargo, *verb*, 2) to tie a ship up at a port or wharf or boat landing 3) to deduct from someone's wages

doc-tor [dŏk′tər] *noun*, a person trained and licensed to practice medicine

doc-trine [dŏk′trǐn] *noun*, 1) that which is taught 2) principle of faith or religion, a tenet

doc-u-ment [dŏk′yə-mənt] *noun*, 1) an original or official paper relied upon as basis 2) any writing or book that conveys information, *verb*, 3) to write down or put on paper

dodge [dŏj] *verb*, to avoid something by moving out of the way, to evade, to escape

doe [dō] *noun*, a female deer

does [dŭz] *verb*, the part of the verb **do** used with he, she and it

does-n't [dŭz′ənt] the *contraction* of the words **does** and **not**

dog [dôg] *noun*, an animal with four legs and a tail that eats meat

dog-ma [dôg′mə] *noun*, 1) positive statements not based on fact, 2) doctrine

dol-drums [dōl′drəmz′] *noun*, listlessness, feeling the blues, an inactive period, depression

doll [dŏl] *noun*, a toy made to look like a baby or a miniature person

dol-lar [dŏl′ər] *noun*, currency used in the United States and Canada, valued at 100 cents

dol-lar sign *noun*, a symbol used to mean dollars : $

dol-phin [dŏl′fǐn] *noun*, any of various small cetaceans with the snout in the shape of a beak and the neck vertebrae partially fused like a whale

do-main [dō-mān'] *noun*, specialty, realm, jurisdiction, field

dome [dōm] *noun*, a roof that resembles a half circle

do-mes-tic [də-měs'tĭk] *adjective*, found in the home or in one's home land

dom-i-nant [dŏm'ə-nənt] *adjective*, governing, controlling, bossy

dom-i-nate [dŏm'ə-nāt'] *verb*, to have control over someone

dom-i-no [dŏm'ə-nō'] *noun*, a game made with uniform blocks each showing any number of dots, 1-6

do-nate [dō'nāt'] *verb*, to make a contribution, to bestow, to give

do-na-tion [dō-nā'shən] *noun*, a gift of money or goods given to a charitable organization

done [dŭn] *adjective*, 1) finished or complete 2) cooked, *verb*, 3) the *past tense* of **do**

don-key [dŏng'kē] *noun*, an animal like a small horse with long ears, a mule, a burro, an ass

do-nor [dō'nər] *noun*, a person who gives money or something of value to a cause

don't [dōnt] the *contraction* of the words **do** and **not**

doom [dōōm] *noun,* 1) a grim fate, *verb*, 2) to destroy or to ruin

door [dôr] *noun*, the moveable piece of wood or metal, standing at an entrance, a portal

door-way [dôr'wā'] *noun*, the opening for an entrance to a building or a room

dor-mant [dôr'mənt] *adjective*, sleeping, inactive, inert

dor-mi-to-ry [dôr'mĭ-tôr'ē] *noun*, a building containing a series of beds and rooms for sleeping

dor-sal [dôr'səl] *adjective*, hair on the back of an animal

dose [dōs] *noun*, the amount of medicine that is to be taken at one time to be effective

dot [dŏt] *noun*, a very small round spot or mark, a point, a fleck

dou-ble [dŭb'əl] *adjective*, 1) twice as much, multiplied by two 2)made up of two like parts *noun*, 3) a stand in or look alike 4) a two-base hit in baseball, *verb*, 5) to make twice as much

doubt [dout] *noun*, 1) a question, a feeling of wonder, mistrust, *verb*, 2) to question, to wonder 3) to hesitate, to ponder

doubt-ful [dout'fəl] *adjective*, questionable, unlikely

dough [dō] *noun*, a mixture of mostly flour and water to be baked to make bread or cake

dough-nut [dō'nŭt'] *noun*, a fried cake rolled in sugar

douse [dous] *verb*, to drench, to saturate, to extinguish a light

dove [dŭv] *noun*, 1) a white pigeon, *verb*, [dōv] 2) *past tense* of **dive**

down [doun] *adverb*, 1) to apply toward the ground, *noun*, 2) soft feathers, *preposition*, 3) in or to a lower place

down-stairs [doun'stârz'] *noun*, *plural*, the landing of the story of a building, a lower floor

downtown [doun'toun'] *noun*, the main business area of a city or town

down-ward [doun′wərd] *adverb*, descending from a higher to a lower level, moving toward the ground

doze [dōz] *verb*, to take a nap, to sleep for a brief time

doz-en [dŭz′ən] *noun*, twelve

Dr. the abbreviation of doctor

drab [drăb] *adjective*, dull in color, dreary, monotonous, dingy

draft [drăft] *noun*, 1) the first unedited copy of a manuscript, *verb*, 2) an order from one person to another directing the payment of money to a person named in the document

drag [drăg] *verb*, to pull along with effort, to pass slowly, to haul

drag-on [drăg′ən] *noun*, a mythical creature usually picture as a huge serpant with wings and large claws that breathes fire

drain [drān] *noun*, 1) a pipe used to transport water into a sewer, *verb*, 2) to relieve the pressure of fluid in a wound 3) to become dry as water flows away

dra-ma [drä′mə] *noun*, 1) a story that can be performed on stage 2) excitement

dra-mat-ic [drə-măt′ĭk] *adjective*, theatrical, exciting, stirring

dra-mat-i-cal-ly [drə-măt′ĭk-lē] *adverb*, in a sudden or alarming way, impressively

drank [drăngk] *past tense* of **drink**

drape [drāp] *verb*, to cover with a piece of cloth

dra-per-y [drā′pə-rē] *noun*, a heavy curtain or fabric designed to cover a window

dras-tic [drăs′tĭk] *adjective*, acting with force, severe, extreme

draw [drô] *noun*, 1) a tie, *verb*, 2) to get, to take, to understand 3) to pull along 4) to make a picture with a writing instrument 5) to choose at random 6) to formulate, to prepare 7) to suck

draw-bridge [drô′brĭj′] *noun*, a bridge that can be raised or turned to allow boats and ships to pass

drawl [drôl] *noun*, a manner of speaking that stretches out the sound of a word or vowels

drawn [drôn] *adjective*, thin, gaunt

dread [drĕd] *noun*, 1) fear, horror, terror, *verb*, 2) to anticipate with anxiety, to fear greatly

dream [drēm] *noun*, 1) what your subconscious sees while you sleep 2) what you imagine you want to do and accomplish, *verb*, 3) to picture something with your imagination 4) to see images while sleeping

drear-y [drîr′ē] *adjective*, gloomy, causing sadness, depressing

dredge [drĕj] *verb*, 1) to try to remove something in deep water 2) to sprinkle food with sugar

drench [drĕnch] *verb*, to make completely wet, to soak

dress [drĕs] *noun*, 1) a piece of clothing that covers the bust, waist and thighs 2) a general term for clothing or attire, *verb*, 3) to put clothes on to wear in front of people 4) to adorn

dress-er [drĕs′ər] *noun*, a chest of drawers, a bureau

drew [drōō] *past tense* of **draw**

drift [drĭft] *verb*, to float

drill [drĭl] *noun*, 1) pieces of pipe used to make holes deep into the earth 2) a practice, an exercise, *verb*, 3) to make a hole in something with a special machine 4) to be trained

drink [drĭngk] *noun*, 1) liquid refreshment, a beverage, *verb*, 2) to consume liquid through the mouth, to swallow, to imbibe

drip [drĭp] *verb*, to fall in drops

drive [drīv] *noun*, 1) a short journey or trip made in a vehicle, *verb*, 2) to steer and control a moving vehicle 3) to force to go, to push through, to demand 4) to make, to compel, to urge

driv-er [drī′vər] *noun*, the person steering and controlling a moving vehicle

droll [drōl] *adjective*, queer, odd, strange and amusing

drone [drōn] *noun*, 1) a worker bee, *verb*, 2) to talk dully, to buzz or murmur like a bee

droop [drōōp] *verb*, to hang down

drop [drŏp] *noun*, 1) a small amount of water, *verb*, 2) to let something fall 3) to decrease

dross [drŏs] *noun*, waste, garbage, trash, rubbish

drought [drout] *noun*, a long period without rain

drown [droun] *verb*, to suffocate by immersion under water

drow-sy [drou′zē] *adjective*, sleepy, very tired

drudg-er-y [drŭj′ə-rē] *noun*, dull work that no one wants to do

drug [drŭg] *noun*, a substance that changes the body's chemistry, medicine, a narcotic

drug-gist [drŭg′ĭst] *noun*, a pharmacist, one who is licensed to sell drugs

drum [drŭm] *noun*, 1) a percussion instrument consisting of a hollow cylinder with a taut covering 2) a cylindrical metal container, *verb*, 3) to make music or beat on a drum

drunk [drŭngk] *adjective*, the effect of too much alcohol so a person cannot control his movement or speech

dry [drī] *adjective*, 1) without moisture or water, *verb*, 2) to expel or remove moisture

dry-er [drī′ər] *noun*, a machine that adds heat and tumbles clothes to evaporate and remove moisture

dual [dōō′əl] *adjective*, to have two purposes, double

duch-ess [dŭch′ĭs] *noun*, the wife or widow of a duke

duck [dŭk] *noun*, 1) a bird that swims and is hunted for its eggs and meat, *verb*, 2) to crouch to avoid something, to dodge

duct [dŭkt] *noun*, a tube or channel, a ventilation shaft

due [dōō] *adjective*, 1) something owed, unpaid 2) expected, *adverb*, 3) caused by, because of, *noun*, 4) a charge or fee for membership

duel [dōō′əl] *noun*, 1) a fight or competition, *verb*, 2) to invite a competition, to challenge, to dare to fight

du-et [doo-ĕt′] *noun*, a song or a piece of music for two people

duke [dook] *noun*, the title of a man who is a member of the royal family

dull [dŭl] *adjective*, 1) not sharp, blunt 2) boring, slow, dim

dumb [dŭm] *adjective*, 1) unable to speak, mute, silent 2) unintelligent, stupid

dump [dŭmp] *noun*, 1) a public place where garbage is collected and buried, *verb*, 2) to leave, to drop or throw away

dune [doon] *noun*, a rounded hill of sand heaped up by the wind

dun-geon [dŭn′jən] *noun*, a close, dark prison, an underground cell

dupe [doop] *verb*, to fool, to hoodwink, to trick, to deceive

du-pli-cate [doo′pli-kĭt] *adjective*, 1) alike, twin, *noun*, 2) a copy, a reproduction, *verb*, 3) to reproduce, to remake

du-ra-ble [door′ə-bəl] *adjective*, able to endure, lasting

du-ra-tion [doo-rā′shən] *noun*, the length of time something continues or exists

dur-ing [door′ĭng] *preposition*, in the course of time, throughout the course of

dusk [dŭsk] *noun*, the time at the end of the day just before dark

dust [dŭst] *noun*, 1) fine particles or any powder of matter, *verb*, 2) to remove the powder by cleaning or wiping

du-ty [doo′tē] *noun*, 1) an obligation or a commitment 2) the right conduct or action

dwarf [dwôrf] *noun*, 1) a small person like a midget or a pygmy, *verb*, 2) to make something look small by comparison

dwell [dwĕl] *verb*, 1) to inhabit, to live in 2) to emphasize, to think about for a long time

dwell-ing [dwĕl′ĭng] *noun*, a home to live, in a residence, an abode

dwin-dle [dwĭn′dl] *verb*, to reduce, to gradually become smaller

dye [dī] *noun*, 1) a liquid used to color cloth, *verb*, 2) fixing colors permanently in the fibers of materials, to change the color

dy-na-mic [dī-năm′ĭk] *adjective*, forces producing motion, forceful, energetic, vigorous

dy-na-mite [dī′nə-mīt′] *noun*, an explosive of nitroglycerin absorbed in porous matter

dy-nas-ty [dī′nə-stē] *noun*, a term in power, a reign of rulers

dys-en-ter-y [dĭs′ən-tĕr′ē] *noun*, a disease of the bowels

dys-func-tion-al [dĭs-fŭngk′shən-əl] *adjective*, functioning abnormally

E

each [ēch] *adjective*, every, all

ea-ger [ē′gər] *adjective*, anxious to do something, enthusiastic

ea-gle [ē′gəl] *noun*, a very large keen-sighted bird of prey

ear [îr] *noun*, 1) the organ, one on each side of the head, used to hear 2) the sleeve in a cereal plant used to hold seeds or flowers

ear-ly [ûr′lē] *adjective*, 1) near the beginning of a time period, *adverb*, 2) before the set time

earn [ûrn] *verb*, 1) to obtain money in return for working, 2) to achieve by one's own efforts

ear-nest [ûr′nĭst] *adjective*, eager, passionate, fervent, serious

ear-ring [îr′rĭng, îr′ĭng] *noun*, ornaments worn in the ear

Earth [ûrth] *noun*, 1) the name of the planet on which we live, the third planet in the solar system, **earth** *noun*, 2) soil, dirt, ground

earth-quake [ûrth′kwāk′] *noun*, a shaking movement of the earth as the ground settles

ease [ēz] *noun*, 1) relaxation, relief, *verb*, 2) to move with grace 3) to relieve stress, worry, or pain

ea-sel [ē′zəl] *noun*, a tripod frame used to support an artist's canvas

eas-i-ly [ē′zə-lē] *adverb*, without difficulty, effortlessly

east [ēst] *noun*, one of the four points of a compass pointing left of north, where the sun rises

east-ern [ē′stərn] *adjective*, 1) belonging to or characteristic of the east 2) oriental

eas-y [ē′zē] *adjective*, 1) effortless, not difficult 2) relaxed, comfortable, satisfying, familiar

eat [ēt] *verb*, 1) to put food in the mouth, chewing it and swallowing it 2) to corrode

eaves [ēvz] *noun*, *plural*, the lower edges of a roof which overhang a building

ebb [ĕb] *noun*, 1) the time when the tide recedes 2) to fall back

eb-on-y [ĕb′ə-nē] *adjective*, 1) the color black, *noun*, 2) a hard, heavy and durable wood

ec-cen-tric [ĭk-sĕn′trĭk] *adjective*, 1) off-center from the regular or normal, *noun*, 2) a circular piece of apparatus pivoted at one side from its center 3) a person who has unusual habits and tastes

ec-cle-si-as-tic [ĭ-klē′zē-ăs′tĭk] *adjective*, pertaining to the Christian church or clergy

ech-o [ĕk′ō] *noun*, 1) a series of reflected sound waves that repeat as they bounce back from an object, repeating the sound *verb*, 2) the sound of something bouncing off a surface

e-clipse [ĭ-klĭps′] *noun*, 1) a period of time when the light from the sun is blocked by the moon during the day and there is complete darkness, *verb*, 2) overshadowing causing a loss of light

e-col-o-gy [ĭ-kŏl′ə-jē] *noun*, the study of how plants and animals interact in the environment

e-co-nom-ic [ĕk′ə-nŏm′ĭk] *adjective*, related to business, finances and the economy

e-co-nom-i-cal [ĕk′ə-nŏm′ĭ-kəl] *adjective*, affordable, a good investment, inexpensive

e-co-nom-ics [ĕk′ə-nŏm′ĭks] *noun*, the study of the way people try to secure goods and services and the necessities of life

e-con-o-my [ĭ-kŏn′ə-mē] *noun*, 1) the careful or thrifty management of resources 2) a system for the management of resources, money, and jobs

ec-o-sys-tem [ĕk′ō-sĭs′təm] *noun*, living things that work together affecting each other

ec-sta-sy [ĕk′stə-sē] *adverb*, intense delight, bliss

ec-ze-ma [ĕk′sə-mə] *noun*, a rash, inflammatory disease of the skin

edge [ĕj] *noun*, 1) the place or point where something begins or ends 2) the rim or brink of something 3) the sharp blade of a cutting instrument such as a knife 3) the point farthest from the middle of a surface 4) an advantage, *verb*, 5) to take small, cautious steps 6) to put a border on something

ed-i-ble [ĕd′ə-bəl] *adjective*, fit to be eaten as food

ed-i-fice [ĕd′ə-fĭs] *noun*, a large, imposing building or structure

ed-i-fy [ĕd′ə-fī′] *verb*, to educate, to instruct, to guide spiritually

ed-it [ĕd′ĭt] *verb*, to revise and correct, to prepare something for publication, to rewrite

e-di-tion [ĭ-dĭsh′ən] *noun*, a book, newspaper or magazine printed as a special issue

ed-i-tor [ĕd′ĭ-tər] *noun*, one who reads and prepares books or newspapers before they are published

ed-u-cate [ĕj′ə-kāt′] *verb*, to teach people, to provide knowledge

ed-u-ca-tion [ĕj′ə-kā′shən] *noun*, *no plural*, teaching and learning, acquired knowledge

eel [ēl] *noun*, a long fish that looks like a snake that lives in the sea

ee-rie [îr′ē] *adjective*, spooky

ef-fect [ĭ-fĕkt′] *noun*, a result, consequence, outcome

ef-fec-tive [ĭ-fĕk′tĭv] *adjective*, having a positive result, helpful

ef-fer-ves-cence [ĕf′ər-vĕs′sĭns] *adjective*, 1) gas escaping in innumerable small bubbles 2) irrepressible excitement

ef-fi-ca-cious [ĕf′ĭ-kā′shəs] *adjective*, producing a desired result, effective, efficient

ef-fi-cien-cy [ĭ-fĭsh′ən-sē] *noun*, bringing the required result with little waste

ef-fi-cient [ĭ-fĭsh′ənt] *adjective*, working well with no waste of money or effort, capable

ef-fi-gy [ĕf′ə-jē] *noun*, a likeness in the form of a dummy, an image of a hated person usually hung

ef-fort [ĕf′ərt] *noun*, an attempt, an action, strenuous exertion

ef-fu-sion [ĭ-fyōō′zhən] *noun*, a pouring forth, an ejection

e.g. for example

egg [ĕg] *noun*, a shell covered fetus of a baby bird, fish or insect containing an embryo

e-go [ē′gō, ĕg′ō] *noun*, a person's self-importance, self-esteem

e-go-tism [ē′gə-tĭz′əm] *adjective*, , self-praise, vanity, self-absorbed

e-go-tist [ē′gə-tĭst] *noun*, excessive interest in one's self

eight [āt] *noun*, *adjective*, the number 8, one plus seven

eight-een [ā-tēn′] *noun*, *adjective*, the number 18, ten plus eight

eight-eenth [ā-tēnth′] *noun*, *adjective*, the number eighteen in order, 18th

eighth [ātth] *noun, adjective*, the number eight in order, 8th

eight-i-eth [ā′tē-ĭth] *noun, adjective*, the number 80 in order, 80th

eight-y [ā′tē] *adjective, noun*, the number 80, eight times ten

ei-ther [ē′thər, ī′thər] *adjective*, 1) being one or the other of two, each, *conjunction*, 2) used before words or phrases the last of which follows "or" to show that they are choices, *pronoun*, 3) the one or the other

e-ject [ĭ-jĕkt′] *verb*, to force or throw out, to expel

e-lab-o-rate [ĭ-lăb′ər-ĭt] *adjective*, 1) made with great care, highly finished, carefully planned, complicated, *verb*, 2) to develop or explain in detail

e-lapse [ĭ-lăps′] *verb*, to pass

e-las-tic [ĭ-lăs′tĭk] *adjective*, 1) that which returns to its original size and shape after it has been stretched completely, *noun*, 2) fabric with fine strings of rubber woven in it

e-la-tion [ĭ-lā′ shən] *noun*, a rise in spirits, excitement, joy

el-bow [ĕl′bō′] *noun*, the part of the arm that bends in the middle

eld-er [ĕl′dər] *adjective*, 1) older, more aged, *noun*, 2) one who is older than others 3) someone people look up to because of age and experience

e-lect [ĭ-lĕkt′] *verb*, to choose by vote, to decide a course of action or representative

e-lec-tion [ĭ-lĕk′shən] *noun*, a time when people make choices by voting

e-lec-tor-ate [ĭ-lĕk′tər-ĭt] *noun*, the voters in an election

e-lec-tric [ĭ-lĕk′trĭk] *adjective*, a source of power

e-lec-tri-cal [ĭ-lĕk′trĭ-kəl] *adjective*, a form of energy used for light and heat

e-lec-tri-cian [ĭ-lĕk-trĭsh′ən] *noun*, one who is licensed to handle electrical equipment

e-lec-tric-i-ty [ĭ-lĕk-trĭs′ĭ-tē] *noun*, the combined effects of electrons and protons

e-lec-tro-cute [ĭ-lĕk′trə-kyōōt′] *verb*, to kill by electric shock

e-lec-trode [ĭ-lĕk′trōd′] *noun*, a conductor by which an electric unit enters or leaves a vacuum

e-lec-tron [ĭ-lĕk′trŏn′] *noun*, a particle smaller than an atom having a negative electric charge

el-e-gant [ĕl′ĭ-gənt] *adjective*, marked by finished or tasteful selection, gorgeous, refined

el-e-ment [ĕl′ə-mənt] *noun*, 1) a basic substance of which the universe is composed 2) the forces that constitute the weather

el-e-men-ta-ry [ĕl′ə-mĕn′tə-rē] *adjective*, the first principles of anything, basic facts of a subject

el-e-phant [ĕl′ə-fənt] *noun*, a huge, herbivore mammal that lives in Africa and India, weighs 5 or 6 tons, stands 10-11 feet high, has thick skin, two tusks and a long nose called a trunk

el-e-vate [ĕl′ə-vāt′] *verb*, to raise to a higher position or level, to lift up

elevation

el-e-va-tion [ĕl′ə-vā′shən] *noun*, the altitude, the height

el-e-va-tor [ĕl′ə-vā′tər] *noun*, a machine with hoisting equipment for conveying persons, goods, etc. to or from different levels

e-lev-en [ĭ-lĕv′ən] *adjective, noun*, the number 11, one plus ten

e-lev-enth [ĭ-lĕv′ənth] *noun*, *adjective*, number 11 in order, 11th

elf [ĕlf] *noun*, an imaginary person who lives in the mountains, a pixie or fairy, a goblin

elic-it [ĭ-lĭs′ĭt] *verb*, to draw forth, to evoke a response

el-i-gi-ble [ĕl′ĭ-jə-bəl] *adjective*, qualified to be chosen

e-lim-i-nate [ĭ-lĭm′ə-nāt′] *verb*, to get rid of, to remove or leave out, to exclude

e-lite [ĭ-lēt′] *adjective*, the best, choice, select, superior

elk [ĕlk] *noun*, an animal similar to a moose, a large deer

elm [ĕlm] *noun*, a hardwood shade tree, similar to an oak

e-lope [ĭ-lōp′] *verb*, to run away to marry secretly

el-o-quence [ĕl′ə-kwəns] *adjective*, able to use the language, fluent

else [ĕls] *adverb*, 1) otherwise, instead of 2) more, in addition

else-where [ĕls-hwâr′] *adverb*, in some other place

e-lude [ĭ-lōōd′] *verb*, to get away from, to avoid, to escape

e-lu-sive [ĭ-lōō′sĭv] *adjective*, baffling, hard to grasp, slippery

em-a-nate [ĕm′ə-nāt′] *verb*, to flow out, to radiate, to come out

e-man-ci-pate [ĭ-măn′sə-pāt′] *verb*, to set free from restrictions

e-man-ci-pa-tion [ĭ-măn′sə-pā′shən] *noun*, freeing from oppression or slavery

em-balm [ĕm-bäm′] *verb*, to mummify or preserve a corpse

em-bar-go [ĕm-bär′gō] *noun*, a law that prevents a country from trading with another

em-bark [ĕm-bärk′] *verb*, 1) to board an airplane or boat 2) to begin an undertaking

em-bar-rass [ĕm-băr′əs] *verb*, to confuse, to humiliate, to make someone ill at ease, to chagrin

em-bar-rass-ment [ĕm-băr′əs-mənt] *noun*, *no plural*, feeling self-conscious, distressed, ashamed

em-bas-sy [ĕm′bə-sē] *noun*, the building in which people work to represent their own country in another country

em-bed [ĕm-bĕd′] *verb*, to bury into something, to fix firmly

em-bel-lish [ĕm-bĕl′ĭsh] *adjective*, to adorn, to make beautiful

em-bez-zle [ĕm-bĕz′əl] *verb*, to steal money from a company or savings accounts

em-blem [ĕm′bləm] *noun*, a badge or symbol of something

em-bod-y [ĕm-bŏd′ē] *verb*, to include as a whole

em-brace [ĕm-brās′] *verb*, 1) to hold closely in one's arms, to hug 2) to accept something, to include 3) to take up seriously

em-broi-der [ĕm-broi′dər] *verb,*
1) to sew ornamental patterns on
fabric 2) to embellish a story

em-bry-o [ĕm′brē-ō′] *noun,* the
first stage of development of an
animal or plant organism

em-bry-on-ic [ĕm′brē-ŏn′ĭk]
adjective, in the initial stage of
development of an organism

em-er-ald [ĕm′ər-əld] *noun,* a
precious stone of rich, deep
green color

e-merge [ĭ-mûrj′] *verb,* to come out
or become visible, to appear

e-mer-gen-cy [ĭ-mûr′jən-sē] *noun,*
an unexpected event that calls
for immediate action 2) a
pressing necessity, a crisis

em-i-grant [ĕm′ĭ-grənt] *noun,* a
refugee leaving one country to
live in another

em-i-grate [ĕm′ĭ-grāt′] *verb,* to
leave one country to go and live
in another

em-i-nent [ĕm′ə-nənt] *adjective,*
high, lofty, noted, famous

e-mis-sion [ĭ-mĭsh′ən] *noun,*
substances given off from other
processes, such as air pollution

e-mo-tion [ĭ-mō′shən] *noun,*
feelings, passion, excitement

e-mo-tion-al [ĭ-mō′shə-nəl]
adjective, excitable, passionate,
showing emotions, fervent

em-pa-thize [ĕm′pə-thīz′] *verb,* to
relate to the feelings of others

em-pa-thy [ĕm′pə-thē] *noun,* being
able to share the feelings of
another person, by imagining
yourself as that person

em-per-or [ĕm′pər-ər] *noun,* the
male ruler of a country or empire

em-pha-sis [ĕm′fə-sĭs] *noun,* to
give something more attention
by highlighting it

em-pha-size [ĕm′fə-sīz′] *verb,* to
call to someone's attention, to
bring out clearly, to stress

em-pire [ĕm′pīr′] *noun,* a group of
nations or states united under a
single sovereign power

em-pir-i-cal [ĕm-pîr′ĭ-kəl] *adjective,*
based on observation and
experience rather than theory

em-ploy [ĕm-ploi′] *verb,* 1) to hire
and provide a livelihood for 2) to
make use of, to engage

em-ploy-ee [ĕm-ploi′ē] *noun,* one
who is employed by or works for
another for wages, a worker

em-ploy-er [ĕm-ploi′ər] *noun,* a
person or business that employs
or hires one or more persons

em-ploy-ment [ĕm-ploi′mənt]
noun, no plural, a job, an
occupation

em-pow-er [ĕm-pou′ər] *verb,* to
authorize, to enable

emp-ty [ĕmp′tē] *adjective,*
1) having nothing inside,
something that is without
content or meaning, *verb,* 2) to
remove the contents completely,
to make vacant, to unload

em-u-late [ĕm′yə-lāt′] *verb,* to
follow or copy in the hope of
equalling or surpassing

en-a-ble [ĕ-nā′bəl] *verb,* to make
possible, to empower

en-act [ĕn-ăkt′] *verb,* 1) to make
into law 2) to act, to perform

e-nam-el [ĭ-năm′əl] *noun*, 1) a smooth glossy surface finish 2) the outer covering of the teeth

en-chant [ĕn-chănt′] *verb*, to delight, to bewitch, to captivate

en-chant-ment [ĕn-chănt′mənt] *noun*, fascination, appeal, charm

en-clave [ĕn′klāv′] *noun*, territory enclosed in a foreign land

en-close [ĕn-klōz′] *verb*, to put something inside of something else, shut on all sides

en-clo-sure [ĕn-klōz′hər] *noun*, something that surrounds on all sides

en-com-pass [ĕn-kŭm′pəs] *verb*, to enclose in a circle, to surround

en-core [ŏn′kôr′] *noun*, an additional performance given in response to calls from the audience, as by applause

en-coun-ter [ĕn-koun′tər] *verb*, to meet, especially unexpectedly or briefly

en-cour-age [ĕn-kûr′ĭj] *verb*, to give praise or hope to someone to help them, to urge

en-cour-age-ment [ĕn-kûr′ĭj-mənt, -kŭr′-] *noun*, support, assurance, helpfulness, praise

en-croach-ment [ĕn-krōch′mənt] *noun*, crowding, overstepping boundaries or someone's rights

en-cy-clo-pe-di-a [ĕn-sī′klə-pē′dē-ə] *noun*, a comprehensive summary of knowledge on many subjects written in a set of volumes

end [ĕnd] *noun*, 1) the farthest point or edge of anything 2) when something has been finished or completed 3) the aim, reason, design, *verb*, 4) to finish

en-dan-ger [ĕn-dān′jər] *verb*, to put something in danger of extinction, to hazard

en-deav-or [ĕn-dĕv′ər] *noun*, 1) an attempt, *verb*, 2) to try, to make an earnest effort, to strive

end-less [ĕnd′lĭs] *adjective*, measureless, infinite, ceaseless

en-dorse [ĕn-dôrs′] *verb*, 1) to write on the back of 2) to sanction, to declare approval of

en-dow [ĕn-dou′] *verb*, to provide or bestow income or property

en-dow-ment [ĕn-dou′mənt] *noun*, giving of money to provide a regular income

en-dur-ance [ĕn-door′əns] *noun*, *no plural*, the ability to survive, the will to persevere

en-dure [ĕn-door′] *verb*, to bear, to put up with, to tolerate, to suffer

en-e-my [ĕn′ə-mē] *noun*, a hostile person or country that wants to harm you, a foe, a rival

en-er-gize [ĕn′ər-jīz′] *verb*, to make active, to invigorate

en-er-gy [ĕn′ər-jē] *noun*, 1) the capacity for performing work 2) power that makes things work

en-fee-ble [ĕn-fē′bəl] *verb*, to weaken, to make helpless

en-fold [ĕn-fōld′] *verb*, to envelope, to enclose

en-force [ĕn-fôrs′] *verb*, to impose by force, to require compliance

en-gage [ĕn-gāj′] *verb*, 1) to put in gear 2) to fight 3) to hire 4) to become involved 5) to keep busy or active 6) to promise

en-gage-ment [ĕn-gāj′mənt] *noun*,
1) an encounter 2) an action 3) a
proposal for marriage

en-gen-der [ĕn-jĕn′dər] *verb*, to
cause, to give rise to

en-gine [ĕn′jĭn] *noun*, 1) a
mechanism, a device, a tool 2) a
locomotive 3) a source of power

en-gi-neer [ĕn′jə-nîr′] *noun*, 1) one
who designs machines or
structures 2) one who drives an
engine, *verb*, 3) to bring about
an event, to direct, to manage

Eng-lish [ĭng′glĭsh] *adjective*,
1) the language of the people of
Britain and America 2) relating
to or belonging to the British
people

en-grave [ĕn-grāv′] *verb*, 1) to cut
in 2) to leave an impression

en-gulf [ĕn-gŭlf′] *verb*, to cover
completely, to close over

en-hance [ĕn-hăns′] *verb*, to make
better, to increase the value or
quality of something

e-nig-ma [ĭ-nĭg′mə] *noun*, one of a
kind, very unusual, an oddity, a
puzzle, a mysterious person

en-joy [ĕn-joi′] *verb*, to take
pleasure in

en-joy-a-ble [ĕn-joi′ŭb-əl]
adjective, , pleasant, beneficial

en-joy-ment [ĕn-joi′mənt] *noun*,
delight, satisfaction, happiness

en-large [ĕn-lärj′] *verb*, to increase,

en-light-en [ĕn-līt′n] *verb*, to
educate, to inform, to instruct

en-list [ĕn-lĭst′] *verb*, to sign up, to
enroll in the armed forces

en-mi-ty [ĕn′mĭ-tē] *noun*, hatred

e-nor-mi-ty [ĭ-nôr′mĭ-tē] *noun*,
immensity, great size

e-nor-mous [ĭ-nôr′məs] *adjective*,
monstrous, huge, immense

e-nough [ĭ-nŭf′] *adjective*, 1) all
that is needed, ample, *adverb*,
2) adequate, sufficient, *noun*,
3) the required amount

en-rage [ĕn-rāj′] *verb*, to anger, to
inflame, to make furious

en-rich [ĕn-rĭch′] *verb*, to improve,
to make rich, to enhance

en-roll [ĕn-rōl′] *verb*, to register in
a list, to become a member

en-roll-ment [ĕn-rōl′mənt] *noun*,
the total number registered

en-sem-ble [ŏn-sŏm′bəl] *noun*, the
whole, all the parts taken
together as a group

en-sue [ĕn-sōō′] *verb*, to make sure
or certain, to follow up

en-sure [ĕn-shōōr′] *verb*, to make
sure or certain or guarantee

en-ter [ĕn′tər] *verb*, to go into

en-ter-prise [ĕn′tər-prīz′] *noun*, a
bold undertaking, a project or
business activity, a venture

en-ter-tain [ĕn′tər-tān′] *verb*, 1) to
hold the interest of the people
you are talking to or performing
for, 2) to amuse, to delight
3) to have in mind

en-ter-tain-ment [ĕn′tər-tān′mənt]
noun, a performance that
interests an audience

en-thrall [ĕn-thrôl′] *verb*, to
fascinate, to mesmerize

en-thu-si-asm [ĕn-thōō′zē-ăz′əm]
noun, *no plural*, eagerly wanting
to do something

en-thu-si-as-tic [ĕn-thōō′zē-ăs′tĭk] *adjective*, passionate and devoted to a belief or idea

en-tice [ĕn-tīs′] *verb*, to lure, to attract, to tempt, to persuade

en-tire [ĕn-tīr′] *adjective*, whole, complete, with no parts missing

en-ti-tle [ĕn-tīt′l] *verb*, to be given as a right from birth

en-ti-ty [ĕn′tĭ-tē] *noun*, something, a being, an individual

en-trance [ĕn′trəns] *noun*, 1) the place to go into, the means of entry [ĕn′trăn-s] *verb*, 2) to mesmerize, to fill with wonder

en-tree [ŏn′trā] *noun*, the main dish

en-tre-pre-neur [ŏn′trə-prə-nûr′] *noun*, a businessman

en-trust [ĕn-trŭst′] *verb*, to confide, to give something to someone for safekeeping

en-try [ĕn′trē] *noun*, 1) entrance, a way in 2) an item in a log

e-nun-ci-ate [ĭ-nŭn′sē-āt′] *verb*, to pronounce, to state clearly

en-vel-op [ĕn-vĕl′əp] *verb*, to cover, to enclose, to surround

en-vel-ope [ĕn′və-lōp′] *noun*, a folded paper cover like a pocket for a letter

en-vi-ous [ĕn′vē-əs] *adjective*, jealous, full of envy

en-vi-ron-ment [ĕn-vī′rən-mənt] *noun*, the surroundings or place where something lives

en-vy [ĕn′vē] *noun*, *no plural*, 1) the feeling of anger or bitterness because someone has more of something, or a better life, *verb*, 2) to want something someone has, to be jealous

e-on [ē′ŏn′, ē′ən] *noun*, a thousand years, a long time

e-phem-er-al [ĭ-fĕm′ər-əl] *adjective*, short-lived, fleeting

ep-ic [ĕp′ĭk] *noun*, a long poem about a hero and his deeds

ep-i-dem-ic [ĕp′ĭ-dĕm′ĭk] *adjective*, 1) affecting many in the community, widespread, *noun*, 2) a contagious disease infecting the population

ep-i-der-mis [ĕp′ĭ-dûr′mĭs] *noun*, the outer layer of the skin

ep-i-gram [ĕp′ĭ-grăm′] *noun*, a pithy phrase, a witty saying e.g. *Fish and guests stink after three day. Ben Franklin, 1736*

ep-i-lep-sy [ĕp′ə-lĕp′sē] *noun*, a disorder of the nervous system accompanied by convulsions

ep-i-logue [ĕp′ə-lôg′] *noun*, a short conclusion added to the end of a book

ep-i-sode [ĕp′ĭ-sōd′] *noun*, 1) a prominent occurrence 2) a separate incident in a longer story or poem

ep-i-taph [ĕp′ĭ-tăf′] *noun*, an inscription in memory of a dead person on a monument

e-pit-o-me [ĭ-pĭt′ə-mē] *noun*, the smallest amount of something representing the whole

ep-och [ĕp′ək] *noun*, 1) a time period characterized by a distinctive development 2) a date marking the beginning of a new period of history

e-qual [ē′kwəl] *adjective*, 1) the same as 2) alike, *noun*, 3) something valued as the same

amount 4) a symbol = that shows
both sides have the same value,
verb, 5) to be as good as
someone else

e-qual-i-ty [ĭ-kwŏl′ĭ-tē] *noun*, the
state or condition of being equal,
the same as

e-qual-ize [ē′kwə-līz′] *verb*, to
make even, to balance

e-qual-ly [ē′kwə-lē] *adverb*, as
well as

e-qua-tion [ĭ-kwā′shən] *noun*,
1) the act or process of being
equal 2) a statement that the
values of two mathematical
expressions are equal (=)

e-qua-tor [ĭ-kwā′tər] *noun*, the
imaginary band circling the
middle of the earth, an equal
distance between the North and
South Poles

e-ques-tri-an [ĭ-kwĕs′trē-ən] *noun*,
a rider on horseback

e-qui-lib-ri-um [ē′kwə-lĭb′rē-əm]
noun, a state of balance between
opposing forces

e-qui-lat-er-al tri-angle
[ē′kwə-lăt′ər-əl trī′ăng′gəl]
noun, a triangle with 60 degree
angles and equal sides

e-qui-nox [ē′kwə-nŏks′] *noun*, the
time of year when night and day
are of equal length: March 21;
October 21

e-quip [ĭ-kwĭp′] *verb*, to provide
what is needed, to outfit

e-quip-ment [ĭ-kwĭp′mənt] *noun*,
no plural, tools, supplies,
whatever is needed for a job or a
special activity

eq-ui-ta-ble [ĕk′wĭ-tə-bəl] *noun*,
fair, just, impartial

eq-ui-ty [ĕk′wĭ-tē] *noun*, 1) money
paid on the balance of a loan
2) justice, impartiality, fair

e-quiv-a-lent [ĭ-kwĭv′ə-lənt] *noun*,
1) equal in value or worth
2) similar in force, measure,
meaning or effect, identical

e-quiv-o-cate [ĭ-kwĭv′ə-kāt′] *verb*,
to evade, to avoid, to waffle

e-ra [îr′ə, ĕr′ə] *noun*, a period of
time, or history, an epoch

e-rad-i-cate [ĭ-răd′ĭ-kāt′] *verb*, to
destroy completely

e-rase [ĭ-rās′] *verb*, to rub out, to
remove all traces of

e-ras-er [ĭ-rā′sər] *noun*, instrument
of rubber or cloth that removes
marks made by chalk or pencil

e-rect [ĭ-rĕkt′] *adjective*,
1) standing straight, upright,
verb, 2) to build or construct

er-mine [ûr′mĭn] *noun*, a small
rodent with white fur and black
spots

e-rode [ĭ-rōd′] *verb*, 1) to wear
away gradually 2) to cause
something to disintegrate

e-ro-sion [ĭ-rō′zhən] *noun*, the
process of wearing down,
breaking up, and carrying off
from the land, soil, and rock, by
wind and water

err [ûr, ĕr] *verb*, to be mistaken

er-rand [ĕr′ənd] *noun*, a short trip
to accomplish a task

er-rat-ic [ĭ-răt′ĭk] *adjective*,
irregular, inconsistent

er-ro-ne-ous [ĭ-rō′nē-əs] *adjective*,
misleading, false, incorrect

er-ror [ĕr′ər] *noun*, 1) a belief in what is untrue 2) an act involving a departure from truth and accuracy, a mistake

er-satz [er′säts] *adjective*, 1) made or used as a substitute 2) not real or genuine

er-u-dite [ĕr′yə-dīt] *adjective*, learned, well-educated, scholarly

e-rupt [ĭ-rŭpt′] *verb*, 1) to burst out suddenly, to explode 2) to become active like lava

es-ca-la-tor [ĕs′kə-lā′tər] *noun*, moving stairs that can carry you up or down or a long distance

es-cape [ĭ-skāp′] *noun*, 1) a way out, *verb*, 2) to free yourself

es-chew [ĕs-chōō′] *verb*, to avoid, to keep away from, to evade

es-cort [ĕs′kôrt′] *noun*, 1) a guide, an attendant, *verb*, 2) to accompany someone, to go with them as protection or an honor

es-crow [ĕs′krō′] *noun*, money set aside in a special bank account that can only be used for a specified purpose

Es-ki-mo [ĕs′kə-mō′] *noun*, a person whose ancestors have been living in Russia, Alaska, Canada, and Greenland for hundreds of years

e-soph-a-gus [ĭ-sŏf′ə-gəs] *noun*, the tube that leads from the pharynx to the stomach

es-o-ter-ic [ĕs′ə-tĕr′ĭk] *adjective*, understood only by a chosen few

es-pe-cial-ly [ĕ-spĕsh′ə-lē] *adverb*, particularly, to a great extent

es-pi-o-nage [ĕs′pē-ə-näzh′] *noun*, a case of spying

es-prit de corps [ĕ-sprē′ də kôr′] *noun*, team spirit

es-say [ĕs′ā′, ĕ-sā′] *noun*, 1) a literary composition expressing personal observations, *verb*, 2) to try out, to test, to attempt

es-sence [ĕs′əns] *noun*, 1) the real character or spiritual nature of something 2) perfume

es-sen-tial [ĭ-sĕn′shəl] *adjective*, 1) indispensable, most important, **essential**s *noun*, *plural* 2) something that cannot be done without, necessities

es-tab-lish [ĭ-stăb′lĭsh] *verb*, to make stable or firm, to settle

es-tab-lish-ment [ĭ-stăb′lĭsh-mənt] *noun*, a place of business or residence, an organization

es-tate [ĭ-stāt′] *noun*, 1) property planned as a unit 2) material things, property left at someone's death

es-teem [ĭ-stēm′] *noun*, 1) respect, honor, *verb*, 2) to regard favorably

es-ti-mate [ĕs′tə-māt′] *noun*, 1) an educated guess, an approximation, *verb*, 2) to calculate the value or amount of something 3) to give an approximate rather than an exact answer, to guess

et cet-er-a [ĕt sĕt′ər-ə] abbr. **etc.**, and so on

etch [ĕch] *verb*, to engrave by biting out with an acid

e-ter-nal [ĭ-tûr′nəl] *adjective*, endless, infinite, ceaseless

e-ter-ni-ty [ĭ-tûr′nĭ-tē] *noun*, forever

eth-i-cal [ĕth′ĭ-kəl] *adjective*, honest, humane, doing the right thing

eth-ics [ĕth′ĭks] *noun*, *plural*, rules of moral duty governing a group

eth-nic [ĕth′nĭk] *adjective*, relating to different races, ancestry, or culture, moral principles

eth-nol-o-gy [ĕth-nŏl′ə-jē] *noun*, the study of man

et-i-quette [ĕt′ĭ-kĕt′] *noun*, good manners, polite behavior, courtesy

et-y-mol-o-gy [ĕt′ə-mŏl′ə-jē] *noun*, the study of the origin and derivation of words

eu-gen-ics [yoō-jĕn′ĭks] *adjective*, regarding the improvement of race by controlling breeding

eu-lo-gy [yoō′lə-jē] *noun*, praise lauding a deceased person

eu-phe-mism [yoō′fə-mĭz′əm] *noun*, a mild expression to replace a blunt remark

eu-re-ka [yoō-rē′kə] *noun*, an expression of triumph concerning a discovery

Eu-rope [yoōr′əp] *noun*, one of the seven continents containing 12% of the world's population

eu-tha-na-sia [yoō′thə-nā′zhə] *noun*, mercy killing

e-vac-u-ate [ĭ-văk′yoō-āt′] *verb*, to abandon, to empty out

e-vade [ĭ-vād′] *verb*, to avoid, to escape by cleverness

e-val-u-ate [ĭ-văl′yoō-āt′] *verb*, to measure the value, to appraise

e-vap-o-rate [ĭ-văp′ə-rāt′] *verb*, 1) to turn from solid or liquid into vapor, to dry up 2) to vanish

e-vap-o-ra-tion [ĭ-văp′ə-rā′shən] *noun*, the process in which water disappears from the surface as it becomes vapor, the loss of water into the air

e-va-sive [ĭ-vā′sĭv] *adjective*, trying to avoid something, elusive

eve [ēv] *noun*, the night before

e-ven [ē′vən] *adjective*, 1) a flat, smooth surface 2) any number divided by two 3) equal, *adverb*, 4) more than expected 5) still to go on 6) actually, also

eve-ning [ēv′nĭng] *noun*, the time between the end of the afternoon and bedtime, dusk

e-ven-ly [ē′vən-lē] *adverb*, to give the same number to each

e-vent [ĭ-vĕnt′] *noun*, something that takes place, an activity

e-ven-tu-al-ly [ĭ-vĕn′choō-ə-lē] *adverb*, at long last, after the passing of some time

ev-er [ĕv′ər] *adverb*, 1) always, at all times 2) at any time

ev-er-green [ĕv′ər-grēn′] *noun*, a plant that keeps its leaves all year long, including holly, rhododendrons, pine trees

eve-ry [ĕv′rē] *adjective*, each one, inclusive of all, all possible

eve-ry-bod-y [ĕv′rē-bŏd′ē] *pronoun*, every person

eve-ry-day [ĕv′rē-dā′] *adjective*, ordinary, commonplace, normal

eve-ry-one [ĕv′rē-wŭn′] *pronoun*, each person, everybody

eve-ry-thing [ĕv′rē-thĭng′] *noun*, all together

eve-ry-where [ĕv′rē-hwâr′] *adverb*, in all places

ev-i-dence [ĕv′ĭ-dəns] *noun*, *no plural*, 1) one or more facts or items 2) testimony

ev-i-dent [ĕv′ĭ-dənt] *adjective*, clear, apparent, obvious

e-vil [ē'vəl] *noun*, very bad, corrupt, immoral, wicked

e-voke [ĭ-vōk'] *verb*, 1) to call forth, 2) to bring about

ev-o-lu-tion [ĕv'ə-lōō'shən] *noun*, the natural process of growth and change, development

e-volve [ĭ-vŏlv'] *verb*, to come into being, to develop gradually

ex-ac-er-bate [ĭg-zăs'ər-bāt'] *verb*, to make worse, to embitter

ex-act [ĭg-zăkt'] *adjective*, precise, correct, completely accurate

ex-act-ly [ĭg-zăkt'lē] *adverb*, precisely, specifically

ex-ag-ger-ate [ĭg-zăj'ə-rāt'] *verb*, to enlarge beyond the bounds of truth, to overstate

ex-alt [ĭg-zôlt'] *verb*, to praise, to elevate, to rise in rank

ex-am-i-na-tion [ĭg-zăm'ə-nā'shən] *noun*, scrutiny, inquiry

ex-am-ine [ĭg-zăm'ĭn] *verb*, 1) to inspect carefully, to look at every point 2) to scrutinize

ex-am-ple [ĭg-zăm'pəl] *noun*, a model, a sample of what others are like, an illustration

ex-as-per-ate [ĭg-zăs'pə-rāt'] *verb*, to irritate, to provoke, to enrage

ex-ceed [ĭk-sēd'] *verb*,1) to pass beyond the measure of something, 2) to surpass

ex-cel [ĭk-sĕl'] *verb*, to improve on, to be superior, to be better than ever at something

ex-cel-lence [ĕk'sə-ləns] *noun*, the quality of being exceptional

ex-cel-lent [ĕk'sə-lənt] *adjective*, superior, very good, near the standard or model

ex-cel-si-or [ĕk'sə-l'sē'ər] *adjective*, higher, always upward

ex-cept [ĭk-sĕpt'] *preposition*, apart from, to take out, to exclude

ex-cep-tion [ĭk-sĕp'shən] *noun*, 1) not conforming to the general rule 2) leaving out, an omission

ex-cep-tion-al [ĭk-sĕp'shə-nəl] *adjective*, superior, above average, unusually good

ex-cerpt [ĕk'sûrpt'] *noun*, something that has been removed from its original text

ex-cess [ĭk-sĕs'] *adjective*, extra, more than enough, more than expected, exceeding a limit

ex-ces-sive [ĭk-sĕs'ĭv] *adjective*, greater than the usual amount, very great, too much

ex-change [ĭks-chānj'] *noun*, 1) the process of settling accounts between parties, *verb*, 2) to change something for something else, to replace, to trade

ex-cise [ĕk'sīz'] *adjective*, an inland fee levied upon specific articles with a country

ex-cite [ĭk-sīt'] *verb*, to be energized, full of feeling

ex-cite-ment [ĭk-sīt'mənt] *noun*, 1) ado, commotion 2) a thrill

ex-cit-ing [ĭk-sī'tĭng] *adjective*, feeling energized, stimulating

ex-claim [ĭk-sklām'] *verb*, to speak suddenly and with strong feeling

ex-cla-ma-tion mark [ĕk'sklə-mā'shən märk] *noun*,

the sign ! used to show surprise, humor, or strong feeling

ex-clude [ĭk-sklo͞od′] *verb*, to keep someone or something out

ex-clu-sive [ĭk-sklo͞o′sĭv] *adjective*, restricted to one or a few people

ex-clu-sive-ly [ĭk-sklo͞o′sĭv-l ē] *adverb*, without exception, only

ex-cru-ci-at-ing [ĭk-skro͞o′shē-ā′tĭng] *adjective*, very painful, agonizing

ex-cur-sion [ĭk-skûr′zhən] *noun*, a journey chiefly for recreation, a brief tour, an outing

ex-cuse [ĭk-skyo͞oz′] *noun*, 1) a reason, a defense, *verb*, 2) to free, to allow an interruption

ex-e-cute [ĕk′sĭ-kyo͞ot′] *verb*, 1) to kill someone 2) to take action

ex-e-cu-tion [ĕk′sĭ-kyo͞o′shən] *noun*, 1) accomplishment, implementation 2) putting to death a condemned person

ex-ec-u-tive [ĭg-zĕk′yə-tĭv] *noun*, 1) a business person, a manager 2) one of the three branches of government, the other two are the judicial and the legislative branches

ex-em-pla-ry [ĭg-zĕm′plə-rē] *adjective*, commendable, serving as a desirable model

ex-em-pli-fy [ĭg-zĕm′plə-fī′] *adjective*, to show or illustrate by example, to typify

ex-empt [ĭg-zĕmpt′] *verb*, to excuse, to make immune

ex-er-cise [ĕk′sər-sīz′] *noun*, 1) an activity used to train the body, *verb*, 2) to train the body to make it stronger and healthy

ex-ert [ĭg-zûrt′] *verb*, to force upon

ex-er-tion [ĭg-zûr′shən] *noun*, hard work, a strong effort

ex-hale [ĕks-hāl′] *verb*, to breathe out, to blow, to puff, to emit

ex-haust [ĭg-zôst′] *verb*, to be thorough covering all points

ex-haus-tion [ĭg-zôs′chən] *noun*, the feeling of being very tired

ex-haus-tive [ĭg-zô′stĭv] *adjective*, complete, thorough, attending to every detail, comprehensive

ex-hib-it [ĭg-zĭb′ĭt] *noun*, 1) a display of a collection of something, *verb*, 2) to display

ex-hi-bi-tion [ĕk′sə-bĭsh′ən] *noun*, a public display of art

ex-hil-a-ra-tion [ĭg-zĭl′ə-rā′shən] *noun*, a feeling of excitement

ex-i-gen-cy [ĕk′sə-jən-sē] *noun*, a pressing necessity, a case demanding immediate action

ex-ile [ĕg′zīl′] *noun*, 1) one who is not allowed to live in his own country as a form of punishment, *verb*, 2) to be forced out of a place not allowed to return

ex-ist [ĭg-zĭst′] *verb*, 1) to be, to live, to be present, to occur

ex-ist-ence [ĭg-zĭs′təns] *noun*, *no plural*, 1) state of being actual, life 2) an actual occurrence, reality 3) that which lives

ex-it [ĕg′zĭt] *noun*, 1) a way to go out, the door leading out of a building, *verb*, 2) to leave, to go out, to depart, to go away

ex-o-dus [ĕk′sə-dəs] *noun*, mass departure, migration

ex-or-cise [ĕk′sôr-sīz′] *verb*, to drive out evil spirits

ex-o-skel-e-ton [ĕk′sō-skĕl′ĭ-tn]
noun, outer structure that provides
protection to an insect or shell fish

ex-ot-ic [ĭg-zŏt′ĭk] *adjective*,
strange, interesting, or unusual

ex-pand [ĭk-spănd′] *verb*, to grow
or make larger, to extend

ex-panse [ĭk-spăns′] *noun*, a space,
a region, an area of land

ex-pa-tri-ate [ĕk-spā′trē-āt′] *noun*,
someone who left the land where
he was born

ex-pect [ĭk-spĕkt′] *verb*, to wait for
something that is supposed to
happen, to anticipate

ex-pec-ta-tion [ĕk′spĕk-tā′shən]
noun, anticipation, calculation

ex-pe-di-en-cy [ĭk-spē′dē-ən-sē]
adjective, that which is practical,
advantageous, right or just

ex-pe-di-tion [ĕk′spĭ-dĭsh′ən]
noun, a journey or excursion for
a specific purpose, a trip

ex-pel [ĭk-spĕl′] *verb*, to drive out,
to eject, to force to leave

ex-pend-i-ture [ĭk-spĕn′də-chər]
noun, a laying out, as of money
or labor, the cost

ex-pense [ĭk-spĕns′] *noun*, 1) cost,
payment 2) responsibility,
obligation 3) sacrifice

ex-pen-sive [ĭk-spĕn′sĭv] *adjective*,
costing a lot of money

ex-pe-ri-ence [ĭk-spîr′ē-əns] *noun*,
1) encountering, adventuring
2) knowledge gained by
observation or trial, *verb*, 3) to
live, to feel by participating

ex-per-i-ment [ĭk-spĕr′ə-mənt]
noun, 1) a test done to see if
something is valid or true, *verb*,

2) to carry out a plan in order to
test a prediction

ex-pert [ĕk′spûrt′] *noun*, the most
knowledgeable person in his
particular field, an authority

ex-per-tise [ĕk′spûr-tēz′] *noun*,
specialized knowledge, aptitude

ex-pi-ra-tion [ĕk′spə-rā′shon]
noun, 1) a coming to a close
2) the death, the end

ex-pire [ĭk-spīr′] *verb*, to end, to
finish, to cease to exist

ex-plain [ĭk-splān′] *verb*, to clarify
or to give a reason, to account for

ex-pla-na-tion [ĕk′splə-nā′shon]
noun, act of interpreting an
action or a description

ex-ple-tive [ĕk′splĭ-tĭv] *noun*, a
curse word, an obscenity

ex-plic-it [ĭk-splĭs′ĭt] *adjective*,
plain, definite, expressed

ex-plode [ĭk-splōd′] *verb*, to
release energy with a loud noise

ex-ploit [ĕk′sploit′] *noun*, 1) a deed
or act, especially one of renown,
verb, 2) to make use of for one's
own profit, to act unfairly

ex-plo-ra-tion [ĕk′splə-rā′shon]
noun, a search, an investigation

ex-plore [ĭk-splôr′] *verb*, 1) to look
into closely, to examine or
investigate 2) to go into new or
unknown places, to find new
things and learn about them

ex-plor-er [ĭk-splôr′ər] *noun*,
someone who travels to a region
previously unknown to learn
more about it, a pioneer

ex-plo-sion [ĭk-splō′zhən] *noun*, a
blast, a bomb bursting

ex-plo-sive [ĭk-splō′sĭv] *adjective*, flammable, combustible

ex-port [ĭk-spôrt′] *noun*, 1) something sold abroad as part of international trade, *verb*, 2) merchandise conveyed from one country to another

ex-pose [ĭk-spōz′] *verb* 1) to make something easy to see 2) to reveal, to make known 3) to leave unprotected

ex-po-si-tion [ĕk′spə-zĭsh′ən] *noun*, a statement of intent, a public exhibition or display

ex-po-sure [ĭk-spō′zhər] *noun*, not sheltered from the elements such as cold, sun, rain, wind, etc.

ex-press [ĭk-sprĕs′] *adjective*, 1) the fastest means available, *verb*, 2) to indicate, to show

ex-pres-sion [ĭk-sprĕsh′ən] *noun*, 1) the look on someone's face, a facial gesture 2) a remark, a cliché 3) a way of showing feelings or emotion through actions or words

ex-pul-sion [ĭk-spŭl′shən] *noun*, a driving or forcing out

ex-punge [ĭk-spŭnj′] *verb*, to cancel, to remove

ex-quis-ite [ĕk′skwĭ-zĭt] *adjective*, 1) carefully crafted or executed, 2) delicately beautiful

ex-tend [ĭk-stĕnd′] *verb*, 1) to stretch out, to lengthen 2) to offer

ex-ten-sion [ĭk-stĕn′shən] *noun*, a part added on

ex-ten-sive [ĭk-stĕn′sĭv] *adjective*, having wide reach, broad

ex-tent [ĭk-stĕnt′] *noun*, the area covered or affected by something, the amount

ex-ten-u-ate [ĭk-stĕn′yōō-āt′] *adjective*, out of reach, beyond your grasp or control

ex-te-ri-or [ĭk-stûr′ē-ər] *noun*, 1) external, outward 2) on the surface or appearance

ex-ter-mi-nate [ĭk-stûr′mə-nāt′] *verb*, to drive out, to destroy completely, to kill

ex-ter-nal [ĭk-stûr′nəl] *adjective*, 1) on or relating to the outside of the subject 2) the outer part

ex-tinct [ĭk-stĭngkt′] *adjective*, no longer existing or living

ex-tin-guish [ĭk-stĭng′gwĭsh] *verb*, to put out as a fire or a light

ex-tort [ĭk-stôrt′] *verb*, to get money by making threats and by intimidation or coercion

ex-tor-tion [ĭk-stôr′shən] *noun*, demanding money by making threats, blackmail

ex-tra [ĕk′strə] *adjective*, 1) a surplus, more than needed, *adverb*, 2) in addition

ex-tract [ĭk-străkt′] *verb*, 1) to take out 2) to distill or remove

ex-tra-di-tion [ĕk′strə-dĭsh′ən] *noun*, the surrender of a prisoner by another authority

ex-tra-ne-ous [ĭk-strā′nē-əs] *adjective*, not essential, foreign

ex-traor-di-nar-y [ĭk-strôr′dn-ĕr′ē, ĕk′strə-ôr′-] *adjective*, unusual, rare, exceptional, uncommon

ex-tra-ter-res-tri-al [ĕkstrətə′restrēəl] *adjective*, of or from outside the earth's atmosphere

ex-trav-a-gant [ĭk-străv′ə-gənt] *adjective*, wasteful, exceeding reasonable limits, spending too much, very expensive

ex-treme [ĭk-strēm′] *adjective*, 1) to the highest or a very high degree 2) the utmost limit or degree, radical 3) final, last

ex-trem-i-ty [ĭk-strĕm′ĭ-tē] *noun*, 1) the end or the boundary 2) an appendage such as a hand or foot

ex-tri-cate [ĕk′strĭ-kāt′] *verb*, to clear away, to free, to untangle

ex-trin-sic [ĭk-strĭn′sĭk] *adjective*, external, from outside

ex-u-ber-ant [ĭg-zōō′bər-ənt] *adjective*, joyful, high spirits

ex-ude [ĭg-zōōd′] *verb*, to discharge, to ooze, to give forth

ex-ul-ta-tion [ĕk′səl-tā′shən] *noun*, joy, celebration, showing delight

eye [ī] *noun*, 1) the organ located on the head used to see 2) the tiny hole at the end of a needle

eye-lash [ī′lăsh′] *noun*, one of the hairs growing on the eyelid

eye-lid [ī′lĭd′] *noun*, the skin which covers the eye

F

fa-ble [fā′bəl] *noun*, a fictitious short story with a moral in which animals talk

fab-ric [făb′rĭk] *noun*, 1) any cloth that is woven or knit from fibers 2) make up, substance

fab-ri-cate [făb′rĭ-kāt′] *verb*, 1) to build, to construct 2) to lie

fab-u-lous [făb′yə-ləs] *adjective*, marvelous, wonderful, amazing

fa-cade [fə-säd′] *noun*, the decorated entrance of a building

face [fās] *noun*, 1) the front of the head, with the eyes, nose and mouth 2) the front of something 3) a flat surface of a space figure, *verb*, 4) to stand in front of or turn to confront

fac-et [făs′ĭt] *noun*, 1) one of the polished plane surfaces of a precious stone 2) an aspect of a subject, a side, a feature

fa-ce-tious [fə-sē′shəs] *adjective*, joking, sportive, kidding around, not serious, flippant, glib

fac-ile [făs′əl] *adjective*, easy

fa-cil-i-tate [fə-sĭl′ĭ-tāt′] *verb*, to make easier, to aid, to help

fac-sim-i-le [făk-sĭm′ə-lē] *noun*, an exact copy or likeness of anything, a reproduction

fact [făkt] *noun*, an indisputable piece of information, something that is true, a certainty

fac-tion [făk′shən] *noun*, a dissenting party, an inner circle, a clique, a group within a group

fac-tious [făk′shəs] *adjective*, broken into groups

fac-tor [făk′tər] *noun*, 1) a condition that causes a result 2) the numbers multiplied or divided to give a product

fac-to-ry [făk′tə-rē] *noun*, a building where products are manufactured

fac-ul-ty [făk′əl-tē] *noun*, 1) the professors in an institution of higher learning 2) a physical or mental ability, talent, capacity

fad [făd] *noun*, temporary fashion or style, the "in" thing

fade [fād] *verb*, to lose light or color, to become dull or dim

Fahr-en-heit [făr′ən-hīt′] *adjective*, of or relating to the temperature scale in which the freezing point of water is 32 degrees and the boiling point of water is 212 degrees under normal atmospheric pressure

fail [fāl] *verb*, to not accomplish what you set out to

fail-ure [fāl′yər] *noun*, lack of success, a debacle

faint [fānt] *adjective*, 1) hard to see, dim 2) inaudible, indistinct, soft, *verb*, 3) to feel weak or dizzy, to collapse unconscious

fair [fâr] *adjective*, 1) treating everyone equally, not favoring one more than another, 2) so-so, mediocre 3) pale, light-skinned complexion, *noun*, 4) a gathering for buying and selling goods

fair-y [fâr′ē] *noun*, an imaginary person with magical powers

faith [fāth] *noun*, 1) a belief in something that is not seen, relying on trust 2) religion

faith-ful [fāth′fəl] *adjective*, loyal, trustworthy, full of faith

fake [fāk] *noun*, 1) forgery, imitation, *verb*, 2) to pretend, to falsify 3) to counterfeit

fall [fôl] *noun*, 1) the season of autumn, *verb*, 2) to come down, to tumble 3) to lose power

fal-lacy [făl′ə-sē] *noun*, an error in reasoning, an erroneous idea

fal-li-ble [făl′ə-bəl] *adjective*, liable to err, vulnerable

fall line *noun*, a place where the height of the land drops

suddenly and is usually marked by many waterfalls and rapids

false [fôls] *adjective*, 1) not true 2) artificial, not real

false-hood [fôls′hŏŏd′] *noun*, a lie

fal-ter [fôl′tər] *verb*, to hesitate

fame [fām] *noun*, *no plural*, wide recognition, to become a household name, celebrity

fa-mil-iar [fə-mĭl′yər] *adjective*, to have some knowledge of something or someone

fam-i-ly [făm′ə-lē] *noun*, 1) a parent or parents and children 2) all of a person's relatives

fam-ine [făm′ĭn] *noun*, 1) general scarcity of food 2) extreme scarcity of something

fam-ish-ed [făm′ĭshd] *adjective*, extremely hungry

fa-mous [fā′məs] *adjective*, well known, celebrated, noted

fan [făn] *noun*, 1) any device designed to circulate air 2) an ardent follower or supporter, *verb*, 3) to move air, to cool with a fan 4) to spread out

fa-nat-ic [fə-năt′ĭk] *adjective*, 1)carrying an interest or enthusiasm to extremes, *noun*, 2) a person with a very strong interest in something

fan-cy [făn′sē] *adjective*, elaborate, unusually decorated, not plain

fang [făng] *noun*, a long sharp tooth, the incisor

fan-tas-tic [făn-tăs′tĭk] *adjective*, imaginary, fanciful, unreal

fan-ta-sy [făn′tə-sē] *noun*, a daydream, an imaginary situation, make believe

far [fär] *adjective*, 1) at a great distance, not near or close by, *adverb*, 2) exceedingly 3) a certain point

fare [fâr] *noun*, 1) money paid for a trip by bus or train for example 2) food and drink *verb* 3) to get along well

fare-well [fâr-wĕl′] *noun*, goodbye, adieu, parting

farm [färm] *noun*, buildings or land in the country where people grow food as cash crops or raise animals to sell

farm-er [fär′mər] *noun*, a person who owns and or cultivates crops on a portion of land

far-ther [fär′thər] *preposition*, to be far away, in a remote place

fas-ci-nate [făs′ə-nāt′] *verb*, to attract and hold interest, to captivate, to excite

fas-ci-na-tion [făs′ə-nā′shən] *noun*, personal attraction, infatuation

fash-ion [făsh′ən] *noun*, 1) clothing or articles that are in style, in vogue, *verb*, 2) to shape or form

fast [făst] *adjective*, 1) quick, rapid 2) fixed in place, cannot move, *verb*, 3) to go for an extended period of time without eating

fas-tid-i-ous [făst'ĭd'ē-əs]*adjective*, hard to please, very critical

fat [făt] *adjective*, 1) with a wide round shape 2) thick, *noun*, 3) an oily substance from cooked meat

fa-tal [fāt′l] *adjective*, in a manner that results in death, lethal

fate [fāt] *noun*, an invisible hand that seems to cause certain things to happen over which people have no control, destiny

fa-ther [fä′thər] *noun*, the male parent

fa-ther-in-law [fä′thər-ĭn-lô′] *noun*, the father of your wife or husband

fath-om [făth′əm] *verb*, to understand, to comprehend

fa-tigue [fə-tēg′] *noun*, 1) the condition of being very tired or exhausted, *verb*, 2) to become or make tired

fau-cet [fô′sĭt] *noun*, a fixture for drawing liquid from a pipe and regulating the flow

fault [fôlt] *noun*, 1) mistake, failure to perform a duty 2) a fracture in the earth's surface causing a displacement 3) a defect

fa-vor [fā′vər] *noun*, 1) an act of kindness, *verb*, 2) to prefer, to patronize, to fancy, to like

fa-vor-ite [fā′vər-ĭt] *adjective*, something preferred above others, liked the best

fawn [fôn] *noun*, a young deer less than a year old

fear [fîr] *noun*, 1) expecting harm, *verb*, 2) to dread or be afraid of

fear-ful [fîr′fəl] *adjective*, afraid, timid, frightened, scared

fea-si-ble [fē′zə-bəl] *adjective*, capable of being done, practical, possible, achievable

feast [fēst] *noun*, 1) an elaborate meal, a banquet, *verb*, 2) to eat large amounts of food and drink

feat [fēt] *noun*, an accomplishment, a remarkable achievement

feath-er [fĕth′ər] *noun*, a soft thing which covers birds, like a toothpick covered with fine hairs

fea-ture [fē′chər] *noun*, 1) a distinctive aspect on a person, a characteristic 2) an article in the newspaper

Feb-ru-ar-y [fĕb′rōō-ĕr′ē] *noun*, the second month in the year, having 28 days but 29 every four years on the leap year

fed [fĕd] *past tense* of **feed**

fed-er-al [fĕd′ər-əl] *adjective*, belonging or pertaining to a state formed by the consolidation of several states

fed-er-a-tion [fĕd′ə-rā′shən] *noun*, an alliance, an association

fee [fē] *noun*, money charged for something, the cost or expense

fee-ble [fē′bəl] *adjective*, weak, frail, unsteady, infirm

feed [fēd] *noun*, 1) food for animals, *verb*, 2) to give someone food, to nourish

feel [fēl] *verb*, 1) to perceive in a way other than your five senses, to sense with emotions 2) to acknowledge with the sense of touch 3) to form an opinion

feign [fān] *verb*, to pretend

feist-y [fī′stē] *adjective*, very active or spirited, exuberant

fe-lic-i-ty [fĭ-lĭs′ĭ-tē] *adjective*, the state of being happy

fell [fĕl] *adjective*, 1) cruel, deadly, *verb*, 2) *past tense* of **fall**

fel-low [fĕl′ō] *noun*, 1) a man 2) a companion, a comrade

fel-on [fĕl′ən] *noun*, a person convicted of a serious crime

fel-o-ny [fĕl′ə-nē] *noun*, any offense that is punishable by confinement in prison

felt [fĕlt] *noun*, 1) a soft synthetic fabric without a nap 2) the *past tense* of **feel**

fe-male [fē′māl′] *adjective*, 1) referring to a woman, feminine, *noun*, 2) attributes of a woman and characteristics

fem-i-nine [fĕm′ə-nĭn] *adjective*, of or pertaining to the female sex

fence [fĕns] *noun*, 1) a partition made of wood or wire 2) a person who receives stolen goods, *verb*, 3) to put in an enclosure 4) to duel with a sword

fend-er [fĕn′dər] *noun*, the strip of metal over the wheels of a car, bicycle, or other vehicle to protect against mud or water

fer-ment [fər-mĕnt′] *verb*, to ripen, to acidify with bacteria

fern [fûrn] *noun*, a bright green plant that does not have flowers, and grows in shady places

fe-ro-cious [fə-rō′shəs] *adjective*, savage, fierce, cruel

fer-ret [fĕr′ĭt] *noun*, 1) a small rodent like a weasel, *verb*, 2) to drive or hunt out of hiding

fer-ry [fĕr′ē] *noun*, a boat used to transport people or goods from one side to the other

fer-tile [fûr′tl] *adjective*, fruitful, able to produce abundantly

fer-ti-lize [fûr′tl-īz′] *verb*, to make ready for planting by providing proper amounts of nitrogen, potash, oxygen, etc.

fer-vent [fûr′vənt] *adjective*, hot, showing intense feeling

fes-ter [fĕs′tər] *verb*, to become worse due to infection or anger

fes-ti-val [fĕs/tə-vəl] *noun*, 1) a big party that takes place on a holiday or at another special time 2) a program of cultural events 3) a carnival

fetch [fĕch] *verb*, 1) to go after something and bring it back, to retrieve 2) to bring in

fet-ish [fĕt/ĭsh] *noun*, 1) a quirky habit based on a belief in magic 2) a compulsion

feud [fyōōd] *noun*, 1) a long bitter quarrel, especially between two families, verb, 2) to fight like enemies, to quarrel

fe-ver [fē/vər] *noun*, when the body temperature rises above 98.6 degrees the body is trying to fight off bacteria and infection

few [fyōō] *adjective*, three or more, not many, hardly any

fi-an-cé [fē/än-sā/, fē-än/sā/] *noun*, a man engaged to be married

fi-an-cée [fē/än-sā/] *noun*, a woman engaged to be married

fi-as-co [fē-ăs/kō] *noun*, a situation that is a total failure, an upheaval

fi-ber [fī/bər] *noun*, 1) thread-like structures combined with others to create tissue 2) a filament

fick-le [fĭk/əl] *adjective*, moody, flighty, liable to change, inconsistent, not loyal

fic-tion [fĭk/shən] *noun*, that which is not true or based in fact, make-believe

fic-ti-tious [fĭk-tĭsh/əs] *adjective*, imaginary, unreal, false

fid-dle [fĭd/l] *noun*, 1) another word for a violin, *verb*, 2) to tinker with something

fid-dler [fĭd/l-ər] *noun*, a person who plays a hand-held stringed instrument

fi-del-i-ty [fĭ-dĕl/ĭ-tē, fī-] *noun*, faithfulness, loyalty, honesty

field [fēld] *noun*, 1) land put to special use for sports or farming 2) a sphere or range of activities, *verb*, 3) to receive a ball in a game and pass it to another player

fiend [fēnd] *noun*, a person who is wicked, cruel, or malicious

fierce [fîrs] *adjective*, furious, violent or aggressive

fier-y [fîr/ē] *adjective*, burning, heated, impassioned

fif-teen [fĭf-tēn/] *adjective*, *noun*, the number 15, ten plus five

fif-teenth [fĭf-tēnth/] *adjective*, *noun*, number 15 in order, 15th

fifth [fĭfth] *adjective*, *noun*, number five in order, written 5th

fif-ti-eth [fĭf/tē-ĭth] *adjective*, the ordinal of fifty, written 50th

fif-ty [fĭf/tē] *adjective*, *noun*, the number 50, 5 times 10

fig [fĭg] *noun*, a small fruit grown in warm climates, full of seeds

fight [fīt] *noun*, 1) the use of the body or weapons against someone, *verb*, 2) to engage in a physical struggle 3) to quarrel 4) to struggle in any way

fig-ment [fĭg/mənt] *noun*, an imaginary thing

fig-ur-a-tive [fĭg/yər-ə-tĭv] *adjective*, not literal, metaphorical

fig-ure [fĭg/yər] *noun*, 1) the shape or outline of something 2) numbers or digits in an equation

or used as measurements, *verb*,
3) to calculate numbers 4) to try
to simplify something that is
hard to understand

fig-ure-head [fĭg′yər-hĕd′] *noun*,
1) a carved figure set at the front
of a sailing ship 2) a leader in
name only who has no power or
authority

fil-a-ment [fĭl′ə-mənt] *noun*, a
slender thread-like structure

filch [fĭlch] *verb*, to steal

file [fīl] *noun*, 1) a drawer for papers
2) a metal tool with a rough edge
for making things smooth 3) a line
of people 4) a stored collection of
data, *verb*, 5) to organize papers
or other items into a system 6) to
make a surface smooth through
friction with a tool 7) to walk in a
straight and orderly line 8) to
register

fil-i-al [fĭl′ē-əl] *adjective*, referring
to the duty of a son or daughter

fill [fĭl] *verb*, 1) to occupy with as
much as possible 2) to make full

fil-let [fĭl′ĭt] *noun*, 1) a boneless
piece of meat or fish, *verb*, 2) to
remove bones from

film [fĭlm] *noun*, 1) a solid layer that
floats on liquid 2) a sensitized
coating which receives photographic
impressions 3) a movie shown in a
theater or on television, *verb*, 4) to
photograph something on film 5) to
make a movie using a camera

filth-y [fĭl′thē] *adjective*, 1) very
dirty, nasty, squalid 2) obscene

fil-tra-tion [fĭl-trā′shən] *noun*,
purification, processing

fin [fĭn] *noun*, part of a fish that
sticks out, may be fan-shaped,
used for steering or balance

fi-nal [fī′nəl] *adjective*, the last

fi-na-le [fə-năl′ē] *noun*, the last part
of a show or a musical
composition, the culmination

fi-nal-ly [fī′nə-lē] *adverb*, at last,
reaching the end, ultimately,
eventually, concluding

fi-nance [fə-năns′] *noun*, 1) the
management of money, *verb*,
2) to use money to pay for
something through a loan

find [fīnd] *verb*, to uncover or see
something after a search

fine [fīn] *adjective*, 1) pleasant,
nice, refined, *noun*, 2) a penalty,
verb, 3) to punish someone by
making them pay money

fin-esse [fə-nĕs′] *noun*, 1) a delicate
skill, dexterity 2) tact

fin-ger [fĭng′gər] *noun*, 1) one of
five digits on a hand, *verb*, 2) to
handle, to touch or feel

fin-ger-nail [fĭng′gər-nāl′] *noun*,
the claw or talon on the end of a
finger or foot

fin-ick-y [fĭn′ĭ-kē] *adjective*, very
fussy, hard to please, picky

fin-ish [fĭn′ĭsh] *noun*, 1) a surface
appearance, *verb*, 2) to end or
complete 3) to use up

fi-nite [fī′nīt′] *adjective*, 1) limited
in number 2) having boundaries

fir [fûr] *noun*, an evergreen tree

fire [fīr] *noun*, 1) animation,
excitement, inspiration 2) things
that are burning, *verb*, 3) to
discharge or dismiss from a job

firm [fûrm] *adjective*, 1) solid,
hard, *noun*, 2) a company
managing a business

firm-ly [fûrm′lē] *adverb*, solidly,
resolutely, unwaveringly

first [fûrst] *adjective*, 1) the most important 2) initial, beginning

first aid [fûrst/ād/] *noun*, the emergency help given someone who has been hurt

fish [fǐsh] *noun*, 1) any animal that lives in the sea, seafood, *verb*, 2) to try to catch a fish or get something from someone

fish-er-man [fǐsh/ər-mən] *noun*, someone who catches fish for sport or as a living

fis-sure [fǐsh/ər] *noun*, a crack, a narrow groove, a cleft

fist [fǐst] *noun*, the hand with fingers closed tightly together

fit [fǐt] *adjective*, 1) healthy, active, vibrant 2) good enough, *verb*, 3) to be able to wear the right size 4) to fit in place

five [fǐv] *adjective, noun*, the number 5, one plus four

fix [fǐks] *noun*, 1) a dilemma, a predicament, *verb*, 2) to repair 3) to set, to arrange, to put in place 4) to focus on

fix-ture [fǐks/chər] *noun*, that which is attached to something permanently, apparatus

flac-cid [flăk/sǐd, flăs/ǐd] *adjective*, flabby, hanging loosely

flag [flăg] *noun*, 1) cloth with a special pattern on it, used as the sign of a country, club, etc., *verb*, 2) to wave to someone as a signal to stop 3) to droop

flag-rant [flā/grənt] *adjective*, plainly wicked, unashamed

flail [flāl] *verb*, to move wildly as if beating something, to thrash

flair [flâr] *noun*, an aptitude or talent for doing something well

flake [flāk] *noun*, 1) a small sliver of something, *verb*, 2) to peel off in shavings or tiny bits

flam-boy-ant [flăm-boi/ənt] *adjective*, showy, flashy

flame [flām] *noun*, burning gas seen in a fire

flam-ma-ble [flăm/ə-bəl] *adjective*, capable of igniting easily and burning quickly

flank [flăngk] *noun*, a side, the lateral edge, on either side

flan-nel [flăn/əl] *noun*, a soft woolen cloth of loose texture

flap [flăp] *noun*, 1) the covering of an opening, hanging loose, *verb*, 2) to wave to and fro

flare [flâr] *noun*, 1) a torch or brilliant light, *verb*, 2) to shine, to burn brightly 3) to spread out like a fan, to expand outward

flash [flăsh] *noun*, 1) a split second, a gleam, an instant, *verb*, 2) to show a bright light

flash-light [flăsh/līt/] *noun*, a hand held light operated by batteries

flat [flăt] *adjective*, a level, horizontal surface, smooth

flat-ter-y [flăt/ə-rē] *noun, no plural*, excessive compliments

flaunt [flônt] *verb*, to boast, to wave under one's nose

fla-vor [flā/vər] *noun*, 1) the seasoning of food, the taste of something, *verb*, 2) to add something to make food taste better 3) to savor

flaw [flô] *noun*, a defect, a weak spot, an imperfection

flax [flăks] *noun*, a slender plant with blue flowers whose stalks are used to make thread

flay [flā] *verb*, to whip, to beat with a strap or stick, to hit

flea [flē] *noun*, a jumping insect the size of a pinhead that sucks blood from people and animals

fled [flĕd] *past tense* of **flee**

flee [flē] *noun*, to run away

fleece [flēs] *noun*, 1) the wool on sheep, *verb*, 2) to shake someone down for money 3) to swindle

fleet [flēt] *noun*, a group of ships sailing together, or vehicles, or aircraft under one command

fleet-ing [flē/tǐng] *adjective*, swift in motion, passing quickly

flex [flĕks] *verb*, to bend

flex-i-ble [flĕk/sə-bəl] *adjective*, 1) able to bend easily 2) willing to change a course of action, easily adaptable

flick-er [flĭk/ər] *verb*, 1) to change unsteadily from bright to dim light 2) a brief sensation

flight [flīt] *noun*, 1)the act of flying 2) an airplane trip

flim-sy [flĭm/zē] *adjective*, 1) weak and easily broken 2) ineffectual, not well thought out

flinch [flĭnch] *verb*, to move back suddenly in fear, to make a nervous movement

flint [flĭnt] *noun*, stone, slate

flip [flĭp] *adjective*, 1) sarcastic, kidding, joking 2) a somersault in the air, *verb*, 3) to turn suddenly or swiftly in the air

flip-pant [flĭp/ənt] *adjective*, glib, sarcastic, disrespectful

flirt [flûrt] *verb*, to tease affectionately

float [flōt] *noun*, 1) a portable display in a parade, *verb*, 2) to be supported by liquid or gas

flock [flŏk] *noun*, 1) a company of birds, animals, or people 2) to gather, to congregate, to huddle

flood [flŭd] *noun*, 1) water that stays on land that is below sea level, *verb*, 2) when water overflows or invades a dry area

floor [flôr] *noun*, 1) the part of the room you walk on 2) a level of a building, *verb*, 3) to knock down

flo-ral [flôr/əl] *adjective*, decorated with flowers

Flor-i-da [flôr/ĭ-də] *noun*, on of the 50 United States, located in the Southeast, the capital is Tallahassee. The **Florida** state flower is the orange blossom, the motto is "In God we trust."

floss [flôs] *noun*, 1) a soft silky substance found in the husks of some plants 2) a thin silk thread used for embroidery 3) a thread covered with wax used to clean between the teeth, *verb*, 4) to run thread through the teeth to remove food and tarter

flo-til-la [flō-tĭl/ə] *noun*, a small fleet of ships

floun-der [floun/dər] *noun*, 1) any member of the flatfish family, *verb*, 2) to flail about

flour [flour] *noun*, the finely ground meal of wheat or other cereal plants like oats, rice, corn

flour-ish [flûr/ĭsh] *noun*, 1) a handwritten decoration, *verb*, 2) to prosper, to grow well

flout [flout] *verb*, to scoff at, to ridicule, to disobey a law

flow [flō] *noun*, *no plural*, 1) a current, a steady stream, *verb*, 2) the movement of gases or fluid

flow-er [flou′ər] *noun*, the part of a plant that bears the seeds, the blossom where fruit develops

flu [floo] *noun*, an illness characterized by fever, chills and coughing

fluc-tu-ate [flŭk′choo-āt′] *verb*, 1) to move back and forth like a wave, 2) to vary in degree or value

fluc-tu-a-tion [flŭk′choo-ā′ shən] *noun*, 1) wavering 2) the change of the level or degree of something

flu-ent [floo′ənt] *adjective*, spoken smoothly and easily

flu-id [floo′ĭd] *adjective*, 1) not fixed or settled, able to flow, *noun*, 2) a liquid that flows

fluo-res-cent [floo-rĕs′ənt] *adjective*, absorption of light from another source

flur-ry [flŭr′ē] *noun*, a brief excitement, quick movement

flush [flŭsh] *adjective*, 1) even, *noun*, 2) in poker, five cards of the same suit, *verb*, 3) to turn red in the face 4) to wash out

flust-er [flŭs′tər] *verb*, to confuse, to embarrass, to agitate

flute [floot] *noun*, a musical instrument like a pipe with keys to control the flow of air that can make different sounds

flut-ter [flŭt′ər] *verb*, to flap, to move wings hurriedly

flux [flŭks] *noun*, a flowing series of continuous changes

fly [flī] *noun*, 1) a small flying insect, *verb*, 2) to move through the air 3) to move very fast

foam [fōm] *noun*, froth formed on the surface of liquids by agitation, a mass of small bubbles

fo-cal [fō′kəl] *adjective*, central, midway, of or at a focus

fo-cus [fō′kəs] *noun*, 1) a central point, *verb*, 2) to concentrate, to pay attention to something

foe [fō] *noun*, an enemy, an adversary, an opponent

fog [fôg] *noun*, a cloud of fine water drops of vapor found close to the ground or sea

foil [foil] *noun*, 1) a paper-like sheet of metal, *verb*, 2) to throw off someone's plans

fo-li-age [fōlē′ĭj] *noun*, the leaves of trees and other plants

fold [fōld] *noun*, 1) a part of something that has been wrapped or enclosed *verb*, 2) to double or bend over, to envelope 3) to withdraw or quit

fold-er [fōl′dər] *noun*, a cardboard cover to protect papers and keep them together

folk [fōk] *noun*, people

fol-low [fŏl′ō] *verb*, 1) to go after, to pursue 2) to come after or behind 3) to pay close attention

fol-ly [fŏl′ē] *noun*, foolishness

fond [fŏnd] *adjective*, to have feelings of affection

fon-dle [fŏn′dl] *adverb*, to caress, to stroke lovingly

food [food] *noun*, a source of nourishment, what you eat

fool [fool] *noun*, 1) a silly person, *verb*, 2) to trick or deceive

fool-ish [foo′lĭsh] *adjective*, crazy, silly, senseless, ridiculous

foot [fŏŏt] *noun, plural,* **feet,** 1) the part of the leg that touches the ground 2) a measure of length equal to 12 inches 3) any base that resembles a foot

foot-ball [fŏŏt'bôl'] *noun,* 1) a game in which two teams of eleven players try to score points by passing a ball to each other in a period of time 2) an oval shaped inflated ball

for [fôr, fər when unstressed] *conjunction,* 1) because, *preposition,* 2) being as 3) to be given or used by 4) at a certain price 5) to be going toward 6) to show the meaning of as in an example

for-age [fôr'ĭj] *verb,* to look for food or supplies

for-ay [fôr'ā'] *noun,* a raid, an attempt at something unfamiliar

for-bear [fôr-bâr'] *verb,* to hold back, to wait, to refrain from

for-bear-ance [fôr-bâr'əns] *adjective,* the act of refraining from anything, self-restraint

for-bid [fər-bĭd'] *verb,* to prohibit, to order someone not to do something

force [fôrs, fōrs] *noun,* 1) power 2) a group of people trained in the same manner, *verb,* 3) to cause to happen with conviction

fore-bod-ing [fôr-bō'dĭng] *noun,* a warning or sign that something bad or evil will happen

fore-cast [fôr'kăst'] *noun,* 1) a prediction that something will happen, *verb,* 2) to anticipate, to figure out and say what will happen before it takes place

fore-close [fôr-klōz'] *verb,* to take away land from a person who owes money on it

for-eign [fôr'ĭn] *adjective,* belonging to another country

for-eign-er [fôr'ə-nər] *noun,* someone from another country

fore-man [fôr'mən] *noun,* 1) the overseer of a set of workmen 2) the spokesman of a jury

fore-most [fôr'mōst'] *adjective,* first in importance, time or place

fo-ren-sic [fə-rĕn'sĭk] *adjective,* information that is a suitable argument in court

fore-see-a-ble [fôr-sē'ə-bəl] *adjective,* predictable, able to be seen in the future

fore-sight [fôr'sīt'] *noun,* the ability to foresee future happenings or needs

for-est [fôr'ĭst] *noun,* a large place where trees grow together

fore-stall [fôr-stôl'] *verb,* to keep safe by taking action first

fore-tell [fôr-tĕl'] *verb,* to predict, to make a prophecy, to forecast

for-ev-er [fôr-ĕv'ər] *adverb,* until the end of time, always

for-feit [fôr'fĭt] *verb,* to lose the right, to surrender, to give up

forge [fôrj] *verb,* 1) to move ahead or progress with difficulty 2) to mold metal with heat and then hammering it into shapes 3) to invent or create something 4) to counterfeit with intent to defraud

for-ger-y [fôr'jə-rē] *noun,* the making of a false or altered document for the purpose of fraud, counterfeit

for-get [fər-gĕt′] *verb*, 1) to fail to recall or remember 2) to neglect

for-give [fər-gĭv′] *verb*, to accept an apology, to get rid of a grudge, to cease to feel angry

for-got [fər-gŏt′] *past tense* of **forget**

for-got-ten [fər-gŏt′n] *adjective*, left behind

fork [fôrk] *noun*, 1) a utensil with prongs on the end used to stab food 2) a place in a road where it separates in different directions

for-lorn [fər-lôrn′] *adjective*, sad and unhappy, lonely

form [fôrm] *noun*, 1) the shape of something 2) a document with questions that need to be answered, *verb*, 3) to shape or create, to develop, to mold

for-mal [fôr′məl] *adjective*, following the customs of an occasion, ceremonious

for-mal-i-ty [fôr-măl′ĭ-tē] *noun*, adherence to established rules of etiquette and convention

for-ma-tion [fôr-mā′shən] *noun*, 1) a creation 2) an arrangement

form-er [fôr′mər] *adjective*, the one mentioned first of two

for-mi-da-ble [fôr′mĭ-də-bəl] *adjective*, very large, alarming

for-mu-la [fôr′myə-lə] *noun*, 1) substances mixed together in specific amounts to create something 2) a mathematical statement

for-mu-late [fôr′myə-lāt′] *verb*, to give form to, to express precisely in a few words

fort [fôrt] *noun*, a place for protecting soldiers

forte [fôrt] *noun*, a strong point or special talent or ability

for-ti-eth [fôr′tē-ĭth] *noun*, the ordinal of forty, 40th

for-ti-fy [fôr′tə-fī′] *verb*, 1) to make stronger or more secure 2) to strengthen against attack 3) to empower, to invigorate

for-ti-tude [fôr′tĭ-tōōd′] *noun*, courage, strength of mind

for-tress [fôr′trĭs] *noun*, a building with strong walls made to be defended against attack, a fort

for-tu-nate [fôr′chə-nĭt] *adjective*, lucky, favorable, advantageous

for-tune [fôr′chən] *noun*, 1) whatever good or bad happens to people, luck 2) a large amount of money, wealth, riches

for-ty [fôr′tē] *noun*, the number 40, four times ten

fo-rum [fôr′əm] *noun*, a public meeting place and market square

for-ward [fôr′wərd] *adjective*, 1) near or at the front part, ahead 2) onward 3) bold, confident

fos-sil [fŏs′əl] *noun*, the impression in a rock of the petrified remains of an organism or animal

fos-ter [fô′stər] *verb*, to help, to raise, to encourage

fought [fôt] *past tense* of **fight**

foul [foul] *adjective*, 1) dirty, filthy, squalid, *noun*, 2) a ball hit outside the first or third base line

found [found] *verb*, 1) to establish, to lay a foundation 2) *past tense* of **find**

foun-da-tion [foun-dā′shən] *noun*, 1) the base of a building 2) the

principles of a program or
agency 3) a charitable institution

found-er [foun′dər] *noun*, 1) the
author or prime mover, *verb*,
2) to take on water, to sink

found-ry [foun′drē] *noun*, a plant
refining metal or glass, a factory

foun-tain [foun′tən] *noun*, water
spraying into the air from a pipe

four [fôr] *adjective*, *noun*, the
number 4, one plus three

four-teen [fôr-tēn′] *noun*, 4+10=14

four-teenth [fôr-tēnth′] *adjective*,
noun, number 14 in order, 14th

fourth [fôrth] *adjective*, *noun*, the
ordinal number after third, one
of four equal parts, 4th

fowl [foul] *noun*, a bird eaten as
food or kept for its eggs

fox [fŏks] *noun*, a wild animal the
size of a dog with a pointed
muzzle and bushy tail

foy-er [foi′ər] *noun*, the entry way
in a hall, a hotel lobby

frac-tion [frăk′shən] *noun*, 1) a
piece of something 2) a number
that names part of a group, not a
whole number

frac-ture [frăk′chər] *noun*, 1) a
break, a fissure, *verb*, 2) to break

frag-ile [frăj′əl, -īl′] *adjective*,
easily broken, delicate

frag-ment [frăg′mənt] *noun*, a
piece broken off, a particle

fra-grance [frā′grəns] *noun*, having
a sweet smell, perfume, aroma

frail [frāl] *adjective*, weak, fragile,
delicate, sickly, infirm, ill

frame [frām] *noun*, 1) the structure
that supports something 2) the
outline of something encasing

something, *verb*, 3) to enclose in
a frame 4) to make someone
appear guilty of a crime by
arranging false evidence

fran-chise [frăn′chīz′] *noun*, a
special right or privilege granted
by a government to do business

frank [frăngk] *adjective*, candid,
honest in expressing your thoughts

fran-tic [frăn′tĭk] *adjective*, wild

fra-ter-ni-ty [frə-tûr′nĭ-tē] *noun*, a
fellowship of men in a college

fraud [frôd] *noun*, 1) an imposter, a
fake 2) criminal deception

fraud-u-lent [frô′jə-lənt] *adjective*,
tricky, deceitful, dishonest

fray [frā] *noun*, 1) a conflict, a
commotion, *verb*, 2) to unravel

freak [frēk] *noun*, odd, a strange
event or person, out of the
ordinary, an aberration

freck-le [frĕk′əl] a small, brownish
spot on the skin

free [frē] *adjective*, 1) able to do as
you please, unrestrained
2) something that doesn't cost
any money 3) available, *verb*,
4) to release something from
captivity, to turn loose

free-dom [frē′dəm] *noun*, liberty,
independence, being free

freeze [frēz] *verb*, to become
chilled by cold, to change to ice

freez-er [frē′zər] *noun*, an
appliance that preserves food by
keeping it frozen

freight [frāt] *noun*, transportation
of products in bulk, cargo

freight-er [frā′tər] *noun*, a ship or
aircraft laden with cargo

fren-zy [frĕn′zē] *noun*, craziness,
wild excitement, fury

fre-quen-cy [frē′kwən-sē] *noun*, a common occurrence, returning often or repeatedly

fre-quent [frē′kwənt] *adjective*, 1) happening often, repeated, many times, *verb*, 2) to go to a place often for visits

fresh [frĕsh] *adjective*, new, unused, arrived recently

fresh-man [frĕsh′mən] *noun*, a person in his first year of high school or college

fret [frĕt] *verb*, to fuss, to fume, to feel anxious, to worry

fric-tion [frĭk′shən] *noun*, 1) the force that resists movement between two things that are in contact with each other 2) an argument caused by differences of opinion, rivalry

Fri-day [frī′dē, -dā′] *noun*, the sixth day of the week

friend [frĕnd] *noun*, a person who knows and likes another

friend-ship [frĕnd- shĭp] *noun*, mutual respect, companionship

fright [frīt] *noun*, sudden fear, alarm, horror, a shock

fright-en [frīt′n] *verb*, to scare or alarm, to fill with fear

frig-id [frĭj′ĭd] *adjective*, very cold

frill [frĭl] *noun*, a doodad, bells and whistles, an unnecessary extra feature or luxury

fringe [frĭnj] *noun*, 1) an edge made of short lengths of thread, used to decorate clothes, curtains, etc. 2) an outside edge

frisk-y [frĭs′kē] *adjective*, playful, active, lively, bouncy

frit-ter [frĭt′ər] *verb*, to waste

friv-o-lous [frĭv′ə-ləs] *adjective*, 1) of little weight or importance 2) given to trifling, silly

frog [frôg] *noun*, 1) a jumping amphibian that can swim in water or live on land 2) a tool designed to sit in the bottom of the bowl or vase to hold flower stems in place for Ikebana

frol-ic [frŏl′ĭk] *verb*, merry making, to play happily about

from [frŭm] *preposition*, 1) used to show a starting point 2) used to show a point of separation 3) used to show material source or cause, given by

front [frŭnt] *adjective*, 1) first, ahead of the others, *noun*, 2) the face or the part of something that faces forward 3) the facade

fron-tier [frŭn-tîr′] *noun*, 1) the line between two countries 2) the outer limits of a settled part of a country 3) the outer limits of knowledge

frost [frôst] *noun*, *no plural*, the frozen water that covers the surface of everything outdoors when the temperature goes below freezing, 32 degrees

froth [frôth] *noun*, 1) fine bubbles, foam, effervescence, *verb*, 2) to cause to foam

frown [froun] *noun*, 1) a stern look, *verb*, 2) to scowl or look unhappy, to show disapproval

froze [frōz] *past tense* of **freeze**

fro-zen [frō′zən] *adjective*, the temperature cooled to the point that liquid turns to solid

fru-gal [frōo′gəl] *adjective*, careful management of resources, stingy

fruit [froot] *noun*, 1) an edible product of a flower, bearing seeds 2) the results, the outcome

fruit-ful [froot'fəl] *adjective*, successful, productive

fru-i-tion [froo-ĭsh'ən] *noun*, fulfillment, realization

frus-trate [frŭs'trāt'] *verb*, to prevent from attaining a purpose

fry [frī] *verb*, to cook in hot oil

fudge [fŭj] *noun*, a soft candy usually made with chocolate

fuel [fyoo'əl] *noun*, something that gives off heat when it is burned

fu-gi-tive [fyoo'jĭ-tĭv] *noun*, one who flees from pursuit

ful-fill [fool-fĭl'] *verb*, 1) to satisfy a contract 2) to complete

full [fool] *adjective*, 1) leaving no empty space, holding as much as possible, *adverb*, 2) entirely

full-ness [fool'nĭs] *noun*, the state of being full or complete

fumes [fyoomz] *noun*, *plural*, gases

fu-mi-gate [fyoo'mĭ-gāt'] *verb*, to apply smoke or vapor as a means of disinfecting or killing bugs

fun [fŭn] *noun*, *no plural*, light-hearted, amusement

func-tion [fŭngk'shən] *noun*, 1) the natural and proper action of anything 2) a service, duty, purpose 3) how something works

fund [fŭnd] *noun*, 1) a sum of money collected to help a special purpose, *verb*, 2) to provide money to support a cause

fun-da-men-tal [fŭn'də-mĕn'tl] *adjective*, 1) of first importance, essential 2) elementary

fu-ner-al [fyoo'nər-əl] *noun*, a ceremony honoring the life of someone who died

fun-gus [fŭng'gəs] *noun*, a plant that lacks chlorophyll, leaves, and a flower, i.e., mushrooms, mold, yeast

fun-nel [fŭn'əl] *noun*, an open cone wide at the top with a smaller end to pour liquids from one container to another

fun-ny [fŭn'ē] *adjective*, 1) amusing, entertaining, to make someone laugh 2) strange, unusual, odd

fur [fûr] *noun*, a soft hair covering on some animals, the pelt

fu-ri-ous [fyoor'ē-əs] *adjective*, 1) very angry 2) very fast, strong or wild, intense

fur-nace [fûr'nĭs] *noun*, 1) an enclosed place in which heat is produced by the combustion of fuel 2) a large structure containing a very hot fire

fur-nish [fûr'nĭsh] *verb*, to supply, to stock, or provide, to give

fur-ni-ture [fûr'nĭ-chər] *noun*, *no plural*, the articles in a home that make it more comfortable, such as beds, chairs, tables, etc.

fu-ror [fyoor'ôr'] *noun*, frenzy, a state of great excitement

fur-ther [fûr'thər] *adjective*, at a greater distance, beyond

fur-tive [fûr tĭv] *adjective*, done in a sneaky way, hoping to be unnoticed

fu-ry [fyoor'ē] *noun*, 1) great anger 2) wild or uncontrolled force

fur-ry [fûr'ē] *adjective*, covered with hair, like fur

fuse [fyōoz] *noun*, 1) tinder, kindling, a wick, *verb*, 2) to combine, to blend, to mix

fu-sion [fyōo′zhən] *noun*, the process of joining two or more things together as one

fuss [fŭs] *noun*, 1) complaint, bother, commotion, *verb*, 2) to behave in an excited, worried way, to be overly emotional

fu-tile [fyōot′l, fyōo′tĭl′] *adjective*, ineffective, fruitless, pointless

fu-ture [fyōo′chər] *adjective*, 1) coming after the present, *noun*, 2) a time that has not yet arrived

G

gadg-et [găj′ĭt] *noun*, a mechanical device or tool, a thingamajig

gag [găg] *noun*, 1) something used to keep people from talking, *verb*, 2) to restrain someone from speaking 3) to choke

gai-e-ty [gā′ĭ-tē] *noun*, fun, merriment, cheerfulness

gain [gān] *noun*, 1) an increase, *verb*, 2) to win, to acquire

gait [gāt] *noun*, a manner of walking, skipping or running

gal-ax-y [găl′ək-sē] *noun*, a group of billions of stars held together by mutual gravitation

gale [gāl] *noun*, a strong wind

gall [gôl] *noun*, 1) bitterness, nerve, *verb*, 2) to annoy, to anger, to goad, to infuriate, to chafe

gal-lant [găl′ənt] *adjective*, brave, bold, daring, chivalrous

gal-le-on [găl′ē-ən] *noun*, a large sailing ship of earlier times

gal-ler-y [găl′ə-rē] *noun*, 1) a long, narrow, room-like corridor or passageway 2) a room for the exhibition of works of art

gal-ley [găl′ē] *noun*, 1) an oblong tray to hold type which has been set 2) the lower part of a ship

gal-lon [găl′ən] *noun*, a unit of liquid measure equal to 4 quarts or 96 ounces

gal-lop [găl′əp] *verb*, to trot, to run like a horse

gal-lows [găl′ōz] *noun*, a public place where criminals are hung

gal-va-nize [găl′və-nīz′] *verb*, to stimulate with shock, to stir up

gam-ble [găm′bəl] *verb*, to try to win money by betting

gam-bler [găm′blər] *noun*, a risk taker, a bettor

game [gām] *noun*, a form of entertainment played with rules that tests skill and knowledge

gam-ut [găm′ət] *noun*, the entire range from beginning to end

gan-der [găn′dər] *noun*, a rooster

gang [găng] *noun*, an organized group of people who intimidate and harass

gang-plank [găng′plăngk′] *noun*, a temporary board or ramp used to board or leave a ship

gan-grene [găng′grēn′] *noun*, the death of a limb such as a leg due to infection

gap [găp] *noun*, a hole, an opening, the distance between two places

ga-rage [gə-räzh′] *noun*, 1) a place for housing automobiles 2) a place where automobiles or buses are repaired

gar-bage [gär′bĭj] *noun*, another word for solid waste, especially household waste, trash

gar-den [gär′dn] *noun*, 1) a plot of land where flowers or vegetables grow, *verb*, 2) to look after the plants by watering and pruning

gar-den-er [gärd′nər] *noun*, one who makes and tends a garden

gar-gle [gär′gəl] *verb*, to rinse the throat with a liquid kept in motion by air forced through it from the lungs

gar-goyle [gär′goil′] *noun*, figures of demons carved as a roof spout

gar-ish [gâr′ĭsh, găr′-] *adjective*, in bad taste, gaudy, showy

gar-lic [gär′lĭk] *noun*, a plant like an onion with a strong taste and smell, used to season food

gar-ment [gär′mənt] *noun*, a piece of clothing

gar-ner [gär′nər] *verb*, to gather up, to accumulate, to amass

gar-net [gär′nĭt] *noun*, a precious stone of a deep red color

gar-nish [gär′nĭsh] *noun*, 1) a decoration for food, *verb*, 2) to decorate or adorn 3) to take from someone a little at a time by order of the court to pay a debt

gar-ri-son [găr′ĭ-sən] *noun*, troops stationed in a fort or fortified town, a military post

gas [găs] *noun*, 1) a state of matter in which a substance like air is capable of infinitely expanding, has no fixed form or volume and takes the shape of its container 2) the short word for gasoline

gas-e-ous [găs′ē-əs, găsh′əs] *adjective*, in vapor form

gas-ket [găs′kĭt] *noun*, a piece of rubber sealing a joint between metal surfaces

gas-o-line [găs′ə-lēn′] *noun*, petroleum or diesel motor fuel

gasp [găsp] *noun*, 1) a short breath, *verb*, 2) to pant with a short, excited breath 3) to wheeze

gas-tron-o-my [gă-strŏn′ə-mē] *noun*, the science of preparing and serving good food

gate [gāt] *noun*, a doorway which closes an opening in a wall or fence with a hinged door

gath-er [găth′ər] *verb*, 1) to bring together or accumulate, to harvest 2) to infer from what someone has said

gauche [gōsh] *adjective*, clumsy, lacking social grace, awkward

gaud-y [gô′dē] *adjective*, showy in a tasteless way, ostentatious

gauge [gāj] *noun*, 1) an instrument used to find the exact measurement, *verb*, 2) to measure or estimate

gaunt [gônt] *adjective*, thin, lean

gauze [gôz] *noun*, 1) a cotton bandage 2) a thin fabric

gaze [gāz] *verb*, to stare, to look at steadily and intently, to gawk

gaz-et-teer [gaezə′tir] *noun*, a geographical dictionary

gear [gîr] *noun*, 1) a toothed wheel that passes motion and power along 2) equipment, supplies

gel-a-tin [jĕl′ə-tn] *noun*, a tasteless, transparent, substance, obtained from connective animal tissue such as skin, hooves, or horns

gem [jĕm] *noun*, a precious stone
used as an ornament

gen-der [jĕn′dər] *noun*, a set of
categories applied to nouns as
masculine or feminine

ge-ne-al-o-gy [jē′nē-ŏl′ə-jē] *noun*,
a record of family history

gen-er-al [jĕn′ər-əl] *adjective*, 1) of
or from the whole 2) not
particular or definite 3) usual or
customary, common, lacking
detail, *noun*, 2) the title of the
commander in an army

Gen-er-al As-sem-bly *noun*, 1) the
law-making body in the United
States 2) the assembly of the
representatives of the United
Nations

gen-er-al-i-ty [jĕn′ə-răl′ĭ-tē] *noun*,
a vague statement, a theory

gen-er-al-ize [jĕn′ər-ə-līz′] *verb*, to
make a broad statement

gen-er-al-ly [jĕn′ər-ə-lē] *adverb*,
usually, as a rule, normally

gen-er-ate [jĕn′ə-rāt′] *verb*, to
make or create, to produce

gen-er-a-tion [jĕn′ə-rā′shən] *noun*,
1) the average time interval
between the birth of parents and
the birth of their offspring 2) all
of the people born at the same
time 3) the act or process of
producing something

gen-er-a-tor [jĕn′ə-rā′tər] *noun*, a
machine by which mechanical
energy is changed into electrical
energy

gen-er-os-i-ty [jĕn′ə-rŏs′ĭ-tē] *noun*,
given freely, unselfishness

gen-er-ous [jĕn′ər-əs] *adjective*,
liberal, open-handed, abundant

gen-er-ous-ly [jĕn′ər-əs-lē] *adverb*,
plentifully, unselfishly

ge-ni-al [jēn′yəl] *adjective*, kind,
cheerful, gracious friendly

gen-ius [jēn′yəs] *noun*,
extraordinary mental superiority

gen-re [zhän′rə] *noun*, a category,
art or music characterized by a
particular style, class

gen-til-i-ty [jĕn-tĭl′ĭ-tē] *noun*, men
or women who are born to
money, well-bred

gen-tle [jĕn′tl] *adjective*, 1) kind,
courteous, amiable 2) soft and
mild 3) not steep, moderate

gen-tle-man [jĕn′tl-mən] *noun*, a
well-mannered man

gent-ly [jĕnt′lē] *adverb*, carefully

gen-u-flect [jĕn′yə-flĕkt′] *verb*, to
bend one knee and bow in
worship or reverence

gen-u-ine [jĕn′yo͞o-ĭn] *adjective*,
the real thing, not false

gen-u-ine-ly [jĕn′yo͞o-ĭn-lē]
adverb, truly, really

ge-og-ra-phy [jē-ŏg′rə-fē] *noun*,
the science that describes the
surface of the earth, and its
division into continents

ge-ol-o-gist [jē-ŏl′ə-jĭst] *noun*, a
person who studies the structure
of the earth, its physical changes
and the causes producing these
to form a hypothesis

ge-ol-o-gy [jē-ŏl′ə-jē] *noun*, the
science that studies the history
and structure of the earth, its
physical changes, and the causes
producing these changes

ge-om-e-try [jē-ŏm′ĭ-trē] *noun*, the
branch of mathematics that
studies the relations, properties,
and measurement of solids,
surfaces, lines, and angles

Geor-gia [jôr′jə] *noun*, one of the 50 United States located in the Southeast, the capital city is Atlanta. The state flower of **Georgia** is the cherokee rose and the motto is: "Wisdom, justice, and moderation."

ge-ra-ni-um [jə-rā′nē-əm] *noun*, a flower with pink, red, or white blooms and velvety leaves

germ [jûrm] *noun*, a micro-organism that causes disease

ger-mi-nate [jûr′mə-nāt′] *verb*, to cause to grow or develop

ges-ta-tion [jĕ-stā′shən] *noun*, the period of time a new life is in the womb until it is born, pregnancy

ges-tic-u-late [jĕ-stĭk′yə-lāt′] *verb*, to make a motion, to gesture

ges-ture [jĕs′chər] *noun*, 1) any action of the hand or face 2) something done to show one's feelings, *verb*, 2) to make a movement of the arm or hand, to signal

get [gĕt] *verb*, 1) to take, have or buy, to acquire 2) to become

gey-ser [gī′zər] *noun*, a natural hot spring that throws out a spray of steam and water occasionally

ghet-to [gĕt- tō] *noun*, any section of a city in which many members of a minority group live, or to which they are restricted by discrimination

ghost [gōst] *adjective*, the appearance of a dead person, an apparition haunting the living

gi-ant [jī′ənt] *adjective*, 1) very large, huge, *noun*, 2) a legendary person or animal who is a very, very large size

gift [gĭft] *noun*, 1) a present 2) a person's inborn ability to do something well, talent

gi-gan-tic [jī-găn′tĭk] *adjective*, very big, large, huge, titanic

gig-gle [gĭg′əl] *verb*, to laugh in a silly, undignified way

gill [gĭl] *noun*, respiratory organs in fish that separate dissolved oxygen from water

gi-raffe [jə-răf′] *noun*, a tall African animal with a very long neck and very long legs and large brown spots on its coat

gird-er [gûr′dər] *noun*, iron or steel beams to span an opening or carry weight of a building

girl [gûrl] *noun*, a female child

gist [jĭst] *noun*, the main point

give [gĭv] *verb*, 1) to present, to donate, to handover 2) to impart something 3) to provide

giv-en [gĭv′ən] *adjective*, 1) stated, typical 2) habitually inclined

gla-cier [glā′shər] *noun*, a body of ice, formed in a region of perpetual snow, moving slowly down a mountain slope or valley

glad [glăd] *adjective*, 1) feeling pleasure or joy 2) very willing

glad-i-o-lus [glăd′ē-ō′ləs] *noun*, *plural*, **gladioli**, a native bulb from Africa of the iris family, that has sword-shaped leaves stems of brightly colored flowers

glad-ly [glăd′lē] *adverb*, happily, willingly, joyfully, thrilled

glam-or [glăm′ər] *noun*, beauty, attraction, alluring charm

glance [glăns] *noun*, 1) a quick or a brief look, *verb*, 2) to look quickly or briefly, to glimpse

gland [glănd] *noun*, a vital part of a living organism that secretes a substance

glare [glâr] *noun*, 1) a bright light reflecting shine or a gleam 2) a frown or scowl, *verb*, 3) to stare disapprovingly, to glower

glass [glăs] *noun*, *no plural*, 1) a hard clear substance that is easily shattered, used for windows, *plural*, **glasses**, 2) a drinking container 3) specially cut lenses made of plastic or glass held in a frame to adjust vision so that a person can see better, spectacles

glaze [glāz] *noun*, 1) a shiny, glossy surface, *verb*, 2) to cover with a thin shiny surface

gleam [glēm] *noun*, 1) a brief flash of light, *verb*, 2) to shine brightly, to glisten, to glitter

glean [glēn] *verb*, to gather leftovers, to collect

glee [glē] *noun*, joy, merriment

glib [glĭb] *adjective*, flippant, to speak freely, sarcastic

glide [glīd] *verb*, to move very smoothly, effortlessly

glid-er [glī′dər] *noun*, an aircraft that flies without an engine

glimpse [glĭmps] *noun*, 1) a brief look at something, *verb*, 2) to get a quick look at something

glit-ter [glĭt′ər] *noun*, 1) something that sparkles or twinkles, *verb*, 2) to sparkle or twinkle as a reflection of light, to shimmer

gloat [glōt] *verb*, to brag, to boast

globe [glōb] *noun*, 1) the round replica of the Earth marked with countries, continents, oceans 2) anything round like a ball

gloom [gloom] *noun*, 1) darkness 2) a feeling of sadness, dejection

gloom-y [gloo′mē] *adjective*, 1) dark, shadowy 2) despondent, dejected, disheartened

glo-ri-fy [glôr′ə-fī′] *verb*, to exalt, to acclaim, to worship

glo-ri-ous [glôr′ē-əs] *adjective*, exhibiting qualities that deserve or receive glory or praise

glory [glôr′ē] *noun*, praise given to someone who has accomplished something extraordinary

glos-sa-ry [glô′sə-rē] *noun*, a list of specialized words used in the text with definitions

gloss-y [glô′sē] *adjective*, smooth and shiny, gleaming bright

glove [glŭv] *noun*, a covering for the hand with separate sections for the fingers and the thumb

glow [glō] *noun*, 1) gentle radiance, light and warmth, *verb*, 2) to radiate light or heat

glu-cose [gloo′kōs′] *noun*, sugar created in the body by processing starch

glue [gloo] *noun*, 1) a substance used to make things stick together 2) to fasten with an adhesive, to attach closely

glut-ton [glŭt′n] *noun*, someone who is greedy for food, a pig

glyc-er-in [glĭs′ər-ĭn] *noun*, a softening agent for skin

gnarled [närld] *adjective*, twisted in a knotted form

gnat [năt] *noun*, a very small flying insect which bites or sucks

gnaw [nô] *verb*, to bite so as to wear away persistently

go [gō] *noun*, 1) a try, an effort, *verb*, 2) to move along 3) to depart 4) to proceed

goad [gōd] *verb*, to use some object to make an animal move

goal [gōl] *noun*, 1) a personal level of achievement 2) the cage or line at the end of the field which the opposite team tries to penetrate 3) a score when the ball goes into the end zone or cage 4) an objective

goat [gōt] *noun*, a small, horned farm animal that is raised for its milk and fur

gob-ble [gŏb'əl] *verb*, to eat fast

gob-lin [gŏb'lĭn] *noun*, in folklore, an ugly, grotesque, evil creature

god [gŏd] *noun*, a being to whom people pray and has control of the forces in the world

God [gŏd] *proper noun*, the supreme being, the master of the universe

god-dess [gŏd'ĭs] *noun*, a female form of deity

goes [gōz] *present tense* of **go** used in the singular, with *he*, *she* or *it*

gog-gles [gŏg'əlz] *noun*, large spectacles for protecting the eyes

gold [gōld] *adjective*, 1) bright yellow color, *noun*, 2) the precious metal

golf [gŏlf] *noun*, a game played by two or more people in which a small ball is hit into 18 holes arranged on a large piece of land called a course

gone [gôn] *past tense* of **go**,

gong [gông] *noun*, a large piece of metal hung up and hit to make a loud noise

good [gŏŏd] *adjective*, 1) right, not wrong, the opposite of bad 2) appropriate, suitable, useful 3) well-behaved 4) nice, pleasant, *noun*, *no plural*, 5) what is virtuous and kind

good-bye [gŏŏd-bī'] *interjection*, an expression used to acknowledge parting

good-ness [gŏŏd'nĭs] *noun*, generosity, kindness, strength

goods [gŏŏdz] *noun*, merchandise, things made by people

good-will [gŏŏd wĭl] *noun*, the value of kindness in business and relationships

goose [gŏŏs] *noun*, *plural*, **geese**, a water bird like a duck but larger and having a longer neck

gore [gôr] *noun*, blood and guts

gorge [gôrj] *noun*, to stuff yourself

gor-geous [gôr'jəs] *adjective*, strikingly beautiful

go-ril-la [gə-rĭl'ə] *noun*, a very large animal similar to a monkey, the largest ape

gos-pel [gŏs'pəl] *noun*, the word of God according to the Bible

gos-sa-mer [gŏs'ə-mər] *adjective*, sheer, like cobwebs

gos-sip [gŏs'əp] *noun*, 1) rumors about people that are unkind or untrue 2) a person who makes false remarks about others, *verb*, 3) to talk about other people

got [gŏt] *past tense* of **get**

gouge [gouj] *noun*, 1) a chisel with a trough-like blade 2) a groove or hole, *verb*, 3) to cut or tear out

141

gourd [gôrd] *noun*, a large fruit similar to squash, its shell is often made into a container

gour-met [goor-mā′] *adjective*, the finest food and drink

gov-ern [gŭv′ərn] *verb*, to lead or control by authority

gov-ern-ment [gŭv′ərn-mənt] *noun*, the offices, agencies, and people who control a nation, state, or city by making laws and seeing that they are followed

gov-er-nor [gŭv′ər-nər] *noun*, 1) the elected leader who controls a state or country 2) an attachment to an engine for controlling its speed

gown [goun] *noun*, 1) a dress worn by a woman that reaches to the floor 2) a covering over the body so that a doctor can examine you without your clothes on

grab [grăb] *verb*, to seize, to clutch, to snatch suddenly

grace [grās] *noun*, 1) elegance, refinement, charm 2) kindness, 3) holiness, devotion, love 4) mercy, pardon, 5) a blessing, giving thanks, *verb*, 6) to beautify, to enhance, to adorn

grace-ful [grās′fəl] *adjective,* having beauty of movement

grace-ful-ly [grās′fə-lē′] *adverb*, gently, calmly, smoothly

gra-cious [grā′shəs] *adjective*, kindly, pleasing, courteous

grade [grād] *noun*, 1) a rating of evaluation or class, *verb*, 2) to rate in a group, according to size, quantity or aptitude

grad-u-al [grăj′oo-əl] *adjective*, slow progress, step by step

grad-u-al-ly [grăj′oo-ə-lē] *adverb*, progressively, slowly

grad-u-ate [grăj′oo-āt′] *noun*, 1) someone who has received a diploma to document that he has completed a required course of study 2) a cup or cylinder marked for measuring liquid, *verb*, 3) to get one's degree by finishing school

grad-u-a-tion [grăj′oo-ā′shən] *noun*, a ceremony to award diplomas to graduates

graf-fi-to [grə-fē′tō] *noun, plural*, **graffiti**, words written or painted in public places to deface a building or surface

graft [grăft] *verb*, 1) a bud or a slit of a tree inserted into another tree to grow 2) a piece of transplanted living skin or tissue 3) use of one's position to make dishonest gains

grain [grān] *noun*, 1) a very small piece of something, a particle 2) seed from a cereal plant such as wheat, corn or rice 3) the design or pattern of wood

gram [grăm] *noun*, the metric unit of weight used to measure small objects

gram-mar [grăm′ər] *noun, no plural*, rules that govern the correct use of language

grand [grănd] *adjective*, very large and fine, impressive in size

grand-child [grănd′chīld′] *noun, plural* **grandchildren**, the child of someone's son or daughter

grand-daugh-ter [grăn′dô′tər] *noun*, the daughter of a person's son or daughter

gran-deur [grăn′jər] *noun*,
1) splendor, magnificence 2) intellectual greatness, nobility

grand-fa-ther [grănd′fä′thər] *noun*, the father of one of your parents

grand-moth-er [grănd′mŭth′ər] *noun*, the mother of one of your parents

grange[g rānj]noun, 1) a farm with its dwelling house, barns, etc. 2) an association of farmers founded. in 1867 for mutual welfare and advancement

gran-ite [grăn′ĭt] *noun*, a hard igneous rock, consisting of feldspar and quartz used in monuments or buildings

grant [grănt] *noun*, 1) money donated by the government or a charitable foundation, *verb*, 2) to allow, to confer, to give

gran-u-late [grăn′yə-lāt′] *verb*, to process a dense solid into grains

grape [grāp] *noun*, a small, round edible, juicy fruit that grows in clusters on a vine

grape-fruit [grāp′frōōt′] *noun*, a bitter-tasting citrus fruit

graph [grăf] *noun*, 1) a picture indicating measurements to show change 2) a drawing used to show information

graph-ic [grăf′ĭk] *adjective*, 1) pertaining to the arts of painting and drawing 2) clearly and vividly described

graph-ite [grăf′īt′] *noun*, native carbon in the form of crystals

grap-ple [grăp′əl] *verb*, to grab and struggle, to seize in a firm grip

grasp [grăsp] *noun*, 1) a clasp, held in the hand, *verb*, 2) to hold or seize 3) to understand or learn, to comprehend, to take in

grass [grăs] *noun*, 1) a fine ground cover with thin leaves that animals eat 2) as a lawn

grass-hop-per [grăs′hŏp′ər] *noun*, an insect that lives in the grass with strong back legs for jumping and wings for flying

grate [grāt] *noun*, 1) a metal frame with wires laid in the shape of a grid, *verb*, 2) to cut or shave into small pieces, to shred by rubbing on a jagged surface

grate-ful [grāt′fəl] *adjective*, thankful, appreciative

grat-i-fy [grăt′ə-fī′] *verb*, to satisfy, to please, to delight, to indulge

grat-i-tude [grăt′ĭ-tōōd′] *noun*, *no plural*, a feeling of thankfulness

gra-tu-i-ty [grə-tōō′ĭ-tē] *noun*, a tip

grave [grāv] *adjective*, 1) serious, *noun*, 2) a hole in the ground where a dead person is buried

grav-el [grăv′əl] *noun*, a mixture of pebbles and rock fragments

grav-i-ty [grăv′ĭ-tē] *noun*, 1) the force that draws everything to the earth's center 2) solemnity

gra-vy [grā′vē] *noun*, a sauce made with meat drippings

gray [grā] *adjective*, a mixture of white and black

graze [grāz] *verb*, 1) to eat grass or vegetation 2) to rub against something and scraping off the surface

grease [grēs] *noun*, 1) a soft animal fat, any oily matter, *verb*, 2) to lubricate or make slippery with oils, to coat with grease

great [grāt] *adjective*, 1) large in size, quantity, duration 2) eminent, remarkable

great-er [grā′tər] *adjective*, more than something in comparison to it, bigger, larger, better

great-est [grā′tĭst] *adjective*, the most, superior, the best

greed [grēd] *noun, no plural*, the feeling that you want more than you need, desire for wealth

green [grēn] *adjective*, 1) a color made by combining yellow and blue, *noun*, 2) a plot of grass

green-house [grēn′hous′] *noun*, a house made of glass or transparent plastic used to grow plants in a controlled environment, year round

greet [grēt] *verb*, to welcome

greet-ing [grē′tĭng] *noun*, good wishes, a friendly expression

gre-gar-i-ous [grĭ-gâr′ē-əs] *adjective*, very social

grew [grōō] *past tense* of **grow**

grey [grā] see **gray**, *adjective*, a mixture of black and white

grid [grĭd] *noun*, 1) grate, a lead plate with perforations 2) a network of horizontal and vertical lines used to locate a particular point on a map

grid-dle [grĭd′l] *noun*, a flat square pan used for cooking waffles

grief [grēf] *noun*, very deep sorrow, sadness

griev-ance [grē′vəns] *noun*, cause of complaint, a protest

grieve [grēv] *verb*, to be upset about something lost or done, to be filled with sadness

grill [grĭl] *noun*, 1) a cooking unit fueled by a flame with a metal rack used for broiling food 2) a system of bars forming a lattice, *verb*, 3) to broil food 4) to question intensely

grim [grĭm] *adjective*, dreadful, ghastly, stern, unyielding

grin [grĭn] *noun*, 1) a big smile, *verb*, 2) to smile broadly

grind [grīnd] *verb*, to crush something so hard that it becomes powder

grip [grĭp] *noun*, 1) a hold, a grasp, *verb*, 2) to hold or clench something tightly

gris-ly [grĭz′lē] *adjective*, ghastly, horrible, gruesome, shocking

grist [grĭst] *noun*, corn

grit [grĭt] *noun*, 1) courage, fortitude, bravery 2) sand, gravel

grits [grĭts] *noun, plural*, coarsely ground hominy

griz-zly [grĭz′lē] *noun*, a big bear

groan [grōn] *noun*, 1) a moan, a complaint, a cry of distress, *verb*, 2) to emit a low cry in disapproval or pain

gro-cer [grō′sər] *noun*, someone who sells food and sundries

gro-cer-y [grō′sə-rē] *noun*, food and household items

groom [grōōm] *noun*, 1) the husband to be 2) a stable boy, *verb*, 3) to make oneself presentable, neat and trim 4) to train for a career

groove [grōōv] *noun*, 1) a channel or furrow cut into something, *verb*, 2) to move to a rhythm

grope [grōp] *verb*, to feel your way in the dark

gross [grōs] *adjective*, 1) bulky, coarse, unrefined, *noun*, 2) twelve dozen

gro-tesque [grō-těsk'] *adjective*, distorted, abnormal, strange-looking in shape, bizarre

ground [ground] *noun*, *no plural*, 1) the surface of the earth, *verb*, 2) past tense of **grind**

ground-wa-ter [ground'wô'tər] *noun*, water beneath the earth's surface that supplies wells and springs

group [grōōp] *noun*, 1) people or things classed together because of similar qualities 2) people who share the same needs and interests

grove [grōv] *noun*, a group of trees

grov-el [grŏv'əl] *verb*, to crawl or creep, to beg, to cringe

grow [grō] *verb*, 1) to become larger, to increase in size or maturity 2) to change and become 3) to sprout, to germinate, to develop

growl [groul] *noun*, 1) a low angry or unfriendly noise coming from the throat like a dog, *verb*, 2) to make such a sound

grown [grōn] *adjective*, to reach maturity or full growth

growth [grōth] *noun*, *no plural*, 1) development 2) a tumor or other form of swelling

grub [grŭb] *noun*, 1) larva, a maggot, *verb*, 2) to dig, to uproot

grudge [grŭj] *noun*, hatred, spite

gruel-ing [grōō'ə-lĭng] *adjective*, exhausting, very difficult

grue-some [grōō'səm] *adjective*, ghastly, frightening, horrible

grum-ble [grŭm'bəl] *verb*, to complain, to grouse, to mutter

grunt [grŭnt] *verb*, to make a guttural sound, to snort like a pig

guar-an-tee [găr'ən-tē'] *noun*, 1) a pledge, *verb*, 2) to honor a promise

guard [gärd] *noun*, 1) a person or thing which forms security or a defense, *verb*, 2) to watch over, to prevent an escape

guard-i-an [gär'dē-ən] *noun*, someone who takes care of another or their property

guer-ril-la [gə-rĭl'ə] *noun*, a revolutionary

guess [gĕs] *noun*, 1) a conjecture, *verb*, 2) to imagine, to think or believe 3) to give an answer without really knowing

guest [gĕst] *noun*, 1) a person invited to stay with someone else, a visitor 2) a customer

guid-ance [gīd'ns] *noun*, *no plural*, help, the act or result of guiding

guide [gīd] *noun*, 1) someone or something that shows the way to find something, *verb*, 2) to show the way or lead 3) to counsel

guile [gīl] *noun*, the ability to trick others, crafty, cunning, deceit

guilt [gĭlt] *noun*, *no plural*, knowing you have done wrong, accepting liability and blame for a wrongdoing

guilt-y [gĭl′tē] *adjective*, 1) legally charged with a crime 2) full of shame, feeling remorse

guin-ea [gĭn′ē] *noun*, Italian money

guise [gīz] *noun*, general, external appearance, manner, pretense

gui-tar [gĭ-tär′] *noun*, a musical instrument with six strings, a fretted neck, and a pear-shaped sound box

gulf [gŭlf] *noun*, 1) part of a sea partially surrounded by land, larger and deeper than a bay 2) a large empty space, or chasm

gul-li-ble [gŭl′ə-bəl] *adjective*, easily tricked or cheated

gul-ly [gŭl′ē] *noun*, a small gulch

gulp [gŭlp] *noun*, 1) a mouthful, *verb*, 2) to swallow quickly 3) to choke or gasp nervously

gum [gŭm] *noun*, 1) the tissue in your mouth above the teeth to protect the roots of each tooth 2) a sticky substance used to hold things together, like glue 3) a soft substance, flavored like candy that can be chewed but it does not dissolve in the mouth

gun [gŭn] *noun*, a weapon consisting of a metal tube used to shoot bullets under pressure

gust [gŭst] *noun*, a sudden rush of wind or strong blast of air

gut [gŭt] *noun*, a stomach, the belly

gut-ter [gŭt′ər] *noun*, drainage ditch, drainpipe, the lower ground or channel along the side of the road

gym-na-si-um [jĭm-nā′zē-əm] *noun*, a place where athletic exercises are performed

gym-nas-tics [jĭm-năs′tĭks] *noun*, *plural*, exercises designed to stretch and strengthen the body

gyp-sy [jĭp′sē] *noun*, a person who belongs to a wandering group of people that came from India to Europe, 4,000 years ago, and now live all over the world

H

hab-it [hăb′ĭt] *noun*, a tendency to do something as part of a routine

hab-i-tat [hăb′ĭ-tăt′] *noun*, the natural environment in which an animal or plant lives

hab-i-ta-tion [hăb′ĭ-tā′shən] *noun*, home, abode, dwelling

hab-it-u-al [hə-bĭch′ōō-əl] *adjective*, customary regular, according to habit

hack-les [hăk′əlz] *noun*, hairs on the back and neck of a dog

hack-neyed [hăk′nēd] *adjective*, common, dull, worn-out

had [hăd] *past tense* of **have**

hag-gle [hăg′əl] *verb*, to argue about price, to bargain

hai-ku [hī′kōō] *noun*, a form of poetry consisting of 3 lines of 5, 7, and 5 syllables each

hail [hāl] *noun*, 1) frozen rain drops, *verb*, 2) to shout with enthusiasm, to call out in a greeting, to welcome

hair [hâr] *noun*, a strand of filament that grows out of the skin of animals

hal-cy-on [hăl′sē-ən] *adjective*, calm, peaceful, happy

hale [hāl] *adjective*, 1) healthy, strong, *verb*, 2) to haul, to draw

half [hăf] *noun*, 1) one of two equal parts of a whole 2) partial

half-dol-lar [hăf′dŏl′ər] *noun*, a coin equal to fifty cents

half-way [hăf′wā′] *adjective*, midway between two points

hal-i-but [hăl′ə-bət] *noun*, the largest species of marine flatfish

hall [hôl] *noun*, 1) the corridor in a building 2) a large room for entertaining 3) a building at a college or university

hal-low [hăl′ō] *verb*, to bless or make holy or sacred

hal-lu-ci-na-tion [hə-lōō′sə-nā′shən] *noun*, something that is seen or heard but is not actually happening, a delusion, a vision

halt [hôlt] *verb*, to cause to stop

halt-er [hôl′tər] *noun*, 1) a rope for a horse, a rein 2) a woman's top tied behind the neck across the back 3) a hangman's noose

halve [hăv] *verb*, to divide in two

ham [hăm] *noun*, the meat from a pig's leg, preserved with salt or smoke 2) a comedian

ham-burg-er [hăm′bûr′gər] *noun*, ground beef served as a patty on a sandwich or alone

ham-mer [hăm′ər] *noun*, 1) an instrument used for driving nails, *verb*, 2) to hit or pound with a hammer or mallet, to strike

ham-mock [hăm′ək] *noun*, a bed of canvas cord hanging between two supports

ham-per [hăm′pər] *noun*, 1) a basket for dirty clothes, *verb*, 2) to obstruct, to hinder

ham-ster [hăm′stər] *noun*, a rodent similar to a rat, with large cheek pouches often kept as a pet

hand [hănd] *noun*, 1) the part of the arm used to hold something, *verb*, 2) to deliver or pass along

hand-i-cap [hăn′dē-kăp′] *noun*, an impediment, a physical defect

hand-ker-chief [hăng′kər-chĭf] *noun*, a cloth carried for wiping hands, the face or nose

han-dle [hăn′dl] *noun*, 1) the part of a tool or instrument held in the hand, *verb*, 2) to touch, feel, or examine with the hand 3) to control, manage or deal with

han-dle-bar [hăn′dl-bär′] *noun*, the part of a bicycle used to steer when you ride it

hand-shake [hănd′shāk′] *noun*, a greeting made by shaking hands

hand-some [hăn′səm] *adjective*, good-looking, attractive, manly

hand-writ-ing [hănd′rī′tĭng] *noun*, penmanship, writing done with a pen usually in cursive

hand-y [hăn′dē] *adjective*, convenient, easy to use

hang [hăng] *verb*, 1) to fasten so that something can swing freely, to suspend 2) to suspend from a rope until dead 3) to linger, to wait 4) to be contingent

hang-ar [hăng′ər] *noun*, a building for storing aircraft

hang-er [hăng′ər] *noun*, a device on which something hangs

hap [hăp] *noun*, chance, luck

hap-haz-ard [hăp-hăz′ərd] *adjective*, random, by chance

hap-less [hăp′lĭs] *adjective*, unfortunate, unlucky

happen

hap-pen [hăp′ən] *verb,* 1) to occur by chance, 2) to come to pass

hap-pi-ly [hăp′ə-lē] *adverb,* good-natured, graciously, kindly

hap-pi-ness [hăp′ē-nĭs] *noun,* conscious enjoyment of good fortune, contentment

hap-py [hăp′ē] *adjective,* content, delighted or pleased

ha-rangue [hə-răng′] *noun,* a loud address to a multitude, a noisy, ranting speech

ha-rass [hăr′əs] *verb,* to trouble by repeated attacks, to annoy

har-bor [här′bər] *noun,* 1) a port where ships or boats can dock safely, *verb,* 2) to hide, to protect, or conceal

hard [härd] *adjective,* 1) difficult to understand 2) firm, solid, impenetrable 3) difficult, fatiguing, challenging, *adverb,* 4) intensely, diligently

hard drive [hard-drīv] *noun,* the memory on a computer

hard-en [här′dn] *verb,* 1) to make solid 2) to become unfeeling

hard-ly [härd′lē] *adverb,* barely, almost never, scarcely

hard-ship [härd′shĭp′] *adjective,* suffering, extreme privation

hard-ware [härd′wâr′] *noun,* 1) the pieces of equipment that make up a computer 2) the items such as nuts and bolts needed to put something together

hare [hâr] *noun,* an animal similar to a rabbit, that is larger, unlike a rabbit, it does not burrow and its young are born covered with fur

harm [härm] *noun,* 1) injury, damage, *verb,* 2) to hurt or injure

harm-ful [härm′fəl] *adjective,* something that could cause damage or injury

harm-less [härm′lĭs] *adjective,* causing no harm or damage

har-mon-i-ca [här-mŏn′ĭ-kə] *noun,* a hand-held rectangular musical instrument played by exhaling and inhaling through a row of metal reeds

har-mo-ni-ous [här-mō′nē-əs] *adjective,* symmetrical, congruous, in agreement

har-mo-nize [här′mə-nīz′] *verb,* to sing in harmony, to cooperate

har-mo-ny [här′mə-nē] *noun,* 1) in tune 2) agreement, accord

har-ness [här′nĭs] *noun,* the working gear of a horse

harp [härp] *noun,* 1) a large stringed instrument, played with fingers, *verb,* 2) to nag

har-poon [här-pōōn′] *noun,* 1) a barbed spear, like a missile, *verb,* 2) to stab with a harpoon

har-row [hăr′ō] *noun,* 1) a tractor accessory used to turn over the soil, *verb,* 2) to cultivate

harsh [härsh] *adjective,* 1) hard to bear, severe, cruel 2) inharmonious, jarring, dissonant

har-vest [här′vĭst] *noun,* 1) the season's crop, the amount gathered, *verb,* 2) to gather or pick a crop, to reap, to gain

has [hăz] the part of the *verb* **have** used with *he, she,* and *it*

haste [hāst] *noun,* hurry

hast-i-ly [hāst′ĭ-lē] *adverb,* hurriedly, quickly, fast

hast-y [hā′stē] *adjective,* done in a hurry, quickly, swiftly hat [hăt] *noun*, a clothing accessory used to cover the head

hatch [hăch] *noun*, 1) a small doorway or opening in a wall or a floor, *verb*, 2) to emerge from an egg when born

hatch-et [hăch′ĭt] *noun*, a small ax with a short handle

hate [hāt] *verb*, to dislike, to detest

hate-ful [hāt′fəl] *adjective*, full of hate, feeling a strong dislike

ha-tred [hā′trĭd] *noun*, a strong feeling of dislike, hostility

haugh-ty [hô′tē] *adjective*, snobbish, proud, arrogant

haul [hôl] *verb*, to pull or draw with force, to drag

haunt [hônt] *noun*, 1) a deserted place, *verb*, 2) to intrude upon continually, to annoy

have [hăv] *verb*, 1) to hold possession of or keep 2) a word used to indicate the *past tense* in a description 3) need to, must

ha-ven [hā′vən] *noun*, a safe place

hav-oc [hăv′ək] *noun*, widespread destruction of life and property, devastation, confusion

Ha-waii [hə-wä′ē] *noun*, one of the 50 United States located in the Pacific ocean, the capital city is Honolulu. The state flower of **Hawaii** is the red hibiscus, and the motto is "The life of the land is preserved in righteousness."

hawk [hôk] *noun*, 1) a large bird of prey that kills birds and small animals for food, *verb*, 2) to peddle, to sell on the street

hay [hā] *noun*, dry grass eaten by farm animals such as cows, goats, horses

haz-ard [hăz′ərd] *noun*, a danger

haz-ard-ous [hăz′ər-dəs] *adjective*, dangerous, risky

haze [hāz] *noun*, *no plural*, widely dispersed dust particles in the air that cloud the view

ha-zel [hā′zəl] *adjective*, a color of gray, blue, and brown

he [hē] *pronoun*, *plural*, **they**, the male person or animal, who is the subject of the sentence

head [hĕd] *noun*, 1) the boney structure at the top of humans and animals that contains the brain, the face, the ears, eyes, mouth, nose 2) the seat of reason 3) the leader 4) the top of something, *verb*, 5) to move in the direction of something 6) to lead, to be the chief of

head-ache [hĕd′āk′] *noun*, 1) a dispute, a nuisance 2) pain in the head behind the eyes, a migraine

head-ing [hĕd′ĭng] *noun*, the title or caption at the top of a piece of writing

head-line [hĕd′līn′] *noun*, 1) a line of type displayed conspicuously at the top of a page or column of a newspaper, etc.

head-quar-ters [hĕd′kwôr′tərz] *noun*, the center of operations from which orders are issued

heal [hēl] *verb*, to cure, to treat an illness, to remedy, to repair

health [hĕlth] *noun*, 1) the condition of the body or mind free from sickness or disease

health-y [hĕl′thē] *adjective*, 1) having a sound mind and body, strong, vibrant, robust 2) wholesome, nutritious

heap [hēp] *noun*, 1) a pile of things, *verb*, 2) to put into a large pile, to stack

hear [hîr] *verb*, 1) to take in sounds 2) to listen 3) to receive news

heard [hûrd] *past tense* of **hear**

hear-ing [hîr'ĭng] *noun*, an opportunity to be heard before a judge or peers

hearse [hûrs] *noun*, a car used to carry a coffin to the cemetery

heart [härt] *noun*, 1) the hollow muscular organ that pumps blood to the arteries and receives blood from the veins 2) the center of someone's feelings and spirit

hearth [härth] *noun*, 1) the bottom of a furnace where metal that is melted into a liquid, collects 2) a fireplace or fireside

heat [hēt] *noun*, *no plural*, 1) the feeling of warmth or something hot caused by friction, *verb*, 2) to raise the temperature to become hot, to make warmer

heat-er [hē'tər] *noun*, an appliance that warms a room

heath-en [hē'thən] *noun*, an atheist, someone who has no god

heave [hēv] *verb*, to lift or pull something of great weight

heav-en [hĕv'ən] *noun*, 1) the home of God, where the saved go to live after death 2) a state of supreme happiness or bliss

heav-en-ly [hĕv'ən-lē] *adjective*, divine, very nice or pleasant

heav-i-er [hĕv'ē-ər] *adjective*, more weight

heav-y [hĕv'ē] *adjective*, 1) a lot of weight 2) burdensome

heck-ler [hĕk'əl- ər] *noun*, a person who shouts and bothers a public speaker by asking impertinent questions

hec-tic [hĕk'tĭk] *adjective*, unsettled, confused, frenzied

hedge [hĕj] *noun*, a dense row of shrubs or bushes to make a wall or border forming a boundary

heed [hēd] *verb*, to notice, to pay careful attention to

heel [hēl] *noun*, 1) the back part of a foot, below the ankle 2) the back, bottom part of a shoe 3) an unpleasant person, *verb*, 4) to follow obediently or closely

heif-er [hĕf'ər] *noun*, a young cow

height [hīt] *noun*, distance upwards, how tall something is

hei-nous [hā'nəs] *adjective*, hateful, odious, flagrant

heir [âr] *noun*, one who inherits, or has the right to inherit, property after the death of its owner

heir-ess [âr'ĭs] *noun*, a woman who has inherited a fortune

heir-loom [âr'lo͞om'] *noun*, a family possession that is given to a new generation

held [hĕld] *past tense* of **hold**

hel-i-cop-ter [hĕl'ĭ-kŏp'tər] *noun*, a flying machine propelled by action of screws and propeller blades rotating on a vertical axis without supporting wings

he-li-um [hē'lē-əm] *noun*, an inert, gaseous element occurring principally in the atmosphere of the sun and stars

hell [hĕl] *noun*, a place where Satan lives, and where the wicked go when they die

hel-lo [hĕ-lō′] a greeting

helm [hĕlm] *noun*, 1) the wheel or tiller by which a ship is steered 2) a position of control

hel-met [hĕl′mĭt] *noun*, a protective head covering

help [hĕlp] *noun*, 1) assistance, support, guidance, *verb*, 2) to do something for someone, to assist

help-ing [hĕl′pĭng] *adjective*, 1) willing to assist, *noun*, 2) a portion of food, a serving

help-less [hĕlp′lĭs] *adjective*, unable to help oneself

help-ful [hĕlp′fəl] *adjective*, ready to give help, useful

hem [hĕm] *noun*, 1) the edge of a garment that has been sewn to keep it from fraying and to hang on the body at the correct length, *verb*, 2) to sew the bottom of a garment 3) to confine, to enclose

hem-i-sphere [hĕm′ĭ-sfîr′] *noun*, one of the halves of the Earth

hem-or-rhage [hĕm′ər-ĭj] *noun*, any discharge of blood from the blood vessels, heavy bleeding

hen [hĕn] *noun*, a female bird

her [hər, ər; hûr when stressed] *pronoun*, belonging to a female

her-ald [hĕr′əld] *noun*, 1) a bearer of news, *verb*, 2) to announce publicly, to proclaim

herb [ûrb] *noun*, a plant used for medicine or adding flavor to food or aromatherapy

her-biv-o-rous [hûr-bĭv′ər-əs] *adjective*, plant or grain-eating

herd [hûrd] *noun*, 1) a group of animals eating, traveling, or living together, *verb*, 2) to control or steer a group of animals 3) to congregate

here [hîr] *adverb*, referring to at this place or going to this place

he-red-i-tar-y [hə-rĕd′ĭ-tĕr′ē] *adjective*, an inherited trait passed on through genes

he-red-i-ty [hə-rĕd′ĭ-tē] *noun*, inherited characteristics or traits, from parents to offspring

her-e-sy [hĕr′ĭ-sē] *noun*, nonconformity, dissenting view

her-e-tic [hĕr′ĭ-tĭk] *noun*, a nonbeliever, a skeptic

her-it-age [hĕr′ĭ-tĭj] *noun*, rights and or possessions handed down from ancestors, traditions

her-mit [hûr′mĭt] *noun*, a recluse

her-mit-age [hûr′mĭ-tĭj] *noun*, home of one who lives alone

he-ro [hîr′ō] *noun*, a person of great strength or courage

her-o-ine [hĕr′ō-ĭn] *noun*, a woman admired for her brave deeds

her-o-ism [hĕr′ō-ĭz′əm] *noun*, courage, boldness, bravery

her-on [hĕr′ən] *noun*, a large bird that lives in the wetlands

her-ring [hĕr′ĭng] *noun*, a small fish abundant in the ocean and occasionally in fresh water

hers [hûrz] *pronoun*, something belonging to a woman or girl

her-self [hûr-sĕlf′] *pronoun*, *plural*, **themselves**, the reflexive form of **she**, belonging to her

hes-i-tate [hĕz′ĭ-tāt′] *verb*, to act slowly, as if in doubt, to be uncertain because of fear

hes-i-ta-tion [hĕz′ĭ-tā′shən] *noun*, doubt, uncertainty, skepticism

hew [hyōō] *verb*, to cut to pieces

hex-a-gon [hĕk′sə-gŏn′] *noun*, a plane figure having six sides

hi-a-tus [hī-ā′təs] *noun*, a gap, an interval in time, a pause

hi-ber-nate [hī′bər-nāt′] *verb*, to sleep during the winter

hic-cup [hĭk′əp] *noun*, a small belch

hid-den [hĭd′n] *adjective*, concealed from sight

hide [hīd] *noun, no plural*, 1) the skin of an animal, *verb*, 2) to put out of sight or keep a secret

hid-e-ous [hĭd′ē-əs] *adjective*, horribly ugly, shocking

hi-er-ar-chy [hī′ə-rär′kē] *noun*, a pecking order, ranking

hi-er-o-glyph-ics [hī′ər-ə-glĭf′ĭks] *noun, plural*, characters in the picture writing of the ancient Egyptians

high [hī] *adjective*, 1) how tall something is from the ground 2) expensive, above average 3) a shrill pitch 4) elated, happy

high-land [hī′lənd] *noun*, land that is high up in the hills

high-ness [hī′nĭs] *noun*, a title of a member of the royal family

high-way [hī′wā′] *noun*, the main public road that usually connects cities and towns

hi-jack [hī′jăk′] *verb*, to force the driver of a vehicle to go somewhere, to commandeer

hike [hīk] *noun*, 1) a walk, a journey, a trek, *verb*, 2) to go on an extended walk

hi-lar-i-ous [hĭ-lâr′ē-əs] *adjective*, very funny, boisterously merry

hill [hĭl] *noun*, a rise in land making the ground higher than the earth's surface

him [hĭm] *pronoun*, a man or boy

him-self [hĭm-sĕlf′] *pronoun*, the subject mentioned later in the sentence, *plural*, **themselves**

hind [hīnd] *adjective*, situated in the rear, back, posterior

hind-er [hĭn′dər] *verb*, to make it more difficult for someone to do something, to obstruct

hind-rance [hĭn′drəns] *noun*, something that slows you down, an impediment

Hin-du [hĭn′dōō] *noun*, 1) a native of India 2) a religion in the culture of India that embraces ancient philosophy and traditions

hinge [hĭnj] *noun*, 1) a joint on which a door or gate turns or swings, *verb*, 2) to hang on or attach to something

hint [hĭnt] *noun*, 1) a suggestion, a clue 2) a slight trace

hip [hĭp] *noun*, the part of your body below the stomach where your legs are connected to the rest of your body

hip-po-pot-a-mus [hĭp′ə-pŏt′ə-məs] *noun*, a beast with thick skin that measures 4-5 feet high, 12 feet long, and weighs 4 tons that lives in Africa

hire [hīr] *verb*, to pay someone to do work or perform services

his [hĭz] *pronoun*, belonging to a man or a boy, the possessive case of **he**

hiss [hĭs] *noun*, 1) a sound made by forcing air through the teeth, *verb*, 2) to make a sound like a defensive cat

his-tol-o-gy [hĭ-stŏl′ə-jē] *noun*, the science which treats of the minute structure of animal and vegetable tissues

his-to-ri-an [hĭ-stôr′ē-ən] *noun*, an authority or author of past events in history

his-tor-ic [hĭ-stôr′ĭk] *adjective*, something embodying people or things of the past

his-tor-ic-al [hĭ-stôr′ĭ-kəl] *adjective*, information or articles belonging to the past

his-to-ry [hĭs′tə-rē] *noun*, the story and knowledge of the past

hit [hĭt] *noun*, 1) a blow 2) when the ball comes in contact with the bat 3) a smash, a wow, a success, *verb*, 4) to bring something down hard onto something else, to smack

hitch [hĭch] *verb*, to tie something to something else, to attach

hive [hīv] *noun*, 1) any place busy with activity 2) a variety of allergic reactions marked by itching welts or a rash

hoard [hôrd] *noun*, 1) a large amount of something, *verb*, 2) to collect and save, to accumulate, to set aside and hide away

hoarse [hôrs] *adjective*, sounding husky and rough or harsh

hoax [hōks] *noun*, a practical joke, a deceitful trick, a prank

hob-by [hŏb′ē] *noun*, a favorite activity that holds someone's interest that one likes to do

hock-ey [hŏk′ē] *noun*, a game played on ice or grass by two teams of several players who use curved sticks to hit a puck or a ball into a net or goal

hoe [hō] *noun*, 1) a tool used to lift and loosen the earth, *verb*, 2) to turn the soil over with a hoe

hog [hôg, hŏg] *noun*, a large pig

hoist [hoist] *verb*, to lift up, especially with a pulley

hold [hōld] *noun*, 1) the part of the ship that holds cargo, *verb*, 2) to grasp 3) to contain 4) to host an event 5) to restrain, to keep back 6) to balance something

hole [hōl] *noun*, an empty space in something, a gap, a cavity

hol-i-day [hŏl′ĭ-dā′] *noun*, 1) a period of time that involves relaxation, not work 2) a day on which a particular event is celebrated

hol-low [hŏl′ō] *adjective*, 1) not solid, an empty space inside 2) feeling unworthy, small, *noun*, 3) a small valley

hol-ly [hŏl′ē] *noun*, an evergreen, with bright red berries and prickly and glossy leaves

hol-o-caust [hŏl′ə-kôst′] *noun*, great or total destruction by fire, extermination, genocide

ho-ly [hō′lē] *adjective*, of God or the gods, divine, sacred

hom-age [hŏm′ĭj] *noun*, to honor, to pay tribute and respect to a person publicly

home [hōm] *noun*, a place where someone lives, a residence

home-ly [hōm′lē] *adjective*, unattractive, plain looking

ho-me-o-path [hō′mē -ə-pǎth′]
noun, one who combats disease
with remedies producing a
similar complaint on a healthy
person

home run [hōm rŭn′] *noun*, a run
scored on a hit which enables the
batter to touch all four bases

home-stead [hōm′stěd′] *verb*, to
make a place one's home

home-work [hōm′wûrk′] *noun*,
material from school to be
completed at home

hom-i-cide [hŏm′ĭ-sīd′] *noun*, the
killing of one human being by
another

hom-i-ly [hŏm′ə-lē] *noun*, a
sermon, a moral lesson

ho-mo-ge-neous [hō′mə-jē′nē-əs]
adjective, 1) of the same kind or
nature 2) consisting of similar
parts throughout

ho-mog-e-nize [hə-mŏj′ə-nīz′]
verb, to reduce particles in
liquids such as milk or paint to
the same size

hom-o-phone [hŏm′ə-fōn′] *noun*, a
word that sounds the same as
another, but is spelled
differently, and has a different
meaning,e.g. *see* and *sea*

hone [hōn] *verb*, to sharpen

hon-est [ŏn′ĭst] *adjective*, truthful,
not deceitful, trustworthy

hon-est-ly [ŏn′ĭst-lē] *adverb*,
truthfully, openly, candidly

hon-es-ty [ŏn′ĭ-stē] *noun*, telling
the truth, integrity, fairness,
truthfulness, sincerity

hon-ey [hŭn′ē] *noun*, a syrup made
by bees with treacle gathered
from flowers

hon-ey-comb [hŭn′ē-kōm′] *noun*,
six-sided wax cells made by bees
to store honey or eggs

hon-ey-moon [hŭn′ē-mo͞on′] *noun*,
a holiday taken by a couple who
have just been married

hon-or [ŏn′ər] *noun*, 1) fame or a
rank of high esteem, *verb*, 2) to
hold in high respect 3) to keep an
agreement

hon-or-a-ble [ŏn′ər-ə-bəl]
adjective, respectable

hood [ho͝od] *noun*, 1) a covering for
the head attached to a jacket or a
robe 2) an engine cover

hoof [ho͞of] *noun*, *plural*, **hooves**,
the horny, hard covering on the
foot of a horse, cow, or goat

hook [ho͝ok] *noun*, 1) a bent piece
of metal or plastic used to catch
something, *verb*, 2) to clasp, to
hang or fasten with a hook

hoop [ho͞op] *noun*, a circular band
such as a hoola hoop or a
basketball hoop

hop [hŏp] *noun*, 1) a short jump,
verb, 2) to move on one foot
3) to jump with both feet
together to move along like a
bird or rabbit

hope [hōp] *noun*, 1) a feeling of
confidence, expectation *verb*, 2)
to look forward to something
believing that it will happen

horde [hôrd] *noun*, crowd, a mob, a
large multitude, a swarm

ho-ri-zon [hə-rī′zən] *noun*, where
the earth appears to intersect the
sky, the skyline

hor-i-zon-tal [hôr′ĭ-zŏn′tl]
adjective, 1) parallel to the
horizon 2) flat, level

horn [hôrn] *noun*, 1) one of two hard growths growing out from the heads of some animals such as deer, elk, etc. 2) the device on a car or bicycle used to warn people 3) a musical wind instrument in the brass family with a set of keys to regulate the flow of air

hor-net [hôr′nĭt] *noun*, a bee similar to a wasp

hor-o-scope [hôr′ə-skōp′] *noun*, an astrological prediction

hor-ri-ble [hôr′ə-bəl] *adjective*, shocking, hideous, terrible

hor-rid [hôr′ĭd] *adjective*, vile, disgusting, vulgar, frightful

hor-ri-fy [hôr′ə-fī′] *verb*, to frighten, to terrify, to petrify

hor-ror [hôr′ər] *noun*, great fear and shock caused by something terrible, a feeling of terror

horse [hôrs] *noun*, an animal with long legs that eats grass and can pull a cart or carry people

horse-back [hôrs′băk′] *noun*, to ride on the back of a horse

horse-pow-er [hôrs′pou′ər] *noun*, a unit of power, 33,000 foot pounds of work per minute

horse-shoe [hôrs′shōō′] *noun*, a strip of iron conformed to the rim of a horse's hoof

hor-ti-cul-ture [hôr′tĭ-kŭl′chər] *noun*, the study of growing fruits, vegetables, flowers

hose [hōz] *noun*, 1) a rubber tube used to transfer liquid, *verb*, 2) to water with a hose

hos-pi-tal [hŏs′pĭ-tl] *noun*, a place that provides medical and surgical care for the sick and injured

hos-pi-tal-i-ty [hŏs′pĭ-tăl′ĭ-tē] *noun*, *no plural*, a warm welcome or reception to visitors

host [hōst] *noun*, 1) a person who entertains guests 2) an organism harboring a parasite 3) the communion wafer, representing the bread of the last supper, *verb*, 4) to receive guests, to entertain 5) to serve as a master of ceremonies

hos-tage [hŏs′tĭj] *noun*, a person or thing taken prisoner for ransom

host-ess [hō′stĭs] *noun*, the woman of the house, the one who entertains the guests

hos-til-i-ty [hŏ-stĭl′ĭ-tē] *noun*, unfriendliness, antagonism, showing ill will, rancor

hot [hŏt] *adjective*, 1) a very high temperature 2) a very strong, burning taste, spicy

ho-tel [hō-tĕl′] *noun*, a large building that offers lodging and food to paying customers, etc.

hound [hound] *noun*, a breed of dog used for hunting or racing

hour [our] *noun*, 1)sixty minutes, a measure of time, 1/24th of a day 2) a particular time of day

house [hous] *noun*, 1) a building or part of a building where people live or dwell 2) a residence

house-hold [hous′hōld′] *noun*, all the people who live in a house together as a unit

House of Representatives *noun*, the lower house of the legislature of the United States

house-wife [hous′wīf′] *noun*, the spouse who maintains the house for the family

hov-el [hŭv′əl] *noun*, a shack

hov-er [hŭv′ər] *verb*, to stay in one place floating, flying, or fluttering in the air

hov-er-craft [hŭv′ər-krăft′] *noun*, a sort of boat that travels over land or water by floating on air pushed out by its engines

how [hou] *adverb*, 1) used in questions about time, amount, or size 2) used to ask about health 3) used to show surprise or pleasure 4) used to show the way something is done

how-ev-er [hou-ev′ər] *adverb*, nonetheless, despite, though, yet, after all, in whatever manner

howl [houl] *verb*, 1) to make a long wailing cry like a wolf, 2) to moan loudly in pain

hub [hŭb] *noun*, 1) the center of activity 2) the center position of a wheel or propeller

hud-dle [hŭd′l] *verb*, to gather together, to cluster

hue [hyōō] *noun*, a shade of a color

hug [hŭg] *noun*, 1) a close embrace with arms around the body, *verb*, 2) to put arms around someone, to hold them closely, to embrace

huge [hyōōj] *adjective*, very large

hulk [hŭlk] *noun*, a bulky body, a big clumsy person or thing

hull [hŭl] *noun*, 1) the body of a ship 2) a shell, a husk, a case

hum [hŭm] *noun*, 1) to make a low, buzzing sound like a bee, *verb*, 2) to sing a melody to yourself with lips closed

hu-man [hyōō′mən] *noun*, pertaining to mankind, a person

hu-mane [hyōō-mān′] *adjective*, kind, compassionate, loving

hu-man-i-ty [hyōō-măn′ĭ-tē] *noun*, mankind, people, the human race

hum-ble [hŭm′bəl] *adjective*, 1) modest, not proud, 2) poor, plain, small 3) respectful, courteous, *verb*, 4) to bring down to defeat, to humiliate

hu-mid-i-ty [hyōō-mĭd′ĭ-tē] *noun*, moisture, warm, damp air

hu-mil-i-ate [hyōō-mĭl′ē-āt′] *verb*, to treat in a way that takes away a person's pride or self-respect

hu-mil-i-a-tion [hyōō-mĭl′ē-ā′shən] *noun*, embarrassment, shame

hu-mil-i-ty [hyōō-mĭl′ĭ-tē] *noun*, the state of being modest, meek

hu-mor [hyōō′mər] *noun*, *no plural*, 1) being able to make others laugh, *verb*, 2) to indulge

hu-mor-ous [hyōō′mər-əs] *adjective*, full of or characterized by humor, funny, comical

hump [hŭmp] *noun*, a round lump

hu-mus [hyōō′məs] *noun*, organic material consisting of decayed vegetable matter that provides nutrients for plants

hunch [hŭnch] *noun*, 1) a premonition or suspicion, *verb*, 2) to sit or stand with the back bent or arched

hun-dred [hŭn′drĭd] *noun*, the number 100, ten times ten

hun-dreth [hŭn′drĭdth] *noun*, 1) the ordinal number after 99th 2) the number two digits to the right of the decimal point

hung [hŭng] *past tense* of **hang**

hun-ger [hŭng′gər] *noun*, 1) wanting food, craving, *verb*, 2) to feel a strong desire for something 3) to yearn

hun-gry [hŭng′grē] *adjective*, to desire something like food

hunt [hŭnt] *verb*, to seek out to capture for food or sport, 2) to search for

hur-dle [hûr′dl] *noun*, 1) a barrier, an obstacle to overcome, *verb*, 2) to jump over, to vault

hurl [hûrl] *verb*, to throw hard

hur-ri-cane [hûr′ĭ-kān′] *noun*, a severe storm or cyclone that develops in tropical areas

hur-ry [hûr′ē] *noun*, 1) confusion, *verb*, 2) to move quickly, to hasten, to rush, to be quick

hurt [hûrt] *verb*, 1) to injure, to inflict pain or damage 2) to offend 3) to ache, to throb

hur-tle [hûr′tl] *verb*, to move at great speed, to rush headlong

hus-band [hŭz′bənd] *noun*, the male spouse in a marriage

hus-band-ry [hŭz′bən-drē] *noun*, home management, careful attention to spending money

hush [hŭsh] *noun*, 1) quiet, calm, silence, *verb*, 2) to muffle, to make silent, to quiet down

hus-tle [hŭs′əl] *verb*, 1) to hurry, move, or do quickly 2) to trick someone in business

hut [hŭt] *noun*, a small, usually wooden building, a shack

hy-brid [hī′brĭd] *noun*, two species fertilized to create a new breed

hy-drant [hī′drənt] *noun*, a pipe located at a curb with a valve for fire engines to connect a hose to a source of water to fight fire

hy-drau-lic [hī-drô′lĭk] *noun*, operated or moved by means of water under pressure

hy-dro-gen [hī′drə-jən] *noun*, a gas that is tasteless, odorless, and inflammable used in rockets

hy-dro-pho-bi-a [hī′drə-fō′bē-ə] *noun*, 1) fear of water 2) a symptom of rabies

hy-dro-plane [hī′drə-plān′] *noun*, 1) a boat which glides on the surface of water, *verb*, 2) to lose control when a vehicle does not have traction and loses the ability to stop

hy-e-na [hī-ē′nə] *noun*, a large, carnivorous wild dog that lives in Africa and Australia

hy-giene [hī′jēn′] *noun*, practices that keep people clean and healthy, cleanliness

hymn [hĭm] *noun*, a religious song

hy-per-bo-le [hī-pûr′bə-lē] *noun*, an expression used as a figure of speech, *i.e.*, *I would give him the shirt off my back.*

hy-phen [hī′fən] *noun*, the mark -, used to separate syllables, as at the end of a line e.g. *composition,* or to join the parts of compound words e.g. *self-confidence, self-reliant*

hyp-no-tize [hĭp′nə-tīz′] *verb*, to put in a trance in which the subject is susceptible to suggestion, to mesmerize

hy-po-chon-dri-ac [hī′pə-kŏn′drē-ăk′] *noun*, a person always worrying about the state of their health

hy-poc-ri-sy [hĭ-pŏk′rĭ-sē] *adjective*, the act or practice of feigning to be what one is not, especially the false assumption of virtue or religion

hyp-o-crite [hĭp′ə-krĭt′] *noun*, someone who says one thing and does another, a fraud

hy-poth-e-sis [hĭ-pŏth′ĭ-sĭs] *noun*, an educated guess

hy-po-thet-i-cal [hī′pə-thĕt′ĭ-kəl] *adjective*, an assumption based on theory or conjecture

hys-ter-i-cal [hĭ-stĕr′ĭ-kəl] *adjective*, excitable, nervous

I

I [ī] *pronoun, plural,* **we**, first person singular, the individual speaking

ice [īs] *noun,* 1) frozen water, *verb,* 2) to put frosting on a cake

ice-berg [īs′bûrg′] *noun,* a large mass of ice floating in the water

ice cream [īs′krēm′] *noun,* cream, milk, flavoring, and eggs mixed together in a pot of ice and salt to make frozen cream

i-ch-thy-ol-o-gy [ĭk′thē-ŏl′ə-jē] *noun,* the study of fish

i-cing [ī′sĭng] *noun,* frosting

i-ci-cle [ī′sĭ-kəl] *noun,* a piece of ice in the form of a spike made by water dripping and freezing

i-con [ī′kŏn′] *noun,* 1) a religious image 2) a symbol that represents something

i-cy [ī′sē] *adjective,* 1) very cold, freezing, chilly 2) slippery

I'd [īd] the *contraction* of the words: **I** and **would**

Ida-ho [ī′də-hō′] *noun,* one of the 50 United States located in the Northwest, the capital city is Boise. The state flower of **Idaho** is the syringa, and the motto is "May she endure forever."

i-de-a [ī-dē′ə] *noun,* 1) a concept, a thought, an impression 2) a belief, a judgment, a view, an opinion 3) a goal or intention

i-de-al [ī-dē′əl] *adjective,* 1) perfect, ultimate, imaginary, *noun,* 2) an image of something in its perfect form, a model

i-den-ti-cal [ī-dĕn′tĭ-kəl] *adjective,* the very same in every way

i-den-ti-fi-ca-tion [ī-dĕn′tə-fĭ-kā′shən] *noun,* the means to verify who someone is

i-den-ti-fy [ī-dĕn′tə-fī′] *verb,* 1) to classify or name 2) to equate

i-den-ti-ty [ī-dĕn′tĭ-tē] *noun,* who someone is, individuality

i-de-ol-o-gy [ī′dē-ŏl′ə-jē] *noun,* beliefs, opinions, political ideas

id-i-om [ĭd′ē-əm] *noun,* a group of words which when used together have a special meaning, i.e., *raining cats and dogs*

id-i-o-syn-cra-sy [ĭd′ē-ō-sĭng′krə-sē] *noun,* a peculiar habit, a quirk

id-i-ot [ĭd′ē-ət] *noun,* a nitwit, jerk

i-dle [īd′l] *adjective,* not busy, unoccupied, inactive, lazy

i-dol [īd′l] *noun,* a figure that people worship or honor

i.e. an abbreviation, for example…

if [ĭf] *conjunction*, 1) on condition that 2) whether

ig-loo [ĭg′lōō] *noun*, a home made of ice and snow used by Eskimos

ig-ne-ous [ĭg′nē-əs] *adjective*, produced by fire, volcanic

ig-nite [ĭg-nīt′] *verb*, 1) to set on fire, to burn 2) to inspire

ig-no-ble [ĭg-nō′bəl] *adjective*, unworthy, of low character

ig-no-rance [ĭg′nər-əns] *noun*, *no plural*, lack of knowledge

ig-no-rant [ĭg′nər-ənt] *adjective*, not knowing something

ig-nore [ĭg-nôr′] *verb*, to take no notice of, to pretend something is not there, to disregard

ill [ĭl] *adjective*, feeling unhealthy or badly, uneasy, not well

I'll [īl] the *contraction* of the words **I** and **will**

il-le-gal [ĭ-lē′gəl] *adjective*, against the law, prohibited, criminal

il-leg-i-ble [ĭ-lĕj′ə-bəl] *adjective*, so poorly written that it is unreadable

il-le-git-i-mate [ĭl′ĭ-jĭt′ə-mĭt] *adjective*, unlawful, illegal

il-lic-it [ĭ-lĭs′ĭt] *adverb*, not according to law, not permitted or allowed, forbidden

il-lim-it-a-ble [ĭ-lĭm′ĭ-tə-bəl] *adjective*, infinite, vast, limitless

Il-li-nois [ĭl′ə-noi′] *noun*, one of the 50 United States located in the Midwest, the capital is Springfield. The state flower of **Illinois** is the meadow violet, and the motto is "State sovereignty, national union."

il-lit-er-ate [ĭ-lĭt′ər-ĭt] *adjective*, unable to read and write

ill-ness [ĭl′nĭs] *noun*, a sickness, not feeling well, an ailment

il-log-i-c-al [ĭ-lŏj′ĭ-kəl] *adjective*, not observing the rules of correct reasoning, unreasonable

il-lu-mi-nate [ĭ-lōō′mə-nāt′] *verb*, 1) to fill with light 2) to make understandable, to clarify

il-lu-sion [ĭ-lōō′zhən] *noun*, a fantasy, a fallacy, a delusion

il-lu-sive [ĭ-lōō′sĭv] *adjective*, deceptive, false, evasive

il-lus-trate [ĭl′ə-strāt′] *verb*, 1) to explain by examples or verbal images 2) to provide with pictures or designs, to show

il-lus-tra-tion [ĭl′ə-strā′shən] *noun*, 1) an explanation 2) a diagram that provides a design or pictures

il-lus-tri-ous [ĭ-lŭs′trē-əs] *adjective*, distinguished, well known, acclaimed, celebrated

I'm [īm] the *contraction* of the words, **I** and **am**

im-age [ĭm′ĭj] *noun*, 1) a picture in the mind, or in a mirror 2) a figure made of stone, wood, etc.

im-ag-i-nar-y [ĭ-măj′ə-nĕr′ē] *adjective*, unreal, fanciful, existing only in the mind

im-ag-i-na-tion [ĭ-măj′ə-nā′shən] *noun*, fantasy, dreams

im-ag-ine [ĭ-măj′ĭn] *verb*, 1) to form a notion or mental picture 2) to conceive, to suppose

im-bue [ĭm-byōō′] *verb*, 1) to teach, to inspire 2) to saturate with moisture or color

im-i-tate [ĭm′ĭ-tāt′] *verb*, to do something the same way as someone else, to emulate

im-i-ta-tion [ĭm′ĭ-tā′shən] *adjective*, 1) artificial, synthetic, counterfeit, *noun*, 2) a copy of something, a fake, a replica

im-mac-u-late [ĭ-măk′yə-lĭt] *adjective*, without stain or blemish, spotlessly clean

im-ma-te-ri-al [ĭm′ə-tîr′ē-əl] *adjective*, not applicable, irrelevant, not important

im-ma-ture [ĭm′ə-tyо̄о̆r′] *adjective*, 1) not grown up 2)using poor judgment, childish

im-meas-ur-a-ble [ĭ-mĕzh′ər-ə-bəl] *adjective*, vast, limitless, endless, infinite

im-me-di-ate [ĭ-mē′dē-ĭt] *adjective*, right away

im-me-di-ate-ly [ĭ-mē′dē-ĭt-lē] *adverb*, at once, without interval of time or delay, now

im-mense [ĭ-mĕns′] *adjective*, huge, very large, enormous

im-men-si-ty [ĭ-mĕn′sĭ-tē] *noun*, enormity, endlessness

im-mi-grant [ĭm′ĭ-grənt] *noun*, a settler from another country

im-mi-gra-tion [ĭm′ĭ-grā′shən] *noun*, a journey or voyage to another place

im-mi-grate [ĭm′ĭ-grāt′] *verb*, to go to a country of which one is not a native to live as a resident

im-mi-nent [ĭm′ə-nənt] *adjective*, impending, near at hand, likely to occur at any moment

im-mo-bil-i-ty [ĭ-mō-bĭl′-ĭ-tē] *noun*, the state of being unable to move, motionless

im-mortal [ĭ-môr′tl] *adjective*, undying, everlasting, perpetual

im-mor-tal-i-ty [ĭm′ôr-tăl′ĭ-tē] *noun*, eternal life, unending fame

im-mune [ĭ-myoōn′] *adjective*, 1) exempt from an obligation 2) not affected by disease

im-mu-ni-ty [ĭ-myо̄о̄′nĭ-tē] *noun*, freedom from natural or usual liability

im-mu-nize [ĭm′yə-nīz′] *verb*, to protect the body from incurable diseases by injecting an inactive form of the virus or bacteria

im-mu-ni-za-tion [ĭm′yə-nĭ-zā′shən] *noun*, an injection that protects the body from a virus or disease

im-pact [ĭm′păkt′] *noun*, 1) the striking of one object by another 2) a forceful impression, a strong effect

im-pair [ĭm-pâr′] *verb*, to weaken, to diminish in value, make worse

im-pale [ĭm-pāl′] *verb*, to pierce

im-part [ĭm-pärt′] *verb*, to tell, to reveal, to make known, to give

im-par-tial [ĭm-pär′shəl] *adjective*, fair, just, not favoring one more than another, not biased

im-pas-sa-ble [ĭm-păs′ə-bəl] *adjective*, a place that cannot be passed or traversed

im-passe [ĭm′păs′] *noun*, an inescapable situation

im-pa-tience [ĭm-pā′shəns] *noun*, *no plural*, anticipation, anxious expectancy, feeling anxious

im-pa-tiens [ĭm-pā′shəns] *noun*,
a flower with white or red
flowers that bloom in summer

im-pa-tient [ĭm-pā′shənt]
adjective, not wanting to wait

im-peach [ĭm-pēch′] *verb*, to
charge with a crime while in
office, to discredit

im-pec-ca-ble [ĭm-pĕk′ə-bəl]
adjective, faultless, perfect

im-pede [ĭm-pēd′] *verb*, to hinder
or obstruct, to slow action

im-ped-i-ment [ĭm-pĕd′ə-mənt]
noun, hindrance, obstacle

im-pel [ĭm-pĕl′] *verb*, to push, to
motivate, to drive forward

im-pend-ing [ĭm-pĕn′dĭng]
adjective, nearing, approaching

im-per-a-tive [ĭm-pĕr′ə-tĭv]
adjective, 1) expressive of
command 2) binding, obligatory

im-per-fect [ĭm-pûr′fĭkt] *adjective*,
flawed, defective, faulty

im-per-fec-tion [ĭm′pər-fĕk′shən]
noun, 1) a flaw, a defect 2) a
failing, a shortcoming, a fault

im-pe-ri-al [ĭm-pîr′ē-əl] *adjective*,
relating to an empire and its ruler

im-per-il [ĭm-pĕr′əl] *verb*, to put in
danger or peril

im-pe-ri-ous [ĭm-pîr′ē-əs]
adjective, bossy, domineering

im-per-me-a-ble
[ĭm-pûr′mē-ə-bəl] *adjective*,
unable to pass through
something, impervious

im-per-son-al [ĭm-pûr′sə-nəl]
adjective, detached, withdrawn

im-per-son-ate [ĭm-pûr′sə-nāt′]
verb, to pretend to be someone

by assuming their appearance

im-per-ti-nent [ĭm-pûr′tn-ənt]
adjective, 1) rude, insolent, pert
2) irrelevant

im-per-vi-ous [ĭm-pûr′vē-əs]
adjective, 1) impenetrable, not
admitting entrance or passage
through 2) unaffected

im-pet-u-ous [ĭm-pĕch′ōō-əs]
adjective, impulsive, rash

im-pe-tus [ĭm′pĭ-təs] *noun*, a
driving force, motive, incentive

im-plau-si-ble [ĭm-plô′zə-bəl]
adjective, unlikely, unbelievable

im-ple-ment [ĭm′plə-mənt] *noun*,
1) a tool, a device, *verb*, 2) to
complete, to accomplish, to
make good on an idea

im-pli-ca-tion [ĭm′plĭ-kā′shən]
noun, something that is implied,
hinted or suggested

im-plic-it [ĭm-plĭs′ĭt] *adjective*,
1) implied, unsaid, understood
2) trusting without reserve

im-plore [ĭm′plôr] *verb*, to plead
with or beg for with much
feeling

im-ply [ĭm-plī′] *verb*, to suggest a
meaning not spoken, to hint

im-po-lite [ĭm′pə-līt′] *adjective*,
rude, discourteous, disrespectful

im-port [ĭm-pôrt′] *noun*,
1) merchandise brought into
another country 2) the
significance, the meaning, *verb*,
3) to bring goods into another
country 4) to signify

im-por-tance [ĭm-pôr′tns] *noun*, *no
plural*, of great significance or
consequence, prominence

161

im-por-tant [ĭm-pôr′tnt] *adjective*, significant, having a great value

im-por-tu-nate [ĭm-pôr′chə-nĭt] *adjective*, urging, demanding

im-pose [ĭm-pōz′] *verb*, 1) to take unfair advantage of 2) to force on others

im-pos-si-ble [ĭm-pŏs′ə-bəl] *adjective*, unrealizable, unattainable, hopeless

im-pos-tor [ĭm-pŏs′tər] *noun*, one who practices deception, a cheat

im-po-tent [ĭm′pə-tənt] *adjective*, weak, ineffective, powerless

im-prac-ti-cal [ĭm-prăk′tĭ-kəl] *adjective*, not easily done, infeasible, not useful

im-pre-cate [ĭm′prĭ-kāt′] *verb*, to hope evil will happen, to curse

im-preg-na-ble [ĭm-prĕg′nə-bəl] *adjective*, invulnerable, unbeatable, irrefutable

im-press [ĭm-prĕs′] *verb*, to affect someone's feelings or mind in a positive way

im-pres-sion [ĭm-prĕsh′ən] *noun*, 1) a sensation, a feeling 2) an imprint on the mind, an effect

im-pres-sive [ĭm-prĕs′ĭv] *adjective*, touching, stirring, convincing

im-print [ĭm-prĭnt′] *noun*, 1) a deep impression left by pressure, *verb*, 2) to leave a mark by applying pressure

im-pris-on [ĭm-prĭz′ən] *verb*, to hold someone in jail

im-prob-a-ble [ĭm-prŏb′ə-bəl] *adjective*, unlikely to happen

im-prop-er [ĭm-prŏp′ər] *adjective*, something that is considered rude or inappropriate

im-pro-pri-e-ty [ĭm′prə-prī′ĭ-tē] *noun*, an inappropriate expression or act

im-prove [ĭm-prōōv′] *verb*, to make better, to uplift, to refine

im-prove-ment [ĭm-prōōv′mənt] *noun*, an act, a state, or result of improving or bettering

im-pro-vise [ĭm′prə-vīz′] *verb*, 1) to make do with whatever is at hand 2) to perform without preparation at a moment's notice

im-pru-dent [ĭm-prōōd′nt] *adjective*, injudicious, indiscreet

im-pu-dent [ĭm′pyə-dənt] *adjective*, shameless, bold

im-pulse [ĭm′pŭls′] *noun*, a sudden desire to do something, a whim

im-pul-sive [ĭm-pŭl′sĭv] *adjective*, spontaneous, unexpected

im-pu-ni-ty [ĭm-pyōō′nĭ-tē] *noun*, freedom from punishment

im-pu-ri-ty [ĭm-pyōōr′ĭ-tē] *noun*, something mixed with something else that is usually not as good, contamination

im-pute [ĭm-pyōōt′] *verb*, to credit, to attribute, to ascribe

in [ĭn] *adjective*, 1) being inside, *preposition*, 2) showing where 3) at or inside of something 4) during, before the end 5) wearing 6) at hand or on hand 7) near

in-a-bil-i-ty [ĭn′ə-bĭl′ĭ-tē] *adjective*, not able

in-ac-ces-si-ble [ĭn′ăk-sĕs′ə-bəl] *adjective*, 1) out of reach, unattainable 2) hard to contact

in-ac-cu-ra-cy [ĭn-ăk′yər-ə-sē]
noun, inexactness, the state of
being incorrect, an error

in-ad-e-quate [ĭn-ăd′ĭ-kwĭt]
adjective, not sufficient,
deficient, not enough

in-ad-ver-tent-ly [ĭn′əd-vûr′tnt-lē]
adverb, not purposely,
unintentionally, accidentally

in-al-ien-a-ble [ĭn-āl′yə-nə-bəl]
adjective, not to be taken away

in-an-i-mate [ĭn-ăn′ə-mĭt]
adjective, lifeless, dull

in-ar-tic-u-late [ĭn′är-tĭk′yə-lĭt]
adjective, lacking the ability to
express oneself in speech

in-au-gu-rate [ĭn-ô′gyə-rāt′] *verb*,
1) to commence or enter upon, to
begin 2) to install into office in a
formal manner

in-can-des-cent [ĭn′kən-dĕs′ənt]
adjective, glowing or luminous
with intense heat or brightness

in-can-ta-tion [ĭn′kăn-tā′shən]
noun, singing or chanting of
magic spells

in-ca-pa-ble [ĭn-kā′pə-bəl]
adjective, unable, powerless

in-ca-pac-i-tate [ĭn′kə-păs′ĭ-tāt′]
verb, to disable

in-car-cer-a-tion [ĭn-kär′sə-rā′
shən] *noun*, imprisonment

in-car-nate [ĭn-kär′nĭt] *adjective*,
in the flesh, in human form

in-car-na-tion [ĭn′kär-nā′shən]
noun, the act of assuming a
human body after death

in-cen-di-ar-y [ĭn-sĕn′dē-ĕr′ē]
noun, 1) causing something to
burn 2) tending to inflame
passion, a political agitator

in-cense [ĭn-sĕns′] *noun*, 1) scent,
aroma, fragrance, *verb*, 2) anger

in-cen-tive [ĭn-sĕn′tĭv] *noun*,
encouragement, stimulus

in-ces-sant [ĭn-sĕs′ənt] *adjective*,
continuing without interruption

in-ces-sant-ly [ĭn-sĕs′ənt-lē]
adverb, without stopping,
continual, not ceasing

inch [ĭnch] *noun*, a measure of
length, 1/12 of a foot

in-ci-dent [ĭn′sĭ-dənt] *noun*, a
happening, an event, a casual
occurrence, an episode

in-ci-den-tal-ly [ĭn′sĭ-dĕn′tl-ē]
adverb, casually, by chance

in-cin-er-a-tor [ĭn-sĭn′ə-rā′tər]
noun, a container or facility
where garbage is burned

in-ci-sion [ĭn-sĭzh′ən] *noun*, a cut
or slit into tissue, a gash

in-cite [ĭn-sīt′] *verb*, to stir up, to
stimulate to action, to provoke

in-cite-ment [ĭn-sīt′mĕnt] *noun*, a
stimulus, a catalyst

in-clem-ent [ĭn-klĕm′ənt]
adjective, stormy, unkind

in-cli-na-tion [ĭn′klə-nā′shən]
noun, a preference, a tendency

in-cline [ĭn-klīn′] *verb*, 1) to lean,
to slant 2) likely to, to tend to

in-clude [ĭn-klo͞od′] *verb*, 1) to
have as part of, to comprise, to
contain 2) to count or think of
someone or something as a part
of a group, to involve

in-clu-sive [ĭn-klo͞o′sĭv] *adjective*,
not leaving anyone out

in-cog-ni-to [ĭn′kŏg-nē′tō]
adjective, in disguise, a secret
identity

in-co-her-ent [ĭn′kō-hîr′ənt]
adjective, incomprehensible,
lacking meaningful connection

in-come [ĭn′kŭm′] *noun*, the money
a person earns from work or
investments, salary

in-com-par-able
[ĭn-kŏm′pər-ə-bəl] *adjective*,
excellent, unequaled by
comparison, unrivaled,

in-com-pat-i-ble
[ĭn′kəm-păt′ə-bəl] *adjective*, not
easy to get along with

in-com-pe-tent [ĭn-kŏm′pĭ-tənt]
adjective, unfit, clumsy

in-com-plete [ĭn′kəm-plēt′]
adjective, unfinished, under
construction, lacking

in-con-ceiv-a-ble
[ĭn′kən-sē′və-bəl] *adjective*,
unimaginable, incredible, hard to
believe, unthinkable

in-con-gru-ous [ĭn-kŏng′grŏŏ-əs]
adjective, not fitting, absurd

in-con-se-quen-tial
[ĭn-kŏn′sĭ-kwĕn′shəl] *adjective*,
of little importance or
consequence

in-con-ven-ience [ĭn′kən-vēn′yəns]
adjective, bothersome,
troublesome, bad timing

in-con-ven-ient [ĭn′kən-vēn′yənt]
adjective, untimely, awkward

in-cor-po-rate [ĭn-kôr′pə-rāt′]
verb, 1) to unite into one body
2) to form into a corporation

in-cor-rect [ĭn′kə-rĕkt′] *adjective*,
not right in facts, wrong

in-cor-rect-ly [ĭn′kə-rĕkt′-lē]
adverb, mistakenly, erroneously

in-crease [ĭn-krēs′] *noun*, 1) the
amount by which something gets

larger , *verb*, 2) to become
greater in size or quantity

in-cred-i-ble [ĭn-krĕd′ə-bəl]
adjective, surpassing belief,
unlikely, unbelievable

in-cred-u-lous [ĭn-krĕj′ə-ləs]
adjective, shocking, surprising,
full of disbelief

in-cre-ment [ĭn′krə-mənt, ĭng′-]
noun, an increase, addition

in-crim-i-nate [ĭn-krĭm′ə-nāt′]
verb, to accuse or charge with a
crime or fault

in-crim-i-nat-ing
[ĭn-krĭm′ə-nā′tĭng] *adjective*,
showing proof of guilt

in-cu-bate [ĭn′kyə-bāt′] *verb*, to
apply enough heat or warmth to
make something hatch

in-cum-bent [ĭn-kŭm′bənt]
adjective, 1) required, obligatory,
noun, 2) an officeholder

in-cur [ĭn-kûr′] *verb*, to bring
something unpleasant upon
oneself

in-cur-a-ble [ĭn-kyŏŏr′ə-bəl]
adjective, irreparable, incapable
of being cured, terminally ill

in-cur-sion [ĭn-kûr′zhən] *noun*,
temporary invasion

in-debt-ed [ĭn-dĕt′ĭd] *adjective*,
under obligation to repay money

in-deed [ĭn-dēd′] *adverb*, really,
truly, in fact, certainly

in-def-i-nite [ĭn-dĕf′ə-nĭt]
adjective, uncertain, inexact

in-def-i-nite-ly [ĭn-dĕf′ə-nĭt-lē]
adverb, no set time frame

in-del-i-ble [ĭn-dĕl′ə-bəl] *adjective*,
something that cannot be
removed or blotted out,
permanent

in-dem-ni-ty [ĭn-dĕm′nĭ-tē] *noun*, protection or exemption from loss or damage, security

in-dent [ĭn-dĕnt′] *verb*, 1) to set lines in from the margin 2) to form notches from the edge

in-den-ture [ĭn-dĕn′chər] *verb*, the written work agreement of a servant to a master

in-de-pend-ence [ĭn′dĭ-pĕn′dəns] *noun*, *no plural*, the state of acting alone or separately

in-de-pend-ent [ĭn′dĭ-pĕn′dənt] *adjective*, free, not relying on others, without bias

in-de-pend-ent-ly [ĭn′dĭ-pĕn′dənt-lē] *adverb*, to work alone, self-reliant, self-directed, autonomously

in-de-struct-i-ble [ĭn′dĭ-strŭk′tə-bəl] *adjective*, incapable of being destroyed

in-dex [ĭn′dĕks′] *noun*, a table for finding reference to topics, names, etc.

In-di-an [ĭn′dē-ən] *noun*, a native American

In-di-ana [ĭn′dē-ăn′ə] *noun*, one of the 50 United States located in the Midwest, the capital is Indianapolis. The state flower of **Indiana** is the peony, and the state motto is "The crossroads of America."

in-di-cate [ĭn′dĭ-kāt′] *verb*, to show or describe, to point out

in-di-ca-tion [ĭn′dĭ-kā′shən] *noun*, something that shows a trait

in-di-ca-tor [ĭn′dĭ-kā′tər] *noun*, 1) something that points to a trait 2) an instrument for automatically showing the pressure of the working fluid in an engine

in-dict [ĭn-dīt′] *verb*, to formally charge with a crime

in-dict-ment [ĭn-dīt′mənt] *noun*, a formal accusation charging someone with a crime

in-dif-fer-ent [ĭn-dĭf′ər-ənt] *adjective*, mediocre, unresponsive, unconcerned

in-dig-e-nous [ĭn-dĭj′ə-nəs] *adjective*, native to an area

in-di-gent [ĭn′dĭ-jənt] *adjective*, poor, homeless, destitute

in-di-gest-i-ble [ĭn′dĭ-jĕs′tə-bəl] *adjective*, something that cannot be swallowed or eaten

in-di-ges-tion [ĭn′dĭ-jĕs′chən] *noun*, a stomach ache

in-dig-nant [ĭn-dĭg′nənt] *adjective*, angry because of something that is not right, resentful

in-dig-na-tion [ĭn′dĭg-nā′shən] *noun*, anger, displeasure, rage

in-dig-ni-ty [ĭn-dĭg′nĭ-tē] *noun*, insulting or abusive treatment

in-di-go [ĭn′dĭ-gō′] *noun*, a deep blue-violet dye made from plants

in-di-rect [ĭn′dĭ-rĕkt′] *adjective*, in a roundabout way, circuitous

in-di-rect-ly [ĭn′dĭ-rĕkt′lē] *adverb*, inadvertently, not intentionally

in-dis-creet [ĭn′dĭ-skrēt′] *adjective*, lacking prudence or discretion, unwise, not careful, imprudent

in-dis-pen-sa-ble [ĭn′dĭ-spĕn′sə-bəl] *adjective*, absolutely necessary, essential

in-dis-put-a-ble [ĭn′dĭ-spyōō′tə-bəl] *adjective*, sure, undeniable, certain

in-di-vid-u-al [ĭn′də-vĭj′oō-əl], *adjective*, 1) unique, special *noun* 2) a person, an entity, one

in-di-vid-u-al-ly [ĭn′də-vĭj′oō-əl-lē] *adverb*, separately, one by one, distinctly

in-di-vis-i-ble [ĭn′də- vĭz′ə-bəl] *adjective*, 1) that cannot be divided 2) cannot be divided without leaving a remainder

in-do-lent [ĭn′də-lənt] *adjective*, lazy, inactive, slothful

in-dom-i-ta-ble [ĭn′dŏm′ĭ-tə-bəl] *adjective*, unconquerable, invincible, unyielding

in-door [ĭn′dôr′] *adjective*, located or used inside of a building

in-duce [ĭn-doōs′] *verb*, to persuade, to influence

in-duce-ment [ĭn-doōs′mənt] *noun*, any motive that leads one to act

in-duc-tive [ĭn-dŭk′tĭv] *adjective*, going from specific to general

in-dulge [ĭn-dŭlj′] *verb*, to allow, to give way to, to foster

in-dul-gence [ĭn-dŭl′jəns] *noun*, a giving way to one's desires

in-dus-tri-al [ĭn-dŭs′trē-əl] *adjective*, commercial, related to economy or industry

in-dus-tri-ous [ĭn-dŭs′trē-əs] *adjective*, hard-working

in-dus-try [ĭn′də-strē] *noun*, 1) hard work, being active 2) center of commerce, enterprise, business activity

in-el-i-gi-ble [ĭn-ĕl′ĭ-jə-bəl] *adjective*, not qualified to be chosen for an office

in-ept [ĭn-ĕpt′] *adjective*, out of place, incapable of performing a task, incompetent, unskilled

in-eq-ui-ty [ĭn-ĕk′wĭ-tē] *noun*, unfairness, not equal, bias

in-ert [ĭn-ûrt′] *adjective*, without power, inactive

in-er-tia [ĭ-nûr′shə] *noun*, 1) a state of matter by which it retains its state of rest, inactivity 2) acted upon by an external force

in-ev-i-ta-ble [ĭn-ĕv′ĭ-tə-bəl] *adjective*, not to be evaded or shunned, unavoidable

in-fal-li-ble [ĭn-făl′ə-bəl] *adjective*, unfailing, reliable, certain

in-fa-mous [ĭn′fə-məs] *adjective*, notoriously bad, wicked

in-fan-cy [ĭn′fən-sē] *adjective*, the first part of life, early childhood

in-fant [ĭn′fənt] *noun*, a very young child in the first stage of life, a newborn, a baby

in-fan-try [ĭn′fən-trē] *noun*, the ground forces of the army

in-fat-u-a-tion [ĭ-făch′oō-ā′shən] *adjective*, the state of having very strong love or admiration for someone, an obsession

in-fect [ĭn-fĕkt′] *verb*, to share a virus or infection with others

in-fec-tion [ĭn-fĕk′shən] *noun*, any disease caused by germs

in-fec-tious [ĭn-fĕk′shəs] *adjective*, capable of being spread or communicated, contagious

in-fer [ĭn-fûr′] *verb*, to arrive at a conclusion, to deduce, to guess

in-fer-ence [ĭn′fər-əns] *noun*, the act of drawing a conclusion from certain premises, a guess

in-fe-ri-or [ĭn-fîr′ē-ər] *adjective*, 1) lower 2) of less importance 3) inequality, subordinate

in-fer-nal [ĭn-fûr′nəl] *adjective*, referring to hell, diabolical

in-fest [ĭn-fĕst′] *verb*, to overrun in a troublesome manner

in-fil-tra-tion [ĭn′fĭl-trā′shən] *noun*, 1) the act of penetrating the pores of a substance 2) passing or joining gradually

in-fi-nite [ĭn′fə-nĭt] *adjective*, unending, there is no end

in-fin-i-tes-i-mal [ĭn′fĭn-ĭ-tĕs′ə-məl] *adjective*, immeasurably small

in-fin-i-ty [ĭn-fĭn′ĭ-tē] *noun*, the symbol ~, to continue indefinitely

in-firm [ĭn-fûrm′] *adjective*, weak, sick, in poor health, ill, not well

in-fir-ma-ry [ĭn-fûr′mə-rē] *noun*, a small hospital, a first-aid tent

in-fir-mi-ty [ĭn-fûr′mĭ-tē] *noun*, weakness or ailment

in-flam-ma-ble [ĭn-flăm′ə-bəl] *adjective*, explosive, volatile

in-flam-ma-tion [ĭn′flə-mā′shən] *noun*, swelling, redness

in-flate [ĭn-flāt′] *verb*, to fill with air, to enlarge, to expand

in-fla-tion [ĭn-flā′shən] *noun*, ever higher price levels

in-flict [ĭn-flĭkt′] *verb*, to force, to impose, to deal as a blow

in-flu-ence [ĭn′flōo-əns] *noun*, 1) power, authority or control 2) the gradual or unseen operation of some cause, *verb*, 3) to act upon 4) to persuade

in-flu-en-tial [ĭn′flōo-ĕn′shəl] *adjective*, important, able to change people's thoughts on something, compelling

in-flu-en-za [ĭn′flōo-ĕn′zə] *noun*, no plural, the flu, an epidemic characterized by fever, headaches and fatigue

in-flux [ĭn′flŭks′] *noun*, flowing into or coming in

in-form [ĭn-fôrm′] *verb*, 1) to tell, to notify, to advise 2) to teach

in-for-mal [ĭn-fôr′məl] *adjective*, casual, not dressy or formal

in-form-ant [ĭn-fôr′mənt] *noun*, someone who gives information

in-for-ma-tion [ĭn′fər-mā′shən] *noun*, facts, data, the things you need to know, knowledge

in-frac-tion [ĭn-frăk′shən] *noun*, a violation, a breach of trust

in-fringe-ment [ĭn-frĭnj′mənt] *adjective*, an act that goes further than what is right or fair

in-fu-ri-ate [ĭn-fyŏŏr′ē-āt′] *verb*, to make someone angry, furious

in-fuse [ĭn-fyooz′] *verb*, 1) to instill a principle or quality 2) to steep in water or other fluid without boiling for the purpose of extracting useful qualities

in-fu-sion [ĭn-fyōo′zhən] *noun*, the act of introducing, spreading or dispersion of properties

in-ge-nu-i-ty [ĭn′jə-nōo′ĭ-tē] *noun*, skill in discovering, inventing or planning

in-gest [ĭn-jĕst′] *verb*, to take in food or liquids gradually

in-grat-i-tude [ĭn-grăt′ĭ-tood′] *adjective*, unthankfulness

in-gre-di-ent [ĭn-grē′dē-ənt] *noun*, 1) a component part of any mixture 2) things that are mixed together to make something

in-hab-it [ĭn-hăb′ĭt] *verb*, to live someplace, to make a place a home, to live or dwell in

in-hab-it-ant [ĭn-hăb′ĭ-tənt] *noun*, an animal or person who lives at a location, a resident

in-hale [ĭn-hāl′] *verb*, to take a deep breath, to breathe in

in-her-ent [ĭn-hîr′ənt] *adjective*, part of the nature of a person or thing

in-her-it [ĭn-hĕr′ĭt] *verb*, to receive something, usually from a relative by will or succession

in-her-it-ance [ĭn-hĕr′ĭ-təns] *noun*, the estate, heritage, legacy

in-hib-it [ĭn-hĭb′ĭt] *verb*; 1) to intimidate, to make someone hold back in fear 2) to restrain

in-hi-bi-tion [ĭn′hə-bĭsh′ən] *noun*, feeling painfully self-conscious

in-iq-ui-ty [ĭ-nĭk′wĭ-tē] *noun*, a gross injustice, wickedness

in-i-tial [ĭ-nĭsh′əl] *adjective*, 1) first, *noun*, 2) the first letter of a name, *verb*, 3) to sign with one's initials

in-i-tial-ly [ĭ-nĭsh′ə-lē] *adverb*, at the beginning, starting with

in-i-ti-ate [ĭ-nĭsh′ē-āt′] *verb*, to introduce, to begin or start something, to originate

in-i-ti-a-tion [ĭ-nĭsh′ē-ā′shən] *noun*, an induction, an inaugural

in-i-t-ia-tive [ĭ-nĭsh′ə-tĭv] *noun*, 1) an act that begins 2) to do something because it needs to be done, an introductory step

in-ject [ĭn-jĕkt′] *verb*, 1) to introduce, to insert 2) to give someone a shot 3) to force something into

in-jec-tion [ĭn-jĕk′shən] *noun*, an insertion of medicine through a needle into a vein, a shot

in-jure [ĭn′jər] *verb*, to cause harm of any kind, to wound

in-ju-ri-ous [ĭn-jŏŏr′ē-əs] *adjective*, hurtful, harmful

in-ju-ry [ĭn′jə-rē] *noun*, 1) a wound, harm to the body 2) an offense, harm done

in-jus-tice [ĭn-jŭs′tĭs] *noun*, violation of the rights of another, an unfair or unjust act

ink [ĭngk] *noun*, a colored liquid used with a pen for writing

in-land [ĭn′lənd] *adjective*, away from the shore or water

in-law [ĭn′lô′] *noun*, used after a word to mean a person related through marriage

in-let [ĭn′lĕt′] *noun*, a channel of water extending into the land from the river

in-mate [ĭn′māt′] *noun*, a prisoner or resident of a mental hospital

inn [ĭn] *noun*, a small hotel that serves meals, a tavern

in-nate [ĭ-nāt′] *adjective*, instinct, inherited, natural behavior

in-ner [ĭn′ər] *adjective*, 1) further in, or in the middle, 2) spiritual

in-ning [ĭn′ĭng] *noun*, the period in a baseball game for both teams to get three outs

in-no-cence [ĭn′ə-səns] *noun*, *no plural*, 1) freedom from guilt, 2) simplicity, naive

in-no-cent [ĭn′ə-sənt] *adjective*, 1) not guilty of a crime, blameless 2) pure, harmless

in-no-va-tion [ĭn′ə-vā′shən] *noun*, making changes, modernization

in-no-va-tive [ĭn′ə-vā′tĭv] *adjective*, the introduction of something new, change

in-nu-en-do [ĭn′yoō-ĕn′dō] *noun*, a suggestion, a hint, an insinuation

in-oc-u-late [ĭ-nŏk′yə-lāt′] *verb*, to inject an animal or person with a virus to prevent an infection

in-op-por-tune [ĭn-ŏp′ər-toōn′] *adjective*, poorly timed

in-or-di-nate [ĭn-ôr′dn-ĭt] *adjective*, excessive, extravagant

in-or-gan-ic [ĭn′ôr-găn′ĭk] *adjective*, matter that does not contain carbon atoms in its main structure

in-put [ĭn′poōt′] *noun*, anything put in, feedback, data entry

in-quest [ĭn′kwĕst′] *noun*, judicial inquiry or examination

in-quire [ĭn-kwīr′] *verb*, to ask about, to investigate

in-quir-y [ĭn-kwīr′ē, ĭng′-] *noun*, investigation, study, research

in-qui-si-tion [ĭn′kwĭ-zĭsh′ən, ĭng′] *noun*, formal questioning

in-quis-i-tive [ĭn-kwĭz′ĭ-tĭv] *adjective*, curious, apt to ask questions, given to research

in-sane [ĭn-sān′] *adjective*, crazy, psychotic, mad, deranged

in-sa-tia-ble [ĭn-sā′shə-bəl] *adjective*, incapable of being satisfied or appeased

in-scrip-tion [ĭn-skrĭp′shən] *noun*, an engraved message

in-scru-ta-ble [ĭn-skroō′tə-bəl] *adjective*, unexplainable, mysterious

in-sect [ĭn′sĕkt′] *noun*, a bug with six legs and two pairs of wings

in-sec-ti-cide [ĭn-sĕk′tĭ-sīd′] *noun*, a poison used to kill bugs

in-sen-si-ble [ĭn-sĕn′sə-bəl] *adjective*, not aware, unconscious, unable to feel

in-sep-a-ra-ble [ĭn-sĕp′ər-ə-bəl] *adjective*, incapable of being separated or disjoined

in-sert [ĭn-sûrt′] noun, 1) an extra piece sewn or put n place, *verb*, 2) to put something into something, to inject, to set in place

in-side [ĭn-sīd′] adjective, 1) internal, interior, *noun*, 2) the portion that is in the middle of something, *preposition*, 3) something contained in something else, lying within

in-sid-i-ous [ĭn-sĭd′ē-əs] *adjective*, evil, sly, deceitful, beguiling

in-sight [ĭn′sīt′] *noun*, understanding, perceptiveness

in-sig-ni-a [ĭn-sĭg′nē-ə] *noun*, an emblem, a symbol, a badge

in-sig-nif-i-cance [ĭn′sĭg-nĭf′ĭ-kəns] *noun*, lack of importance, irrelevance

in-sig-nif-i-cant [ĭn′sĭg-nĭf′ĭ-kənt] *adjective*, unimportant, trivial

in-sin-cere [ĭn′sĭn-sîr′] *adjective*, untrue, false, disingenuous

in-sin-u-ate [ĭn-sĭn′yoō-āt′] *verb*, to introduce artfully, to suggest, to imply, to hint slyly

in-sip-id [ĭn-sĭp′ĭd] *adjective*, uninteresting, dull, banal

in-sist [ĭn-sĭst′] *verb*, 1) to say firmly, to demand, to speak with emphasis 2) to assert or demand persistently, to reiterate

in·so·lence [ĭn′sə-ləns] *noun*, arrogance, rudeness, impudence

in·so·lent [ĭn′sə-lənt] *adjective*, rude, insulting, disrespectful

in·sol·u·ble [ĭn-sŏl′yə-bəl] *adjective*, incapable of being dissolved

in·som·ni·a [ĭn-sŏm′nē-ə] *noun*, the inability to sleep

in·spect [ĭn-spĕkt′] *verb*, to examine carefully and critically, to scrutinize, to look over

in·spec·tion [ĭn-spĕk′shən] *noun*, a careful examination

in·spec·tor [ĭn-spĕk′tər] *noun*, someone who investigates

in·spi·ra·tion [ĭn′spə-rā′shən] *noun*, something that stimulates imagination and goodwill

in·spire [ĭn-spīr′] *verb*, to stimulate to think and to act, to influence

in·sta·bil·i·ty [ĭn′stə-bĭl′ĭ-tē] *noun*, lack of stability, unsteadiness

in·stall [ĭn-stôl′] *verb*, 1) to place in possession of an office, rank, or order 2) to put in place

in·stal·la·tion [ĭn′stə-lā′shən] *noun*, hooking up, placement

in·stall·ment [ĭn-stôl′mənt] *noun*, one of a number of parts, an affordable payment

in·stance [ĭn′stəns] *noun*, an example, a case in point

in·stant [ĭn′stənt] *adjective*, 1) happening or done at once, *noun*, 2) a moment, a very short period of time

in·stan·ta·ne·ous [ĭn′stən-tā′nē-əs] *adjective*, done in an instant

in·stant·ly [ĭn′stənt-lē] *adverb*, at once, immediately

in·stead [ĭn-stĕd′] *adverb*, in place of someone or something, a substitute, replacement

in·step [ĭn′stĕp′] *noun*, the arched middle portion of the human foot or of a shoe

in·sti·gate [ĭn′stĭ-gāt′] *verb*, to urge, to start, to provoke

in·still [ĭn-stĭl′] *verb*, 1) to teach, to educate, to inform 2) to saturate, to infuse, to permeate

in·stinct [ĭn′stĭngkt′] *noun*, a natural impulse that makes animals know how to take care of themselves, innate aptitude

in·stinc·tive·ly [ĭn-stĭngk′tĭv-lē] *adverb*, inborn, natural habits

in·sti·tute [ĭn′stĭ-tōot′] *noun*, 1) an organization devoted to research and study, *verb*, 2) to form or establish an organization

in·sti·tu·tion [ĭn′stĭ-tōo′shən] *noun*, 1) a building used to serve a group of people 2) a commitment to a way of life and society's rules or customs

in·struct [ĭn-strŭkt′] *verb*, to teach, to show someone how to do something, to direct

in·struc·tion [ĭn-strŭk′shən] *noun*, teaching, direction, a training manual, knowledge imparted

in·struc·tor [ĭn-strŭk′tər] *noun*, a teacher or mentor

in·stru·ment [ĭn′strə-mənt] *noun*, 1) a device, a mechanism, a tool 2) a device, such as a trumpet, piano, or drum, used to make musical sounds 3) a means

in-stru-men-tal [ĭn′strə-měn′tl] *adjective*, helpful, acting to get something done

in-suf-fer-a-ble [ĭn-sŭf′ər-ə-bəl] *adjective*, painful, not to be endured, agonizing, intolerable

in-suf-fi-cient [ĭn′sə-fĭsh′ənt] *adjective*, not enough

in-su-late [ĭn′sə-lāt′] *verb*, to cover, to wrap, to protect

in-su-la-tion [ĭn′sə-lā′shən] *noun*, a material that prevents passage of heat, electricity, or sound

in-sult [ĭn-sŭlt′] *noun*, 1) an unkind or rude remark, *verb*, 2) to treat with abuse

in-sur-ance [ĭn-shŏŏr′əns] *noun*, *no plural*, money paid to an agency that keeps it as an investment to be returned to pay the cost of an accident, property damage, death, etc.

in-sure [ĭn-shŏŏr′] *verb*, to guarantee, to protect

in-sur-gent [ĭn-sûr′jənt] *noun*, a person who rises up against the law or leadership

in-sur-rec-tion [ĭn′sə-rĕk′shən] *noun*, an uprising, a rebellion

in-tan-gi-ble [ĭn-tăn′jə-bəl] *adjective*, something that cannot be touched, like an idea

in-te-ger [ĭn′tĭ-jər] *noun*, a whole number, an intact unit

in-te-grate [ĭn′tĭ-grāt′] *verb*, to make into a whole, to combine into one unit, to merge

in-teg-ri-ty [ĭn-tĕg′rĭ-tē] *noun*, honesty and virtue

in-tel-lect [ĭn′tl-ĕkt′] *noun*, the power to understand, the mind

in-tel-lec-tu-al [ĭn′tl-ĕk′chōō-əl] *adjective*, smart, intelligent, brainy, able to reason

in-tel-li-gence [ĭn-tĕl′ə-jəns] *noun*, *no plural*, 1) the faculty of understanding 2) information communicated, reasoning

in-tel-li-gent [ĭn-tĕl′ə-jənt] *adjective*, very smart, clever

in-tel-li-gent-ly [ĭn-tĕl′ə-jənt-lē] *adverb*, sensibly, skillfully

in-tel-li-gi-ble [ĭn-tĕl′ĭ-jə-bəl] *adjective*, capable of being understood

in-tend [ĭn-tĕnd′] *verb*, 1) to plan to do something 2) to signify

in-tense [ĭn-tĕns′] *adjective*, a strong feeling, earnest

in-ten-si-fy [ĭn-tĕn′sə-fī′] *verb*, to make greater or more intense

in-ten-si-ty [ĭn-tĕn′sĭ-tē] *noun*, great concentration of energy

in-ten-sive [ĭn-tĕn′sĭv] *adjective*, thorough, involving special effort

in-tent [ĭn-tĕnt′] *noun*, 1) purpose 2) meaning

in-ten-tion [ĭn-tĕn′shən] *noun*, aim, a plan, a purpose, a meaning

in-ten-tion-al-ly [ĭn-tĕn′shə-nəl-lē] *adverb*, to do something on purpose, deliberately

in-tent-ly [ĭn-tĕnt′lē] *adverb*, strongly, purposefully

in-ter [ĭn-tûr′] *verb*, to bury

in-ter-act [ĭn′tər-ăkt′] *verb*, to relate mutually to, to exchange

in-ter-cede [ĭn′tər-sēd′] *verb*, to mediate, to act on one's behalf

in-ter-cept [ĭn′tər-sĕpt′] *verb*, to stop or seize something before it reaches its intended destination

in-ter-est [ĭn′trĭst] *noun*, 1) a
concern or pleasure from
learning something 2) title or
share in a thing 3) a rate percent
of money paid for the use of
money, *verb*, 4) to involve or
gain the attention of someone

in-ter-est-ing [ĭn′trĭ-stĭng]
adjective, able to hold the
attention or curiosity of someone

in-ter-face [ĭn′tər-fās′] *noun*, 1) a
surface forming a common
boundary between adjacent areas
2) in computer science the
software or hardware connecting
one drive or system to another

in-ter-fere [ĭn′tər-fîr′] *verb*, to be
in opposition, to meddle

in-ter-fer-ence [ĭn′tər-fîr′əns]
noun, obstruction, interrupting

in-ter-im [ĭn′tər-ĭm] *noun*, 1) in the
meantime 2) temporary

in-te-ri-or [ĭn-tîr′ē-ər] *adjective*,
1)having to do with the inside
part of something, inside *noun*,
2) the inside part of something

in-ter-lude [ĭn′tər-lōōd′] *noun*, a
pause, a lull, a recess

in-ter-me-di-ate [ĭn′tər-mē′dē-ĭt]
adjective, in the middle

in-ter-ment [ĭn-tûr′mənt] *noun*, the
burial ceremony

in-ter-mi-na-ble [ĭn-tûr′mə-nə-bəl]
adjective, endless, prolonged

in-ter-mis-sion [ĭn′tər-mĭsh′ən]
noun, a break or recess

in-ter-mit-tent [ĭn′tər-mĭt′nt]
adjective, coming and going at
intervals, beginning again

in-tern [ĭn′tûrn′] *noun*, 1) a
doctor's apprentice, *verb*, 2) to
imprison, to confine

in-ter-nal [ĭn-tûr′nəl] *adjective*,
1) inside of something, inner
2) referring to any function
inside the body, in word

in-ter-na-tion-al [ĭn′tər-năsh′ə-nəl]
adjective, in nations all over the
world, universal, global

in-ter-ne-cine [ĭn′tər-nĕs′ēn′]
adjective, mutually destructive

In-ter-net [ĭn′tûr-nĕt] *noun*, the
link to other computers,
connected by a server

in-ter-pret [ĭn-tûr′prĭt] *verb*, to
translate, to define, to explain

in-ter-pre-ta-tion
[ĭn-tûr′prĭ-tā′shən] *noun*, an
explanation of someone's art
work, a translation

in-ter-pret-er [ĭn-tûr′prĭ-tər] *noun*,
1) one who explains or expounds
2) a translator

in-ter-ro-gate [ĭn-tĕr′ə-gāt′] *verb*,
to ask questions of someone

in-ter-rupt [ĭn′tə-rŭpt′] *verb*, to
stop something from continuing,
to disturb, to hinder

in-ter-sect [ĭn′tər-sĕkt′] *verb*, when
two lines meet to form angles

in-ter-sec-tion [ĭn′tər-sĕk′shən]
noun, 1) a figure formed by two
lines 2) a place where two or
more roads meet

in-ter-state [ĭn′tər-stāt′] *adjective*,
pertaining to commerce between
the states

in-ter-val [ĭn′tər-vəl] *noun*, the
space of time between any two
points or events, a break

in-ter-vene [ĭn′tər-vēn′] *verb*, 1) to
come or be between 2) to occur
between points of time or events

in-ter-view [ĭn′tər-vyōō′] *noun*, a
meeting in which people
exchange information

in-ti-ma-cy [ĭn′tə-mə-sē] *noun*,
close familiarity or association,
very personal

in-ti-mate [ĭn′tə-mĭt] *adjective*,
familiar, close, dear, private

in-tim-i-date [ĭn-tĭm′ĭ-dāt′] *verb*,
to make someone afraid, to
threaten someone, to scare

in-to [ĭn′tōō] *preposition*, to go
toward the middle of something

in-tol-er-a-ble [ĭn-tŏl′ər-ə-bəl]
adjective, unbearable,
unendurable, insufferable

in-tol-er-ance [ĭn-tŏl′ər-əns] *noun*,
prejudice, bias, bigoted

in-to-na-tion [ĭn′tə-nā′shən] *noun*,
a tone of voice, speech pattern,
inflection, melody of pitch

in-tox-i-cate [ĭn-tŏk′sĭ-kāt′] *verb*,
1) to enliven, to excite, to elate
2) to feel drugged or drunk

in-trac-ta-ble [ĭn-trăk′tə-bəl]
adjective, unruly, refractory

in-tran-si-tive [ĭn-trăn′sĭ-tĭv]
adjective, a verb in a sentence
that does not need a direct object
to complete its meaning

in-trep-id [ĭn-trĕp′ĭd] *adjective*,
fearless, dauntless

in-tri-cate [ĭn′trĭ-kĭt] *adjective*,
complicated, involved, complex

in-tri-cate-ly [ĭn′trĭ-kĭt-lē] *adverb*,
in a complex or involved manner

in-trigue [ĭn′trēg′] *noun*, 1) a
conspiracy, a strategy, a plan,
verb, 2) to be drawn into, to
fascinate, to arouse curiosity

in-trin-sic [ĭn-trĭn′zĭk] *adjective*,
basic and essential to what
something is

in-tro-duce [ĭn′trə-dōōs′] *verb*,
1) to bring into notice 2) to make
known by formal announcement
3) to bring into use, to lead in

in-tro-duc-tion [ĭn′trə-dŭk′shən]
noun, 1) a piece of writing at the
beginning of a book that
explains something about the
story 2)) the act of formally
making persons known to each
other

in-tro-vert [ĭn′trə-vûrt′] *noun*, a
loner, someone who feels self-
conscious, a shy person

in-trude [ĭn-trōōd′] *verb*, to
trespass, to enter uninvited

in-tru-sion [ĭn-trōō′zhən] *noun*, the
act of forcing in without right or
welcome or permission

in-tu-i-tion [ĭn′tōō-ĭsh′ən] *noun*,
instinctive knowledge or feeling

in-un-date [ĭn′ŭn-dāt′] *verb*, to
overflow, to flood, to overwhelm

in-ure [ĭn-yōōr′] *verb*, to accustom,
to harden by use or exposure

in-vade [ĭn-vād′] *verb*, to enter by
force in order to take over

in-va-lid [ĭn-văl′ĭd] *adjective*,
1) null and void, insupportable,
[ĭn-val′ĭ-d] *noun*, 2) a person
who is feeble, weak, frail, sickly

in-val-i-date [ĭn-văl′ĭ-dāt′] *verb*, to
deprive of legal force

in-var-i-ab-ly [ĭn-vâr′ē-ə-blē]
adverb, without fail, predictably

in-va-sion [ĭn-vā′zhən] *noun*, an
attack, a raid

in-vent [ĭn-vĕnt'] *verb*, to create or design a new product

in-ven-tion [ĭn-vĕn'shən] *noun*, a design or creation of an idea

in-ven-tor [ĭn-vĕn'tər] *noun*, one who finds out or devises a new product or device

in-ven-to-ry [ĭn'vən-tôr'ē] *noun*, an itemized list of stock on hand with their estimated worth

in-verse [ĭn-vûrs'] *adjective*, the opposite, reverse in direction

in-vert [ĭn-vûrt'] *verb*, to overturn

in-ver-te-brate [ĭn-vûr'tə-brĭt] *noun*, animals without a backbone or spinal column, e.g. *a worm*

in-vest [ĭn-vĕst'] *verb*, to empower, to qualify, to devote time

in-ves-ti-gate [ĭn-vĕs'tĭ-gāt'] *verb*, to examine into, to inquire

in-ves-ti-ga-tion [ĭn-vĕs'tĭ-gā'shən] *noun*, an inquiry into a matter, a thorough exploration or inspection, an examination

in-vest-ment [ĭn-vĕst'mənt] *noun*, money used to create a source of income or profit

in-ves-tor [ĭn-vĕs'tər] *noun*, one who makes an investment or spends money on something with a view of obtaining income or profit

in-vet-er-ate [ĭn-vĕt'ər-ĭt] *adjective*, deep-rooted, habitual

in-vig-o-rate [ĭn-vĭg'ə-rāt'] *verb*, to fill with life and energy

in-vin-ci-ble [ĭn-vĭn'sə-bəl] *adjective*, irresistible, unyielding

in-vis-i-ble [ĭn-vĭz'ə-bəl] *adjective*, something that cannot be seen

in-vi-ta-tion [ĭn'vĭ-tā'shən] *noun*, a request asking for someone to attend a gathering

in-vite [ĭn-vīt'] *verb*, to ask someone to do something or come somewhere

in-voke [ĭn-vōk'] *verb*, to call upon, to ask for, to cause

in-vol-un-tar-y [ĭn-vŏl'ən-tĕr'ē] *adjective*, without will or choice, done unwillingly

in-volve [ĭn-vŏlv'] *verb*, to take part in, to make a part of

in-vul-ner-a-ble [ĭn-vŭl'nər-ə-bəl] *adjective*, incapable of injury

in-ward [ĭn'wərd] *adjective*, internal, inner, center

in-ward-ly [ĭn'wərd-lē] *adverb*, 1) toward the middle, halfway 2) internal, privately

i-o-dine [ī'ə-dīn'] *noun*, a nonmetallic element isolated as a crystalline solid

i-o-ta [ī-ō'tə] *noun*, the very smallest quantity or degree, a particle

Io-wa [ī'ə-wə] *noun*, one of the 50 United States located in the Midwest, the capital city is Des Moines. The state flower of **Iowa** is the wild rose and the state motto is "Our liberties we prize and our rights we will maintain."

i-rate [ī-rāt'] *adjective*, angry

ir-i-des-cent [ĭr'ĭ-dĕs'ənt] *adjective*, the rainbow-like play of colors, shimmering

irk-some [ûrk'səm] *adjective*, repetitious, tedious, annoying

i-ron [ī'ərn] *noun*, 1) a gray metal 2) a heated device used to smooth the wrinkles from clothes 3) an element, *verb*, 4) to press with a heated iron to remove wrinkles on clothes

i-ron-ic [ī-rŏn′ĭk] *adjective*, what appears to be false or contradictory, but in fact is true

i-ro-ny [ī′rə-nē, ī′ər-] *noun*, the outcome of events contrary to what was expected

ir-rad-i-ate [ĭ-rā′dē-āt′] *verb*, 1) to illuminate, to enlighten 2) to expose to ultraviolet rays

ir-ra-tion-al [ĭ-răsh′ə-nəl] *adjective*, absurd, unreasonable,

ir-rec-on-cil-a-ble [ĭ-rĕk′ən-sī′lə-bəl] *adjective*, not able to be resolved by compromise, unfriendly

ir-reg-u-lar [ĭ-rĕg′yə-lər] *adjective*,1) not conforming to rule 2) erratic, not uniform

ir-rel-e-vant [ĭ-rĕl′ə-vənt] *adjective*, not applicable or pertinent, not relevant

ir-rep-a-ra-ble [ĭ-rĕp′ər-ə-bəl] *adjective*, permanently damaged

ir-re-proach-a-ble [ĭr′ĭ-prō′chə-bəl] *adjective*, blameless, above reproach

ir-re-sist-i-ble [ĭr′ĭ-zĭs′tə-bəl] *adjective*, tempting

ir-re-triev-a-ble [ĭr′ĭ-trē′və-bəl] *adjective*, lost, gone forever

ir-rev-er-ent [ĭ-rĕv′ər-ənt] *adjective*, lacking proper respect

ir-rev-o-ca-ble [ĭ-rĕv′ə-kə-bəl] *adjective*, fixed, constant

ir-ri-gate [ĭr′ĭ-gāt′] *verb*, to supply with water by causing a stream to flow through, as in channels

ir-ri-tate [ĭr′ĭ-tāt′] *verb*, 1) to annoy, to excite to anger, to bother 2) to make skin sore, red or swollen

ir-ri-ta-tion [ĭr′ĭ-tā′shən] *noun*, 1) something that makes a person mildly angry 2) an allergic reaction, an irritant

is [ĭz] the part of the *verb* to **be** that we use with *he*, *she* and *it*

Is-lam [ĭs-läm′] *noun*, Muslim religion founded in Arabia in the 7th century based on the teaching of Muhammed

is-land [ī′lənd] *noun*, a body of land completely surrounded by water

isle [īl] *noun*, a small island

is-n't [ĭz′ənt] the *contraction* of the words **is** and **not**

i-so-late [ī′sə-lāt′] *verb*, 1)to separate and set apart, to place alone 2) to cut off from others

iso-la-tion [ī′sə-lā′shən] *noun*, solitude, detachment

i-sos-ce-les tri-ang-le [ī-sŏs′ə-lēz′ trī′ăng′gəl] *noun*, a triangle with two equal sides

is-sue [ĭsh′o͞o] *noun*, 1) a subject, a topic 2) the condition or action of being made available, *verb*, 3) to give, to send, or come out

isth-mus [ĭs′məs] *noun*, a neck of land joining two larger portions of land bordered by water

it [ĭt] *pronoun, plural*, **they**, 1) the subject of the sentence 2) used about the weather, time, and dates, and in other phrases

i-tal-ic [ĭ-tăl′ĭk] *noun*, type in which the letters slant

itch [ĭch] *noun*, 1) a tickling feeling in the skin that makes one want to scratch, *verb*, 2) to scratch and rub a sore because it is irritating

i-tem [ī′təm] *noun*, a point, a news article, a single thing in a list

i-tem-ize [ī′tə-mīz′] *verb*, to list or state by particulars

itin-er-ant [ī-tĭn′ər-ənt] *adjective*, traveling from place to place

itin-er-ar-y [ī-tĭn′ə-rĕr′ē] *noun*, the detailed plan for a journey

its [ĭts] of **it**, belonging to **it**, the possessive of the word **it**

it's [ĭts] the *contraction* of the words: **it** and **is**

it-self [ĭt-sĕlf′] *plural*, **themselves**, referring to the subject in the sentence

I've [īv] the *contraction* of the words **I** and **have**

i-vo-ry [ī′və-rē, īv′rē] *noun*, hard, white substance taken from the tusks of elephants

i-vy [ī′vē] *noun*, a leafy vine

J

jack-al [jăk′əl] *noun*, a wild animal like a dog that eats meat

jack-et [jăk′ĭt] *noun*, 1) a covering worn over clothing, similar to a coat, cut below the waist 2) a covering put over something

jade [jād] *noun*, a stone of a green color, but sometimes whitish, capable of a fine polish

jad-ed [jā′dĭd] *adjective*, tired, weary from overwork

jag-uar [jăg′wär′] *noun*, a wild cat with a spotted coat that lives in South America

jail [jāl] *noun*, a prison

jam [jăm] *noun*, *no plural*, 1) a sweet spread, made of boiled fruit and sugar 2) when traffic is stopped because no one can move 3) a predicament, a spot, *verb*, 4) to pack tightly together, to compress, to press or push

jan-i-tor [jăn′ĭ-tər] *noun*, one who takes care of the repairs and cleaning of a public building

Jan-u-ary [jăn′yo͞o-ĕr′ē] *noun*, the first month of the year, having 31 days

jar [jär] *noun*, a container like a bottle with a wide opening

jar-gon [jär′gən] *noun*, language used by a particular group, special terminology

jaun-dice [jôn′dĭs] *noun*, a blood condition that makes the skin look yellow

jaunt [jônt] *noun*, a short trip

jave-lin [jăv′lĭn] *noun*, a spear thrown by hand for distance as in an athletic event

jaw [jô] *noun*, the bony part of the face in which the teeth are set

jazz [jăz] *noun*, a style of music using unusual tone effects on brass instruments improvised by southern blacks in the 1900's

jeal-ous [jĕl′əs] *adjective*, to be unhappy because you want what someone else has, envious

jeal-ous-y [jĕl′ə-sē] *noun*, suspicious fear or watchfulness, resentment of someone's success

jeans [jēnz] *noun*, *plural*, pants made of denim fabric

jeer [jîr] *verb*, to make fun of someone and laugh in their face, to mock, to scorn

jel-ly [jĕl′ē] *noun, no plural,* 1) any substance between liquid and solid 2) a sweet spread made of fruit and pectin

jel-ly-fish [jĕl′ē-fĭsh′] *noun,* a soft gelatinous sea creature that looks like a squishy blob

jeop-ard-ize [jĕp′ər-dīz′] *verb,* to expose to danger or even death

jerk [jûrk] *noun,* 1) a quick pull or twist movement, *verb,* 2) to pull, to push or to twist suddenly

jer-sey [jûr′zē] *noun,* clothing usually made of cotton with elastic that covers the top of the body like a shirt

jest [jĕst] *verb,* to joke, to jeer

jet [jĕt] *noun,* 1) an aircraft propelled through air by an engine, which releases a stream of hot air behind itself as it travels 2) a stream of gas, air, or liquid that spurts from a hole

jet-ti-son [jĕt′ĭ-sən] *verb,* to throw overboard, to discard

jet-ty [jĕt′ē] *noun,* a kind of wall built out into water for protection against the waves

Jew [jōō] *noun,* a member of the Jewish religion

jew-el [jōō′əl] *noun,* a precious stone used as an ornament, a gem of great value

jew-el-er [jōō′ə-lər] *noun,* one who makes, or deals in jewels, gems and precious stones, etc.

jew-el-ry [jōō′əl-rē] *noun, no plural,* ornaments like rings, earrings, necklaces made with precious or semiprecious stones, gems and metals

jilt [jĭlt] *verb,* to break up a romantic interest

jin-gle [jĭng′gəl] *verb,* 1) to make a tingling sound like the sound of bells, *noun,* 2) a catchy verse

job [jŏb] *noun,* 1) a task or responsibility 2) work done for payment 3) a position or duty

jog [jŏg] *verb,* 1) to run for exercise 2) to jolt, to bump, to shake 3) to stir into activity

join [join] *verb,* 1) to connect, to bring together, to meet 2) to become a member

joint [joint] *adjective,* 1) an endeavor involving two or more people, *noun,* 2) the part of the body where bones connect 3) the place where two things are joined or connected 4) a poor quality restaurant or night club

joke [jōk] *noun,* 1) a funny story or funny phrase spoken to amuse people, *verb,* 2) to tell funny stories or say things to make people laugh, to fool around

jolt [jōlt] *noun,* 1) a sudden and unexpected movement, *verb,* 2) to cause a sudden movement

joule [jōōl, joul] *noun,* the measurement of force or energy

jour-nal [jûr′nəl] *noun,* a book that records daily thoughts or transactions, a diary

jour-nal-ism [jûr′nə-lĭz′əm] *noun, no plural,* the job of writing for a publication or broadcasting news

jour-nal-ist [jûr′nə-lĭst] *noun,* a professional writer for a newspaper or magazine

jour-ney [jûr′nē] *noun*, the passage from one place to another, a trip

jo-vi-al [jō′vē-əl] *adjective*, joking in a friendly manner

joy [joi] *noun*, *no plural*, great happiness, delight, elation

joy-ful [joi′fəl] *adjective*, full of joy, very glad, happy

ju-bi-lant [jōō′bə-lənt] *adjective*, thrilled, rejoicing, great joy

Ju-da-ism [jōō′dē-ĭz′əm] *noun*, the religion of people descended from the Israelites in the Bible

judge [jŭj] *noun*, 1) one with legal authority who applies the rules of law in a court 2) the person designated to decide the winner in a contest, *verb*, 3) to make a decision, to form an opinion

judg-ment [jŭj′mənt] *noun*, 1) a decision or opinion 2) intelligence, understanding

ju-di-cial [jōō-dĭsh′əl] *adjective*, of the administration of justice by judges, laws and the courts

ju-di-cious [jōō-dĭsh′əs] *adjective*, sound in judgment, wise

ju-do [jōō′dō] *noun*, *no plural*, a form of martial arts that uses the body as a weapon, and bans throws and blows

jug [jŭg] *noun*, a container with a handle for holding liquids

jug-ger-naut [jŭg′ər-nôt′] *noun*, an overpowering or crushing force

jug-gle [jŭg′əl] *verb*, to throw several things into the air and keep them moving by throwing and catching them as a trick

jug-u-lar [jŭg′yə-lər] *noun*, one of the large veins returning the blood from the head

juice [jōōs] *noun*, the natural fluid that comes from fruit, vegetables or cooked meat

Ju-ly [jōō-lī′] *noun*, the seventh month of the year, having 31 days

jump [jŭmp] *noun*, 1) a spring or leap, *verb*, 2) to lift the feet off the ground into the air, to leap 3) to rise suddenly

jump-er [jŭm′pər] *noun*, a sleeveless dress worn over a blouse

junc-tion [jŭngk′shən] *noun*, a place where roads or lines of railways meet and cross

June [jōōn] *noun*, the sixth month of the year, having 30 days

jun-gle [jŭng′gəl] *noun*, land thickly overgrown with vines, bushes, trees, etc.

jun-ior [jōōn′yər] *adjective*, 1) a student in his third year of a four-year course 2) the son named after his father, abbreviated **Jr**. after the name

jun-i-per [jōōn′pər] *noun*, an evergreen bush or tree related to cypresses, with tiny bluish cones that look like berries

junk [jŭngk] *noun*, stuff regarded as unwanted or useless

jun-ket [jŭng′kĭt] *noun*, a trip, a short pleasure excursion

ju-ris-dic-tion [jŏŏr′ĭs-dĭk′shən] *noun*, the legal authority of a court, to apply the law

ju-ror [jŏŏr′ər, -ôr′] *noun*, one of a group of peers chosen to judge someone accused of a crime

ju-ry [jŏŏr′ē] *noun*, a group of people chosen to decide if a person is guilty or innocent of a crime in a court of law

just [jŭst] *adjective*, 1) fair and right 2) normal, *adverb*, 3) to the amount needed, but no more 4) a very short time ago, by a short time 5) only 6) used to compare two things

jus-tice [jŭs′tĭs] *noun*, *no plural*, fairness, equity, impartiality

jus-ti-fy [jŭs′tə-fī′] *verb*, 1) to show or prove to be right 2) to make even or true, as lines of type by proper spacing

ju-ve-nile [jōō′və-nīl′] *adjective*, 1) young, youthful, *noun*, 2) a youth, anyone under the legal adult age of 18

K

ka-lei-do-scope [kə-lī′də-skōp′] *noun*, an optical instrument exhibiting an endless variety of colored forms

kan-ga-roo [kăng′gə-rōō′] *noun*, a marsupial, an animal that eats only plants, lives in Australia, jumps around on its large back legs

Kan-sas [kăn′zəz] *noun*, one of the 50 United States, located in the Midwest, the capital city is Topeka. The state flower of **Kansas** is the sunflower, and the motto is "To the stars through difficulties."

kar-at [kăr′ət] *noun*, the unit measure of the purity of gold

kay-ak [kī′ăk′] *noun*, a canoe that holds one person

keel [kēl] *noun*, the bottom of the boat, the underside

keen [kēn] *adjective*, 1) having a sharp edge 2) mentally aware, quick to understand 3) showing a strong interest, eager

keep [kēp] *verb*, 1) to hold something 2) to store something, to stock it 3) to make someone stay someplace 4) to prevent someone from doing something 5) to continue on

ken-nel [kĕn′əl] *noun*, a place to keep and care for dogs

Ken-tucky [kən-tŭk′ē] *noun*, one of the 50 United States, located in the South, the capital is Frankfort. The state flower of **Kentucky** is the goldenrod, and the state motto is "United We Stand, Divided We Fall."

kept [kĕpt] *past tense* of **keep**

ker-chief [kûr′chĭf] *noun*, a woman's scarf worn as a covering for the head

ker-nel [kûr′nəl] *noun*, a grain or seed, the edible part of a nut

ker-o-sene [kĕr′ə-sēn′] *noun*, thin oil used to ignite a fire

ket-tle [kĕt′l] *noun*, a large metal pot with a lid and a handle and a spout

key [kē] *noun*, 1) a metal instrument used for locking and unlocking things 2) a lever on a musical instrument that affects the sound 3) the piece of information needed to make something understood

key-board [kē′bôrd′] *noun*, a common input device

khak-i [kăk′ē] *noun*, a tan or brownish cotton cloth

kick [kĭk] *noun*, 1) an impulsive move with the foot, *verb*,

2) to hit something with the foot, to move the foot suddenly

kid [kĭd] *noun*, 1) a young goat 2) a child, *verb*, 3) to tease or joke

kid-nap [kĭd′năp′] *verb*, to take someone against their will

kid-nap-per [kĭd′nă-pər] *noun*, someone who takes a person against their will

kid-ney [kĭd′nē] *noun*, one of two parts inside the body which remove waste liquid from the blood and excretes uric acid

kill [kĭl] *verb*, to make someone die

kill-er [kĭl′ər] *noun*, a person who murders or takes the life from another

kil-o-gram [kĭl′ə-grăm′] *noun*, a unit of weight equal to 1,000 grams

kil-o-me-ter [kĭ-lŏm′ĭ-tər] *noun*, a distance equal to 1,000 meters, or 3,281feet

kil-o-watt [kĭl′ə-wŏt′] *noun*, a unit of power equal to 1,000 watts

kin [kĭn] *noun*, *no plural*, family members, one's relatives

kind [kīnd] *adjective*, 1) nice, good, helpful, willing to please, *noun*, 2) a sort or particular one

kin-der-gar-ten [kĭn′dər-gär′tn] *noun*, a preschool for children before the first grade

kin-dle [kĭn′dl] *verb*, 1) to set fire, to ignite 2) to arouse, to inspire

kind-ness [kīnd′nĭs] *noun*, *no plural*, goodness, compassion

kin-dred [kĭn′drĭd] *adjective*, 1) similar in nature, comparable, alike 2) related by family

ki-net-ic [kĭ-nět′ĭk] *adjective*, the changes of motion produced by forces, the force of motion

king [kĭng] *noun*, a male ruler of a country, especially one who comes from a family of rulers

king-dom [kĭng′dəm] *noun*, the lands owned by a monarchy, an empire, an independent realm

kin-ship [kĭn′shĭp′] *noun*, a relationship or connection between family members

kiss [kĭs] *noun*, 1) a slight touch with the lips, *verb*, 2) to touch with the lips as a sign of affection or love

kit [kĭt] *noun*, a set of instruments needed for doing something or going somewhere

kitch-en [kĭch′ən] *noun*, the room where food is prepared and cooked and eaten

kite [kīt] *noun*, a light frame covered with fabric, plastic or paper, flown in the air at the end of a long string

kit-ten [kĭt′n] *noun*, a baby cat

knap-sack [năp′săk′] *noun*, a supply bag of canvas or leather with shoulder straps, worn on the back

knave [nāv] *noun*, an unscrupulous, dishonest person, a rascal

knead [nēd] *verb*, to manipulate by pressing and stretching to mix and remove air, to massage

knee [nē] *noun*, the joint between the thigh and the lower leg connecting the femur and tibia

kneel [nēl] *verb*, to stand on your knees, to bend down so that your calf is laying on the ground

knew [nōō] *past tense* of **know**

knife [nīf] *noun, plural, knives*, a cutting instrument consisting of a sharp blade fastened to a handle

knight [nīt] *noun*, 1) an armed soldier in the Middle Ages 2) a chess piece in the shape of a horse's head 3) an honorary title in Britain

knit [nĭt] *verb*, to join wool into a cloth using long needles

knob [nŏb] *noun*, a handle in the shape of a round lump

knock [nŏk] *noun*, 1) a noise made when something is hit *verb*, 2) to hit or strike with the fist 4) to make a noise by hitting on the surface

knoll [nōl] *noun*, a small round hill

knot [nŏt] *noun*, 1) one or more threads tied together, *verb*, 2) to tie something securely with rope or cord so that it cannot come untied

know [nō] *verb*, 1) to understand clearly, to learn 2) to be skilled in 3) to be acquainted with someone

knowl-edge [nŏl′ĭj] *noun*, something learned and remembered, gained by experience

know-ledge-a-ble [nŏl′ĭ-jə-bəl] *adjective*, familiar with, learned

known [nōn] past participle of **know**

knuck-le [nŭk′əl] *noun*, one of the joints in the finger or toe

Ko-ran [kə-răn′, -rän′, kô-, kō-] *noun*, the holy book of Islam containing the revelations of Muhammed

L

la-bel [lā′bəl] *noun*, 1) a slip of paper indicating contents, ownership, etc. *verb*, 2) to affix a tag or sticker on something naming what it is

la-bor [lā′bər] *noun*, 1) work 2) the pain a woman feels birthing a baby, *verb*, 3) to work hard

lab-o-ra-to-ry [lăb′rə-tôr′ē] *noun*, a place devoted to experiments in natural science

la-bo-ri-ous [lə-bôr′ē-əs] *adjective*, hardworking, tiresome, tedious

lab-y-rinth [lăb′ə-rĭnth′] *noun*, a series of intricate passageways, a maze, an inexplicable difficulty

lace [lās] *noun*, 1) a string used to tie a shoe 2) frilly threadwork used as edging or as an embroidery design, *verb*, 3) to thread through holes with a string or cord

lac-er-a-tion [lăs′ə-rā′shən] *noun*, a cut, a tear made with a knife

lack [lăk] *noun*, 1) not enough, a shortage, *verb* 2) to be without

lack-a-dai-si-cal [lăk′ə-dā′zĭ-kəl] *adjective*, half-hearted, lazy, indifferent, listless

la-con-ic [lə-kŏn′ĭk] *adjective*, brief and to the point

lac-quer [lăk′ər] *noun*, a liquid that hardens into a clear tough film

lac-tic acid [lăk′tĭk-ăs′ĭd] *noun*, a colorless, syrupy liquid, occurring in sour milk

lad [lăd] *noun*, a boy

lad-der [lăd′ər] *noun*, a device used for climbing designed with two side pieces joined together with shorter pieces called rungs that form steps

lad-en [lād′n] *adjective*, heavily loaded, full of cargo

la-dle [lād′l] *noun*, a cup-shaped spoon with a long handle, used for serving or dipping

la-dy [lā′dē] *noun*, 1) a polite and considerate woman 2) in Britain, a title for the wife or daughter of a Lord

lag [lăg] *verb*, to move at a slower pace to stay behind the others

la-goon [lə-gōōn′] *noun*, a shallow body of water near a sea, or pond-like body of water

laid [lād] *past tense* of **lay**

lair [lâr] *noun*, a den, a hole, a hideout for a wild animal

la-i-ty [lā′ĭ-tē] *noun*, someone not connected to the clergy, a layman practicing religion

lake [lāk] *noun*, a body of water with land all around

lamb [lăm] *noun*, a baby sheep

lame [lām] *adjective*, limping or disabled in a hand or joint

la-ment [lə-mĕnt′] *verb*, to mourn, to express grief or regret

lam-i-nat-ed [lăm′ə-nā′tĭd] *adjective*, arranged in layers of plastic one upon the other

lamp [lămp] *noun*, a device used to project artificial light

lam-poon [lăm-pōōn′] *noun*, a cartoon or satire in print

lance [lăns] *noun*, 1) a spear, a harpoon, a javelin, *verb*, 2) to puncture 3) to cut into with a surgical knife

land [lănd] *noun*, *no plural*, 1) the dry part of the earth, not covered by water 2) a country, region, area, *verb*, 3) to arrive on the ground from the air or water

land-fill [lănd′fĭl′] *noun*, a large outdoor site where solid waste is covered by soil

land-lord [lănd′lôrd′] *noun*, a person who owns a building which he lets others use or live in, in return for money

land-mark [lănd′märk′] *noun*, 1) an achievement, a turning point, benchmark 2) a distinctive building or geographical marker like a bend in the river

land-scape [lănd′skāp′] *noun*, a view of an area with its natural surroundings

lane [lān] *noun*, a short road

lan-guage [lăng′gwĭj] *noun*, 1) any means of conveying or communicating ideas 2) human speech 3) dialect, jargon, vocabulary used by a group

lan-guish [lăng′gwĭsh] *verb*, to lose strength, to weaken

lan-guor [lăng′gər] *noun*, a feeling of depression, nervous exhaustion, weakness of mind or body

lank [lăngk] *adjective*, 1)thin, lean 2) hanging straight or limp

lan-tern [lăn′tərn] *noun*, a portable lamp, with a protective case for lights, carried with a handle often used outdoors for light

lap [lăp] *noun*, 1) the seat created by the legs and chest when sitting 2) the distance once around a track 3) a place of rest, *verb*, 4) to drink water with a tongue, like an animal

lapel [lə-pĕl′] *noun*, the flat hanging appendage on a jacket

lapse [lăps] *noun*, 1) an interval of time in which something stops and then starts again 2) termination of policy because of non-payment of premium when due 3) a slip or error

lar-ce-ny [lär′sə-nē] *noun*, taking and carrying away of things, stealing, robbery

lard [lärd] *noun*, grease, fat

large [lärj] *adjective*, greater than average in size, number, etc., immense, big

large-ly [lärj′lē] *adverb*, mostly

lar-ger [lär′jər] *adjective*, bigger

lar-gess [lär-zhĕs′] *noun*, the generous giving of gifts

lar-va [lär′və] *noun*, *plural*, *larvae or larvas*, the second stage of an insect that goes through complete metamorphosis

lar-yn-gi-tis [lăr′ən-jī′tĭs] *noun*, infection of the larynx or upper end of the windpipe

lar-ynx [lăr′ĭngks] *noun*, the voice organ in the upper throat

la-ser [lā′zər] *noun*, a narrow beam of high-energy light that is all exactly the same color

lash [lăsh] *noun*, 1) a whip, a cane, *verb*, 2) to hit hard 3) to secure something in place with a rope

las-so [lăs′ō] *noun*, 1) a rope with a loop at the end used for catching cattle, *verb*, 2) to catch by throwing the noose of a rope around the neck or torso

last [lăst] *adjective*, 1) the end, final 2) an instance when something happens 3) to come after all others 4) to be a part of for a long time, to endure, *adverb*, 5) to stay in good condition, to be preserved 6) most recently

last-ing [lăs′tĭng] *adjective*, the length of time something continues, enduring

latch [lăch] *noun*, 1) a small bolt used to fasten a door, *verb*, 2) to fasten something shut

late [lāt] *adjective*, 1) after the time agreed on 2) near the end of a day, year, etc. 3) a missed opportunity 4) deceased

late-ly [lāt′lē] *adverb*, recently, a short time ago, not long since

la-tent [lāt′nt] *adjective*, hidden, present but not visible

lat-er [lāt′ər] *adverb*, after a certain amount of time, subsequently

lat-er-al [lăt′ər-əl] *adjective*, coming from the side

lat-est [lā′tĭst] *adjective*, most recent, newest, popular

lath-er [lăth′ər] *noun*, *no plural*, the bubbles created by soap

lat-i-tude [lăt′ĭ-tood′] *noun*, *no plural*, 1) vertical lines on a map that indicate north and south 2) freedom to act and to choose

lat-ter [lăt′ər] *adjective*, the second of two things, the last mentioned

laud [lôd] *verb*, to praise, to commend

laugh [lăf] *verb*, to make a sound showing something is funny, to express happiness

laugh-ter [lăf′tər] *noun*, *no plural*, a giggle, mirth, a chuckle

launch [lônch] *noun*, 1) a small boat propelled by an engine, *verb*, 2) to send off, to put a boat into water or send a space ship into orbit 3) to plunge boldly into action

laun-dry [lôn′drē] *noun*, dirty clothes washed together

lau-rel [lôr′əl] *noun*, an evergreen shrub with pink flowers

la-va [lä′və] *noun*, very hot melted rock that bursts out of a volcano

lav-a-to-ry [lăv′ə-tôr′ē] *noun*, a room with a toilet and a sink

lav-en-der [lăv′ən-dər] *noun*, a fragrant plant with purple flowers, classified as an herb

lav-ish [lăv′ĭsh] *adjective*, 1) extravagant, excessive, 2) giving in great amounts, *verb*, 3) to give generously

law [lô] *noun*, a written rule of conduct made and enforced by the government

law-mak-er [lô′mā′kər] *noun*, a person in government who is part of the legislature

lawn [lôn] *noun*, a piece of ground where grass is grown and is kept short in front of a building

law-yer [lô′yər] *noun*, a professional who practices law, an attorney, barrister

lax-a-tive [lăk′sə-tĭv] *noun*, a medicine or herb that induces a bowel movement

lay [lā] *verb*, 1) to place something in a horizontal position 2) when an adult female produces eggs 3) to remain, to rest

lay-er [lā′ər] *noun*, a covering that is set between two levels

la-zy [lā′zē] *adjective*, disinterested in work, slow-moving, indolent

lead [lĕd] *noun*, 1) a heavy, metallic element, having a bright luster 2) the part of a pencil used for writing, [lēd] *verb*, 3) to guide, to show someone the way

lead-er [lē′dər] *noun*, the chief, the person in charge, the ruler

lead-er-ship [lē′dər-shĭp′] *noun*, the quality of those who take charge or show the way

leaf [lēf] *noun*, *plural*, *leaves*, 1) a green part of a stem that grows off a branch of a plant 2) a sheet of paper

leaf-let [lē′flĭt] *noun*, a brochure or a small sheet of printed matter

league [lēg] *noun*, 1) an association or alliance with other individuals 2) sports teams organized to compete against each other 3) three nautical miles

leak [lēk] *noun*, 1) a crack or hole through which liquid, gas or light can ooze, *verb*, 2) to let something escape through an unintended opening such as a crack or hole

lean [lēn] *adjective*, 1) thin, without fat, *verb*, 2) to bend or incline toward 3) to rest on something for support to keep from falling, to recline

leap [lēp] *noun*, 1) a jump, *verb*, 2) to jump or spring over a hurdle or through space

leap year *noun*, a year, once every four years, in which February has 29 days instead of 28

learn [lûrn] *verb*, 1) to acquire knowledge or skill through study or experience 2) to find out, to commit to memory

learn-ing [lûr′nĭng] *noun*, knowledge, education

leash [lēsh] *noun*, a restraining device in the form of a rope or cord, chain or strap

least [lēst] *adjective*, the smallest amount, degree or size

leath-er [lĕth′ər] *noun*, skin of a dead animal prepared for use

leave [lēv] *noun*, 1) permission to be absent, *verb*, 2) to go away from, to depart, to exit 3) to allow something to remain untouched 4) to bequeath

lech-er-ous [lĕch′ər-əs] *adjective*, impure in thought and act, lustful, perverted

lec-tern [lĕk′tərn] *noun*, a podium for public speaking

lec-ture [lĕk′chər] *noun*, 1) a speech on any subject, *verb*, 2) to speak, to talk, to teach

ledge [lĕj] *noun*, 1) a narrow shelf, such as at the bottom of a window 2) a narrow, flat piece of rock, on which you can stand

ledg-er [lĕj′ər] *noun*, a book of accounts listing names and debits and credits

left [lĕft] *adjective*, 1) the side of a person or thing that is to the west when facing north, the opposite of the right side, *verb*, 2) the *past tense* of **leave**

leg [lĕg] *noun*, 1) the body part attached to the foot or paw, used for walking 2) one of at least three pieces on a piece of furniture, used to make it stand

leg-a-cy [lĕg′ə-sē] *noun*, a gift of personal property left in a will

le-gal [lē′gəl] *adjective*, based upon or in accordance with law

leg-end [lĕj′ənd] *noun*, 1) a story about someone that has no factual basis 2) a description or key accompanying a map or picture 3) an admirable person

leg-er-de-main [lĕj′ər-də-mān′] *noun*, slight of hand, magic

leg-i-ble [lĕj′ə-bəl] *adjective*, capable of being read with ease

le-gion [lē′jən] *noun*, a large group of soldiers, a multitude of people

leg-is-la-ture [lĕj′ĭ-slā′chər] *noun*, persons in a state elected by the people, having power to make, repeal, or change laws

le-git-i-mate [lə-jĭt′ə-mĭt] *adjective*, 1) lawful, conforming to accepted standards 2) genuine

leg-ume [leg′yōōm] *noun*, a plant of the pea family grown as a crop

lei-sure [lē′zhər] *noun*, *no plural*, free time from the demand of work or duty, spare time

lem-ming [lĕm′ĭng] *noun*, a small rat-like animal known for its mass migration into the sea

lem-on [lĕm′ən] *noun*, a yellow citrus fruit with a very sour taste, grown in tropical climates

185

lem-on-ade [lĕm′ə-nād′] *noun*, a beverage of sweetened lemon juice and water

lend [lĕnd] *verb*, 1) to let someone use something for a while, until they return it 2) to offer help

length [lĕngkth] *noun*, 1) the distance from end to end 2) the amount of time

length-en [lĕngk′thən-] *verb*, to make longer, drawn out

le-ni-ent [lē′nē-ənt] *adjective*, mild, merciful, not strict

lens [lĕnz] *noun*, 1) a piece of curved glass or other transparent material by which light rays meet to form an image 2) a part of the eye that focuses light rays to form an image on the retina

leop-ard [lĕp′ərd] *noun*, one of the big cats with a spotted coat, that weighs 100-175 pounds, is seven feet long and lives in Africa

le-sion [lē′zhən] *noun*, a wound, a sore, diseased tissue

less [lĕs] *adjective*, a smaller amount, a reduction

less-en [lĕs′ən] *verb*, to make smaller, to become less

les-son [lĕs′ən] *noun*, something that must be learned or studied

let [lĕt] *verb*, 1) to allow, to permit, to grant 2) to rent, to lease

le-thal [lē′thəl] *adjective*, deadly

le-thar-gy [lĕth′ər-jē] *noun*, fatigue, without energy, feeling tired, drowsy, sluggish

let's [lĕts] the *contraction* of the words **let** and **us**

let-ter [lĕt′ər] *noun*, 1) a message in writing, often sent through the mail 2) one of the characters of the alphabet

let-tuce [lĕt′əs] *noun*, a plant having succulent leaves that are used in salads

lev-ee [lĕv′ē] *noun*, an embankment, dike, dam

lev-el [lĕv′əl] *adjective*, 1) flat, with no higher or lower places, *noun*, 2) a place or position of a particular height, *verb*, 3) to make a surface flat or even

lev-er [lĕv′ər] *noun*, 1) a bar used to exert pressure or sustain weight 2) a means to an end

lev-i-ty [lĕv′ĭ-tē] *noun*, lightness of mind, high spirits, happiness

le-vy [lĕv′ē] *verb*, to put a judgment on property to collect money, to impose a tax

lewd [lōōd] *adjective*, obscene, indecent, lecherous, lustful

lex-i-cog-ra-pher [lĕk′sĭ-kŏg′rə-fər] *noun*, a person who writes dictionaries

lex-i-con [lĕk′sĭ-kŏn′] *noun*, a dictionary, a word list

li-a-bil-i-ty [lī′ə-bĭl′ĭ-tē] *noun*, 1) a responsibility, 2) a monetary obligation, a debt

li-a-ble [lī′ə-bəl] *adjective*, 1) legally responsible 2) likely, prone, inclined, apt

li-ai-son [lē′ā-zŏn′] *noun*, cooperation that facilitates a close working relationship between people or organizations

li-ar [lī′ər] *noun*, someone who does not tell the truth

li-bel [lī′bəl] *noun*, 1) a false statement made about someone

lib-er-al [lĭb′ər-əl] *adjective*,
1) progressive reform
2) generous, ample, abundant

lib-er-ate [lĭb′ə-rāt′] *verb*, to free
from constraints, to release

lib-er-a-tion [lĭb′ə-rā′shən] *noun*,
freeing, emancipation

lib-er-ty [lĭb′ər-tē] *noun*, freedom,
independence from foreign rule

li-bi-do [lĭ-bē′dō] *noun*, emotional
urges behind human activity

li-brar-i-an [lī-brâr′ē-ən] *noun*,
one who has charge or care of
the materials in a library

li-brar-y [lī′brĕr′ē] *noun*, 1) a place
where books are kept for reading
and research 2) a room or
building where books and
documents are preserved for
future generations

lib-ret-to [lĭ-brĕt′ō] *noun*, the text
of an opera or other dramatic
musical work

li-cense [lī′səns] *noun*, 1) authority
given to do any act 2) excess of
liberty, *verb*, 3) to authorize

li-cen-tious [lī-sĕn′shəs] *adjective*,
unprincipled, unrestrained

li-chen [lī′kən] *noun*, a fungus that
grows like algae on rocks and
tree bark

lick [lĭk] *noun*, 1) a stroke of the
tongue over something *verb*,
2) to touch with the tongue 3) to
defeat

lic-o-rice [lĭk′ər-ĭs] *noun*, a dried
root, or an extract used in candy

lid [lĭd] *noun*, a cover for a pot or
pan, or some other container

lie [lī] *noun*, 1) the intentional
statement of an untruth, *verb*,

2) to put oneself in a reclining
position 3) to say something that
is not true, to fib

lien [lēn] *noun*, a legal claim upon
real or personal property

lieu [loo] *preposition*, in place, as
in the phrase "in lieu of"

lieu-ten-ant [loo-tĕn′ənt] *noun*,
1) a substitute for another in
performance of any duty 2) a
commissioned officer in the
military below captain

life [līf] *noun*, 1) the ability of
living organisms to grow and
reproduce 2) the period between
birth and death

lift [lĭft] *noun*, 1) a free ride in a car
or other form of transportation 2)
the means by which something is
raised, *verb*, 3) to pick up, to
raise, to elevate 4) to go up, to
ascend

lig-a-ment [lĭg′ə-mənt] *noun*, a
tough band of fibrous tissue
connecting bones, or to support
an organ in place in the body

light [līt] *adjective*, 1) buoyant, not
heavy 2) carefree, *noun*, *no
plural*, 3) the opposite of
darkness, *verb*, 4) to ignite a fire
5) to illuminate, to brighten

light-ning [līt′nĭng] *noun*, *no
plural*, flashing of light by
electricity discharged from one
cloud to another during a storm

like [līk] *adjective*, 1) similar to,
equal to, *preposition*,
2) resembling, typical of, *verb*,
3) to enjoy, to find agreeable

like-ly [līk′lē] *adjective*,
1) expected, probable, suitable,
adverb, 2) probably, certainly

like-ness [līk′nĭs] *noun*,
1) similarity, resemblance, 2) a
picture of a person, a portrait

like-wise [līk′wīz′] *adverb*, in the
same way, also, similarly

li-lac [lī′lək] *noun*, a shrub in the
olive family that has fragrant
white or purple flowers

lil-li-pu-tian [lĭl′ə-pyōō′shən]
adjective, extremely small

lil-y [lĭl′ē] *noun*, a plant with
trumpet-like flowers and thick
tubular roots

limb [lĭm] *noun*, 1) a part of the
body such as an arm or leg or
wing 2) the large branch of a
tree

lim-ber [lĭm′bər] *adjective*,
flexible, elastic, easy to bend

lim-bo [lĭm′bō] *noun*, 1) a place
near heaven or hell where certain
souls are kept 2) a West Indian
dance 3) a state of oblivion

lime [līm] *noun*, a green citrus
fruit, similar to a lemon, grown
on trees in warm climates

lim-it [lĭm′ĭt] *noun*, 1) that which
terminates or confines 2) the full
extent, *verb*, 3) to set
boundaries, to restrict

lim-i-ta-tion [lĭm′ĭ-tā′shən] *noun*,
shortcomings, restriction

lim-ou-sine [lĭm′ə-zēn′] *noun*, a
long automobile with many seats
and a permanent top

limp [lĭmp] *adjective*, 1) without
stiffness, not firm, *noun*, 2) the
way you walk when one leg is
hurt, *verb*, 3) to walk carefully
without putting too much
pressure on the leg or foot

lim-pid [lĭm′pĭd] *adjective*, crystal
clear, transparent

line [līn] *noun*, 1) the distance
between 2 points 2) people or
things arranged in a row 3) a
business or vocation 4) a streak
made with a pencil and ruler 5) a
collection of points along a straight
path that goes on and on in
opposite directions 6) a boundary
marker 7) dialogue in performance,
verb, 8) to put a layer between the
container and the product 9) to
collect in a row, to watch or wait
for something 10) to draw a thin
mark

lin-en [lĭn′ən] *noun*, a thread or
cloth made of flax

lin-er [lī′nər] *noun*, a first class
passenger ship or airplane

lin-ger [lĭng′gər] *verb*, 1) to stay on
longer than expected, to dawdle
2) to persist

lin-ge-rie [län′zhə-rā′] *noun*, lacey,
silk or cotton underwear worn by
women, undergarments

lin-guis-tic [lĭng-gwĭs′tĭk]
adjective, pertaining to language

lin-i-ment [lĭn′ə-mənt] *noun*, a
liquid applied to soothe the skin

lin-ing [lī′nĭng] *noun*, a fold, an
overlay that forms a covering

link [lĭngk] *noun*, 1) a connection,
a bond, a tie, *verb*, 2) to be
joined together, to unite

li-no-le-um [lĭ-nō′lē-əm] *noun*, a
washable floor covering having
a base of solidified oil on a
canvas foundation

lint [lĭnt] *noun*, dirt, dust, fuzz

li-on [lī′ən] *noun*, one of the big
cats that lives in Africa

lip [lĭp] *noun*, one of the soft red rims round the mouth

liq-uid [lĭk′wĭd] *adjective*, 1) flowing like water, *noun*, 2) fluid, not solid or gas

liq-ui-date [lĭk′wĭ-dāt′] *verb*, to sell possessions for cash

liq-uor [lĭk′ər] *noun*, any liquid substance, an alcoholic beverage

lisp [lĭsp] *noun*, a speech impediment making it difficult to make an **S** sound

list [lĭst] *noun*, 1) names of different things written down one after another, *verb*, 2) to order things by writing them down in a sequence

lis-ten [lĭs′ən] *verb*, to pay attention to what someone is saying, to heed

list-less [lĭst′lĭs] *adjective*, lacking spirit or energy, languid, tired

lit [lĭt] *past tense* of **light**

lit-a-ny [lĭt′n-ē] *noun*, a prayer in the form of a religious song alternating with the congregation

li-ter [lē′tər] *noun*, a unit in the metric system used to measure liquid equal to 1.057 quarts

lit-er-ac-y [lĭt′ə-r′ə-sē] , *noun* the ability to listen, speak, read, write, view, represent, compute, and solve problems at levels of proficiency necessary to function in the family, in the community, and on the job

li-ter-ate:*adjective,* able to read and write

lit-er-al-ly [lĭt′ər-ə-lē] *adverb*, word for word, as stated, verbatim

lit-er-a-ture [lĭt′ər-ə-chōōr′] *noun*, *no plural*, 1) the study of writing and books 2) printed material

lithe [lĭth] *adjective*, pliable, bendable, limber, supple

lith-i-um [lĭth′ē-əm] *noun*, a soft, silver-white metallic element of the alkali group

lit-i-gant [lĭt′ĭ-gənt] *noun*, a person who wants to sue another

lit-i-ga-tion [lĭt′ĭ-gā′ shən] *noun*, the subject of a lawsuit, any matter before the court or a judge

lit-mus [lĭt′məs] *noun*, a coloring matter obtained from lichens, that is violet-blue, and turns red by an acid, and is restored to blue by an alkali

lit-ter [lĭt′ər] *noun*, 1) newborn baby animals, offspring 2) trash, small pieces of garbage thoughtlessly discarded, *verb*, 3) to throw trash on the ground, not in a garbage can

lit-tle [lĭt′l] *adjective*, 1) a small amount or size 2) a brief period of time

live [līv, lĭv] *adjective*, 1) performed unrehearsed, broadcasted directly 2) full of energy, *verb*, 3) to be, to exist 4) to reside in, to inhabit 5) to be active and to satisfy an ideal

live-li-hood [līv′lē hōōd] *noun*, the means of supporting oneself

live-ly [līv′lē] *adverb*, full of energy, very active, full of life

liv-er [lĭv′ər] *noun*, an organ inside the body which cleans the blood

live-stock [līv′stŏk′] *noun*, animals raised on a farm or ranch

liv-id [lĭv/ĭd] *adjective*, 1) angry 2) lead colored, black and blue

liv-ing [lĭv/ĭng] *adjective*, 1) alive, *noun*, 2) a manner of existence

liz-ard [lĭz/ərd] *noun*, a cold blooded reptile with four short legs and a scaly body

lla-ma [lä/mə] *noun*, a South American ruminant animal, related to the camel and raised for its fleecy wool

load [lōd] *noun*, 1) something carried by a person, vehicle or animal 2) a burden, *verb*, 3) to put cargo or materials on a truck or vehicle 4) to fill a gun with ammunition 5) to burden

loaf [lōf] *noun*, 1) the whole amount of bread before it is sliced, *verb*, 2) to take it easy, to idle away time, to waste time

loan [lōn] *noun*, 1) a thing, especially money, lent to another person, *verb*, 2) to let someone borrow something, to lend

loathe [lōth] *verb*, to detest, to feel intense dislike for, to abhor

lob-by [lŏb/ē] *noun*, 1) an entrance way, a foyer, *verb*, 2) to influence, to pressure, to sway

lob-ster [lŏb/stər] *noun*, an edible sea animal with a shell, a tail, and ten legs

lo-cal [lō/kəl] *adjective*, regional, in a confined area or place

lo-cal-i-ty [lō-kăl/ĭ-tē] *noun*, a position, a situation, a place

lo-cate [lō/kāt/] *verb*, to find, to search out, to come across

lo-ca-tion [lō-kā/shən] *noun*, where something is, a certain place

lock [lŏk] *noun*, 1) a device for fastening things like doors that can only be opened or closed with a key or secret code 2) a strand of hair, *verb*, 3) to close and make secure 4) to fasten with a lock

lock-er [lŏk/ər] *noun*, a box often with a lock, for keeping things

lo-co-mo-tive [lō/kə-mō/tĭv] *noun*, a railroad engine that pulls trains

lo-cust [lō/kəst] *noun*, an insect like a grasshopper that travels in swarms and eats plants

lode [lōd] *noun*, vein-like deposit of mineral ore

lodge [lŏj] *noun*, 1) a place to stay where you pay for the service, *verb*, 2) to pay to stay someplace, like a motel or inn 3) to become embedded in something

loft [lôft] *noun*, an upper room, a storage area, an attic

loft-y [lôf/tē] *adjective*, very high

log [lôg, lŏg] *noun*, 1) a piece of wood that was part of the trunk or limb of a tree 2) a book of records, a register of a trip

log-ic [lŏj/ĭk] *noun*, the art of correct reasoning, especially of inference, sound judgment

log-i-cal [lŏj/ĭ-kəl] *adjective*, in a way that is orderly and makes sense, reasonable

lo-go [lō/gō/] *noun*, a slogan or symbol that represents a name

loi-ter [loi/tər] *verb*, to hang around, to linger aimlessly, to proceed slowly, to waste time

loll [lŏl] *verb*, to lounge about

lone [lōn] *adjective*, single, one, standing apart, isolated

lone-li-ness [lōn′lē-nĕs] *noun*, feeling alone, solitude

lone-ly [lōn′lē] *adjective*, 1) a sad, depressing feeling because one is alone 2) desolate, remote

long [lông] *adjective*, 1) measuring a distance from one end to another 2) farther or taller than the others 3) an extended period or time, *adverb*, 4) a time in the distant past 5) conditionally, *verb*, 6) to want something badly, to pine, to yearn

lon-gev-i-ty [lŏn-jĕv′ĭ-tē] *noun*, to live the duration of a long life

long-ing [lông′ĭng] *verb*, desiring, wanting, yearning, craving

lon-gi-tude [lŏn′jĭ-tōōd′] *noun, no plural,* imaginary lines shown on maps that go from east to west

look [lŏŏk] *noun*, 1) a glance, a gaze 2) appearance, *verb*, 3) to stare, or glance at 4) to seem or appear 5) to expect, to await

loom [lōōm] *noun*, 1) a machine used to weave cloth from yarn, *verb*, 2) to hover over 3) likely to will happen in the future

loop [lōōp] *noun*, a ring made by a rope or string crossing itself

loop-hole [lōōp′hōl′] *noun*, a way of getting around a law or legal agreement, a means of escape

loose [lōōs] *adjective*, 1) not fastened tightly or securely 2) free, not confined or shut in

loos-en [lōō′sən] *verb*, to become unfastened, to make less tight

loot [lōōt] *noun*, 1) money or spoils, *verb*, 2) to steal, to rob, to pillage and plunder

lope [lōp] *verb*, to run with long, slow steps, to gallop slowly

lop-sid-ed [lŏp′sī′dĭd] *adjective*, hanging over to one side, uneven

lo-qua-cious [lōkwā′shəs] *adjective*, talkative, chatty, fond of talking

lord [lôrd] *noun*, 1) a person having great power or authority, a master or ruler 2) the title for a judge, bishop, or nobleman

lore [lôr] *noun*, a lesson, a teaching

lose [lōōz] *verb*, 1) to fail to win 2) to misplace something

loss [lôs] *noun*, 1) damage, injury, ruin 2) deprivation

lost [lôst] *adjective*, 1) confused, bewildered, disoriented 2) wasted 3) the *past tense* of **lose**

lot [lŏt] *noun*, 1) a piece of land or property 2) a large quantity of something 3) a kind or sort

lo-tion [lō′shən] *noun*, a liquid put on the skin to add moisture and make it soft or clean

lot-tery [lŏt′ə-rē] *noun*, a game of gambling in which a prize is given to the person with the matching digits drawn from a pool of numbers

lo-tus [lō′təs] *noun*, a water plant with white, yellow, blue or pink flowers, a water lily

loud [loud] *adjective*, 1) noisy, making or having a loud sound 2) offensively bright colors

Lou-i-si-ana [lōō-ē′zē-ăn′ə] *noun*, one of the 50 United States located in the South, the capital is Baton Rouge. The state flower of **Louisiana** is the flower of the magnolia tree, and the motto is "Union, justice, confidence."

lounge [lounj] *noun*, 1) a couch with little or no back 2) a room in a bar where people relax and listen to music, *verb*, 3) to rest or recline comfortably

louse [lous] *noun*, *plural*, **lice**, a small insect without wings that lives on the skin of animals, birds, and people

lous-y [lou′zē] *adjective*, bad

lov-a-ble [lŭv′ə-bəl] *adjective*, able to be loved, warm, deep feeling

love [lŭv] *noun*, 1) a very strong warm feeling or deep concern for someone, commitment 2) darling, sweetheart, *verb*, 3) to feel passion, or a strong affection for someone or something

love-ly [lŭv′lē] *adjective*, beautiful, having pleasing qualities

lov-ing [lŭv′ĭng] *adjective*, showing affection or fondness

low [lō] *adjective*, 1) close to the ground, not very high 2) quiet, not very loud 3) coarse, mean 4) inferior quality

low-er [lou′ər] *adjective*, 1) inferior 2) less 3) under, *verb*, 4) to make something closer to the ground 5) to diminish the volume of sound 6) to demean, to debase in estimation

loy-al [loi′əl] *adjective*, true and faithful to a person

loy-al-ty [loi′əl-tē] *adjective*, allegiance, faithfulness

lu-bri-cant [lōō′brĭ-kənt] *noun*, grease or gear oil

lu-bri-cate [lōō′brĭ-kāt′] *verb*, to make smooth or slippery

lu-cent [lōō′sənt] *adjective*, glowing with light, shining

lu-cid [lōō′sĭd] *adjective*, 1) bright, 2) easily understood, clear

luck [lŭk] *noun*, 1) something good or bad that happens by chance or accident 2) good fortune

luck-y [lŭk′ē] *adjective*, fortunate

lu-cra-tive [lōō′krə-tĭv] *adjective*, profitable, money making

lu-cre [lōō′kər] *noun*, money

lu-di-crous [lōō′dĭ-krəs] *adjective*, funny, ridiculous, absurd

lug-gage [lŭg′ĭj] *noun*, *no plural*, bags, suitcases, trunks and traveling accessories, baggage

luke-warm [lōōk′wôrm′] *adjective*, not very warm, but not cold, room temperature, tepid

lull [lŭl] *noun*, 1) a quiet interval, a pause, *verb*, 2) to soothe, to calm

lull-a-by [lŭl′ə-bī′] *noun*, a soft song that puts someone to sleep

lum-ba-go [lŭm-bā′gō] *noun*, rheumatic pain in the loins and small of the back

lum-ber [lŭm′bər] *noun*, 1) logs or timber cut into planks or boards, *verb*, 2) to move along slowly and heavily, to plod

lu-mi-nous [lōō′mə-nəs] *adjective*, 1) well lit, bright, radiating light 2) glowing in the dark

lump [lŭmp] *noun*, 1) a swelling on the body that becomes a mass 2) a hard form without a special shape or size, *verb*, 3) to put together in a pile

lu-nar [lōō′nər] *adjective*, having to do with the moon

lu-na-tic [lōō′nə-tĭk] *adjective*, 1) crazy, mad, psychotic, *noun*, 2) an insane person

lunch [lŭnch] *noun*, the meal you eat in the middle of the day

lunch-eon [lŭn′chən] *noun*, a midday meal enjoyed with others

lung [lŭng] *noun*, one of the two sac-like organs inside the chest with which we breathe

lunge [lŭnj] *verb*, to leap, to charge at someone suddenly

lu-pine [lōō′pən] *noun*, a plant related to peas and beans with tall spikes of flowers

lure [lōōr] *noun*, 1) something appealing, *verb*, 2) to entice, to bait, to attract, to tempt

lu-rid [lōōr′ĭd] *adjective*, horrible, sensational, shocking, gruesome

lurk [lûrk] *verb*, to wait in hiding, to watch someone without their knowledge, to go furtively

lus-cious [lŭsh′əs] *adjective*, appealing to taste or smell

lust [lŭst] *noun*, 1) a strong feeling of wanting something that would be selfish 2) sexual appetite

lus-ter [lŭs′tər] *noun*, shine, brightness, sheen or gloss

lux-u-ri-ous [lŭg-zhōōr′ē-əs] *adjective*, fine and expensive, very nice, sumptuous

lux-u-ry [lŭg′zhə-rē] *noun*, an item that adds to pleasure and comfort

lynch [lĭnch] *verb*, to hang someone in an act of mob violence

lyr-ic [lĭr′ĭk] *noun*, a poem intended to be a song

M

ma-ca-bre [mə-kä′brə] *adjective*, horrible, gruesome, ghastly

mac-a-ro-ni [măk′ə-rō′nē] *noun*, a flour paste dried in long slender tubes, pasta

ma-chet-e [mə-shĕt′ē] *noun*, a long, curved knife used for cutting tall vegetation

mach-i-na-tion [măk′ə-nā′shən] *noun*, a scheme, a plot, or plan

ma-chine [mə-shēn′] *noun*, a device made to consume energy to produce another form of energy to do some kind of work

ma-chin-er-y [mə-shē′nə-rē] *noun*, *no plural*, equipment, devices

mack-er-el [măk′ər-əl] *noun*, a fish found traveling in schools in the North Atlantic

mad [măd] *adjective*, 1) insane, crazy, out of your right mind, reckless 2) angry, enraged

mad-am [măd′əm] *noun*, a respectful way of speaking to or writing to a woman

made [mād] *past tense* of **make**

mad-ri-gal [măd′rĭ-gəl] *noun*, a folk song or ballad of the 16th and 17th century

mael-strom [māl′strəm] *noun*, a powerful whirlpool

maes-tro [mīs′trō] *noun*, an eminent composer, a master

mag-a-zine [măg′ə-zēn′] *noun*, a pamphlet or news summary published periodically

mag-ic [măj′ĭk] *noun*, *no plural*, 1) the art or pretended art of controlling forces through the use of secret charms 2) clever tricks done to outsmart people, using sleight of hand

mag-i-cal [măj′ĭ-kəl] *adjective*, enchanting, marvelous

ma-gi-cian [mə-jĭsh′ən] *noun*, one who uses tricks to do things that seem impossible

mag-is-trate [măj′ĭ-strāt′] *noun*, a public civil officer charged with administering the law, a justice of the peace

mag-nan-i-mous [măg-năn′ə-məs] *adjective*, great of mind, generous and forgiving

mag-nate [măg′nāt′] *noun*, a person of prominence and influence, a tycoon

mag-ne-si-um [măg-nē′zē-əm] *noun*, a ductile silver-white metallic element

mag-net [măg′nĭt] *noun*, 1) a combination of metals that have a force that attract metals like iron 2) something that attracts

mag-net-ic [măg-nĕt′ĭk] *adjective*, 1) possessing the properties of the magnet 2) exerting a moral attractive force or charm

mag-nif-i-cence [măg-nĭf′ĭ-səns] *adjective*, grandeur, splendor

mag-nif-i-cent [măg-nĭf′ĭ-sənt] *adjective*, wonderful, impressive

mag-ni-fy [măg′nə-fī′] *verb*, to increase the size of what you are looking at, to enlarge

mag-ni-fy-ing glass [măg′nə-fī′ĭng glăs] *noun*, a lens that enlarges something small so that it is easier to see

mag-ni-tude [măg′nĭ-tōōd′] *noun*, great in size, or extent

ma-hog-a-ny [mə-hŏg′ə-nē] *noun*, a hard, reddish wood from tropical American trees

maid [mād] *noun*, a female servant

maid-en name [mād′n-nām] *noun*, a woman's last name before she is married

mail [māl] *noun*, letters and packages sent through and delivered by the postal service

maim [mām] *verb*, to injure so that part of the body is useless

main [mān] *adjective*, chief, most important, the center piece

Maine [mān] *noun*, one of the 50 United States located in New England, the capital is Augusta. The state flower of **Maine** is the eastern white pine cone and tassel, and the motto is "Direct or guide."

main-spring [mān′sprĭng′] *noun*, the principal spring in a mechanism, as in a watch

main-stay [mān′stā′] *noun*, 1) pillar, support 2) strength

main-tain [mān-tān′] *verb*, 1) to hold or keep in any particular condition, to preserve 2) to support, to insist

main-te-nance [măn′tə-nəns] *noun*, *no plural*, support, sustenance, upkeep, preservation

maize [māz] *adjective*, 1) pale yellow, *noun*, 2) corn

ma-jes-ti-cal-ly [mə-jĕs′tĭk-lē] *adverb*, with grandeur and stately dignity, regal

maj-es-ty [măj′ĭ-stē] *noun*, the title of a sovereign power, grandeur

ma-jor [mā′jər] *adjective*, 1) chief, most important, the largest, *noun*, 2) an officer in the military between Lieutenant Colonel and General

ma-jor-i-ty [mə-jôr′ĭ-tē] *noun*, the greater number, more than half

make [māk] *verb*, 1) to create or produce 2) to force 3) to earn or win 4) to appoint or name

mal-aise [mă-lāz′] *noun*, uneasiness, distress, fatigue

mal-a-prop-ism [măl′ə-prŏp-ĭz′əm] *noun*, comic misuse of a word, *i.e., a verbal agreement isn't worth the paper it is written on*

ma-lar-i-a [mə-lâr′ē-ə] *noun*, a disease produced by the bite of a mosquito causing a fever

male [māl] *adjective*, 1) belonging to men and boys, *noun*, 2) a grown boy, masculine, manly

ma-lev-o-lent [mə-lĕv′ə-lənt] *adjective*, wishing evil or harm

mal-ice [măl′ĭs] *noun*, a desire to inflict bad feelings on another

ma-lign [mə-līn′] *verb*, to say bad things about someone

ma-lig-nant [mə-lĭg′nənt] *adjective*, tending or threatening to produce death as a tumor

mall [môl] *noun*, a public walk, a marketplace, a shopping center

mal-le-a-ble [măl′ē-ə-bəl] *adjective*, 1) capable of being extended or shaped by beating with a hammer 2) easy to influence

mal-let [măl′ĭt] *noun*, a stick, a club, a hammer-like tool

mal-nu-tri-tion [măl′nōō-trĭsh′ən] *noun*, *no plural*, undernourishment caused by poor diet

mam-mal [măm′əl] *noun*, any warm-blooded vertebrate that nurses its young

mam-moth [măm′əth] *adjective*, very large, gigantic, huge

man [măn] *noun*, *plural*, **men**, 1) a full grown male person 2) speaking of mankind in general *verb*, 3) to supply with people

man-age [măn′ĭj] *verb*, 1) to succeed in getting something done, to reach a goal 2) to control or have power over something 3) to function

man-age-a-ble [măn′-ĭ-jə-bəl] *adjective*, governable, submitting to control

man-age-ment [măn′ĭj-mənt] *noun*, people who manage any enterprise, an administration

man-ag-er [măn′ĭ-jər] *noun*, a person who looks after and directs the activities of a business, a supervisor

man-date [măn′dāt′] *noun*, an order, a command

man-da-to-ry [măn′də-tôr′ē] *adjective*, required, imperative

man-drel [măn′drəl] *noun*, an axis inserted to support something

mane [mān] *noun*, the long hair on the back of the neck of an animal, such as a horse or lion

ma-neu-ver [mə-nōō′vər] *noun*, 1) a military movement 2) a scheme or manipulation, *verb*, 3) to move in a skillful way

mange [mānj] *noun*, a contagious skin disease of animals marked by itching and hair loss

man-go [măng′gō] *noun*, a juicy fruit with a large seed that grows on a tree in tropical climates

ma-ni-ac [mā′nē-ăk′] *noun*, a madman, an insane person

man-i-cure [măn′ĭ-kyōōr′] *noun*, a grooming of hands and fingernails

man-i-fest [măn′ə-fĕst′] *adjective*, 1) something obvious or clear, *noun*, 2) an invoice of a ship's, or truck's, or plane's cargo, *verb*, 3) to show, to display, to reveal

man-i-fes-to [măn′ə-fĕs′tō] *noun*, a statement of policy and plans

man-i-fold [măn′ə-fōld′] *adjective*, having many and various kinds or features

ma-nip-u-late [mə-nĭp′yə-lāt′] *verb*, 1) to treat or work with the hands 2) to control the action of something by management

man-kind [măn′kīnd′] *noun*, no *plural*, all human beings

man-ner [măn′ər] *noun*, 1) the way in which something is done or happens, prevailing customs *plural*, 2) ways of behaving in social situations

man-or [măn′ər] *noun*, a large house, a mansion,

man-sion [măn′shən] *noun*, a large, impressive house, a castle

man-tel [măn′tl] *noun*, 1) the shelf above a fireplace 2)the earth's interior layer between the crust and the core

man-u-al [măn′yōō-əl] *adjective*, 1) done by hand, *noun*, 2) an instruction guide

man-u-fac-ture [măn′yə-făk′chər] *verb*, to assemble parts into a product in a factory

ma-nure [mə-nōōr′] *noun*, livestock excrement

man-u-script [măn′yə-skrĭpt′] *noun*, a copy of a work intended for publication

man-y [mĕn′ē] *adjective*, containing a large number of something

map [măp] *noun*, 1) a representation of the physical features of the earth, *verb*, 2) to plan, to chart, to outline

ma-ple [mā′pəl] *noun*, a hardwood tree that produces sweet sap, boiled into syrup

mar [mär] *verb*, to damage, to hurt, to disfigure, to impair, to injure

mar-ble [mär′bəl] *noun*, no *plural*, 1) a hard stone that polishes well, used in buildings 2) a small glass or stone ball used in a game

March [märch] *noun*, the third month of the year, having 31 days

march [märch] *noun*, 1) the distance of a walk using regular steps 2) a piece of music to which soldiers walk, *verb*, 3) to walk with measured steps, to advance in this manner

mare [mâr] *noun*, a female horse

mar-ga-rine [mär′jər-ĭn] *noun*, *no plural*, a butter substitute made from vegetable oils

mar-gin [mär′jĭn] *noun*, 1) the border of empty space around the text on paper 2) a deposit of a portion of the value of stocks or bonds

mar-gin-al [mär′jə-nəl] *adjective*, borderline, passable

mar-i-gold [măr′ĭ-gōld′] *noun*, a yellow or orange flower that is a member of the daisy family

ma-rine [mə-rēn′] *noun*, having to do with the sea

mar-i-on-ette [măr′ē-ə-nĕt′] *noun*, a puppet manipulated by strings

mar-i-time [măr′ĭ-tīm′] *adjective*, bordering on the sea, nautical

mark [märk] *noun*, 1) a spot or a scratch, imprint 2) a point or the total of points indicating how one has scored on an examination or in a class 3) a distinguishing feature, *verb*, 4) to indicate a line or boundary 5) to leave a spot or write something 6) to distinguish

mar-ket [mär′kĭt] *noun*, 1) a store, an outlet, a shop 2) supply and demand, commodity market, *verb*, 3) to develop ways to sell a product by making it attractive

mar-ma-lade [mär′mə-lād′] *noun*, a preserve or jam made of the pulp and rind of fruit

ma-roon [mə-rōōn′] *verb*, to leave someone isolated

mar-riage [măr′ĭj] *noun*, a wedding ceremony, the state of being married

mar-row [măr′ō] *noun*, soft tissue filling in cavities of bones

mar-ry [măr′ē] *verb*, to take someone as a husband or wife

marsh [märsh] *noun*, a low, wet place where the ground is damp and swampy, a bog

mar-shal [mär′shəl] *noun*, 1) the master of ceremonies of a parade or an event 2) an officer of justice similar to a sheriff, *verb*, 3) to manage, to direct

mar-su-pi-al [mär-sōō′pē-əl] *noun*, animals that nurse their offspring in a pouch

mar-tial [mär′shəl] *adjective*, pertaining to military affairs, war or battle, inclined to war

mar-tyr [mär′tər] *noun*, one who suffers persecution, torture or death for a cause or beliefs

mar-vel [mär′vəl] *noun*, 1) something that is remarkable or exciting, *verb*, 2) to be filled with wonder or astonishment

mar-vel-ous [mär′və-ləs] *adjective*, astonishing, causing wonder

Mary-land [mĕr′ə-lənd] *noun*, one of the 50 United States located on the East Coast, the capital is Annapolis. The state flower of **Maryland** is the black-eyed susan and the motto is "Manly deeds, womanly words."

mas-cot [măs′kŏt′] *noun*, an animal or person that represents a school, team, or other group

mas-cu-line [măs′kyə-lĭn] *adjective*, of or pertaining to the male sex, manly

mash [măsh] *verb*, to make into a pulpy mass or mush

mask [măsk] *noun*, 1) a covering that hides a face or its expressions *verb*, 2) to disguise or hide, to cover, to conceal

ma-son [māʹsən] *noun*, a person
who builds with brick, stone, etc.

mas-quer-ade [măsʹkə-rādʹ] *noun*,
1) a party at which people wear
masks and sometimes costumes,
verb, 2) to dress in costume, to
wear a disguise, to conceal

mass [măs] *noun*, 1) a quantity or
amount of something with no
particular shape or number
2) the quantity of matter that a
body contains as measured by
the force of gravity on it
3) a large gathering of people
4) in the Catholic church the
celebration of the Eucharist

Mas-sa-chu-setts [măsʹə-chōōʹsĭts]
noun, one of the 50 United
States, located on the East Coast,
the capital is Boston. The state
flower of **Massachusetts** is the
mayflower, and the motto is "By
the sword we seek peace, but
peace only under liberty."

mas-sa-cre [măsʹə-kər] *noun*,
1) the killing of many
defenseless people, *verb*, 2) to
murder a large number of people

mas-sage [mə-säzhʹ] *noun*, 1) a
rubdown, kneading of the body
with the hands, *verb*, 2) to rub to
stimulate circulation, to caress

mas-sive [măsʹĭv] *adjective*, heavy,
weighty, consisting of a large
amount, immense, huge

mast [măst] *noun*, a tall pole that
holds up a ship's sail

mas-ter [măsʹtər] *noun*, 1) an
accomplished artist, a teacher ,
one in control 2) the title of
address in front of a boy's name,
verb, 3) to accomplish a difficult
task

mas-ter-piece [măsʹtər-pēsʹ] *noun*,
a fine example of skill or
excellence

mat [măt] *noun*, 1) a floor covering
made of straw, rubber, vinyl,
etc. 2) a tangled hair mass

match [măch] *noun*, 1) a slender
piece of wood or cardboard
tipped with a flammable mixture
2) a sports contest, 3) a person
or thing equal to another, *verb*,
4) to sort similar objects, to pair
up

mate [māt] *noun*, 1) an associate
2) one of a male and female pair
of animals, birds, etc., *verb*, 3) to
bring together a male and
female animal to reproduce

ma-te-ri-al [mə-tîrʹē-əl] *adjective*,
1) pertaining to or composed of
matter 2) of consequence,
substantial, *noun* 3) data 4) cloth,
the fabric something is made of
5) possessions, real things not
spiritual things

ma-te-ri-al-ize [mə-tîrʹē-ə-līzʹ]
verb, to take form, to emerge

ma-ter-nal [mə-tûrʹnəl] *adjective*,
motherly, relating to a mother

ma-ter-ni-ty [mə-tûrʹnĭ-tē] *noun*,
motherhood, the time during
pregnancy, or for a mother to
care for a newborn infant

math-e-mat-ics [măthʹə-mătʹĭks]
noun, *plural*, the study of
numbers, arithmetic

ma-tin-ee [mătʹn-āʹ] *noun*, an
afternoon theater performance

ma-tri-arch [māʹtrē-ärkʹ] *noun*, the
woman who rules a family or
social group, clan or tribe

mat-ri-mo-ny [măt′rə-mō′nē] *noun*, marriage, wedlock

ma-trix [mā′trĭks] *noun*, 1) an environment from which something develops or originates 2) a mold or die

mat-ter [măt′ər] *noun, no plural*, 1) the substance of which something is made 2) a situation or affair *verb*, 3) to be of importance, to signify

mat-tress [măt′rĭs] *noun*, a stuffed cushion to sleep on

ma-ture [mə-tyŏŏr′] *adjective*, fully grown or developed, adult

ma-tu-ri-ty [mə-tyŏŏr′ĭ-tē] *noun*, adulthood, full development

maud-lin [môd′lĭn] *adjective*, tearfully sentimental

maul [môl] *verb*, to handle roughly, to lacerate

mau-so-le-um [mô′sə-lē′əm] *noun*, a stately tomb, an above ground structure for burials

mav-er-ick [măv′ər-ĭk] *noun*, a rebel, a nonconformist

max-im [măk′sĭm] *noun*, a proverb, wise words

max-i-mum [măk′sə-məm] *adjective*, 1) greatest attainable, most, *noun*, 2) the largest amount possible

May [mā] *noun*, the fifth month of the year, having 31 days

may [mā] *verb*, 1) used to show that something is possible 2) to be allowed or permitted to 3) showing a hope that something will happen

may-be [mā′bē] *adverb*, perhaps, possibly

may-hem [mā′hĕm′] *noun*, commotion, disturbance

may-on-naise [mā′ə-nāz′] *noun*, a dressing for salad or sandwiches, usually made from eggs and oil

may-or [mā′ər, mâr] *noun*, the chief magistrate and leader of a city or town

me [mē] *pronoun*, the form of "I" used as an object, used by a speaker to refer to himself or herself

mead-ow [mĕd′ō] *noun*, a field of grass or flowers, grassland

mea-ger [mē′gər] *adjective*, small in amount, scant

meal [mēl] *noun*, food served and eaten, the occasion for eating

mean [mēn] *adjective*, 1) unkind, cruel, uncaring 2) the middle, *verb*, 3) to be the same as, to have as a definition 4) to intend to do something 5) to denote

me-an-der [mē-ăn′dər] *verb*, to wander around, to follow a path with twists and turns

mean-ing [mē′nĭng] *noun*, 1) the definition of a word 2) the significance of an action , signs or symbols

means [mēnz] *noun, plural*, resources that help us to do what we want to do, wealth

meant [mĕnt]*verb, past tense* of **mean**

mean-while [mēn′hwīl′] *adverb*, at the same time

mea-sles [mē′zəlz] *noun, plural*, a contagious disease marked by distinct red circular spots, accompanied by a fever

meas-ure [mĕzh′ər] *noun*, 1) a standard or test 2) dimension, size, capacity 3) step, action, proceeding, maneuver, *verb*, 4) to take measurements, to gauge, to scale, to calibrate

meas-ure-ment [mĕzh′ər-mənt] *noun*, the size, weight, distance, mass, etc. of something

meat [mēt] *noun*, the parts of an animal's body used as food

me-chan-ic [mĭ-kăn′ĭk] *noun*, 1) one who repairs cars or other machines, 2) *plural*, **mechanics**, the inner workings or things, the study of how machines work

mech-an-ism [mĕk′ə-nĭz′əm] *noun*, 1) the parts of a machine taken collectively 2)any system that works like a machine

med-al [mĕd′l] *noun*, a small, award, often embossed, given for achievement

med-dle [mĕd′l] *verb*, to interfere

me-di-a [mē′dē-ə] *plural, noun*, see **medium**

me-di-an [mē′dē-ən] *adjective*, 1) halfway, central, *noun*, 2) the middle, center, mid-point

me-di-ate [mē′dē-āt′] *verb*, to settle a dispute as an impartial third party

med-i-cal [mĕd′ĭ-kəl] *adjective*, healing, therapeutic

me-dic-i-nal [mĭ-dĭs′ə-nəl] *adjective*, having properties that cure disease or sickness

med-i-cine [mĕd′ĭ-sĭn] *noun*, 1) the study and treatment of illness 2) a remedy to treat disease and restore good health

me-di-e-val [mē′dē-ē′vəl] *adjective*, characteristic of the Middle Ages from about the fifth to the fifteenth centuries

me-di-oc-ri-ty [mē′dē-ŏk′rĭ-tē] *noun*, the average, commonplace, ordinary

med-i-tate [mĕd′ĭ-tāt′] *verb*, to think deeply

med-i-ta-tion [mĕd′ĭ-tā′shən] *noun*, deep, solemn reflection

me-di-um [mē′dē-əm] *adjective*, 1) sized between large and small, *noun*, 2) conditions, the environment 3) a means to communicate with the dead *plural*, **media,** 4) a network of communication such as television, radio, newspapers , etc. that reaches the public

med-ley [mĕd′lē] *noun*, a mixture

meek [mēk] *adjective*, humble, gentle, patient, quiet

meet [mēt] *verb*, to come face to face, to approach, to encounter

meet-ing [mē′tĭng] *noun*, 1) a gathering to discuss an issue or idea 2) point of contact

meg-a-phone [mĕg′ə-fōn′] *noun*, a cone or electric device that amplifies the sound of a voice

mel-an-chol-y [mĕl′ən-kŏl′ē] *adjective*, 1) filled with sorrow, very sad, *noun*, 2) a feeling of sadness characterized by depression

mel-low [mĕl′ō] *adjective*, 1) good-natured, relaxed, *verb*, 2) to soften with maturity

mel-o-dy [mĕl′ə-dē] *noun*, a harmonious tune, music

mel-on [měl′ən] *noun*, a large juicy fruit growing on vines

melt [mělt] *verb*, to change to a liquid state by applying heat

mem-ber [měm′bər] *noun*, 1) one of the limbs of the body, such as a leg, arm, hand, etc. 2) a part of a group, a component 3) one that belongs to a certain group

mem-ber-ship [měm′bər-shǐp′] *noun*, the state of belonging to a certain group

mem-brane [měm′brān′] *noun*, a thin piece or layer of animal or vegetable tissue that serves to cover separate cells, organs, etc.

me-men-to [mə-měn′tō] *noun*, a token as a reminder, a souvenir

mem-oir [měm′wär′, -wôr′] *noun*, an autobiography

mem-o-ra-ble [měm′ər-ə-bəl] *adjective*, worthy of remembrance, remarkable

mem-o-ran-dum [měm′ə-răn′dəm] *noun*, a note made as a reminder, a written message, a memo

me-mo-ri-al [mə-môr′ē-əl] *noun*, a monument, a marker, a statue

mem-o-rize [měm′ə-rīz′] *verb*, to commit to memory, to learn by heart, to remember exactly

mem-o-ry [měm′ə-rē] *noun*, something a person remembers, the ability to remember

men-ace [měn′ĭs] *noun*, a danger

me-nag-er-ie [mə-năj′ə-rē] *noun*, a varied group or collection of animals, persons, or things that are strange, odd, or startling

mend [měnd] *verb*, 1) to repair or fix, to sew a tear, or patch a hole

2) to heal, to improve, to recover from illness 3) to set right a dispute

me-ni-al [mē′nē-əl] *adjective*, suitable for servants, low

men-in-gi-tis [měn′ĭn-jī′tĭs] *noun*, an infection of the three membranes that envelop the brain and spinal cord, that could be fatal

men-tal [měn′tl] *adjective*, of or existing in the mind

men-tal-ly [měn′tl-lē] *adverb*, in or using the mind, intellectually

men-tion [měn′shən] *noun*, 1) a reference, a notice, *verb*, 2) to remark, to refer to briefly

men-tor [měn′tôr′] *noun*, a trusted teacher, a wise counselor

men-u [měn′yōō] *noun*, 1) a list indicating the choice of meals at a restaurant 2) a list of options in a computer program

mer-chan-dise [mûr′chən-dīz′] *noun*, 1) commodities, wares, goods, *verb*, 2) to sell, to promote, to market

mer-chant [mûr′chənt] *noun*, a person who buys and sells goods

mer-ci-ful [mûr′sĭ-fəl] *adjective*, compassionate, kind-hearted

mer-ci-less [mûr′sĭ-lĭs] *adjective*, cruel, unkind, showing no compassion, inhumane

mer-cu-ry [mûr′kyŏŏrē] *noun, no plural,* a heavy, silver-white metallic chemical element, liquid at ordinary temperatures

mer-cy [mûr′s *noun*, the act of sparing someone punishment shown to an offender or enemy

mere [mîr] *adjective*, not more than what is specified, only

mere-ly [mîr′lē] *adverb*, simply, hardly, barely

merge [mûrj] *verb*, to combine into one, to blend gradually

me-rid-i-an [mə-rĭd′ē-ən] *noun*, an imaginary great circle on the surface of the earth, passing through the poles and any given place on the earth's surface

mer-it [mĕr′ĭt] *noun*, 1) worth, good qualities, *verb*, 2) to deserve

mer-maid [mûr′mād′] *noun*, a mythical creature that lives in the water, having the head and torso of a woman and the tail of a fish

mer-ry [mĕr′ē] *adjective*, happy, full of laughter, joyful

me-sa [mā′sə] *noun*, a flat-topped hill with steep sides, common in dry areas

mesh [mĕsh] *noun*, 1) a net-like material of cords or wires that interlock with spaces between them, like a screen, *verb*, 2) to blend, to intertwine

mes-mer-ize [mĕz′mə-rīz′] *verb*, to hypnotize, to spellbind

mess [mĕs] *noun*, 1) many things mixed up, a jumble, *verb*, 2) to make something dirty or untidy 3) to make a muddle of

mes-sage [mĕs′ĭj] *noun*, 1) a letter or note 2) a communication transmitted from one person to another 3) a central theme

mes-sen-ger [mĕs′ən-jər] *noun*, one who does an errand, a carrier

mes-sy [mĕs′ē] *adjective*, untidy

met [mĕt] *past tense* of **meet**

me-tab-o-lism [mĭ-tăb′ə-lĭz′əm] *noun*, the chemical and physical changes in living cells that process energy in the body

met-al [mĕt′l] *noun*, 1) a chemical element such as silver or gold, or any combination of metals, such as brass, copper, or nickel

met-al-lic [mə-tăl′ĭk] *adjective*, resembling, pertaining to or consisting of metal

meta-mor-phic [mĕt′ə-môr′fĭk] *adjective*, changed in form

met-a-mor-pho-sis [mĕt′ə-môr′fə-sĭs] *noun*, a change from one form of development or shape into another

met-a-phor [mĕt′ə-fôr′] *noun*, a figure of speech in which one thing is spoken of in terms of another, *i.e., the road was a ribbon of moonlight...*

met-a-phys-ics [mĕt′ə-fĭz′ĭks] *noun*, speculative philosophy

me-te-or [mē′tē-ər′] *noun*, a fragment of solid matter that enters the earth's atmosphere and burns, leaving a bright streak in the sky

me-te-or-ite [mē′tē-ə-rīt′] *noun*, the metallic or stony remains of a body that has entered the earth's atmosphere

me-te-or-ol-o-gist [mē′tē-ə-rŏl′ə-jĭst] *noun*, a person who studies the weather

me-ter [mē′tər] *noun*, 1) a machine used for measuring 2) a measure of length equal to 100 centimeters or approximately 39 inches 3) the rhythm in a poem

meth-od [mĕth′əd] *noun*, a process or system used to do something

me-thod-i-cal [mə-thŏd′ĭ-kəl] *adjective*, systematic, orderly

Me-tho-dist [mə-thŏdĭst] *noun*, a member of the Protestant Christian denomination that developed from the teachings of John and Charles Wesley

me-tic-u-lous [mĭ-tĭk′yə-ləs] *adjective*, extremely careful

met-ric sys-tem *noun*, a system that measures length in centimeters, meters, and kilometers; capacity in liters, mass in grams and kilograms; temperature in degrees Celsius

me-trop-o-lis [mĭ-trŏp′ə-lĭs] *noun*, the chief city of a country, state or region, an urban center

met-tle [mĕt′l] *noun*, courage, spirit, fortitude, character

Mich-i-gan [mĭsh′ĭ-gən] *noun*, one of the 50 United States, located in the North, the capital is Lansing. The state flower of **Michigan** is the apple blossom and the motto is "If you seek a pleasant peninsula, look about you."

mi-cro-cosm [mī′krə-kŏz′əm] *noun*, a small world

mi-cro-or-gan-ism [mī′krō-ôr′gə-nĭz′əm] *noun*, an animal , bacterium, plant, or virus that can only be seen with a microscope

mi-cro-phone [mī′krə-fōn′] *noun*, a device that converts sounds into electrical energy signals that can then be amplified

mi-cro-scope [mī′krə-skōp′] *noun*, an optical instrument with a lens that makes it possible to see things too small to be seen with the naked eye

mi-cro-scop-ic [mī′krə-skŏp′ĭk] *adjective*, too small to be seen with a naked eye

mid [mĭd] *adjective*, halfway

mid-dle [mĭd′l] *adjective*, 1) halfway between two points, median, central, *noun*, 2) the center, the point equally distant from two sides

mid-get [mĭj′ĭt] *noun*, a small person who is fully mature

mid-night [mĭd′nīt′] *noun*, 12 o'clock at night

midst [mĭdst, mĭtst] *noun*, the middle, the central point

might [mīt] *noun*, *no plural*, 1) strength or force, *verb*, 2) *past tense* of **may**

might-y [mī′tē] *adjective*, strong, powerful, important, great

mi-grant [mī′grənt] *adjective*, one who travels from place to place in search of work

mi-grate [mī′grāt′] *verb*, 1) to move from one land to settle in another 2) to make a journey each year from one place to another at the same time of year

mi-gra-tion [mī-grā′shən] *noun*, movement to leave one area to live in another

mild [mīld] *adjective*, 1) gentle in feeling 2) not severe or extreme

mile [mīl] *noun*, a measure of length equal to 5,280 feet, 1,760 yards, or 1.6 kilometers

mile-age [mī′lĭj] *noun*, 1) an allowance for the cost per mile traveled 2) the total amount of miles traveled in a given time

mile-stone [mīl′stōn′] *noun*, a turning point, a great achievement or significant event

mil-i-tant [mĭl′ĭ-tənt] *adjective*, 1) warlike, combative, *noun*, 2) a revolutionary, a violent objector, a terrorist

mil-i-tar-y [mĭl′ĭ-tĕr′ē] *noun*, the armed forces of a country

milk [mĭlk] *noun*, 1) the liquid produced by a female mammal to feed her young, *verb*, 2) to force milk from a female mammal

mill [mĭl] *noun*, a building or group of buildings in which resources such as lumber, cotton, corn and steel, are processed

mil-len-ni-um [mə-lĕn′ē-əm] *noun*, a span of one thousand years

mil-li-liter [mĭl′ə-lē′tər] *noun*, a liquid measure equal to 1/1000 of a liter, or less than .04 fluid ounces

mil-li-me-ter [mĭl′ə-mē′tər] *noun*, a measure of length, 1/1000 of a meter, or about .04 inches

mil-lion [mĭl′yən] *noun*, ten times one hundred thousand, 1,000,000

mil-lion-aire [mĭl′yə-nâr′] *noun*, a person whose property and assets total $1,000,000

mime [mīm] *noun*, 1) a performer who only uses motion and facial expression, but no voice, to communicate, *verb*, 2) to use actions instead of speech to communicate

mim-e-o-graph [mĭm′ē-ə-grăf′] *noun*, a stenciling copying device fed by an ink drum

mim-ic [mĭm′ĭk] *verb*, to copy or repeat what someone does or says, to imitate playfully

mim-ic-ry [mĭm′ĭ-krē] *noun*, an imitation, an impersonation

min-a-ret [mĭn′ə-rĕt′] *noun*, a lofty slender tower attached to a mosque

mince [mĭns] *verb*, 1) to cut into very small pieces 2) to moderate or restrain words

mind [mīnd] *noun*, 1) thoughts, a person's reason or way of feeling, *verb*, 2) to watch over, to pay attention to, to obey

mine [mīn] *noun*, 1) a large hole or tunnel deep in the ground made to reach and remove metals or other natural resources from the earth, *pronoun*, 2) something that belongs to the person who is speaking, *verb*, 3) digging into land to remove metal ore or other valuable resources

min-er-al [mĭn′ər-əl] *noun*, any inorganic substance occurring naturally in the earth

min-gle [mĭng′gəl] *verb*, to join or mix together, to blend

min-i-a-ture [mĭn′ē-ə-chŏŏr′] *adjective*, 1) very small, done on a small scale, *noun*, 2) a very small painting, especially a portrait

min-i-mum [mĭn′ə-məm] *adjective*, 1) smallest, *noun*,

2) the least quantity possible or allowable in a given case

min-is-ter [mĭn′ĭ-stər] *noun*,

1) the head of a government department in some countries 2) a representative to a foreign country ranking under an ambassador 3) a Christian member of the clergy, *verb*, 4) to give help to someone, to care for, to serve, to attend

min-is-try [mĭn′ĭ-strē] *noun*, 1) a department of government headed by a minister 2) the work of the clergy 3) the act of serving

Min-ne-so-ta [mĭn′ĭ-sō′tə] *noun*, one of the 50 United States located in the North, the capital is St. Paul. The state flower of **Minnesota**, is the pink and white lady slipper and the motto is,"The star of the north."

min-now [mĭn′ō] *noun*, a small fish

mi-nor [mī′nər] *adjective*,

1) smaller, not very important *noun*, 2) a person younger than the legal age

mi-nor-i-ty [mə-nôr′ĭ-tē] *noun*, a number, part or amount forming less than half of a whole

mint [mĭnt] *noun*, 1) a large amount of money 2) any of the plants of the mint family, used to flavor food, gum, etc.3) a candy

mi-nus [mī′nəs] *preposition*, less

min-ute [mĭn′ĭt] *adjective*, 1) very small, *noun*, 2) a period of time equal to sixty seconds

mi-nu-ti-a [mĭ-nōō′shē-ə] *noun*, *plural*, **minutiae**, very small, petty details, trifling matters

mir-a-cle [mĭr′ə-kəl] *noun*, an occurrence that is unusual and goes beyond the laws of nature

mi-rac-u-lous [mĭ-răk′yə-ləs] *adjective*, supernatural, awesome, incredible

mi-rage [mĭ-räzh′] *noun*, an unreal reflection, an optical illusion

mire [mīr] *noun*, 1) a bog, a swamp, *verb*, 2) to immerse in swampy ground, to entangle

mir-ror [mĭr′ər] *noun*, 1) a looking glass in which one can see reflections 2) something that reflects

mirth [mûrth] *noun*, laughter, glee

mis-be-have [mĭs′bĭ-hāv′] *verb*, to behave badly, to be disruptive

mis-cel-la-ne-ous [mĭs′ə-lā′nē-əs] *adjective*, consisting of several different kinds, mixed, various

mis-chief [mĭs′chĭf] *noun*, an annoying action on the part of a person, havoc, harm or damage

mis-chie-vous [mĭs′chə-vəs] *adjective*, causing petty injury or annoyance to others

mis-con-cep-tion [mĭs′kən-sĕp′shən] *noun*, a misunderstanding, a mistake

mis-con-duct [mĭs-kŏn′dŭkt] *noun*, wrongdoing, improper behavior

mis-con-strue [mĭs′kən-strōō′] *verb*, to misunderstand

mis-cre-ant [mĭs′krē-ənt] *noun*, an evil person, a lawbreaker

mis-de-mean-or [mĭs′dĭ-mē′nər] *noun*, a crime less than a felony, any minor misdeed

mi-ser [mī′zər] *noun*, a tightwad, one who hoards money

mis-er-a-ble [mĭz′ər-ə-bəl] *adjective*, unhappy, causing misery

mis-er-y [mĭz′ə-rē] *noun*, great unhappiness, or distress

mis-for-tune [mĭs-fôr′chən] *noun*, bad luck, something bad that happens, an unlucky event

mis-giv-ing [mĭs-gĭv′ĭng] *noun*, a feeling of worry, doubt about a future event, apprehension

mis-hap [mĭs′hăp′] *noun*, an unfortunate accident

mis-in-ter-pret [mĭs′ĭn-tûr′prĭt] *verb*, to confuse, to mistake

mis-judge [mĭs-jŭj′] *verb*, to mistake, to underestimate

mis-lead [mĭs-lēd′] *verb*, to lead astray, to deceive, to lie

mis-lead-ing [mĭs-lē′dĭng] *adjective*, deceptive, vague, unclear, not accurate

mis-no-mer [mĭs-nō′mər] *noun*, the wrong name, an incorrect identification

mis-place [mĭs-plās′] *verb*, to lose temporarily, to mislay

Miss [mĭs] *noun*, the title before the name of an unmarried woman

miss [mĭs] *noun*, 1) a failure, an error, a slip, *verb*, 2) to fail to make the most of an opportunity or moment 3) to notice someone's absence, to feel sad someone is gone 4) to avoid, to fail to hit, to escape 5) to omit

mis-sile [mĭs′əl] *noun*, a weapon capable of being hurled

mis-sion [mĭsh′ən] *noun*, a special service, task or duty

mis-sion-ar-y [mĭsh′ə-nĕr′ē] *noun*, one whose work is to teach others about his religion

Mis-sis-sip-pi [mĭs′ĭ-sĭp′ē] *noun*, one of the 50 United States, located in the South, the capital is Jackson. The state flower of **Mississippi** is the magnolia blossom, and the motto is "By valor and arms."

mis-sive [mĭs′ĭv] *noun*, a letter

Mis-sou-ri [mĭ-zŏŏr′ē] *noun*, one of the 50 Unieted States, located in the Midwest, the capital is Jefferson City. The state flower of **Missouri** is the hawthorn, and the motto is:"Let the welfare of the people be the supreme law."

mis-spell [mĭs-spĕl′] *verb*, to spell incorrectly

mist [mĭst] *noun*, a thin cloud of fine droplets of water vapor or fog near the ground

mis-take [mĭ-stāk′] *noun*, an error, the wrong answer or action

Mis-ter [mĭs′tər] *noun*, a title for a man, abbreviated, 'Mr.'

mis-tle-toe [mĭs′əl-tō′] *noun*, a plant that lives off another, with evergreen leaves and white berries used at Christmas for decoration

mis-tress [mĭs′trĭs] *noun*, a woman who has control over others

mis-trust [mĭs-trŭst′] *noun*, 1) suspicion, uncertainty, doubt, lack of confidence, *verb*, 2) to doubt, to question, to challenge

mist-y [mĭs′tē] *adjective*, hazy, damp, and foggy, consisting of a mass of fine droplets of water vapor in the atmosphere

mis-un-der-stand
[mĭs′-ŭn-dər-stănd′] *verb*, to
misconstrue, to mistake

mite [mīt] *noun*, 1) a small amount
2) a small, usually parasitic
arachnid, related to the tick

mi-ter [mī′tər] *verb*, to bevel the
ends for the purpose of matching
together, as a picture frame

mitt [mĭt] *noun*, a baseball glove
with padding to protect the hand

mit-ten [mĭt′n] *noun*, a glove with
a section for covering the thumb
and a larger section to cover all
four fingers

mix [mĭks] *verb*, to combine
different things to make one

mix-ture [mĭks′chər] *noun*, a
combination of different
substances

moan [mōn] *verb*, to make a sound
of pain or suffering, to complain

mob [mŏb] *noun*, a crowd, a horde

mo-bile [mō′bəl, *adjective* 1)
moveable, portable, [mō′-bīl′]
noun, 2) objects attached by a
wire that move airily

mo-bi-lize [mō′bə-līz′] *verb*, to put
into action, to move

moc-ca-sin [mŏk′ə-sĭn] *noun*, a
heelless shoe of deerskin or
other soft leather

mock [mŏk] *adjective*, 1) not real,
pretended, 2) *verb*, to make fun
of, to treat with ridicule

mode [mōd] *noun*, 1) manner
2) fashion, style

mod-el [mŏd′l] *noun*, 1) a person
who sets a good example 2) a
version of something that can be
studied, a copy, 3) someone

who wears clothes for
advertisement, *verb*, 3) to make
a copy of something 4) to wear
clothes as part of an exhibit

mod-er-a-tion [mŏd′ər-ā′shən]
noun, restraint, a happy medium

mod-ern [mŏd′ərn] *adjective*,
representative of the current
trends and fashions

mod-est [mŏd′ĭst] *adjective*,
1) humble, decent 2) simple

mod-es-ty [mŏd′ĭ-stē] *noun*, no
plural, humility, decency

mod-i-cum [mŏd′ĭ-kəm] *noun*, a
moderate or small amount

mod-i-fi-ca-tion
[mŏd′ə-fĭ-kā′shən] *noun*, a slight
change in form, an alteration

mod-i-fy [mŏd′ə-fī′] *verb*, to
change, to adjust, to alter

mod-u-late [mŏj′ə-lāt′] *verb*, to
adjust the sound

mo-gul [mō′gəl] *noun*, a powerful
person, especially in business

moist [moist] *adjective*, a little wet,
damp, humid, soggy

mois-ture [mois′chər] *noun*, no
plural, small drops of water,
wetness caused by water

mo-lar [mō′lər] *noun*, one of
several back teeth used for
chewing

mo-las-ses [mə-lăs′ĭz] *noun*, the
thick brown syrup drained from
sugar in the refining process

mold [mōld] *noun*, 1) a cast for
shapes into which a substance is
poured 2) a furry substance that
grows on clothing or food if left
damp and away from the sun, itis
a kind of small fungus, *verb*, 3)

to shape something into a form
or a mold

mold-ing [mōl′dĭng] *noun*, a plane
or curved narrow surface, used
for ornamentation

mole [mōl] *noun*, 1) a small
insectivore that makes and lives
in holes underground 2) a small
round brown mark on the skin

mol-e-cule [mŏl′ĭ-kyoōl′] *noun*,
1) a unit of matter consisting of
two or more atoms combined
2) a particle

mo-lest [mə-lĕst′] *verb*, 1) to touch
without permission 2) to annoy

molt [mōlt] *verb*, to shed the outer
skin, feathers or exoskeleton

mol-ten [mōl′tən] *adjective*, made
liquid by great heat, melted

mo-ment [mō′mənt] *noun*, a very
brief period of time, an instant

mo-men-tar-y [mō′mən-tĕr′ē]
adjective, fleeting, passing

mo-men-tous [mō-mĕn′təs]
adjective, very important

mo-men-tum [mō-mĕn′təm] *noun*,
the force or speed of movement

mon-arch [mŏn′ərk, -ärk′] *noun*, a
king or queen 2) an orange and
black butterfly

mon-ar-chy [mŏn′ər-kē] *noun*, a
government headed by one
person, a king or a queen

mon-as-ter-y [mŏn′ə-stĕr′ē] *noun*,
a residence in which monks live

Mon-day [mŭn′dē, -dā′] *noun*, the
second day of the week

mon-e-tar-y [mŏn′ĭ-tĕr′ē]
adjective, of or pertaining to
money or currency, pecuniary

mon-ey [mŭn′ē] *noun*, currency
and coins issued by a
government to use in exchange
for goods or services

mon-i-tor [mŏn′ĭ-tər] *noun*, 1) a
pupil or student given authority
2) a video screen attached to a
computer, *verb*, 3) to keep watch
over, to supervise, to check

monk [mŭngk] *noun*, a religious
man who lives with other men
who devote their lives to prayer
and religion, a member of a
religious order

mon-key [mŭng′kē] *noun*, a
primate that usually has a long
tail and lives in trees

mon-o-gram [mŏn′ə-grăm′] *noun*,
initials consisting of two or more
letters interwoven, usually
written, embroidered, or
engraved

mon-o-logue [mŏn′ə-lôg′] *noun*, a
long speech, a soliloquy

mo-nop-o-lize [mə-nŏp′ə-līz′]
verb, to have exclusive
possession of something

mo-nop-o-ly [mə-nŏp′ə-lē] *noun*,
exclusive control of the supply
of any commodity

mon-o-the-ism [mŏn′ə-thē-ĭz′əm]
noun, belief in one God

mon-o-tone [mŏn′ə-tōn′] *noun*, a
voice that never changes in pitch

mo-not-o-nous [mə-nŏt′n-əs]
adjective, 1) unvarying in pitch
2) repetitiously dull, being the
same all the time

mon-soon [mŏn-soōn′] *noun*,
heavy rains and strong winds in
southern Asia

mon-ster [mŏn′stər] *noun*, 1) an animal or person with a strange or unusual shape 2) an imaginary and usually frightening creature

Mon-tana [mŏn-tăn′ə] *noun*, one of the 50 United States located in the Northwest, the capital is Helena. The state flower of **Montana** is the bitterroot, and the motto is "Gold and silver."

month [mŭnth] *noun*, a period of 30 days, four weeks, or 1/12 of the year

mon-u-ment [mŏn′yə-mənt] *noun*, a statue or building erected as a tribute to someone who has died

mon-u-men-tal [mŏn′yə-mĕn′tl] *adjective*, massive, enormous, exceptionally great

mood [mo͞od] *noun*, the way a person feels, state of mind

moon [mo͞on] *noun*, the large body in the sky that shines at night, a satellite of a planet

moose [mo͞os] *noun*, a large North American and Eurasian deer

moot [mo͞ot] *adjective*, questionable, debatable

mop [mŏp] *noun*, 1) a tool with a long handle and a sponge or fabric on the bottom used for cleaning floors, *verb*, 2) to clean the floor, usually with soap and water or to remove spilled liquids

mor-al [môr′əl] *adjective*, ethical, honest, concerned with right conduct, virtuous, upright

mo-rale [mə-răl′] *noun*, spirit, confidence, resolve,

mo-ral-i-ty [mə-răl′ĭ-tē] *noun*, virtue, righteousness, ethics

mor-al-ly [môr′ə-lē] *adverb*, in accordance with ethical duty, righteously, respectfully

mor-a-to-ri-um [môr′ə-tôr′ē-əm] *noun*, 1) a suspension, an interval 2) a legal delay of payment

mor-bid [môr′bĭd] *adjective*, depressing, having to do with disease, gruesome

more [môr] *adjective*, 1) larger in amount or number, *adverb*, 2) again, in addition

mo-res [môr′āz′] *noun, plural*, the customs of a society, the code of acceptable conduct

morgue [môrg] *noun*, a place where dead bodies go to be identified before burial

Mor-mon [môr mŭn] *noun*, a member of the Church of Jesus Christ of Latter-day Saints founded in the U.S. in 1830 by Joseph Smith

morn-ing [môr′nĭng] *noun*, the time from when the sun rises until noon or the time from midnight to noon

mo-rose [mə-rōs′] *adjective*, sad, dejected, gloomy or sullen

mor-phine [môr′fēn′] *noun*, a narcotic derived from opium

morse code [môrs kōd] *noun, no plural*, a way of sending messages using flashing lights or a pattern of sounds

mor-sel [môr′səl] *noun*, a little piece of food, a crumb

mor-tal [môr′tl] *adjective*, 1) destructive to life, fatal 2) human and therefore able to die

mor-tal-i-ty [môr-tăl′ĭ-tē] *noun*, the condition of being mortal

mor-tar [môr′tər] *noun*, 1) a mixture made of cement, lime, sand, and water, used to hold bricks or stone together 2) a short cannon used to fire shells at a high angle 3) a bowl in which substances are crushed to a powder with a pestle

mort-gage [môr′gĭj] *noun*, a legal agreement regarding property between a borrower and a lender

mor-ti-cian [môr-tĭsh′ən] *noun*, an undertaker, a funeral director

mor-ti-fy [môr′tə-fī′] *verb*, to shame, to belittle, to humiliate

mor-tu-ar-y [môr′chōō-ĕr′ē] *noun*, a place where dead bodies are kept until burial or cremation

mo-sa-ic [mō-zā′ĭk] *noun*, a surface decoration made by inlaying small pieces of variously colored material such as tile, marble, or glass

Mos-lem [mŏz′ləm] *noun*, *adjective*, see Muslim

mosque [mŏsk] *noun*, a holy building where Muslims worship

mos-qui-to [mə-skē′tō] *noun*, an insect capable of puncturing the skin of humans and other animals and sucking blood

moss [môs, mŏs] *noun*, a bright green plant that grows flat on wet ground and stones

most [mōst] *adjective*, 1) greatest in amount or number, *adverb*, 2) very, *noun*, 3) the largest amount or number

mote [mōt] *noun*, a small speck

moth [môth] *noun*, an insect with four wings, like a butterfly that flies at night

moth-er [mŭth′ər] *noun*, the female parent

moth-er-in-law [mŭth′ər-ĭn-lô′] *noun*, the mother of one's spouse

mo-tion [mō′shən] *noun*, the action or process of moving

mo-tion-less [mō′shən-lĭs] *adjective*, not moving

mo-ti-vate [mō′tə-vāt′] *verb*, to provide the inner drive that causes one to act

mo-ti-va-tion [mō′tə-vā′shən] *noun*, the feeling that moves one to act

mo-tive [mō′tĭv] *noun*, the reason for doing something, incentive

mo-tor [mō′tər] *noun*, a compact engine that makes things move or work

mo-tor-boat [mō′tər-bōt′] *noun*, a small boat with an engine

mo-tor-cy-cle [mō′tər-sī′kəl] *noun*, a two-wheeled vehicle without a top, moved by a motor

mot-tled [mŏt′ld] *adjective*, spotted with different colors

mot-to [mŏt′ō] *noun*, a saying or catch-phrase that expresses a belief, a guiding principle

mound [mound] *noun*, 1) a heap of earth or a small hill, a knoll 2)any small pile

mount [mount] *noun*, 1) a mountain named after someone, *verb*, 2) to rise 3) to get on 4)to ascend 5)to prepare by putting in place

moun-tain [moun′tən] *noun*, a large hill or raised part of land with steep sides that rises from the earth's surface

mourn [môrn] *verb*, to feel sadness at a loss or death of someone

mourn-ing [môr′nĭng] *noun*, *no plural*, a period of grief to honor someone who died

mouse [mous] *noun*, *plural*, **mice**, 1) a small rodent, similar to a rat 2) a hand-held device that guides the cursor or pointer on a computer

mouth [mouth] *noun*, 1) the opening through which all animals consume food 2) an entry

move [mo͞ov] *verb*, 1) to change position or cause something to change position 2) to go to live in a new place 3) to affect, to arouse, or influence 3) to budge, to change opinion, to go on

move-ment [mo͞ov′mənt] *noun*, 1) an inclination, a tendency toward change 2) action, transit, change

mov-ie [mo͞o′vē] *noun*, a film, a motion picture

mov-ing [mo͞o′vĭng] *adjective*, touching, emotional

mow [mō] *verb*, to cut grass

Mr. *noun*, a title put before a man's name

Mrs. *noun*, a title put before a married woman's name

Ms. *noun*, a title put before a woman's name

much [mŭch] *adjective*, 1) of a large amount, great. *adverb*, 2) often, to a great degree *noun*, 3) a large amount or a lot of

mucus [myo͞o′kəs] *noun*, spit, of or relating to mucus, a shiny substance coating the inside of the body cavities

mud [mŭd] *noun*, wet soft earth

mud-dy [mŭd′ē] *adjective*, 1) murky, dirty 2) not clear or pure

muf-fin [mŭf′ĭn] *noun*, a quick bread made from batter and baked in a cup-shaped pan

muf-fle [mŭf′əl] *verb*, to cover or suppress a sound, to tone down

muf-fler [mŭf′lər] *noun*, 1) any of various devices to deaden the noise of escaping gases or vapors 2) a scarf worn around the neck

mug [mŭg] *noun*, 1) a cup with a handle, *verb*, 2) to assault someone, usually to rob them

mug-gy [mŭg′ē] *adjective*, warm and very damp, humid

mulch [mŭlch] *noun*, a protective layer of decomposing organic matter

mule [myo͞ol] *noun*, an animal whose parents were a female horse and a male donkey

mul-ti-form [mŭl′tə-fôrm′] *adjective*, having many shapes or forms

mul-ti-ple [mŭl′tə-pəl] *adjective*, 1) many 2) having many parts

mul-ti-pli-ca-tion [mŭl′tə-plĭ-kā′shən] *noun*, *no plural*, the process of finding the number or quantity by repeated addition of a specified number

mul-ti-ply [mŭl′tə-plī′] *verb*, to increase by a number of times

mul-ti-tude [mŭl′tĭ-to͞od′] *noun*, a crowd, a large gathering

mum-my [mŭm′ē] *noun*, a dead body preserved with chemicals and wrapped in cloth

mumps [mŭmps] *noun*, *plural*, an illness that causes fever and swelling in the neck and throat

mun-dane [mŭn-dān′] *adjective*, commonplace, worldly as opposed to spiritual

mu-nic-i-pal [myōō-nĭs′ə-pəl] *adjective*, having a local self-government, as in a city or town

mu-ral [myōōr′əl] *noun*, a picture painted directly on a wall

mur-der [mûr′dər] *noun*, 1) the act of killing someone, either after first planning to do so or during another crime, such as a robbery, *verb*, 2) to kill a person intentionally

mur-der-er [mûr′dər-ər] *noun*, a person who intends to kill someone and does so

mur-mur [mûr′mər] *verb*, to speak very softly, to whisper

mus-cle [mŭs′əl] *noun*, an organ whose special function is to produce movement

muse [myōōz] *noun*, 1) one's inner voice as a source of creativity, **Muse**, 2) one of the nine Greek goddesses, responsible for art, music, and literature, *verb*, 3) to ponder, to meditate, to think, to reflect

mu-se-um [myōō-zē′əm] *noun*, a building that exhibits a collection of things of interest

mush-room [mŭsh′rōōm′] *noun*, a plant that is not green and is a fungus, it usually has a stem with a cap on it

mu-sic [myōō′zĭk] *noun*, the art or science of harmonic sounds

mu-si-cian [myōō-zĭsh′ən] *noun*, one who plays a musical instrument and performs

mus-ket [mŭs′kĭt] *noun*, a gun with a long barrel, used by soldiers before rifles were invented

musk-rat [mŭs′krăt′] *noun*, a rodent that lives in or near water

Mus-lim [mŭz′ləm] *noun*, a follower of the religion that believes in Allah as the one god and the teachings of Muhammed as written in the Koran

must [mŭst] *verb*, 1) used with another verb to show what is necessary or what has to be done 2) showing what is sure or likely

mus-tache [mŭs′tăsh′] *noun*, the hair covering an upper lip

mus-tang [mŭs′tăng′] *noun*, a wild horse of the American plains

mus-tard [mŭs′tərd] *noun*, 1) a plant with yellow flowers and a seed in pods 2) powder or sauce made of ground mustard seed

mu-ta-ble [myōō′tə-bəl] *adjective*, 1) changing in form 2) fickle

mute [myōōt] *adjective*, voiceless, speechless, unable to speak

mu-ti-late [myōōt′l-āt′] *verb*, to cut off or remove an essential part

mu-ti-nous [myōōt′n-əs] *adjective*, unruly, rebellious, insurgent

mu-ti-ny [myōōt′n-ē] *noun*, a revolt, a rebellion, an uprising

mu-tual [myōō′chōō-əl] *adjective*, having a common interest

muz-zle [mŭz′əl] *noun*, 1) a fastening on the mouth of an animal to prevent it from biting or eating 2) the open end of a gun from which the bullet comes out when the gun is fired, *verb*, 3) to gag, to silence, to hush, to restrain speech and expression

my [mī] *adjective*, belonging to the person speaking

my-op-ic [mī-ŏp′ĭk] *adjective*, 1) nearsighted 2) self-centered, small minede

myr-i-ad [mĭr′ē-əd] *noun*, a very large number

my-self [mī-sĕlf′] *pronoun*, the same person as the one speaking

mys-ter-y [mĭs′tə-rē] *noun*, something not explained or beyond human comprehension

mys-ti-cal [mĭs′tĭ-kəl] *adjective*, beyond understanding, spiritual

mys-ti-fy [mĭs′tə-fī′] *verb*, to confuse, to puzzle, to bewilder

myth [mĭth] *noun*, 1) a legendary story used to explain nature or religion and often featuring heroes or supernatural beings

my-thol-o-gy [mĭ-thŏl′ə-jē] *noun*, the study of stories and legends

N

nag [năg] *verb*, to pester, to annoy, to bother, to find faults

nail [nāl] *noun*, 1) a piece of pointed metal to be put into wood with a hammer 2) the horny tip on the ends of fingers and toes, *verb*, 3) to fasten with a hammer and nail

na-ive [nä-ēv′] *adjective*, unsophisticated, innocent

na-ked [nā′kĭd] *adjective*, without clothes, uncovered, nude

name [nām] *noun*, 1) a word given to someone or something for identification, *verb*, 2) to decide what someone or something will be called, to specify

name-ly [nām′lē] *adverb*, that is

name-sake [nām′sāk′] *noun*, one having the same name as another one, named after another

nap [năp] *noun*, 1) a short rest 2) a textured surface on fabric

nap-kin [năp′kĭn] *noun*, a piece of cloth or paper used at a meal to keep one's clothes, hands, and mouth clean while eating

nar-cot-ic [när-kŏt′ĭk] *noun*, a drug that relieves pain, blunts the senses and produces sleep

nar-rate [năr′āt′] *verb*, to make known, to tell a story, to recount

nar-ra-tive [năr′ə-tĭv] *noun*, a story, a history, a recital

nar-ra-tor [năr′āt′ ər] *noun*, a person who relates a story

nar-row [năr′ō] *adjective*, limited in width, breadth, or scope

na-sal [nā′zəl] *adjective*, associated with the nose

nas-cent [năs′ənt] *adjective*, coming into existence, the beginning

na-tion [nā′shən] *noun*, a group of people living in a particular territory under the same government, a federation

na-tion-al [năsh′ə-nəl] *adjective*, of or pertaining to a country or territory

na-tion-al-i-ty [năsh′ə-năl′ĭ-tē]
noun, the identification of a
person based on the country
where he or she was born or
holds citizenship

na-tive [nā′tĭv] *adjective*,
1) belonging to the place where
a person is born, indigenous,
noun, 2) a person born and
raised in a certain place

nat-u-ral [năch′ər-əl] *adjective*,
1) produced by nature, not man-
made 2) inborn, an innate ability

nat-u-ral-ly [năch′ər-ə-lē] *adverb*,
as one would expect

na-ture [nā′chər] *noun*, *no plural*,
1) that which is not man-made,
the physical world 2) traits
characteristic of someone,
disposition, personality

nau-se-a [nô′zē-ə] *noun*, 1) a
stomach sickness accompanied
by a desire to vomit 2) disgust

nau-se-ate [nô′zē-āt′] *verb*, to
cause to become sick, to fill with
disgust, causing a sick feeling

nau-ti-cal [nô′tĭ-kəl] *adjective*,
having to do with the sea

na-val [nā′vəl] *adjective*, seafaring,
nautical, relating to the navy

na-vel [nā′vəl] *noun*, the belly
button, where the umbilical cord
was attached before birth

nav-i-gate [năv′ĭ-gāt′] *verb*, to
control in which direction a
vehicle will go on a route

nav-i-ga-tion [năv′ĭ-gā′shən] *noun*,
no plural, the ability to travel
through water or plan and follow
a specific route

nav-i-ga-tor [năv′ĭ-gā′tər] *noun*,
one who directs and controls the
course of a vehicle

na-vy [nā′vē] *noun*, the officers
and personnel of warships

near [nîr] *adverb*, 1) a short
distance away, *preposition*,
2) close to

near-by [nîr′bī′] *adverb*, close at
hand, not far away

near-ly [nîr′lē] *adverb*, almost

near-sight-ed [nîr′sī′tĭd] *adjective*,
able to see only things close by

neat [nēt] *adjective*, clean and
orderly, well arranged

Ne-bras-ka [nə-brăs′kə] *noun*, one
of the 50 United States, centrally
located west of the Missouri
River, the capital is Lincoln.
The state flower of **Nebraska** is
the goldenrod, and the motto is
"Equality before the law."

nec-es-sar-y [něs′ĭ-sĕr′ē] *adjective*,
needing to be done, essential

ne-ces-si-tate [nə-sĕs′ĭ-tāt′] *verb*,
to demand, to require

ne-ces-si-ty [nə-sĕs′ĭ-tē] *noun*,
1) something indispensable or
necessary 2) an urgent matter

neck [něk] *noun*, 1) the part of the
body connecting the head and
the torso 2) a narrow connecting
part, as of a bottle

neck-lace [něk′lĭs] *noun*, jewelry
worn around the neck

nec-tar [něk′tər] *noun*, a sweet
liquid made by flowers

need [nēd] *noun*, 1) a lack, *verb*, 2)
to lack something that is
necessary or essential 3) must

nee-dle [nēd'l] *noun*, an instrument for sewing, made of steel, sharp at one end, with an eyehole for thread at the other end

need-y [nē'dē] *adjective*, poor, penniless, impoverished

neg-a-tive [nĕg'ə-tĭv] *adjective*, 1) expressing denial, refusal or resistance, not positive, *noun*, 2) an image on developed film

ne-glect [nĭ-glĕkt'] *noun, no plural*, 1) disregard, carelessness, lack of care or attention, *verb*, 2) to not look after something, letting it deteriorate, to ignore

neg-li-gence [nĕg'lĭ-jəns] *noun*, failure to give the care the occasion demands

neg-li-gi-ble [nĕg'lĭ-jə-bəl] *adjective*, very small, insignificant

ne-go-ti-a-ble [nĭ-gō'shə-bəl] *adjective*, 1) transferable or able to be assigned to another person 2) able to be compromised about, open to discussion

ne-go-ti-ate [nĭ-gō'shē-āt'] *verb*, to settle, to work out, to deal or bargain with others, to compromise

Ne-gro [nē'grō] *noun*, a member of the human species having black or brown skin color

neigh-bor [nā'bər] *noun*, a person who lives near another

neigh-bor-hood [nā'bər-hŏŏd'] *noun*, a small part of a community shared by a group of people who live near each other

nei-ther [nē'thər] *adjective, pronoun,* 1) not the one nor the other, *conjunction,* 2) not either

ne-on [nē'ŏn'] *noun*, an inert gaseous element often used in display lights

neph-ew [nĕf'yōō] *noun*, the son of a brother or a sister

nep-o-tism [nĕp'ə-tĭz'əm] *noun*, favoritism to a relative in a business situation

nerve [nûrv] *noun*, a thread in the body that carries feelings and messages to and from the brain

nerv-ous [nûr'vəs] *adjective*, very worried and a little scared

nest [nĕst] *noun*, a temporary home built by an animal or insect

nes-tle [nĕs'əl] *verb*, to be very close to someone else in a comfortable way

net [nĕt] *noun* 1) a fabric made of tying threads together in a uniform design to form a mesh 2) remaining income after deducting taxes and expenses

net-tle [nĕt'l] *noun*, 1) a weed with small thorns, *verb*, 2) to annoy, to bother, to irritate, to vex

neu-rot-ic [nŏŏ-rŏt'ĭk] *adjective*, suffering from a mental disorder characterized by anxiety and worry and phobias

neu-ter [nōō'tər] *verb*, to make sterile, to spay or castrate

neu-tral [nōō'trəl] *adjective*, 1) neither positive nor negative 2) not on either side during a war 3) in the middle 4) a position in which gears are not engaged

neu-tron [nōō'trŏn'] *noun*, an uncharged particle in an atomic nucleus

Ne-va-da [nə-văd′ə] *noun*, one of the 50 United States, located in the West, the capital is Carson City. The state flower of **Nevada** is the sagebrush, and the motto is "All for our country."

nev-er [něv′ər] *adverb*, not at any time, absolutely not

nev-er-the-less [něv′ər-thə-lěs′] *adverb*, notwithstanding, in spite of that, yet, however

new [nōō] *adjective*, 1) not having been seen or used before 2) additional

New Hampshire [nōō hămp′shər] *noun*, one of the 50 United States located in New England, the capital is Concord. The state flower of **New Hampshire** is the purple lilac, the motto is "Live free or die."

New Jersey [nōō, jûr′zē] *noun*, one of the 50 United States located in the East, the capital is Trenton. The state flower of **New Jersey** is the purple violet and the motto is "Liberty and prosperity."

New Mexico [nōō měk′sĭ-kō′] *noun*, one of the 50 United States, located in the Southwest, the capital is Santa Fe. The state flower of **New Mexico** is the yucca and the motto is "It grows as it goes."

news [nōōz] *noun*, *plural*, a report of current events, things that happened recently

news-pa-per [nōōz′pā′pər] *noun*, a publication that appears daily or weekly, containing recent information

New York [nōō yôrk] *noun*, one of the 50 United States located in the Northeast, the capital is Albany. The state flower of **New York** is the rose and the motto is "Excelsior", which expresses the idea of still higher, ever upward.

next [někst] *adverb*, 1) in the nearest position, without anything between 2) immediately following

nib-ble [nĭb′əl] *verb*, to eat small bites of something

nice [nīs] *adjective*, 1) kind, good 2) pleasant, socially agreeable

niche [nĭch] *noun*, 1) a small, suitable place 2) the position of a particular population in an ecological community

nick-el [nĭk′əl] *noun*, 1) five cents, equal to five pennies 2) a hard, silvery metallic element of the iron group

nick-name [nĭk′nām′] *noun*, a name given to someone which is not his real name

nic-o-tine [nĭk′ə-tēn′] *noun*, a toxic alkaloid, the addictive property of tobacco

niece [nēs] *noun*, the daughter of one's brother or sister

nig-gard-ly [nĭg′ərd-lē] *adjective*, meanly stingy, meager

night [nīt] *noun*, the time of day after the sun goes down

night-mare [nīt′mâr′] *noun*, a terrifying dream or event

nine [nīn] *adjective*, *noun*, the number 9, eight plus one

nine-teen [nīn-tēn′] *noun*, nine plus ten, written 19

nine-teenth [nīn-tēnth′]*adjective*, the number 19 in order, 19th

nine-ti-eth [nīn′tē-ĭth] *adjective*, the number ninety in order, 90th

nine-ty [nīn′tē] *noun*, written 90, 9 X 10

ninth [nīnth] *noun, adjective*, number 9 in order, 9th

nip-ple [nĭp′əl] *noun*, 1) the tip of the breast where the milk ducts discharge 2) the rubber cap on a baby bottle

nir-va-na [nîr-vä′nə] *noun*, in Buddhist teachings, the ideal state, freedom from pain and worry, perfect bliss

ni-tro-gen [nī′trə-jən] *noun*, a gas necessary for life that is colorless, tasteless and odorless

no [nō] *adjective*, 1) not any, *adverb*, 2) a word we use to answer a question, or indicate disagreement

no-bil-i-ty [nō-bĭl′ĭ-tē] *adjective*, a member of the royal circle

no-ble [nō′bəl] *adjective*, 1) aristocratic, born to wealth 2) unselfish, of good character

no-bod-y [nō′bŏd′ē] *pronoun*, no one, a person of no importance

noc-tur-nal [nŏk-tûr′nəl] *adjective*, active or functioning at night

nod [nŏd] *noun*, 1) an inclination of the head, *verb*, 2) to tip the head slightly to acknowledge something

noise [noiz] *noun*, 1) a loud harsh sound, 2) sound of any kind

no-mad [nō′măd′] *noun*, a person who travels about with a tribe and has no address or home

nom-i-nate [nŏm′ə-nāt′] *verb*, to name for an office or place

nom-i-na-tion [nŏm′ə-nā′shən] *noun*, to submit a name for an award or appointment

nom-i-nee [nŏm′ə-nē′] *noun*, a person selected or appointed to run for a position

non-cha-lant [nŏn′shə-länt′] *adjective*, showing a jaunty coolness, aloof, unconcerned

none [nŭn] *adjective*, not any, nothing, no one, not one

non-ex-ist-ent [nŏn′ĭg-zĭs′tənt] *adjective*, not real

non-fic-tion [nŏn-fĭk′shən] *noun*, a true story

non-sense [nŏn′sĕns′] *noun*, folly, silliness, absurdity, meaningless

noon [nōōn] *noun*, midday, twelve o'clock, daytime

nor [nôr; nər when unstressed] *conjunction*, a word used between two choices after neither

nor-mal [nôr′məl] *adjective*, not unusual, common, every day

nor-mal-ly [nôr′mə-lē] *adverb*, generally, ordinarily, frequently

north [nôrth] *adjective*, 1) facing or coming from the north, *adverb*, 2) to or toward the north *noun*, 3) the direction to the left of one facing east, the compass point opposite to south, **North**, 4) regions or countries north of a point that is mentioned or understood

North America [nôrth-ə-mĕr′ĭ-kə]
noun, one of the seven continents
containing 7.9% of the world's
population

North Carolina [nôrth kăr′ə-lī′nə]
noun, one of the 50 United States
located in the Southeast, the
capital is Raleigh. The state flower
of **North Carolina** is the
dogwood, the state motto is "To
be rather than to seem."

North Dakota [nôrth də-kō′tə]
noun, a state located in the
Northwest of the United States,
the capital is Bismark. The state
flower of **North Dakota** is the
wild prairie rose, and the state
motto is "Liberty and union,
now and forever, one and
inseparable."

north-east [nôrth-ēst′] *noun,* the
point between north and east on
a compass

north-west [nôrth-wĕst′] *noun,* the
point between north and west on
the compass

nose [nōz] *noun,* 1) the part of the
face that contains the nostrils
and the organs of smell 2) the
front of a plane, rocket, etc.

nose-gay [nōz′gā′] *noun,* a small
hand-held fragrant bouquet

nos-tal-gia [nŏ-stăl′jə] *noun,*
homesickness, a sentimental
longing for something in the past

nos-tril [nŏs′trəl] *noun,* either of
the external openings of the nose

not [nŏt] *adverb,* a word that forms
a contradiction in a sentence

no-ta-ble [nō′tə-bəl] *adjective,*
remarkable, worthy of notice

no-ta-ry [nō′tə-rē] *noun,* an officer
who witnesses and verifies
commercial papers and is
authorized to take affidavits

no-ta-tion [nō-tā′shən] *noun,* notes
or marks to indicate something

notch [nŏch] *noun,* 1) a nick, a
dent, *verb,* 2) to make a notch

note [nōt] *noun,* 1) an observation
stuck in the mind 2) a bill, IOU,
paper money 3) a single sound
of music on the scale 4) a brief
written message, *verb,* 5) to
make an observation and commit
it to memory 6) to mention

noth-ing [nŭth′ĭng] *adjective,* not
anything, zero, insignificant

no-tice [nō′tĭs] *noun,* 1) a written
warning indicating that
something has happened or will
happen, *verb,* 2) to observe, to
become aware of

no-tice-a-ble [nō′tĭ-sə-bəl]
adjective, capable of being
observed, conspicuous

no-ti-fi-ca-tion [nō′tə-fĭ-kā′shən]
noun, something made known to
someone

no-ti-fy [nō′tə-fī′] *verb,* to inform
someone of something, to give
notice

no-tion [nō′shən] *noun,* a general or
vague idea, a thought, a whim

no-to-ri-e-ty [nō′tə-rī′ĭ-tē] *noun,*
fame, renown, celebrity status

no-to-ri-ous [nō-tôr′ē-əs]
adjective, widely or commonly
known, usually unfavorable

not-with-stand-ing
[nŏt′wĭth-stăn′dĭng] *adverb,* 1)
all the same, nevertheless,
preposition, 2) in spite of

noun [*noun*] *noun*, a word that refers to a person, place, or thing

nour-ish [nûr′ĭsh] *verb*, 1) to feed or foster the growth of something 2) to keep healthy, to nurture, to strengthen

nour-ish-ment [nûr′ĭsh-mənt] *noun*, food that strengthens the body and helps it grow

nov-el [nŏv′əl] *adjective*, 1) new, unusual, *noun*, 2) a work of fiction or romance

nov-el-ist [nŏv′ə-lĭst] *noun*, a person who writes fictional stories

nov-el-ty [nŏv′əl-tē] *noun*, something new

No-vem-ber [nō-vĕm′bər] *noun*, the eleventh month of the year, having 30 days

nov-ice [nŏv′ĭs] *noun*, 1) a beginner 2) a person admitted to a religious order before taking the final vows

now [nou] *adverb*, at the present time, immediately

no-where [nō′hwâr′] *adverb*, 1) not anywhere, *noun*, 2) a place that does not exist

nox-ious [nŏk′shəs] *adjective*, harmful, deadly fumes

noz-zle [nŏz′əl] *noun*, a short outlet or pipe as of a hose, a spout

nu-ance [nōō′äns′] *noun*, a shade or trace of differences in meaning, or color, or tone

nu-cle-ar [nōō′klē-ər] *adjective*, using the very great power made by splitting an atom or joining atoms in nuclear fission

nu-cle-us [nōō′klē-əs] *noun*, 1) a central mass of an atom about which matter is concentrated 2) the seed of the cell is the

positively charged central region of an atom

nude [nōōd] *adjective*, naked, bare

nudge [nŭj] *noun*, 1) a gentle push, *verb*, 2) to push someone lightly

nug-get [nŭg′ĭt] *noun*, a small lump, especially of natural gold

nui-sance [nōō′səns] *noun*, anything which annoys or gives trouble, a disturbance

nul-li-fy [nŭl′ə-fī′] *adjective*, to render invalid or void

numb [nŭm] *adjective*, 1) insensitive from excessive cold 2) unable to feel pain

num-ber [nŭm′bər] *noun*, 1) a symbol representing a quantity 2) abundance, *verb*, 3) to identify with a number 4) to amount to

nu-mer-al [nōō′mər-əl] *noun*, a symbol used to represent a number

nu-mer-ous [nōō′mər-əs] *adjective*, very many, myriad, several

nun [nŭn] *noun*, a woman who gives her life to serving God

nup-tial [nŭp′shəl] *adjective*, marriage vows, a wedding

nurse [nûrs] *noun*, 1) a person who takes care of the sick and dying, *verb*, 2) to look after and attend to the needs of the sick

nurs-er-y [nûr′sə-rē] *noun*, 1) a child's room or playroom 2) a greenhouse for young plants

nur-ture [nûr′chər] *verb*, to care for, to feed and protect

nut [nŭt] *noun*, 1) the edible kernel of a seed 2) a small piece of metal with a hole in it, used to hold a bolt in place

nut-meg [nŭt′mĕg′] *noun*, the aromatic seed of an East Indian tree used as a spice

nu-tri-ent [noo′trē-ənt] *noun*, vitamins and minerals derived from food, proteins

nu-tri-tion [noo-trĭsh′ən] *noun*, a healthy diet, that nourishes the body

nu-tri-tious [noo-trĭsh′əs] *adjective*, wholesome, healthy

ny-lon [nī′lŏn′] *noun*, a strong man-made elastic thread

nymph [nĭmf] *noun*, 1) the young of an insect that has not fully developed the structural proportions and size of an adult 2) a fairy

O

oaf [ōf] *noun*, a clumsy person

oak [ōk] *noun*, a tree of hard wood bearing the acorn as a fruit

oar [ôr, ōr] *noun*, a paddle with a flat blade at the end, used to row or push a boat

o-a-sis [ō-ā′sĭs] *noun*, a fertile spot in a desert having a spring

oat [ōt] *noun*, a member of the cereal family having edible seeds such as rice, wheat or corn

oath [ōth] *noun*, a promise, a commitment, a solemn appeal

o-be-di-ence [ō-bē′dē-əns] *noun*, *no plural*, compliance with authority, submission

o-be-di-ent [ō-bē′dē-ənt] *adjective*, willing to do what one is told to do

ob-e-lisk [ŏb′ə-lĭsk] *noun*, a tall 4-sided column tapering and ending in a pyramid apex

o-bese [ō-bēs′] *adjective*, very fat

o-bey [ō-bā′] *verb*, to follow or carry out an order or law

ob-fus-cate [ŏb′fə-skāt′] *verb*, to confuse, to muddle, to obscure

o-bit-u-ar-y [ō-bĭch′oo-ĕr′ē] *noun*, an announcement of someone's death often with a short biography

ob-ject [ŏb′jĭkt] *noun*, 1) a thing, *verb*, 2) to oppose, to disagree

ob-jec-tion [əb-jĕk′shən] *noun*, a statement, a complaint

ob-jec-tive [əb-jĕk′tĭv] *adjective*, 1) not influenced by emotions, fair, *noun*, 2) a purpose or goal

ob-jur-gate [ŏb′jər-gāt′] *verb*, to scold, to rebuke severely

ob-li-ga-tion [ŏb′lĭ-gā′shən] *noun*, 1) the binding power of a promise 2) a debt of gratitude

ob-lig-a-to-ry [ə-blĭg′ə-tôr′ē] *adjective*, necessary, binding in law or conscience, mandatory

o-blige [ə-blīj′] *verb*, 1) to put under obligation, to accommodate 2) to do a favor for, to help, to assist

o-blique [ō-blēk′] *adjective*, 1) sloping or slanting 2) indirect or evasive, not explicit

ob-lit-er-ate [ə-blĭt′ə-rāt′] *verb*, to erase or blot out, to efface

ob-liv-i-on [ə-blĭv′ē-ən] *noun*, being utterly forgotten

ob-liv-i-ous [ə-blĭv′ē-əs] *adjective*, forgetful, lost in thought

ob-long [ŏb′lông′] *adjective*, a flat shape with four straight sides and four equal angles, that is longer than it is wide

ob-nox-ious [ŏb-nŏk′shəs]
adjective, annoying, offensive

ob-scure [ŏb-skyoŏr′] *adjective*,
1) unclear, vague, unknown
2) not easily seen or heard
3) dark, hidden

ob-scu-ri-ty [ŏb-skyoŏr′ĭ-tē] *noun*,
oblivion, no where

ob-se-qui-ous [ŏb-sē′kwē-əs]
adjective, submissive, fawning

ob-ser-vance [əb-zûr′vəns] *noun*,
1) noticing with attention 2)
compliance as a duty or custom

ob-ser-va-tion [ŏb′zər-vā′shən]
noun, the noting and recording
of facts and events

ob-serv-a-to-ry [əb-zûr′və-tôr′ē]
noun, a building equipped with
instruments for observing the
heavenly bodies, such as the
stars, moon, etc.

ob-serve [əb-zûrv′] *verb*, 1) to see
or notice 2) to comment 3) to
mark an event or day

ob-ses-sion [əb-sĕsh′ən, ŏb-] *noun*,
a fixed idea, continued brooding

ob-so-lete [ŏb′sə-lēt′] *adjective*, no
longer in use, outdated

ob-sta-cle [ŏb′stə-kəl] *noun*,
something that stands in the
way, an obstruction

ob-ste-tri-cian [ŏb′stĭ-trĭsh′ən]
noun, a physician specializing in
the delivery of babies

ob-sti-nate [ŏb′stə-nĭt] *adjective*,
not yielding to reason, stubborn

ob-struct [əb-strŭkt′] *verb*, to get
in the way of something or stop
it completely, to hinder

ob-struc-tion [əb-strŭk′shən] *noun*,
a barrier, an interference

ob-tain [əb-tān′] *verb*, to acquire or
win through effort or request

ob-tain-a-ble [əb-tān′ə-bəl]
adjective, capable of being
acquired or won

ob-trude [ŏb-troōd′] *verb*, 1) to get
in on, to meddle, 2) to stick out

ob-tuse [ŏb-toōs′] *adjective*, an
angle greater than 90 degrees

ob-vi-ous [ŏb′vē-əs] *adjective*,
easily seen or understood,
evident, apparent to everyone

ob-vi-ous-ly [ŏb′vē-əs-lē] *adverb*,
apparently, clearly, evidently

oc-ca-sion [ə-kā′zhən] *noun*, 1) an
event, a happening, an important
time 2) a chance, an opportunity

oc-cas-ion-al-ly [ə-kā′zhə-nə-lē]
adverb, once in a while

oc-cult [ə-kŭlt′] *adjective*, the
mysterious, secret, supernatural

oc-cu-pant [ŏk′yə-pənt] *noun*, the
inhabitant, a resident, a tenant

oc-cu-pa-tion [ŏk′yə-pā′shən]
noun, 1) a job, a way of using
time, *no plural*, 2) being in a
certain place or space

oc-cu-py [ŏk′yə-pī′] *verb*, 1) to
make or hold possession of
something 2) to live in a place

oc-cur [ə-kûr′] *verb*, 1) to happen
2) to come to mind, to appear

oc-cur-rence [ə-kûr′əns] *noun*,
something that happens

o-cean [ō′shən] *noun*, a body of
salt water covering three fifths of
the earth

o-cea-nog-ra-phy [ō′shə-nŏg′rə-fē]
noun, the study of life in the sea

o'clock [ə-klŏk′] *adverb*, according to the clock

Oc-to-ber [ŏk-tō′bər] *noun*, the tenth month of the year having 31 days

oc-to-pus [ŏk′tə-pəs] *noun*, a sea creature that has eight long limbs

oc-u-list [ŏk′yə-lĭst] *noun*, a doctor who treats the eyes

odd [ŏd] *adjective*, 1) any number that cannot be evenly divided by two 2) unusual, very different

odds [ŏdz] *noun*, *plural*, uncertain advantages, doubtful chances

odi-ous [ō′dē-əs] *adjective*, hateful, repulsive, repugnant

odi-um [ō′dē-əm] *noun*, repugnance, intense hatred

o-dom-e-ter [ō-dŏm′ĭ-tər] *noun*, a device in a vehicle used to measure distance traveled

o-dor [ō′dər] *noun*, a smell

o-dor-ous [ō′dər-əs] *adjective*, having a strong smell

of [ŭv] *preposition*, 1) belonging to 2) from among 3) made from 4) about 5) in relation to

off [ôf] *adverb*, 1) away from 2) not on or not working, *preposition*, 3) at a distance

of-fend [ə-fĕnd′] *verb*, to hurt someone's feelings, to insult

of-fense [ə-fĕns′] *noun*, a misdeed, an insult, a sense of insult or injury, an act of aggression

of-fen-sive [ə-fĕn′sĭv] *adjective*, 1) insulting, in poor taste, *noun*, 2) the position of an attack

of-fer [ô′fər] *noun*, 1) a proposal, a suggestion, *verb*, 2) to present to be accepted or refused

of-fice [ô′fĭs] *noun*, 1) a place in which business or professional work is done 2) a position that requires certain duties and tasks

of-fi-cer [ô′fĭ-sər] *noun*, a person lawfully invested with a position of trust or authority

of-fi-cial [ə-fĭsh′əl] *adjective*, 1) having the right or permission to do something, *noun*, 2) a person who is in charge of something 3) of or pertaining to the holding of an office

of-fi-ci-ate [ə-fĭsh′ē-āt′] *verb*, to act as an officer in performing a duty such as a minister

off-spring [ôf′sprĭng′] *noun*, descendants, children, progeny

of-ten [ô′fən] *adverb*, many times, frequently, repeatedly

o-ha-na [ōhŏna] *noun*, family, no one left behind

Ohio [ō-hī′ō] *noun*, one of the 50 United States located in the Midwest region, the capital is Columbus. The state flower of **Ohio** is the scarlet carnation and the motto is "With God all things are possible."

ohm [ōm] *noun*, the unit of electrical resistance

oil [oil] *noun*, 1) any liquid substance that does not dissolve in water 2) a nutrient derived from a pressed nut 3) a dark, thick mineral liquid, found deep in the ground 4) a substance used to burn in a lantern or lamp post, *verb*, 5) to lubricate with grease or oil

oint-ment [oint′mənt] *noun*, a lubricant put on the skin, containing medicine

Okla-ho-ma [ō′klə-hō′mə] *noun*, one of the 50 United States located in the Southwest, the capital is Oklahoma City. The state flower of **Oklahoma** is the mistletoe, and the motto is "Labor conquers all things."

old [ōld] *adjective*, 1) not young, having lived a long time 2) the word we use to show our age 3) not new 4) something that has lasted over time, antique

ol-fac-to-ry [ŏl-făk′tə-rē] *adjective*, referring to the nose and the sense of smell

ol-i-gar-chy [ŏl′ĭ-gär′kē] *noun*, a government run by a few leaders

ol-ive [ŏl′ĭv] *noun*, the fruit of an olive tree eaten as a relish

om-i-nous [ŏm′ə-nəs] *adjective*, looking or sounding as if something bad will happen, portending evil or harm

o-mis-sion [ō-mĭsh′ən] *noun*, failure to do something, something that is left out

o-mit [ō-mĭt′] *verb*, to leave something out, to fail to include

om-e-lette [ŏm′ə-lĭt] *noun*, eggs beaten together and cooked in hot fat in a flat pan

om-nip-o-tent [ŏm-nĭp′ə-tənt] *adjective*, having power over all

om-ni-pres-ent [ŏm′nĭ-prĕz′ənt] *adjective*, universally present at the same time, everywhere

om-nis-cient [ŏm-nĭsh′ənt] *adjective*, all-knowing

om-ni-vore [ŏm′nə-vôr′] *noun*, an animal that eats both plants and animals

on [ŏn] *adverb*, 1) to be precise 2) a location 3) to show time or date 4) about 5) to continue, *preposition*, 6) atop

once [wŭns] *adverb*, 1) one time 2) a long time ago

one [wŭn] *adjective*, 1) the number 1 2) the same, unified 3) act as the same, *noun*, 4) a single thing 5) any person 6) a unit

one-self [wŭn-sĕlf′] *pronoun*, the same person who is the subject of the sentence

on-ion [ŭn′yən] *noun*, a garden vegetable having an edible bulb with pungent taste and odor

on-ly [ōn′lē] *adjective*, 1) sole, no one else, nothing more, 2) but, *adverb*, 3) merely, just

on-o-mat-o-poe-ia [ŏn′ə-măt′ə-pē′ə] *noun*, a term used to describe words that actually sound like the noises they depict such as *boom*

on-slaught [ŏn′slôt′] *noun*, a vicious assault, a massacre

on-to [ŏn′tōō′] *preposition*, to a place, to a position on

o-nus [ō′nəs] *noun*, 1) a burden 2) a responsibility, blame

on-ward [ŏn′wərd] *adverb*, moving toward a point ahead, forward

ooze [ōōz] *verb*, to flow out slowly

o-pal [ō′pəl] *noun*, a precious stone of milky hue, with a play of iridescent colors

o-paque [ō-pāk'] *adjective*, not reflecting or giving out light, not transparent; dull, dark

o-pen [ō'pən] *adjective*, 1) not shut or covered 2) ready for business 3) not surrounded by other things 4) unbiased, free thinking, *verb*, 5) to allow entrance 6) to start, to unfold 7) to remove the cover or door 8) to undo, to untie

o-pe-ra [ŏp'ər-ə] *noun*, a dramatic presentation set to music

op-er-ate [ŏp'ə-rāt'] *verb*, to make something work or function

op-er-a-tion [ŏp'ə-rā'shən] *noun*, 1) any methodical action 2) a plan put into action 3) surgery

op-er-a-tor [ŏp'ə-rā'tər] *noun*, one who operates equipment or works as a manager

o-pi-ate [ō'pē-ĭt] *noun*, a drug which causes rest or sleep, a narcotic or sedative

o-pin-ion [ə-pĭn'yən] *noun*, a belief stronger than an impression, a point of view, a conclusion

o-pos-sum [ə-pŏs'əm] *noun*, a rodent that sleeps upside down hanging from a tree during the day, and travels at night

op-po-nent [ə-pō'nənt] *noun*, a member of the other team, an antagonist, an adversary

op-por-tune [ŏp'ər-tōōn'] *adjective*, timely, well chosen

op-por-tun-ist [ŏp'ər-tōō'nĭst] *noun*, an individual who takes advantage of a situation and ignores ethics to achieve a goal

op-por-tu-ni-ty [ŏp'ər-tōō'nĭ-tē] *noun*, a fit or convenient time, occasion, a good chance

op-pose [ə-pōz'] *verb*, to go against, to defend, to combat, to object, to resist, to fight

op-po-site [ŏp'ə-zĭt] *adjective*, 1) facing, set over against 2) one that is contrary to another

op-po-si-tion [ŏp'ə-zĭsh'ən] *noun*, against, hostile or contrary action, a negative response

op-press [ə-prĕs'] *verb*, to overpower, to subdue, to treat harshly, to hold down

op-pres-sion [ə-prĕsh'ən] *noun*, hardship, cruelty, persecution

op-pres-sive [ə-prĕs'ĭv] *adjective*, 1) unreasonably burdensome 2) tyrannical, cruel, unfair

op-tic [ŏp'tĭk] *adjective*, pertaining to the eye, light and vision

op-ti-cal [ŏp'tĭ-kəl] *adjective*, visual, seeing

op-ti-cian [ŏp-tĭsh'ən] *noun*, a person who makes glasses, a dealer in optical goods

op-ti-mal [ŏp'tə-məl] *adjective*, most favorable condition

op-ti-mist [ŏp'tə-mĭst] *noun*, one who looks on the bright side of things, positive thinking

op-ti-mis-tic [ŏp'tə-mĭs-tĭk] *adjective*, having a positive outlook, upbeat, hopeful

op-tion [ŏp'shən] *noun*, 1) a choice 2) a stock available at a certain time at a stated price

op-tom-e-trist [ŏp-tŏm'ĭ-trĭst] *noun*, one who examines eyes and fits glasses to help a person see better

op-u-lence [ŏp'yə-ləns] *adjective*, having property, wealthy, rich, affluent, luxurious

o-pus [ō′pəs] *noun*, a great creative literary or musical work

or [ôr;] *conjunction*, used to connect words when given a choice

o-ral [ôr′əl] *adjective*, 1) spoken information 2) of the mouth

or-ange [ôr′ĭnj] *adjective*, 1) a color made when red and yellow are mixed together, *noun*, 2) a sweet, juicy fruit grown on trees in tropical climates

or-a-tor [ôr′ə-tər] *noun*, a public speaker of great eloquence

or-a-to-ri-o [ôr′ə-tôr′ē-ō′] *noun*, a dramatic poem set to music

or-a-to-ry [ôr′ə-tôr′ē] *noun*, a moving or powerful speech

or-bit [ôr′bĭt] *noun*, 1) the course an object follows in space around another, *verb*, 2) to travel through space in a cycle

or-chard [ôr′chərd] *noun*, a grove where fruit trees grow for food

or-ches-tra [ôr′kĭ-strə] *noun*, 1) a band of performers on various instruments 2) the lower floor in a theater for instruments

or-chid [ôr′kĭd] *noun*, an exotic tropical flower with three petals

or-dain [ôr-dān′] *verb*, to appoint, to install, to elect, to decree

or-deal [ôr-dēl′] *noun*, an extremely stressful situation

or-der [ôr′dər] *noun*, 1) a careful arrangement 2) a special system used to organize information 3) grouping, placement 4) a request telling someone to do something, *verb*, 5) to arrange carefully 6) to issue a command, to instruct

or-di-nal num-ber [ôr′dn-əl-nŭm′bər] *noun*, a number expressing degree or order in a series such as 1st, 2nd,

or-di-nance [ôr′dn-əns] *noun*, 1) a local law, an edict or decree 2) a prescribed practice or usage

or-di-nar-i-ly [ôr′dn-âr′ə-lē] *adverb*, usually, generally

or-di-nar-y [ôr′dn-ĕr′ē] *adjective*, 1) according to custom or established order 2) regular, commonplace, normal

ore [ôr, ōr] *noun*, a kind of rock in which metal is found

Or-e-gon [ôr′ĭ-gən] *noun*, one of the 50 United States in the Northwest, the capital is Salem. The state flower of **Oregon** is the Oregon grape, and the motto is "The Union."

or-gan [ôr′gən] *noun*, 1) a musical instrument similar to a piano, with long pipes through which air travels through to make the sound 2) a part of the body that serves a specific function

or-gan-ic [ôr-găn′ĭk] *adjective*, grown naturally without pesticides or fertilizers

or-gan-ism [ôr′gə-nĭz′əm] *noun*, any individual life form

or-gan-i-za-tion [ôr′gə-nĭ-zā′shən] *noun*, an association, a group of people working together for a common purpose

or-gan-ize [ôr′gə-nīz′] *verb*, 1) to plan, to arrange and implement in an orderly way 2) to establish or construct a plan or a scheme

o-ri-en-tal [ôr′ē-ĕn′tl] *adjective*, pertaining to the Orient or East Asia

o-ri-en-ta-tion [ôr'ē-ĕn-tā'shən] *noun*, the act of getting your bearings, an introduction

or-i-fice [ôr'ə-fĭs] *noun*, a mouthlike opening, a slit, any hole in a body

or-i-gin [ôr'ə-jĭn] *noun*, the first use or the beginning

o-rig-i-nal [ə-rĭj'ə-nəl] *adjective*, the first of its kind, new and different, inventive

o-rig-i-nal-i-ty [ə-rĭj'ə-năl'ĭ-tē] *noun*, creativeness, innovation

o-rig-i-nate [ə-rĭj'ə-nāt'] *verb*, 1) to bring into existence, to initiate 2) to produce as new

or-na-ment [ôr'nə-mənt] *noun*, a decoration, beautification

or-na-men-tal [ôr'nə-mĕn'tl] *adjective*, decorative, adorned, beautiful, embellished

or-nate [ôr-nāt'] *adjective*, excessively decorated, adorned

or-ni-thol-o-gist [ôr'nə-thŏl'ə-jĭst] *noun*, one who studies birds

or-phan [ôr'fən] *noun*, a child whose parents are dead

or-phan-age [ôr'fə-nĭj] *noun*, a home for children who have no family

or-tho-dox [ôr'thə-dŏks'] *adjective*, accepted, traditional doctrine, very conservative

or-thog-ra-phy [ôr-thŏg'rə-fē] *noun*, correct spelling

os-cil-late [ŏs'ə-lāt'] *verb*, 1) to swing backward and forward, to vibrate 2) varying above or below a mean value

os-si-fy [ŏs'ə-fī'] *verb*, 1) to change or harden cartilage into bone 2) to set into a repeating pattern

os-ten-si-ble [ŏ-stĕn'sə-bəl] *adjective*, shown, professed, apparent, pretended

os-ten-ta-tious [ŏs'tĕn-tā'shəs] *adjective*, pretentious, conspicuous, showiness as an attempt to attract attention

os-tra-cize [ŏs'trə-sīz'] *verb*, to banish, to cast out from social favor or fellowship

os-trich [ŏs'trĭch] *noun*, a very large bird, with long legs, that is black and white, and cannot fly

oth-er [ŭth'ər] *adjective*, 1) not the same, a different one, *adverb*, 2) differently, *pronoun*, 3) someone or something not mentioned specifically

oth-er-wise [ŭth'ər-wīz'] *adverb*, 1) if not 2) except for that 3) of a different kind

ought [ôt] *verb*, should

ounce [ouns] *noun*, 1) a weight measurement equal to 28.35 grams 2) a tiny bit

our [our] *adjective*, belonging to us

our-selves [our-sĕlvz'] *pronoun*, referring to the subjects in a sentence, we or us

oust [oust] *verb*, to dismiss, to kick out, to expel, to force out

out [out] *adverb*, 1) not in or inside, away from 2) absent, away 3) not shining or burning 4) to bring to a conclusion 5) excluded, omitted

out-cast [out'kăst'] *noun*, a displaced person, an exile

out-come [out′kŭm′] *noun*, a possible result as in an experiment, a final product

out-doors [out-dôrz′, -dōrz′] *adjective*, not inside a building, located outside in the open air

out-doors-man [out-dôrz′mən, -dōrz′-] *noun*, a person who enjoys nature and outdoor activities

out-er [ou′tər] *adjective*, on the outside or edge of something

out-fit [out′fĭt′] *noun*, 1) clothing used for a particular purpose, *verb*, 2) to provide clothes or equipment

out-grow [out-grō′] *verb*, to change as one develops

out-growth [out′grōth′] *noun*, development, product

out-ing [ou′tĭng] *noun*, a short trip

out-let [out′lĕt′, -lĭt] *noun*, 1) an opening, 2) manufacturer

out-line [out′līn′] *noun*, 1) a line showing the shape of something 2) the plan or format for a composition, *verb*, 3) to give the main points of

out-live [out-lĭv′] *verb*, to outlast, to live or last longer than

out-rage [out′rāj′] *noun*, 1) anger caused by injury or insult 2) anything that causes resentment or anger, a wicked or brutal act or remark *verb*, 2) to fill with anger of resentment

out-side [out-sīd′] *adverb*, 1) on or to the exterior surface, *noun*, 2) the outer part or surface

out-stand-ing [out-stăn′dĭng] *adjective*, 1) very good, excellent 2) not paid or settled

o-val [ō′vəl] *adjective*, a round flat shape like an egg

ov-en [ŭv′ən] *noun*, an electric appliance like a sealed box used to bake food

o-ver [ō′vər] *adjective*, 1) in control of 2) finished, *adverb*, 3) from start to finish again 4) unused, extra 5) to go beyond the top, *preposition*, 6) covering 7) across, from one side to the other 8) in every part

o-ver-all [ō′vər-ôl′] *adjective*, including all things, as a whole

o-ver-alls [ō′vər-ôlz′] *noun*, a kind of loose trousers with a bib

o-ver-board [ō′vər-bôrd′] *adverb*, over the side of a boat into the water

o-ver-cast [ō′vər-kăst′] *adjective*, cloudy, sunless, gloomy

o-ver-coat [ō′vər-kōt′] *noun*, a coat worn over the other clothing

o-ver-come [ō′vər-kŭm′] *verb*, to beat, to conquer, to lick

o-ver-due [ō′vər-dōō′] *adjective*, delinquent, late, past due

o-ver-es-ti-mate [ō′vər-ĕs′tə-māt′] *verb*, to calculate over the actual amount

o-ver-flow [ō′vər-flō′] *verb*, 1) to pour over the edge of a container when filled 2) to flood

o-ver-haul [ō′vər-hôl′] *verb*, to restore, to repair, to rebuild

o-ver-head [ō′vər-hĕd′] *adjective*, 1) above our heads, in the sky, *noun*, *no plural*, 2) the fixed costs to maintain something

o-ver-lap [ō′vər-lăp′] *verb*, 1) to rest on top of or over something

and cover part of it 2) to
partially coincide

o-ver-look [ō'vər-loŏk'] *verb*, 1) to
fail to notice 2) to excuse

o-ver-night [ō'vər-nīt'] *adjective*,
1) from one day to the next,
adverb, 2) during the night

o-ver-seas [ō'vər-sēz'] *adjective*,
across the ocean, abroad

o-ver-seer [ō'vər-sē'ər] *noun*, one
who watches over an area or
institution, a supervisor

o-ver-sight [ō'vər-sīt'] *noun*, an
error, a blunder, a failure

o-vert [ō-vûrt'] *adjective*, open to
view, not concealed

o-ver-take [ō'vər-tāk'] *verb*, to
chase and catch up to

o-ver-ture [ō'vər-choŏr'] *noun*, a
proposal, an introduction

o-ver-whelm [ō'vər-hwĕlm'] *verb*,
1) to overpower in mind and
feeling 2) to engulf, to inundate

owe [ō] *verb*, 1) to have to give or
pay 2) to feel grateful to
someone for something

owl [oul] *noun*, a large bird that
lives in the forest, it flies at night
and eats small animals

own [ōn] *adjective*, 1) of or
belonging to oneself or itself,
verb, 2) to have or possess

own-er [ō'nər] *noun*, the person
who has the right to something

own-er-ship [ō'nər-shĭp'] *noun*, a
right of possession, a lawful
claim or title

ox [ŏks] *noun*, *plural*, **oxen**,
a full-grown male of domestic
cattle used for farm work

ox-i-dize [ŏk'sĭ-dīz'] *verb*, to
combine with oxygen

ox-y-gen [ŏk'sĭ-jən] *noun*, a
colorless, tasteless, odorless,
chemically active gas, the part of
air that animals, plants and
people need to stay alive

oys-ter [oi'stər] *noun*, an edible
shellfish, a bivalve mollusk

o-zone [ō'zōn'] *noun*, a faintly blue
gas obtained by the silent
discharge of electricity in air

P

pace [pās] *noun*, 1) a steady step or
speed that is maintained, *verb*,
2) to walk back and forth

pach-y-derm [păk'ĭ-dûrm'] *noun*, a
thick-skinned animal

pac-i-fist [pə-sĭf'ĭ-st] *noun*,
someone who looks for peaceful
means, opposed to force

pac-i-fy [păs'ə-fī'] *verb*, to
appease, to calm, to soothe

pack [păk] *noun*, 1) a bundle or a
collection of things 2) a group of
wild animals that hunt together,
verb, 3) to bundle or place a
group of things into a container

pack-age [păk'ĭj] *noun*, a box or
bundle containing one or more
objects

pact [păkt] *noun*, an agreement, a
contract, a treaty, a bargain

pad [păd] *noun*, 1) a soft material
used to protect a wound 2) a
number of sheets of paper stuck
together at one edge

pad-dle [păd'l] *noun*, 1) an ore or
long stick used to row a boat,

verb, 2) to spank, to whack on the bottom 3) to row a boat 4) to move the hands and feet in shallow water

pad-dock [păd'ək] *noun*, a field or enclosure where horses are kept

pad-dy [păd'ē] *noun*, a field for growing rice

pad-lock [păd'lŏk'] *noun*, a portable lock, jointed at one end

page [pāj] *noun*, 1) one side of a piece of paper in a book 2) a person who delivers messages *verb*, 2) to summon or call indirectly through an intercom or paging system

pag-eant [păj'ənt] *noun*, an exhibition, a show, a parade

paid [pād] *verb*, the *past tense* of **pay**

pail [pāl] *noun*, a bucket or container with a handle

pain [pān] *noun*, physical suffering from injury or illness, distress

pain-ful [pān'fəl] *adjective*, a feeling of hurt, aching

paint [pānt] *noun*, 1) a liquid in a variety of colors that is used to cover a surface to protect or decorate it, *verb*, 2) to coat or color with paint 3) to convey an image by creating it using paints

paint-er [pān'tər] *noun*, a person who paints pictures or buildings

paint-ing [pān'tĭng] *noun*, completed pictures done with paint, a representation

pair [pâr] *noun*, 1) two like things put together as a match 2) an instrument with two parts joined together, e.g. *scissors*, *verb*, 3) to separate into groups of two

pa-ja-mas [pə-jä'mə] *noun*, *plural*, loose fitting clothes worn to bed

pal [păl] *noun*, a friend, chum

pal-ace [păl'ĭs] *noun*, a beautiful home of a wealthy person

pal-at-a-ble [păl'ə-tə-bəl] *adjective*, 1) tasty 2) acceptable to the mind, pleasing

pa-la-tial [pə-lā'shəl] *adjective*, suitable for a palace, impressively spacious

pale [pāl] *adjective*, 1) light in color, faded, *noun*, 2) the area enclosed by a boundary

pal-ette [păl'ĭt] *noun*, a board on which a painter mixes paints

pal-i-sade [păl'ĭ-sād'] *noun*, a partition formed as a barrier

pal-lid: [păl'id] *adjective*, pale

palm [päm] *noun*, 1) a tree with a long trunk without branches and large leaves at the top 2) the wide part inside the hand

pal-pa-ble [păl'pə-bəl] *adjective*, 1) capable of being touched or felt, obvious 2) tangible

pal-pi-tate [păl'pĭ-tāt'] *verb*, to throb, to pound with emotion or exertion, to quiver, to shake

pal-try [pôl'trē] *adjective*, insignificant, petty, trifling

pam-per [păm'pər] *verb*, to coddle, to indulge, to baby, to spoil

pam-phlet [păm'flĭt] *noun*, a book of a few sheets of printed matter

pan [păn] *noun*, a round metal pot for cooking things over heat

pan-a-ce-a [păn'ə-sē'ə] *noun*, a solution to all problems or sickness, a cure-all

pan-cre-as [păng′krē-əs] *noun*, a large gland under and behind the stomach that secretes insulin to the intestine to help digest food

pan-da [păn′də] *noun*, a large black and white animal like a bear restricted to China

pan-de-mo-ni-um [păn′də-mō′nē-əm] *noun*, a wild tumult, an uproar, confusion

pan-der [păn′dər] *verb*, to cater to the low desires of others

pane [pān] *noun*, a piece of glass used in windows or doors

pan-el [păn′əl] *noun*, 1) a sunken compartment with raised margins in a ceiling 2) a complete jury 3) a group assembled to judge a contest or conduct a public discussion

pan-ic [păn′ĭk] *noun*, 1) a sudden fear which can spread quickly, *verb*, 2) to feel frightened or alarmed, filled with terror

pan-sy [păn′zē] *noun*, a flower that blooms through the winter

pant [pănt] *verb*, to breathe quickly

pan-ther [păn′thər] *noun*, one of the large wild cats, a cougar

pan-to-mime [păn′tə-mīm′] *noun*, a play in which actors use gestures without speech

pan-try [păn′trē] *noun*, a small room where food is stored

pants [pănts] *noun*, *plural*, trousers

pa-pa [pä′pə] *noun*, a father, a dad

pa-per [pā′pər] *noun*, 1) sheets of thin material used for writing, wrapping, etc. 2) a newspaper 3) paper with writing or printing on it verifying identity

pap-ri-ka [pă-prē′kə] *noun*, a Turkish red pepper

pa-py-rus [pə-pī′rəs] *noun*, a reed plant that grows in the Nile Valley once used as paper

par [pär] *noun*, average, standard, the norm, equality in value

par-a-ble [păr′ə-bəl] *noun*, a short, simple story teaching a lesson

par-a-chute [păr′ə-shōot′] *noun*, 1) a light device made of silk or nylon, designed in an umbrella shape, used to retard the free fall from an aircraft, *verb*, 2) to eject from an aircraft by jumping out equipped with a parachute used to slow something's fall to the ground from a great height

pa-rade [pə-rād′] *noun*, 1) a number of people walking or marching together, a public procession, *verb*, 2) to walk or march with a group of people

par-a-digm [păr′ə-dīm′] *noun*, a model, an example, a pattern

par-a-dise [păr′ə-dīs′] *noun*, a place of complete happiness or beauty, like heaven

par-a-dox [păr′ə-dŏks′] *noun*, a statement that contradicts itself, but is in fact true

par-af-fin [păr′ə-fĭn] *noun*, a white wax used to make candles and seal preserves

par-a-gon [păr′ə-gŏn′] *noun*, a model of perfection

par-a-graph [păr′ə-grăf′] *noun*, a subdivision in writing, beginning with an indent on a new line

par-a-keet [păr′ə-kēt′] *noun*, a small blue or green domestic song bird

par-al-lel [păr′ə-lĕl′] *adjective*,
1) lying in the same direction and always the same distance apart, never intersecting 2) similar

par-al-lel-o-gram
[păr′ə-lĕl′ə-grăm′] *noun*, a four sided plane figure with opposite sides equal and parallel

pa-ral-y-sis [pə-răl′ĭ-sĭs] *adjective*, loss of sensation and motion of body parts, a helpless state

par-a-lyze [păr′ə-līz′] *verb*, to destroy the energy of something, to stop all movement or feeling

pa-ram-e-ter [pə-răm′ĭ-tər] *noun*, the limit, a set standard

par-a-mount [păr′ə-mount′] *adjective*, most important

par-a-noi-a [păr′ə-noi′ə] *noun*, insanity marked by delusions of persecution or glory

par-a-phrase [păr′ə-frāz′] *verb*, to restate a passage in one's own words for clarity

par-a-site [păr′ə-sīt′] *noun*, animal or plant living off another, using it as its only source of food

par-cel [pär′səl] *noun*, 1) a package wrapped in paper, for mailing or carrying, *verb*, 2) to divide into portions to hand out

parch [pärch] *verb*, to dry in the sun, to dehydrate, to scorch

parch-ment [pärch′mənt] *noun*, the skin of a goat or sheep prepared for writing

par-don [pär′dn] *noun*, *no plural*, 1) forgiveness, *verb*, 2) to excuse, to forgive, to absolve

par-don-a-ble [pär′dn-ə-bəl] *adjective*, excusable, forgivable

pare [pâr] *verb*, 1) to peel, to cut off the outer edge, 2) to reduce

par-ent [pâr′ənt] *noun*, 1) the father or mother 2) the source

pa-ren-the-sis [pə-rĕn′thĭ-sĭs] *noun*, one of the curved lines which enclose an aside ( )

par-ish [păr′ĭsh] *noun*, an area served by one church

park [pärk] *noun*, 1) a place in town designated for recreation, rest, and beauty, *verb*, 2) to put a car or vehicle in place

par-ka [pär′kə] *noun*, a heavy winter coat with a hood

par-ley [pär′lē] *noun*, a conference between two opponents, a discussion to settle a dispute

par-lia-ment [pär′lə-mənt] *noun*, an assembly in Great Britain similar to the United States Congress

par-lor [pär′lər] *noun*, a room that is usually used for talking or entertaining, a living room

pa-ro-chi-al [pə-rō′kē-əl] *adjective*, of or pertaining to a parish, a local area

par-o-dy [păr′ə-dē] *noun*, a humorous imitation, satire

pa-role [pə-rōl′] *noun*, an early conditional release from prison, free on probation

par-rot [păr′ət] *noun*, a brightly colored bird with a short curved beak that can mimic speech

par-si-mo-ny [pär′sə-mō′nē] *noun*, selfishness, stinginess

pars-ley [pär′slē] *noun*, an herb with flat or curled leaves used to flavor soups, etc., or to garnish food

par-snip [pär′snĭp] *noun*, a fleshy white root, in the carrot group

part [pärt] *noun*, 1) some of a thing or things 2) a share in an activity 3) a character in a play or film, *verb*, 4) to separate, to leave from 5) to break apart

par-tial-ly [pär′shə-lē] *adverb*, partly, somewhat, slightly

par-ti-al-i-ty [pär′shē-ăl′ĭ-tē] *adjective*, biased, inclined to favor one side

par-tic-i-pate [pär-tĭs′ə-pāt′] *verb*, to have a share in common with others, to take part, to join in

par-tic-i-pa-tion [pär-tĭs′ə-pā′shən] *noun*, the act of getting involved with others

par-ti-ci-ple [pär′tĭ-sĭp′əl] *noun*, one of two forms of a verb forming the past or future tense

par-ti-cle [pär′tĭ-kəl] *noun*, the smallest form of matter

par-tic-u-lar [pər-tĭk′yə-lər] *adjective*, relating to a part, individual, precise, specific

par-tic-u-lar-ly [pər-tĭk′yə-lər-lē] *adverb*, specifically, in detail

par-ti-san [pär′tĭ-zən] *adjective*, an adherent to a party or faction

par-ti-tion [pär-tĭsh′ən] *noun*, 1) a division, a dividing wall, *verb*, 2) to divide, to separate

part-ly [pärt′lē] *adverb*, to some extent, partially, somewhat

part-ner [pärt′nər] *noun*, 1) an associate in business 2) a companion in dancing or sports

par-ty [pär′tē] *noun*, 1) a meeting of friends to enjoy themselves 2) a group of people who have the same interests or goals, *verb*, 3) to celebrate

pass [păs] *noun*, 1) a high mountain road **2)** a paper allowing you to go somewhere or have something 3) a forward gesture, *verb*, 4) to give something to someone else 5) to be allowed to advance to the next level

pas-sage [păs′ĭj] *noun*, 1) an excerpt from a chapter, a clause or quotation 2) the act or process of passing as through time or from place to place, a journey 3) a path 4) a part of a written work or a piece of music

pass-é [pă-sā′] *adjective*, behind the times, old fashioned

pas-sen-ger [păs′ən-jər] *noun*, a traveler in a vehicle such as a car, boat, bus, etc.

pas-sion [păsh′ən] *noun*, a strong feeling or desire, enthusiasm

pas-sion-ate [păsh′ə-nĭt] *adjective*, ardent in feeling or desire

pas-sive [păs′ĭv] *adjective*, to be unresponsive, not active

pass-port [păs′pôrt′] *noun*, an official permission issued to a person allowing travel out of the country and return, a document of identification

past [păst] *adjective*, 1) referring to time and events in history, *noun, no plural*, 2) time gone by, *preposition*, 3) to or on the further side of 4) beyond in time

pas-ta [päs′tə] *noun*, noodles made from flour, water, salt, and eggs

paste [pāst] *noun*, 1) a sticky substance that glues things together, *verb*, 2) to put in place, for example with an adhesive

pas-tel [pă-stĕl′] *noun*, a sketch or drawing of a soft color or hue

pas-teur-ize [păs′chə-rīz′] *verb*, to subject fluids to a high temperature (131 degrees Celsius, 158 degrees Fahrenheit) to prevent the growth of bacteria

pas-time [păs′tīm′] *noun*, something that you do to amuse yourself or to relax, like a hobby

pas-tor [păs′tər] *noun*, a reverend, the clergyman of a church

pas-to-ral [păs′tər-əl] *adjective*, rural, in the country, rustic

pas-try [pā′strē] *noun*, sweet baked foods, such as cakes or pies

pas-ture [păs′chər] *noun*, grassland for grazing livestock

pat [păt] *noun*, 1) a light touch with an open hand, a gentle tap, *verb*, 2) to touch lightly with a hand

patch [păch] *noun*, 1) a small piece of cloth used to mend a tear, *verb*, 2) to put a piece of material on a surface to repair a hole or strengthen a worn area

pat-ent [păt′nt] *adjective*, 1) obvious, plain, *noun*, 2) the right given to an inventor by the government, to exclusively produce a product for a number of years

pa-ter-nal [pə-tûr′nəl] *adjective*, fatherly, kind, devoted

path [păth] *noun*, a track worn by people or animals walking on it

pa-thet-ic [pə-thĕt′ĭk] *adjective*, causing pity or grief, full of pathos and sympathy

pa-thol-o-gy [pă-thŏl′ə-jē] *noun*, the science of treating disease, their natural causes, results, etc.

pa-thos [pā′thŏs′] *noun*, human or animal experience that cause feelings of pity, compassion

pa-tience [pā′shəns] *noun*, *no plural*, the act of waiting calmly for someone or something

pa-tient [pā′shənt] *adjective*, 1) able to bear quietly or wait for something calmly, *noun*, 2) a person under medical treatment

pat-i-o [păt′ē-ō′] *noun*, a paved outdoor space that is used for eating, cooking or relaxing

pa-tri-arch [pā′trē-ärk′] *noun*, 1) father and ruler of a family or tribe 2) a venerable old man

pa-tri-ot [pā′trē-ət] *noun*, one who loves and defends his country

pa-tri-ot-ic [pā′trē-ŏt′ĭk] *adjective*, public-spirited, flag-waving

pa-tri-ot-ism [pā′trē-ə-tĭz′əm] *adjective*, love of one's country

pa-trol [pə-trōl′] *noun*, 1) a small group of soldiers or police officers, *verb*, 2) to follow a route regularly looking for criminal activity

pa-tron [pā′trən] *noun*, 1) someone who provides financial support to a cause 2) a customer

pa-tron-ize [pā′trə-nīz′] *verb*, 1) to act as a supporter 2) to treat with condescension, to look down on

pat-tern [păt′ərn] *noun*, 1) anything used as a guide or model for making things 2) a repeated sequence 3) to make by following a prototype or model

pau-per [pô′pər] *noun*, destitute, a very poor person, a beggar

pause [pôz] *noun*, 1) a break in action of speech, *verb*, 2) to stop briefly, to delay action or speech

pave [pāv] *verb*, 1) to cover the surface with asphalt bricks or cement 2) to prepare the way

pave-ment [pāv′mənt] *noun*, highway surface, asphalt

paw [pô] *noun*, the foot of a four-legged animal that has claws

pawn [pôn] *noun*, 1) a victim 2) a chess piece, *verb*, 3) to sell

pay [pā] *noun*, 1) the compensation received for work, *verb*, 2) to give someone money for goods and services

pay-ment [pā′mənt] *noun*, compensation, installment

pay-roll [pā′rōl′] *noun*, the money used to pay workers

pea [pē] *noun*, a seed from a pod bearing vine that can be eaten

peace [pēs] *noun*, 1) a period of harmony among nations when there is no war 2) a calm ordered condition,

peach [pēch] *noun*, *plural*, **peaches** 1) a juicy yellow-orange fruit with downy skin, from an orchard tree that has pink flowers 2) a yellowish-pink color

pea-cock [pē′kŏk′] *noun*, a bird with a large brightly colored tail

peak [pēk] *noun*, 1) the pointed top of a mountain or ridge 2) the point of greatest development

peal [pēl] *noun*, 1) a loud ringing of bells *verb*, 2) to resound

pea-nut [pē′nŭt′] *noun*, a legume that bears edible pods that taste like nuts

pear [pâr] *noun*, a juicy yellow, brown or green fruit from an orchard tree

pearl [pûrl] *noun*, a precious white substance like a bead found inside an oyster, used in jewelry

peas-ant [pĕz′ənt] *noun*, a person who makes a living from working the soil, especially in poorer countries

peb-ble [pĕb′əl] *noun*, a small stone worn by erosion

pe-can [pĭ-kän′] *noun*, the edible kernel nut from a hickory tree grown in the south

peck [pĕk] *noun*, 1) ¼ of a bushel, *verb*, 2) to cut or lift with the beak

pec-to-ral [pĕk′tər-əl] *adjective*, pertaining to the chest

pe-cu-liar [pĭ-kyo͞ol′yər] *adjective*, oddly different from the usual, strange, distinctive, particular

pe-cu-ni-ar-y [pĭ-kyo͞o′nē-ĕr′ē] *adjective*, pertaining to or consisting of money

ped-a-gogue [pĕd′ə-gŏg′] *noun*, a narrow-minded teacher

ped-al [pĕd′l] *noun*, 1) a lever acted on by the foot, *verb*, 2) to push with the feet when riding a bicycle in order to get the bicycle to move

ped-dle [pĕd′l] *verb*, to sell on the street or door to door

ped-es-tal [pĕd′ĭ-stəl] *noun*, a base or support for a column

pe-des-tri-an [pə-dĕs′trē-ən] *noun*, a person who walks

pe-di-a-tri-cian [pē′dē-ə-trĭsh′ən] *noun*, a doctor who is an expert in children's diseases

ped-i-gree [pĕd′ĭ-grē′] *noun*, a register of a line of ancestors, descent, ancestry, lineage

peek [pēk] *noun*, 1) a glance, *verb*, 2) to look at, to catch sight of

peel [pēl] *noun*, *no plural*, 1) the outside covering of a fruit or vegetable, *verb*, 2) to cut the outside covering off of a fruit or vegetable to eat the food inside

peer [pîr] *noun*, 1) someone your own age or someone with similar interests, an equal, *verb*, 2) to take a closer look

peg [pĕg] *noun*, a wooden or metal object used as a hook when put on a wall or as a fastening

pe-jo-ra-tive [pĭ-jôr′ə-tĭv] *adjective*, a deteriorating effect on the meaning of a word

pel-i-can [pĕl′ĭ-kän′] *noun*, a large water bird with webbed feet and an expandable pouch in the lower beak for holding fish

pelt [pĕlt] *noun*, 1) the furry skin of an animal 2) to strike over and over with repeated blows

pen [pĕn] *noun*, 1) an instrument for writing using ink 2) an enclosure for animals

pe-nal-ize [pē′nə-līz′] *verb*, to scold, to punish, to chastise

pen-al-ty [pĕn′əl-tē] *noun*, the suffering or fine imposed as a punishment for a violation

pen-chant [pĕn′chənt] *noun*, a strong mental leaning or attraction, a decided taste

pen-cil [pĕn′səl] *noun*, a writing instrument with a gray substance in the middle that marks paper

pend-ant [pĕn′dənt] *noun*, a hanging ornament or piece of jewelry, like a necklace

pend-ent [pĕn′dənt] *adjective*, hanging or suspended

pend-ing [pĕn′dĭng] *adjective*, 1) about to happen, imminent, *preposition*, 2) until 3) during

pen-du-lum [pĕn′jə-ləm] *noun*, a weight which swings from side to side under the combined action of gravity and momentum

pen-e-trate [pĕn′ĭ-trāt′] *verb*, 1) to enter into, 2) to affect deeply

pen-guin [pĕng′gwĭn] *noun*, a large flightless sea bird of Antarctica that has flipper-like wings

pen-in-su-la [pə-nĭn′syə-lə] *noun*, a piece of land extending from the mainland, surrounded by water on three sides

pen-i-ten-tia-ry [pĕn′ĭ-tĕn′shə-rē] *noun*, a federal prison

pen-nant [pĕn′ənt] *noun*, a flag or banner with an insignia

pen-ni-less [pĕn′ē-lĭs] *adjective*, without any money, destitute

Penn-syl-va-nia [pĕn′səl-vān′yə] *noun*, one of the 50 United States located in the Northeast, the capital is Harrisburg. The state flower of **Pennsylvania** is the mountain laurel and the motto is "Virtue, liberty, and independence."

pen-ny [pĕn′ē] *noun*, a copper coin worth one cent

pen-sion [pĕn′shən] *noun*, an allowance of money given to someone regularly starting at retirement

pen-sive [pĕn′sĭv] *adjective*, a deep thinker, thoughtful with a hint of sadness, reflective

pen-ta-gon [pĕn′tə-gŏn′] *noun*, a shape with five equal sides

pent-house [pĕnt′hous′] *noun*, an apartment on the top floor of a building

pe-on [pē′ŏn′] *noun*, an unskilled laborer, a menial worker

peo-ple [pē′pəl] *noun*, *plural* of **person**, 1) human beings, men, women, and children 2) a nation

pep-per [pĕp′ər] *noun*, *no plural*, 1) a powder made from seeds of some plants and used to flavor food 2) the fruit of pepper plants, which can be used in cooking or eaten raw in salads

pep-per-mint [pĕp′ər-mĭnt′] *noun*, 1) oil from an herb used to give a taste of mint 2) a candy that tastes like mint

per [pûr] *preposition*, for or in each, according to

per-ceive [pər-sēv′] *verb*, to comprehend, to obtain knowledge through the senses

per-cent [pər-sĕnt′] *noun*, a number in proportion to 100

per-cent-age [pər-sĕn′tĭj] *adjective*, allowance or commission on a hundred

per-cep-ti-ble [pər-sĕp′tə-bəl] *adjective*, capable of being perceived, discernible

per-cep-tion [pər-sĕp′shən] *noun*, viewpoint, observation

perch [pûrch] *noun*, 1) a place for a bird to sit, *verb*, 2) to sit on a narrow branch or stick

per-co-late [pûr′kə-lāt′] *verb*, to pass or trickle through

per-cus-sion [pər-kŭsh′ən] *noun*, the group of musical instruments, such as a drum, whose sound is produced by striking or hitting

per-emp-to-ry [pə-rĕmp′tə-rē] *adjective*, closed to further discussion or action

per-en-ni-al [pə-rĕn′ē-əl] *adjective*, 1) lasting through the year 2) never failing

per-fect [pûr′fĭkt] *adjective*, 1) beyond improvement, having nothing wrong, *verb*, 2) to improve, to make better

per-fec-tion [pər-fĕk′shən] *noun*, the state of being without fault the highest degree of excellence

per-fect-ly [pûr′fĭkt-lē] *adverb*, completely, ideally, flawlessly

per-fi-dy [pûr′fĭd′ē]*noun*, betrayal of trust, treachery

per-fo-rate [pûr′fə-rāt′] *verb*, to make a small hole, to prick

per-form [pər-fôrm′] *verb*, 1) to act in a play or a musical event 2) to do, to carry out a task

per-for-mance [pər-fôr′məns] *noun*, a show or act presented before an audience

per-form-er [pər-fôrm′ər] *noun*, one who entertains before an audience

per-fume [pûr′fyo͞om′] *noun*, *no plural*, liquid that has a sweet smell, a fragrant oil 2) a delightful odor

per-func-to-ry [pər-fŭngk′tə-rē]
adjective, done merely as a duty
with little interest or care

per-haps [pər-hăps′] *adverb*, by
chance, possibly, maybe

per-il [pĕr′əl] *noun*, a danger, a
grave risk, a hazard

pe-rim-e-ter [pə-rĭm′ĭ-tər] *noun*,
the measurement of the distance
around a two-dimensional figure

pe-ri-od [pĭr′ē-əd] *noun*, 1) an
interval of time having a specific
length or character 2) a
punctuation mark used at the end
of a sentence 3) a session in
school

pe-ri-od-i-cal [pĭr′ē-ŏd′ĭ-kəl]
noun, a publication appearing
with a fixed interval between
issues

pe-riph-er-al [pə-rĭf′ər-əl]
adjective, having to do with, or
situated on the outside

pe-riph-er-y [pə-rĭf′ə-rē] *noun*, the
edge, usually of a round surface

per-i-scope [pĕr′ĭ-skōp′] *noun*, an
optical instrument that allows
observation of objects from a
position not in a direct line of
sight, used in submarines

per-ish [pĕr′ĭsh] *verb*, to die

per-ish-a-ble [pĕr′ĭ-shə-bəl]
adjective, likely to spoil,
something that rots in air, it has
to be refrigerated to make it last

per-ju-ry [pûr′jə-rē] *noun*, false
testimony under oath

per-ma-nent [pûr′mə-nənt]
adjective, meant to last without
changing, lasting, durable

per-ma-nent-ly [pûr′mə-nənt-lē]
adverb, lasting indefinitely

per-me-a-ble [pûr′mē-ə-bəl]
adjective, 1) capable of receiving
fluids or transmitting magnetic
effects 2) allowing liquids to
pass through

per-me-ate [pûr′mē-āt′] *verb*, to
pass through the pores of, to
spread throughout, to pervade

per-mis-sion [pər-mĭsh′ən] *noun*,
formal consent, authorization

per-mit [pər-mĭt′] *noun*, 1) a
written notice that allows you to
do something, *verb*, 2) to allow

per-ni-cious [pər-nĭsh′əs]
adjective, destructive, ruinous,
causing insidious harm

per-pen-dic-u-lar
[pûr′pən-dĭk′yə-lər] *adjective*,
1) perfectly upright 2) at right
angles to a horizontal plane

per-pe-trate [pûr′pĭ-trāt′] *verb*, to
do or to perform, to carry
through, to commit

per-pet-u-al [pər-pĕch′ōō-əl]
adjective, never ceasing,
continuing forever, endless

per-pe-tu-ate [pər-pĕch′ōō-āt′]
verb, to continue, indefinitely

per-plex [pər-plĕks′] *verb*, to
embarrass, to bewilder, to cause
to doubt, to puzzle, to baffle

per-plex-i-ty [pər-plĕk′sĭ-tē] *noun*,
the quality or state of being
puzzled, bewilderment

per-se-cute [pûr′sĭ-kyōōt′] *verb*, to
bother, to oppress, to punish

per-se-cu-tion [pûr′sĭ-kyōō′shən]
noun, the act of persistently
injuring, harassing or annoying

per-se-ver-ance [pûr′sə-vîr′əns]
noun, the will to keep working
until the job is done, persistence

per-se-vere [pûr′sə-vîr′] *verb*, to work to accomplish a goal despite difficulty

per-sist [pər-sĭst′] *verb*, to keep doing something, to persevere, to keep trying

per-sist-ence [pər-sĭs′təns] *noun*, the act of continuing to do something, not giving up

per-son [pûr′sən] *noun*, **people**, *plural*, , a human being, such as a man, woman or child, an unnamed individual

per-son-a-ble [pûr′sə-nə-bəl] *adjective*, good-natured, agreeable, charming, attractive

per-son-al [pûr′sə-nəl] *adjective*, of or concerning a particular person, concerning a person's private life, confidential

per-son-al-i-ty [pûr′sə-năl′ĭ-tē] *noun*, distinctive personal character, individuality

per-son-al-ly [pûr′sə-nə-lē] *adverb*, 1) from one's own viewpoint 2) in person

per-son-i-fi-ca-tion [pər-sŏn′ə-fĭ-kā′shən] *noun*, 1) to think of or represent as a person 2) an embodiment or example of someone

per-son-nel [pûr′sə-nĕl′] *noun*, the staff of an organization

per-spec-tive [pər-spĕk′tĭv] *noun*, as one sees it, a point of view

per-spi-ra-tion [pûr′spə-rā′shən] *noun*, sweat, a salty water fluid secreted from the sweat glands

per-spire [pər-spīr′] *verb*, to sweat

per-suade [pər-swād′] *verb*, to influence or gain over by argument or advice

per-sua-sive [pər-swā′sĭv] *adjective*, an influential argument

pert [pûrt] *adjective*, bold, flippant, forward, impertinent, saucy

per-tain [pər-tān′] *verb*, to relate, to be connected as a part

per-ti-nent [pûr′tn-ənt] *adjective*, belonging or relating to the subject at hand, relevant

per-turb [pər-tûrb′] *verb*, to trouble, or disturb, to agitate

pe-ruse [pə-rōōz′] *verb*, to read

per-vade [pər-vā′d] *verb*, to pass through, to permeate

per-va-sive [pər-vā′sĭv] *adjective*, to become spread throughout all parts, universal, permeating

per-ver-sion [pər-vûr′zhən] *noun*, turning from right to wrong

per-vert [pər-vûrt′] *verb*, to make contrary or obstinate remarks

pes-ky [pĕs′kē] *adjective*, annoying, being like a pest

pes-si-mist [pĕs′ə-mĭst] *noun*, one who looks on the dark side of things and expects the worst

pes-si-mism [pĕs′ə-mĭz′əm] *adjective*, belief that life is basically bad or evil

pest [pĕst] *noun*, an annoying and destructive insect or person

pes-ti-cide [pĕs′tĭ-sīd′] *noun*, any substance designed to kill living organisms, such as insects, plants, fungi and rodents

pes-ti-lence [pĕs′tə-ləns] *noun*, any deadly contagious epidemic of a disease or virus

pet [pĕt] *noun*, 1) an animal you look after and keep in your house, *verb*, 2) to stroke

pet-al [pĕt′l] *noun*, one of the parts of the corolla of a flower that are usually colorful

pe-tite [pə-tēt′] *adjective*, small, little, a woman or girl of small size

pe-ti-tion [pə-tĭsh′ən] *noun*, 1) a formal request accompanied by signatures as an endorsement, *verb*, 2) to request change by making a written formal request

pet-ri-fy [pĕt′rə-fī′] *verb*, to turn to stone, to harden, to deaden

pet-ro-le-um [pə-trō′lē-əm] *noun*, mineral oil, a dark brown or green inflammable liquid

pet-ty [pĕt′ē] *adjective*, a small matter of little importance, insignificant, trivial

pew [pyōō] *noun*, one of a number of fixed benches, like a seat, with backs in a church

pew-ter [pyōō′tər] *noun*, an alloy consisting mainly of tin and other metals

phan-tom [făn′təm] *noun*, an apparition, a ghost, a vision

phar-ma-cy [fär′mə-sē] *noun*, a place where medicines are dispensed, a drug store

phase [fāz] *noun*, a stage in a process of development

pheas-ant [fĕz′ənt] *noun*, a game bird with long tail feathers that lives in the woods

phe-nom-e-non [fĭ-nŏm′ə-nŏn′] *noun*, *plural*, **phenomena**, a marvel, an event, a happening

phil-an-throp-ic [fĭl′ən-thrŏp′ĭk] *adjective*, love for mankind, benevolent, generous

phi-lol-o-gist [fĭ-lŏl′ə-jĭst] *noun*, the study of languages, their structure and origin

phi-los-o-pher [fĭ-lŏs′ə-fər] *noun*, a wise person, a reflective thinker, a sage

phi-los-o-phy [fĭ-lŏs′ə-fē] *noun*, the study of moral and mental sciences, metaphysics

pho-bi-a [fō′bē-ə] *noun*, dislike, an irrational fear of something

phone [fōn] *noun*, the short term for telephone

pho-net-ic [fə-nĕt′ĭk] *adjective*, the sound of letters in a word that create speech

pho-no-graph [fō′nə-grăf′] *noun*, a record player

phos-pho-res-cence [fŏs′fə-rĕs′əns] *noun*, any bright and luminous radiating light

phos-pho-rus [fŏs′fər-əs] *noun*, a nonmetallic element used in matches

pho-to-graph [fō′tə-grăf′] *noun*, 1) a picture produced using film and a camera, *verb*, 2) to take a picture with a camera

pho-to-graph-ic [fō′tə-grăf′ĭk] *adjective*, the process of photography

pho-to-syn-the-sis [fō′tō-sĭn′thĭ-sĭs] *verb*, the process by which chlorophyll-containing cells in green plants use light as an energy source to synthesize carbohydrates from carbon dioxide and water

phrase [frāz] *noun*, 1) a part of a sentence 2) an expression, a saying, *verb*, 3) to express in words or a particular way

phys-i-cal [fĭz′ĭ-kəl] *adjective*, of or pertaining to nature or the human body

phy-si-cian [fĭ-zĭsh′ən] *noun*, a doctor of medicine

phys-ics [fĭz′ĭks] *noun*, the science that deals with matter and energy such as light, motion, sound, heat, electricity, or force, and the laws that govern them

phys-i-ol-o-gy [fĭz′ē-ŏl′ə-jē] *noun*, that part of biology which studies living organisms

phy-sique [fĭ-zēk′] *noun*, the development and bodily structure or appearance of a person, the frame, figure

pi-an-ist [pē-ăn′ĭst] *noun*, a skilled performer on the piano

pi-a-no [pē-ăn′ō] *noun*, a musical instrument played by hitting its keys on a keyboard with fingers, which hit felt-covered hammers which strike metal strings

pick [pĭk] *noun*, 1) a choice or selection 2) a tool used to break the earth, *verb*, 3) to choose 4) to hold with your fingers

pick-et [pĭk′ĭt] *noun*, 1) a stake, a peg, *verb*, 2) to strike, to boycott

pick-le [pĭk′əl] *noun*, 1) any food which has been preserved in brine or in vinegar 2) an awkward situation

pic-nic [pĭk′nĭk] *noun*, a pleasure trip with a meal eaten in the open air

pic-to-graph [pĭk′tə-grăf′] *noun*, a rudimentary diagram or symbol used to represent an idea as the primitive writing of stone tablets and cave drawings

pic-ture [pĭk′chər] *noun*, 1) visual respresentation such as a drawing, painting, photograph, etc., *verb*, 2) to imagine in your mind, to visualize

pic-tur-esque [pĭk′chə-rĕsk′] *adjective*, forming a pleasing picture, graphic, quaint

pie [pī] *noun*, a dish made with fruit, meat or vegetables, surrounded by a pastry crust on the bottom and sometimes on top

piece [pēs] *noun*, 1) a fragment or part separated from the whole, *verb*, 2) to put together, to repair

pier [pîr] *noun*, a wharf or dock extending into the water from the land used as a place to moor ships

pierce [pîrs] *verb*, 1) to pass or break through 2) to make a hole through

pierc-ing [pîrsĭng] *adjective*, very loud and shrill

pig [pĭg] *noun*, any swine raised for pork and bacon

pi-geon [pĭj′ən] *noun*, a plump bird with a small head

pig-ment [pĭg′mənt] *noun*, the color of something, a hue

pile [pīl] *noun*, 1) a heap of things laying one on another 2) the nap of a fabric 3) a post in a pier that goes into the soil beneath the water, *verb*, 4) to amass, to gather together in a heap, to accumulate, to stack

pil-grim [pĭl′grəm] *noun*, a person who journeys to pray at a holy place as an act of religious devotion

pil-grim-age [pĭl′grə-mĭj] *noun*, a journey to a holy place

pill [pĭl] *noun*, a small tablet or capsule of medicine that can be easily swallowed

pil-lage [pĭl′ĭj] *verb*, to strip of money, ransack, rob, destroy

pil-lar [pĭl′ər] *noun*, 1) a strong post 2) a column, a vertical support or structure

pil-low [pĭl′ō] *noun*, a cloth case filled with feathers or other soft material put under the head during sleep

pil-low-case [pĭl′ō-kās′] *noun*, the cloth bag put over a pillow to keep it clean

pi-lot [pī′lət] *noun*, someone who is qualified to guide or navigate a ship or a plane

pin [pĭn] *noun*, 1) a slender piece of metal or wood, sharp on one end round on the other, used to join and fasten things together, *verb*, 2) to fasten with a sharp instrument 3) to hold in a spot

pin-cers [pĭn′sərz] *noun*, an instrument or an appendage having two grasping jaws working on a pivot

pinch [pĭnch] *noun*, 1) a very small amount, *verb*, 2) to squeeze between the finger and thumb 3) to make something feel tight

pine [pīn] *noun*, 1) an evergreen tree with leaves that look like needles, *verb*, 2) to want something very badly

pine-ap-ple [pīn′ăp′əl] *noun*, 1) the fruit of a tropical plant, resembling the shape of a pinecone 2) a symbol of hospitality

pin-ion [pĭn′yən] *noun*, a cogwheel with a small number of teeth, designed to interlock its gear with a larger wheel

pink [pĭngk] *adjective*, 1) a pastel color of red and white *noun*, 2) the color made by mixing red and white

pin-na-cle [pĭn′ə-kəl] *noun*, 1) a small turret above the rest of the building 2) a lofty peak

pint [pīnt] *noun*, a liquid measurement equal to 2 cups or 16 fluid ounces

pi-o-neer [pī′ə-nîr′] *noun*, one of the first to explore an area or field of study, an innovator

pi-ous [pī′əs] *adjective*, devout

pipe [pīp] *noun*, 1) a tube, a cylinder used to convey liquid or gas 2) a device with a bowl at the end of a short straw used to smoke tobacco 3) a musical instrument through which air is blown, *verb*, 4) to make a high pitched sound

pi-quant [pē′kənt] *adjective*, 1) pleasantly tart-tasting, spicy 2) provocative

pique [pēk] *noun*, slight anger or resentment, wounded pride

pi-rate [pī′rĭt] *noun*, a person who robs ships at sea

pis-til [pĭs′təl] *noun*, the part of a flower that produces seeds

pis-tol [pĭs′təl] *noun*, a small gun like a revolver, a firearm

pis-ton [pĭs′tən] *noun*, a metal disk which moves up and down in a cylinder exerting pressure

pit [pĭt] *noun*, 1) a deep hole in the ground 2) the seed of a fruit

pitch [pĭch] *noun*, 1) a gummy substance used for sealing roads and caulking 2) a ball thrown or tossed 3) the tone of a voice or a sound, *verb*, 4) to put in place or set up 5) to toss or throw a ball 6) when a boat leans to an angle while traveling in the water

pitch-er [pĭch′ər] *noun*, 1) a container with a lip and a handle to pour liquid 2) the person who throws the ball to the batter in a baseball game

pitch-fork [pĭch′fôrk′] *noun*, a fork with a long handle used to lift and toss hay

pit-fall [pĭt′fôl′] *noun*, a trap, a snare or hazard not easily avoided, a manhole

pith-y [pĭth′ē] *adjective*, concise, short and to the point

pit-i-ful [pĭt′ĭ-fəl] *adjective*, something that causes sympathy

pit-tance [pĭt′ns] *noun*, a small amount, next to nothing

pit-y [pĭt′ē] *noun*, 1) the sadness felt when someone else is hurt or in trouble, *verb*, 2) to feel sorry for someone, to feel sympathy

piv-ot [pĭv′ət] *noun*, a fixed pin on which something turns

piz-za [pēt′sə] *noun*, a baked flat crust covered with cheese, tomato sauce, etc.

pla-cate [plā′kāt′] *verb*, to pacify, to satisfy, to please someone

place [plās] *noun*, 1) a designated area or a particular space, *verb*, 2) to put in a particular spot 3) to identify by recalling

plac-id [plăs′ĭd] *adjective*, calm, peaceful, tranquil, untroubled

pla-gia-rism [plā′jə-rĭz′əm] *noun*, the act of copying someone's work and claiming it as your own

plague [plāg] *noun*, 1) anything troublesome 2) an infectious disease, *verb*, 3) to annoy

plaid [plăd] *adjective*, a pattern on cloth with stripes of different colors that cross each other

plain [plān] *adjective*, 1) open to view, clear, easy to understand, 2) simple, *noun*, 3) a stretch of gently rolling land with few trees

plain-ly [plān′lē] *adverb*, clearly

plain-tiff [plān′tĭf] *noun*, one who commences a legal action or suit

plan [plăn] *noun*, 1) a drawing made to scale of a new building or a project 2) a scheme indicating how to accomplish something, *verb*, 3) to design, to scheme or devise

plane [plān] *noun*, 1) a hand tool used to smooth the surface of wood 2) the short term for airplane 3) any flat or level surface

plan-et [plăn′ĭt] *noun*, a large mass like the earth that revolves around the sun

plank [plăngk] *noun*, a long, flat, thin piece of wood

plank-ton [plăngk′tən] *noun*, plant and animal organisms, generally microscopic, that float and or drift in great numbers in fresh or salt water

plant [plănt] *noun*, 1) a living thing that is not a mineral or animal 2) a factory, a work place, *verb*, 3) to put into the ground to grow

plan-ta-tion [plăn-tā′shən] *noun*, a large estate or farm cultivated by workers living on it

plaque [plăk] *noun*, 1) a marker, a tablet with an inscription 2) tarter calcified on the teeth

plas-ma [plăz′mə] *noun*, the fourth state of matter, composed of broken up atoms

plas-tic [plăs′tĭk] *adjective*, 1) referring to any pliable petroleum based product, *noun*, 2) any of a group of substances made chemically and molded by heat to form, fibers, bottles, etc.

plate [plāt] *noun*, a shallow dish from which food is eaten

pla-teau [plă-tō′] *noun*, 1) an elevated plane 2) inactive period

plat-form [plăt′fôrm′] *noun*, 1) a part of a station where you get on and off trains 2) a raised stage or dais so the audience can see the speaker or performers 3) beliefs, principles

plat-i-tude [plăt′ĭ-tōōd′] *noun*, a commonplace statement

pla-ton-ic [plə-tŏn′ĭk] *adjective*, a friendly relationship

pla-toon [plə-tōōn′] *noun*, a military unit of troops

plat-ter [plăt′ər] *noun*, 1) a large serving of food 2) a large shallow plate for serving food

plau-dits [plô′dĭts] *noun*, *plural*, enthusiastic approval, applause

plau-si-ble [plô′zə-bəl] *adjective*, apparently right, seeming to be reasonable, credible

play [plā] *noun*, 1) the performance of a story on the stage, *verb*, 2) to amuse or entertain 3) to

take part in a game 4) to perform on a musical instrument

plea [plē] *noun*, 1) an appeal, an urgent request 2) a defense

plead [plēd] *verb*, 1) to ask for something that is felt to be very important, to appeal earnestly 2) to respond to a charge by a court of law

pleas-ant [plĕz′ənt] *adjective*, 1) agreeable, amiable, friendly 2) fair, mild, comfortable

pleas-ant-ly [plĕz′ənt-lē] *adverb*, agreeably, politely, kindly

please [plēz] *noun*, 1) a word added to a request to make it sound polite, *verb*, 2) to make someone happy 3) to wish

pleased [plēzd] *adjective*, glad

plea-sur-ably [plĕzh′ər-ə-bəl] *adverb*, capable of affording satisfaction, pleasant

pleas-ure [plĕzh′ər] *noun*, enjoyment or satisfaction

pledge [plĕj] *noun*, 1) a promise, a guarantee, *verb*, 2) to take an oath, to bind by a promise

plen-ty [plĕn′tē] *adjective*, enough, all that is needed

pleth-o-ra [plĕth′ər-ə] *noun*, a large or excessive amount of something, an abundance

pli-ant [plī′ənt] *adjective*, bendable

pli-ers [plī′ərz] *noun*, a small pincers with long jaws, used for bending metal rods or wire, or holding small objects

plod [plŏd] *verb*, to walk heavily, to trudge, to work steadily

plot [plŏt] *noun*, 1) a piece of land 2) a secret plan, usually to

plow

commit a crime 3) the story of a book, film, etc. *verb*, 4) to plan a secret activity, to conspire

plow [plou] *noun*, 1) a farm tool used to overturn earth for planting *verb*, 2) to turn over the soil to prepare the earth for planting 3) to move forcefully

pluck [plŭk] *verb*, to pull off or pull out 2) to grasp or grab

plug [plŭg] *noun*, 1) a stopper used to keep water from draining out of a basin 2) a device put in a wall socket to conduct electricity, *verb*, 3) to put an electric cord into a socket 4) to put a stopper into a drain hole, to fill a hole 5) to persevere

plum [plŭm] *noun*, a sweet juicy red fruit with a pit as a seed

plumb-er [plŭm′ər] *noun*, a tradesman who fits, and repairs gas and water pipes

plumb-ing [plŭm′ĭng] *noun*, *no plural*, all the water pipes, containers, etc. put in a building so that there can be running water and drainage

plume [plo͞om] *noun*, a large or conspicuous feather

plum-met [plŭm′ĭt] *verb*, to fall straight down quickly

plump [plŭmp] *adjective*, a little fat, rounded in form

plunge [plŭnj] *noun*, 1) a sudden dive or fall, *verb*, 2) to fall quickly from a great height 3) to throw oneself into

plung-er [plŭn′jər] *noun*, 1) a rubber suction cup on a handle, used to clear drains 2) a piston-

like part that acts with a thrusting movement in a pump

plu-ral [plo͞or′əl] *adjective*, of or involving more than one

plus [plŭs] *preposition*, in addition to, increased by

ply-wood [plī′wo͞od′] *noun*, sheets of wood arranged in thin layers where the grain is in opposite directions to give it strength

p.m. *adjective*, post meridian, the afternoon, the period between noon and midnight

pneu-mo-nia [no͞o-mōn′yə] *noun*, infection of the lungs caused by bacteria or viruses

poach [pōch] *verb*, 1) to hunt or fish illegally 2) to cook a raw egg in boiling water

pock-et [pŏk′ĭt] *noun*, an envelope made by sewing a piece of cloth into clothing to hold things

pod [pŏd] *noun*, a long, slender part of a vine plant in which the seeds grow, like a pea or bean

po-di-um [pō′dē-əm] *noun*, a pedestal, a raised platform

po-em [pō′əm] *noun*, a composition, usually in verse, with regular rhythm and sounds that express images in powerful or beautiful language

po-et [pō′ĭt] *noun*, one who writes verse in a specific form

po-et-ic [pō-ĕt′ĭk] *adjective*, pertaining to poetry, imaginative

po-et-ry [pō′ĭ-trē] *noun*, *no plural*, words written in verse

poign-ant [poin′yənt] *adjective*, touching, piercing, sharp, keen, affecting the emotions

point [point] *noun*, 1) the tip of something, usually sharp 2) the purpose of a verbal illustration 3) the number of goals for a team 4) at a particular time 5) the location of an ordered pair of numbers, *verb*, 6) to indicate

point-less [pointlĕs] *adjective*, meaningless, senseless

poise [poiz] *noun*, 1) composure, self-assurance, confidence, *verb*, 2) to balance, to keep steady

poi-son [poi′zən] *noun*, 1) any substance dangerous to life and health, *verb*, 2) to kill with toxic chemicals 3) to ruin

poi-son-ous [poi′zə-nəs] *adjective*, deadly, toxic, noxious

poke [pōk] *verb*, to push something pointed into something, to prod

po-lar [pō′lər] *adjective*, 1) of or near the North or South Poles 2) at opposite ends

pole [pōl] *noun*, 1) a round piece of wood, metal or plastic,etc. a rod 2) one end of the earth or at the end of the axis of a sphere

po-lice [pə-lēs′] *noun, no plural*, 1) a department of government that maintains order and enforces the law 2) members of that department

pol-i-cy [pŏl′ĭ-sē] *noun*, 1) a course of action adopted by an individual or group 2) an insurance contract

pol-ish [pŏl′ĭsh] *noun*, 1) a substance such as oil or cleanser used to make a surface shine when rubbed with a cloth, *verb*, 3) rubbing something to make it shine or glossy

po-lite [pə-līt′] *adjective*, showing good manners, courteous

po-lit-i-cal [pə-lĭt′ĭ-kəl] *adjective*, of or pertaining to the conduct of government or politics

pol-i-ti-cian [pŏl′ĭ-tĭsh′ən] *noun*, a person who runs for public office

pol-i-tics [pŏl′ĭ-tĭks] *noun*, the conduct of government

pol-i-ty [pŏl′ĭ-tē] *noun*, form of government of a nation or a state

poll [pōl] *noun*, 1) a survey, *verb*, 2) to interview, to canvass

pol-len [pŏl′ən] *noun*, a fine yellow dust on the anthers of seed bearing flowers, which serves as the male agent in reproduction

pol-li-nate [pŏl′ə-nāt′] *verb*, to fertilize a flower so that it will reproduce itself by conveying pollen to the stigma

pol-lute [pə-lōōt′] *verb*, to contaminate, to make dirty or unusable, to make foul

pol-lu-tion [pə-lōō′shən] *noun*, substances in the air, in water and on land that contaminate the earth

pol-y-gon [pŏl′ē-gŏn′] *noun*, a many sided plane or figure

pomp-ous [pŏm′pəs] *adjective*, arrogant, showing off

pon-cho [pŏn′chō] *noun*, a sleeveless cover for the upper body, a waterproof cloak

pond [pŏnd] *noun*, a small body of water, smaller than a lake

pon-der [pŏn′dər] *verb*, to think about something, to contemplate, to consider carefully

po-ny [pō′nē] *noun*, a small horse

pool [pōōl] *noun*, 1) any amount of still liquid, a basin 2) a collection of things 3) billiards 4) to gather or combine resources for mutual benefit

poor [pŏŏr] *adjective*, 1) not having much money 2) needing kindness or help 3) not good

pop [pŏp] *noun, no plural*, 1) a sudden noise made when air is released 2) part of the popular culture of the 1960s, *verb*, 3) to make a sudden, sharp, explosive sound

pope [pōp] *noun*, the leader of the Roman Catholic Church

pop-py [pŏp′ē] *noun*, a red flower that has seeds that have a narcotic effect

pop-u-lar [pŏp′yə-lər] *adjective*, pleasing to people in general

pop-u-lar-i-ty [pŏp′yə-lăr′ĭ-tē] *noun, no plural*, a measurement of how happy people are with something

pop-u-late [pŏp′yə-lāt′] *verb*, to occupy, to reside in

pop-u-la-tion [pŏp′yə-lā′shən] *noun*, the count of people or organisms who live in an area

pop-u-lous [pŏp′yə-ləs] *adjective*, crowded with people

por-ce-lain [pôr′sə-lĭn] *noun*, a delicate form of china, made with a special clay

porch [pôrch] *noun*, a covered platform in front of a doorway

por-cu-pine [pôr′kyə-pīn′] *noun*, a small animal that has long spikes all over its body for protection

pore [pôr] *noun*, a small hole in the skin for perspiration

pork [pôrk] *noun, no plural*, the meat from a pig used as food

po-rous [pôr′əs] *adjective*, a rough surface, full of holes or texture

por-ridge [pôr′ĭj] *noun*, food made by boiling grain in water or milk until it is very soft, i.e., oatmeal

port [pôrt] *noun*, 1) a harbor or community where ships and boats dock 2) the left handed side of a ship when facing forward 3) a sweet red wine 4) an opening in a ship for cargo

port-a-ble [pôr′tə-bəl] *adjective*, easily carried by hand

por-tal [pôr′tl] *noun*, an entrance, a large doorway, a gate

por-tend [pôr-tĕnd′] *verb*, to foretell, to predict, to forecast

por-tent [pôr′tĕnt′] *noun*, a sign, an omen, forewarning of a threat

por-tent-ous [pôr-tĕn′təs] *adjective*, ominous, serious

por-ter [pôr′tər] *noun*, a person hired to carry luggage in a hotel or a train station or airport

port-fo-li-o [pôrt-fō′lē-ō′] *noun*, a portable case holding papers, prints, drawings, etc.

por-ti-co [pôr′tĭ-kō′] *noun*, a colonnade at the entrance of a building

por-tion [pôr′shən] *noun*, a part of a whole, a segment, a share

port-ly [pôrt′lē] *adjective*, overweight, heavy set, fat

por-trait [pôr′trĭt] *noun*, a painting or photograph of someone

por-tray [pôr-trā′] *verb*, to describe, to represent

pose [pōz] *noun*, 1) a stance or posture, *verb*, 2) to model, to stand 3) to ask a question

po-si-tion [pə-zĭsh'ən] *noun*, 1) a posture, an attitude 2) a job

pos-i-tive [pŏz'ĭ-tĭv] *adjective*, 1) the affirmative, the opposite of negative 2) confident attitude

pos-sess [pə-zĕs'] *verb*, to have as property, to own, to control

pos-ses-sion [pə-zĕsh'ən] *noun*, that which one owns, property

pos-si-bil-i-ty [pŏs'ə-bĭl'ĭ-tē] *noun*, likelihood, probability, chance

pos-si-ble [pŏs'ə-bəl] *adjective*, what may happen or exist

post [pōst] *noun*, 1) a piece of wood or metal placed in the ground as a support 2) a place where a person is stationed, a job or responsibility 3) sending mail through the post office, *verb*, 4) to make known by putting notices or flyers in public places 5) to station at a place

post-age [pō'stĭj] *noun*, *no plural*, the amount of money paid for something sent through the mail

post-card [pōst'kärd'] *noun*, a small note card with a written message sent through the mail

post-er [pō'stər] *noun*, a large advertisement usually hung on a wall in a public place

pos-ter-i-ty [pŏ-stĕr'ĭ-tē] *noun*, the future generation, descendents

post-hu-mous [pŏs'chə-məs] *adjective*, occurring after death

post-pone [pōst-pōn'] *verb*, to put off, to defer to a later time

post-script [pōst'skrĭpt'] *noun*, 1) a note added to a signed letter 2) an addition to a book

pos-tu-late [pŏs'chə-lāt'] *noun*, 1) a self-evident truth, *verb*, 2) to presume, theorize, speculate

pos-ture [pŏs'chər] *noun*, a stance, an unnatural attitude, position

pot [pŏt] *noun*, a container, usually round made for cooking

pot-ash [pŏt'ăsh'] *noun*, wood ashes, potassium hydrochloride

po-tas-si-um [pə-tăs'ē-əm] *noun*, 1) a soft, white metallic element of the alkali group, 2) Vitamin K

po-ta-to [pə-tā'tō] *noun*, an underground starchy vegetable

po-tent [pōt'nt] *adjective*, powerful, effective, mighty

po-ten-tial [pə-tĕn'shəl] *noun*, the capability someone has to accomplish something

po-tion [pō'shən] *noun*, a dose of a liquid with magical powers

pot-pour-ri [pō'pŏŏ-rē'] *noun*, any miscellaneous grouping

pot-ter-y [pŏt'ə-rē] *noun*, *no plural*, ceramic dishes and pots made of baked clay

pouch [pouch] *noun*, 1) a bag, purse 2) a sack on a marsupial for their young

poul-tice [pōl'tĭs] *noun*, a compress used to heal a bruise or sprain

poul-try [pōl'trē] *noun*, any domestic fowl used as food

pounce [pouns] *verb*, to jump on something, to seize eagerly

pound [pound] *noun*, 1) a form of money in Britain and some other countries 2) a measure of weight equal to 2 cups, 16 oz. 3) a place to keep stray animals *verb*, 4) to crush something by striking it over and over

pour [pôr] *verb*, to flow or cause to flow, as in making a stream

pout [pout] *verb*, to make a long face, to sulk, to mope

pov-er-ty [pŏv′ər-tē] *noun*, the state of being poor, destitute, or without money or resources

pow-der [pou′dər] *noun*, *no plural*, fine grains of matter, like dust

pow-er [pou′ər] *noun*, the ability to do something, strength or force

pow-er-ful [pou′ər-fəl] *adjective*, strong, forceful, mighty

prac-ti-cal [prăk′tĭ-kəl] *adjective*, 1) derived from actual use and experience, virtual 2) given to action rather than to speculation

prac-ti-cal-ly [prăk′tĭk-lē] *adverb*, almost, nearly

prac-tice [prăk′tĭs] *noun*, 1) an exercise 2) an habitual action 3) the work of a profession, *verb*, 4) to perform over and over to learn a skill 5) to do something as a principle of living 6) to work at a profession

prag-mat-ic [prăg-măt′ĭk] *adjective*, practical, matter of fact, using common sense

prai-r-ie [prâr′ē] *noun*, a meadow of grassland with no trees

praise [prāz] *noun*, 1) applause, approval, admiration, *verb*, 2) to say nice things about someone, to applaud, to compliment

prance [prăns] *verb*, to frolic, to dance about in a lively manner

prank [prăngk] *noun*, a playful trick, a joke, a caper

prate [prāt] *verb*, to speak foolishly, to boast idly

pray [prā] *verb*, to talk to God or a god, often asking for something or expressing gratitude

pray-er [prâr] *noun*, words spoken aloud or silently to God

preach [prēch] *verb*, to talk to people about how they should live a better life

preach-er [prē′chər] *noun*, a person who talks about living life according to the words of the Bible, a minister

pre-am-ble [prē′ăm′bəl, prē-ăm′-] *noun*, an introduction or preface

pre-car-i-ous [prĭ-kâr′ē-əs] *adjective*, depending on the will of another, uncertain

pre-cau-tion [prĭ-kô′shən] *noun*, something that is done beforehand to guard against harm, danger, mistakes, or accidents

pre-cede [prĭ-sēd′] *adjective*, to go before, to introduce or preface

prec-e-dent [prĕs′ĭ-dənt] *noun*, 1) an authoritative example 2) preceding in time or rank serving as an example

pre-cept [prē′sĕpt′] *noun*, a principle of conduct

pre-cinct [prē′sĭngkt′] *noun*, a district with certain boundaries

pre-cious [prĕsh′əs] *adjective*, 1) of great value, much loved by someone 2) costly, expensive

prec-i-pice [prĕs′ə-pĭs] *noun*, a very high and steep cliff

pre-cip-i-tate [prĭ-sĭp′ĭ-tāt′] *adjective*, 1) headlong, rash, *noun*, 2) a solid formed in a solvent in which it will not dissolve and settles as particles, *verb*, 3) to condense from vapor and fall to the earth's surface like snow or rain

pre-cip-i-ta-tion [prĭ-sĭp′ĭ-tā′shən] *noun*, water that falls as rain or snow from the atmosphere

pre-cip-i-tious [prĭ-sĭp′ĭ- tious] *adjective*, 1) very steep 2) hasty, done without careful thought

pre-cise [prĭ-sīs′] *adjective*, exact, accurate, clearly expressed

pre-ci-sion [prĭ-sĭzh′ən] *adjective*, accuracy, exactness

pre-clude [prĭ-klōōd′] *verb*, to prevent from happening, to hinder, to make impossible

pre-co-cious [prĭ-kō′shəs] *adjective*, showing skills at an earlier age than usual

pre-cur-sor [prĭ-kûr′sər] *noun*, forerunner, predecessor

pred-a-tor [prĕd′ə-tər, -tôr′] *noun*, an animal that eats other animals

pred-e-ces-sor [prĕd′ĭ-sĕs′ər] *noun*, one that comes before another in business or sequence

pre-dic-a-ment [prĭ-dĭk′ə-mənt] *noun*, a stressful position, dilemma, or situation

pre-dict [prĭ-dĭkt′] *verb*, to make a reasonable statement about what might happen, to foretell

pre-dic-tion [prĭ-dĭk′shən] *noun*, foretelling the future, prophecy

pre-di-lec-tion [prĕd′l-ĕk′shən] *noun*, an inclination toward something, preference

pre-dom-i-nance [prĭ-dŏm′ə-nəns] *noun*, power over someone

pre-em-i-nent [prē-ĕm′ə-nənt] *noun*, distinction above others, outstanding, superior

pref-ace [prĕf′ĭs] *noun*, an explanation to the reader at the beginning of a book

pre-fer [prĭ-fûr′] *verb*, to like one thing better than another

pref-er-a-ble [prĕf′ər-ə-bəl] *adjective*, a better choice

pref-er-ence [prĕf′ər-əns] *noun*, one thing that is liked more than another, inclination

pre-fix [prē′fĭks′] *noun*, letters added at the front of a word to alter the meaning

preg-nant [prĕg′nənt] *adjective*, having a child or offspring developing in the body

pre-his-tor-ic [prē′hĭ-stôr′ĭk] *adjective*, a time prior to recorded history

prej-u-dice [prĕj′ə-dĭs] *noun*, a strong feeling against something

prel-ude [prĕl′yōōd′] *noun*, the introduction, forerunner

pre-ma-ture [prē′mə-tyŏŏr′] *adjective*, too early, undeveloped

pre-mier [prĭ-mîr′] *adjective*, a head of government in many nations, such as Russia

pre-miere [prĭ-mîr′] *noun*, the first public showing or display

pre-mi-um [prē′mē-əm] *noun*, 1) a bonus given as an inducement to purchase products 2) the amount usually paid in installments by a policy holder for coverage under a contract

premonition

pre-mo-ni-tion [prē'mə-nĭsh'ən]
noun, a dream or image in which
one sees what is to come

prep-a-ra-tion [prĕp'ə-rā'shən]
noun, the act of doing something
beforehand to get ready for
something

pre-pare [prĭ-pâr'] *verb*, to make
or get ready for something

pre-pon-der-ance
[prĭ-pŏn'dər-əns] *adjective*,
superiority of weight, influence,
overwhelming power

prep-o-si-tion [prĕp'ə-zĭsh'ən]
noun, a word used before a noun
or pronoun to indicate where or
to express connections and
relationships

pre-pos-ter-ous [prĭ-pŏs'tər-əs]
adjective, absurd, ridiculous

pre-req-ui-site [prē-rĕk'wĭ-zĭt]
noun, an essential condition,
required beforehand

pre-rog-a-tive [prĭ-rŏg'ə-tĭv] *noun*,
an exclusive or peculiar right or
privilege of a certain rank

pres-age [prĕs'ĭj] *verb*, to foretell

pre-scribe [prĭ-skrīb'] *verb*, to give
directions to guide someone

pre-scrip-tion [prĭ-skrĭp'shən]
noun, a written direction for
preparation of a medicine

pres-ence [prĕz'əns] *noun*, the act
or state of being present

pre-sent [prĕz'ənt] *adjective*,
1) not past or future, now 2) in
attendance, being in a certain
place *noun*, 3) the present time,
right now 4) a gift, *verb*, 5) to
show, to perform 6) to introduce
one person to another 7) to give
as a gift

pres-en-ta-tion [prĕz'ən-tā'shən]
noun, exhibition, demonstration,
display or lecture

pre-ser-va-tion [prē-zûr-vā'shən]
noun, protection from harm,
danger, or rot

pre-serve [prĭ-zûrv'] *noun*, 1) an
area for the protection of wildlife
and natural resources, *verb*, 2) to
keep something in its original
form, to protect

pre-serves [prĭ-zûrvz'] *noun*
plural, fruits preserved in syrup

pre-side [prĭ-zīd'] *verb*, to control,
to lead, to manage, to oversee

pres-i-dent [prĕz'ĭ-dənt] *noun*,
1) the elected leader of the
United States government 2) the
head of a university, college or
corporation

press [prĕs] *noun*, 1) the newspaper
business or media generated
news stories, *verb*, 2) to push on
steadily with pressure 3) trying
to get someone to do something
4) pushing something to cause
an effect 5) to iron out the
wrinkles in clothing

pres-sure [prĕsh'ər] *noun*, the
action of a force against an
opposing force

pres-tige [prĕ-stēzh'] *noun*,
influence derived from past
success or reputation

pre-sume [prĭ-zoom'] *verb*, to
assume, to conclude, to suppose

pre-sump-tion [prĭ-zŭmp'shən]
noun, an assumption

pre-sup-pose [prē'sə-pōz'] *verb*, to
assume, to consider beforehand

pre-tend [prĭ-tĕnd'] *verb*, to fake,
to make believe, to imitate

pre-tense [prē′tĕns′] *noun*, the act of pretending to be or have something you are not

pre-ten-sion [prĭ-tĕn′shən] *adjective*, when one assumes an air of superiority

pre-text [prē′tĕkst′] *noun*, a false reason to justify an action

pret-ty [prĭt′ē] *adjective*, 1) attractive, pleasing to the eye, *adverb*, 2) close to, quite, fairly

pre-vail [prĭ-vāl′] *verb*, to command, to control, to win

prev-a-lent [prĕv′ə-lənt] *adjective*, widespread, generally accepted

pre-vent [prĭ-vĕnt′] *verb*, to keep something from happening

pre-ven-tion [prĭ-vĕn′shən] *noun*, that which hinders or obstructs

pre-view [prē′vyoō′] *noun*, 1) a look at something before it is reproduced, *verb*, 2) to see in advance

pre-vi-ous [prē′vē-əs] *adjective*, existing or occurring earlier, prior, preceding

pre-vi-ous-ly [prē′vē-əs-lē] *adverb*, formerly, at one time, before

prey [prā] *noun*, any animal that serves as food for another animal

price [prīs] *noun*, the cost of something, what you have to pay

prick [prĭk] *verb*, to make a small hole with something sharp

pride [prīd] *noun*, *no plural*, the feeling of doing your personal best, self-respect

priest [prēst] *noun*, a clergyman of the Catholic Church

prim [prĭm] *adjective*, very precise and formal, exceedingly proper

pri-ma-ry [prī′mĕr′ē] *adjective*, 1) pertaining to elementary education or schools 2) first

prime [prīm] *adjective*, 1) best, choice, excellent, *verb*, 2) to prepare for operation, to make ready

prime min-is-ter [prīm-mĭn′ĭ-stər] *noun*, the head of government in the parliamentary system

prim-i-tive [prĭm′ĭ-tĭv] *adjective*, characterized by the manner of an early time, crude

prince [prĭns] *noun*, 1) the son of a king or queen 2) the ruler of a small country or state

prin-cess [prĭn′sĭs] *noun*, 1) the daughter of a king or queen 2) the wife of a prince

prin-ci-pal [prĭn′sə-pəl] *noun*, 1) occupying the first place or rank, most important 2) the person in charge of a school

prin-ci-ple [prĭn′sə-pəl] *noun*, 1) a source or origin 2) a fundamental truth or doctrine

print [prĭnt] *noun*, 1) the size and style of lettering in a book or poster, *verb*, 2) to press words and pictures on paper or cloth by machine 3) to write

print-er [prĭn′tər] *noun*, a device connected to a computer to make paper copies of a document

pri-or [prī′ər] *adjective*, earlier, before, preceding in time

pri-or-i-ty [prī-ôr′ĭ-tē] *noun*, most important

prism [prĭz′əm] *noun*, 1) a three dimensional object with all sides equal 2) a transparent solid that refracts light

pris-on [prĭz′ən] *noun*, a place where criminals are kept while they are punished

pris-on-er [prĭz′ə-nər] *noun*, a person who is guilty of a crime and must serve a sentence of time away from society, living among other criminals

pris-tine [prĭs′tēn′] *adjective*, 1) part of earlier times, primitive 2) unspoiled, spotless

pri-va-cy [prī′və-sē] *noun*, isolation, seclusion, solitude

pri-vate [prī′vĭt] *adjective*, 1) confidential, personal, secret, *noun*, 2) an enlisted soldier

pri-va-tion [prī-vā′shən] *noun*, hardship, want, lack of comfort

priv-i-lege [prĭv′ə-lĭj] *noun*, a peculiar benefit, favor, or advantage of a group

prize [prīz] *noun*, 1) a reward for winning a contest, *verb*, 2) to cherish, to hold dear

prob-a-bil-i-ty [prŏb′ə-bĭl′ĭ-tē] *adjective*, the chance an event will occur, likelihood

prob-a-ble [prŏb′ə-bəl] *adjective*, likely to occur, plausible

prob-a-bly [prŏb′ə-blē] *adverb*, very likely to prove true

pro-bate [prō′bāt′] *noun*, proof before an officer of the county that a will is valid

pro-ba-tion [prō-bā′shən] *noun*, a trial period, an experiment

probe [prōb] *verb*, to explore with a pointed object or tool

prob-lem [prŏb′ləm] *noun*, 1) a dilemma, something that needs to be solved or worked out

2) a question on a test

pro-ce-dure [prə-sē′jər] *noun*, a manner of proceeding in any action, a process, conduct

pro-ceed [prō-sēd′] *verb*, to go onward or forward, to advance

proc-ess [prŏs′ĕs′] *noun*, 1) a method of operation 2) to change by a special treatment

pro-ces-sion [prə-sĕsh′ən] *noun*, people following one another as part of a ceremony

pro-claim [prō-klām′] *verb*, to announce officially, to declare

proc-la-ma-tion [prŏk′lə-mā′shən] *noun*, an announcement

pro-cliv-i-ty [prō-klĭv′ĭ-tē] *noun*, an inclination, a natural tendency

pro-cras-ti-nate [prō-krăs′tə-nāt′] *verb*, to postpone, to put off doing something, to delay

pro-cras-ti-na-tion [prō-krăs′tə-nā′ shən] *noun*, hesitation, delaying, tarrying

prod [prŏd] *verb*, 1) to poke, to stir up, to jab 2) to urge, to nag

prod-i-gal [prŏd′ĭ-gəl] *adjective*, one who is reckless with money

pro-di-gious [prə-dĭj′əs] *adjective*, marvelous, enormous, huge

prod-i-gy [prŏd′ə-jē] *noun*, a highly gifted child, a marvel

pro-duce [prə-dōōs′] *noun*, 1) fruits of the harvest, *verb*, 2) to create, to bring forth, to provide 3) to present, to display, to exhibit 4) to cause, to make

pro-duc-er [prə-dōō′sər] *noun*, someone in charge of making a movie or film

project

prod-uct [prŏd′əkt] *noun,*
1) something that has an economic
value such as refined natural
resources or a man-made material
2) the outcome of an endeavor
3) the result of multiplying two
numbers together

pro-duc-tion [prə-dŭk′shən] *noun,*
1) creation, construction 2) that
which is the result of effort

pro-fane [prō-fān′] *adjective,* not
sacred, irreverent, secular

pro-fan-i-ty [prō-făn′ĭ-tē] *noun,*
cursing, swearing, expletive

pro-fess [prə-fĕs′] *verb,* to lay
claim to, to affirm, to declare

pro-fes-sion [prə-fĕsh′ən] *noun,* an
employment which needs
specialized study and training

pro-fes-sion-al [prə-fĕsh′ə-nəl]
adjective, 1) engaged in an
activity as a paid job 2) highly
competent

pro-fes-sor [prə-fĕs′ər] *noun,* a
teacher of the highest rank in a
university or college

prof-fer [prŏf′ər] *verb,* to offer, to
give, to present, to submit

pro-fi-cient [prə-fĭsh′ənt]
adjective, advanced in any
occupation or study, skillful

pro-file [prō′fīl′] *noun,* a drawing
or other representation of a view

prof-it [prŏf′ĭt] *noun,* 1) a
monetary gain, *verb,* 2) to be of
advantage, to gain, to benefit

prof-li-gate [prŏf′lĭ-gĭt] *adjective,*
1) wasteful 2) licentious

pro-found [prə-found′] *adjective,*
1) intense, intellectual 2) very
deep, endless, bottomless

pro-fuse [prə-fyōōs′] *adjective,*
existing in large amounts

pro-fu-sion [prə-fyōō′zhən] *noun,*
plenty, abundance

prog-e-ny [prŏj′ə-nē] *noun,*
children, offspring

prog-no-sis [prŏg-nō′sĭs] *noun,*
foretelling the course and
termination of a disease

prog-nos-ti-cate [prŏg-nŏs′tĭ-kāt′]
verb, to predict, to plan for the
future, to foretell

pro-gram [prō′grăm′] *noun,* 1) a
plan of action 2) a written
outline of the order of events in a
public exercise presented as a
leaflet 3) a set of instructions
that tells a computer to do a
specific job 4) a planned
activity, verb 5) to use create a
set of instructions for a computer

pro-gress [prŏg′rĕs′] *noun, no
plural,* 1) a moving toward a
goal, development, *verb,* 2) to go
forward, 3) to advance to a
higher stage, to improve

pro-gres-sion [prə-grĕsh′ən] *noun,*
the act of moving forward

pro-gres-sive [prə-grĕs′ĭv]
adjective, favoring reform,
forward thinking

pro-hib-it [prō-hĭb′ĭt] *verb,* to
prevent, to forbid, to hinder

pro-hi-bi-tion [prō′ə-bĭsh′ən]
noun, a law forbidding the sale
of alcoholic liquors

proj-ect [prŏj′ĕkt′] *noun,* 1) an
activity that requires planning
and materials, *verb,* [prō-jĕkt′]
2) to cause an image to be shown
on a screen 3) to stick out, to
protrude 4) to estimate

253

projectile

pro-jec-tile [prə-jĕk′təl] *noun*, an object propelled through the air

pro-jec-tion [prə-jĕk′shən] *noun*, 1) a bulge, something sticking out 2) a forecast, a prediction

pro-le-tar-i-at [prō′lĭ-tär′ē-ĭt] *noun*, a member of the working class

pro-lif-ic [prə-lĭf′ĭk] *adjective*, abundantly fruitful, plentiful

pro-logue [prō′lôg′] *noun*, an introduction to a play or a story

pro-long [prə-lông′] *verb*, to lengthen, to extend

prom-e-nade [prŏm′ə-nād′] *noun*, 1) a walk in a public place, *verb*, 2) to walk arm in arm

prom-i-nent [prŏm′ə-nənt] *adjective*, important, noticeable

prom-ise [prŏm′ĭs] *noun*, 1) a declaration, binding one to do or not to do a specific act, *verb*, 2) to engage to do, to pledge

prom-is-so-ry [prŏm′ĭ-sôr′ē] *adjective*, containing or implying a promise, an informal contract such as an I.O.U. note

pro-mote [prə-mōt′] *verb*, to further, to encourage, to advance

pro-mo-tion [prə-mō′shən] *noun*, advancement in rank, improvement, support

prompt [prŏmpt] *adjective*, 1) without delay, without too much time passing, *verb*, 2) to cause to act, to urge

prompt-ly [prŏmpt′lē] *adverb*, instantly, quickly, without delay

prone [prōn] *adjective*, 1) inclined to, apt 2) lying face down

prong [prông] *noun*, a sharp projecting part, a spike

pro-noun [prō′noun′] *noun*, a word like **he, she, it, they**, used instead of using a noun again

pro-nounce [prə-nouns′] *verb*, 1) to assert, to declare, to proclaim 2) to speak emphasizing proper sounds and accents, to enunciate

pro-nun-ci-a-tion [prə-nŭn′sē-ā′shən] *noun*, the act of saying words with the proper sound and accent

proof [proof] *noun*, *no plural*, facts which prove something

prop [prŏp] *noun*, 1) an accessory, *verb*, 2) to support, to reinforce 3) to encourage

prop-a-gan-da [prŏp′ə-găn′də] *noun*, an effort to spread an idea

prop-a-gate [prŏp′ə-gāt′] *verb*, to multiply, to spread around

pro-pel [prə-pĕl′] *verb*, to impel, to cause to move or continue in motion, to push forward

pro-pel-ler [prə-pĕl′ər] *noun*, blades turned by a motor to move a boat or airplane through water or air

pro-pen-si-ty [prə-pĕn′sĭ-tē] *noun*, a natural inclination, a tendency

prop-er [prŏp′ər] *adjective*, fitting, appropriate, respectable

prop-er noun [prŏp′ər noun] *noun*, a particular person, place or thing, the first letter is capitalized

prop-er-ly [prŏp′ər-lē] *adverb*, correctly, rightly, absolutely

prop-er-ty [prŏp′ər-tē] *noun*, land, buildings, cars, boats, and other things people own

prop-er-ty tax [prŏp′ər-tē-tăks] *noun*, a certain amount of money paid as a tax on the land, buildings, cars, and boats a person owns

proph-e-cy [prŏf′ĭ-sī′] *noun*, a prediction made after receiving a sign or a vision

proph-e-sy [prŏf′ĭ-sī′] *verb*, to predict, to foretell, to foresee

proph-et [prŏf′ĭt] *noun*, 1) a person who predicts what is going to happen in the future 2) a person who claims to be selected by God to lead

pro-phy-lac-tic [prō′fə-lăk′tĭk] *noun*, a medicine or device that prevents disease, a preventive

pro-por-tion [prə-pôr′shən] *noun*, the relation of one portion to another in size, the ratio

pro-pos-al [prə-pō′zəl] *noun*, a formal suggestion, an offer

pro-pose [prə-pōz′] *verb*, to offer for consideration, to put forward

prop-o-si-tion [prŏp′ə-zĭsh′ən] *noun*, a plan, a scheme

pro-pri-e-tor [prə-prī′ĭ-tər] *noun*, an owner, one who has legal right to a business

pro-pri-e-ty [prə-prī′ĭ-tē] *noun*, established standards of proper behavior, appropriateness

pro-pul-sion [prə-pŭl′shən] *noun*, driving forward, thrust

prose [prōz] *noun*, ordinary language, not poetry or verse

pros-e-cute [prŏs′ĭ-kyōōt′] *verb*, to enforce the law by legal process, to seek justice for a crime

pros-pect [prŏs′pĕkt′] *noun*, 1) an expectation, a possibility, odds 2) a view, an outlook, *verb*, 3) to explore, to search

pros-pec-tus [prə-spĕk′təs] *noun*, a summary of a plan, an outline

pros-per [prŏs′pər] *verb*, to do well, to become rich, to succeed

pros-per-i-ty [prŏ-spĕr′ĭ-tē] *noun*, material well-being, success

pros-per-ous [prŏs′pər-əs] *adjective*, doing well, successful

pros-the-sis [prŏs-thē′sĭs] *noun*, an artificial device to replace a missing part of the body

pros-trate [prŏs′trāt′] *adjective*, 1) lying down *verb*, 2) to stretch out full on the ground

pro-tect [prə-tĕkt′] *verb*, to guard or defend, to keep from harm

pro-tec-tion [prə-tĕk′shən] *noun*, preservation from loss, injury, or annoyance, safe keeping

pro-té-gé [prō′tə-zhā′] *noun*, one under the care and protection of another, often learning a craft under the guidance of a more experienced person

pro-tein [prō′tēn′] *noun*, a nutrient found in amino acids essential to the growth and repair of animal tissue found in eggs, meat, etc.

pro-test [prə-tĕst′] *noun*, 1) a complaint or formal disagreement, *verb*, 2) to disagree strongly, to object

Prot-es-tant [prŏt′ĭ-stənt] *noun*, *adjective*, a person belonging to any of the Christian churches resulting from the Reformation under the leadership of Calvin, Wesley, Luther, etc.

pro-to-col [prō′tə-kôl′] *noun*, decorum, formalities, proper manners, a set of rules

pro-ton [prō'tŏn'] *noun*, a particle
of matter, a part of an atom

pro-to-type [prō'tə-tīp'] *noun*,
original work used as a model by
others, an example

pro-trac-tor [prō-trăk'tər-] *noun*, a
device for measuring angles on a
plane or on paper

pro-trude [prō-trōōd'] *verb*, to
stick out, to cause to jut out

proud [proud] *adjective*,
a feeling of satisfaction over
something one owns, is, or does

proud-ly [proud'lē] *adverb*, feeling
a sense of accomplishment after
doing your personal best

prove [prōōv] *verb*, 1) to show that
something is true with evidence
and tests 2) to turn out

prov-e-nance [prŏv'ə-nəns] *noun*,
place or source of origin

pro-verb [prŏv'ûrb'] *noun*, a short
saying that speaks the truth

pro-vide [prə-vīd'] *verb*, 1) to look
out for in advance 2) to furnish

prov-i-dence [prŏv'ĭ-dəns] *noun*,
divine direction, foresight

prov-ince [prŏv'ĭns] *noun*, part of a
country, with its own
government

pro-vin-cial [prə-vĭn'shəl]
adjective, limited to a small
area, parochial

pro-vi-sion [prə-vĭzh'ən] *noun*,
requirement, a formal stipulation

pro-voke [prə-vōk'] *verb*, 1) to
make someone angry, to annoy,
2) to incite to action

prowl [proul] *verb*, to sneak, to
rove about stealthily

prox-im-i-ty [prŏk-sĭm'ĭ-tē] *noun*,
nearness, neighborhood, vicinity

prox-y [prŏk'sē] *noun*, 1) a person
authorized to act for another 2) a
document giving such consent

pru-dent [prōōd'nt] *adjective*,
cautious, sagacious, advisory

prune [prōōn] *noun*, 1) a dried
plum, *verb*, 2) to cut off
branches of a plant

pry [prī] *verb*, 1) to lift off or open
by pushing with leverage 2) to
snoop, to spy, to search

psalm [säm] *noun*, a sacred song or
poem used to praise God

pseu-do-nym [sōōd'n-ĭm'] *noun*, a
fictitious name, a pen name

psy-che [sī'kē] *noun*, 1) the soul or
spirit 2) the mind's force

psy-chic [sī'k'ĭk] *adjective*, 1)
apprently sensitive to
supernatural forces 2) a person
who predicts the future

psy-chi-a-trist [sĭ-kī'ə-trĭst] *noun*,
a doctor who analyzes and treats
mental problems

psy-cho [sī'kē'ō]*noun*, the mind or
mental processes

psy-chol-o-gy [sī-kŏl'ə-jē] *noun*,
the study of mental states, the
mind and its processes

psy-cho-sis [sī-kō'sĭs] *noun*, a
major mental disorder in which
the personality is very seriously
disorganized and one's sense of
reality is usually altered

pto-maine [tō'mān'] *noun*, an
alkaloid from decomposing
animals that is usually poisonous

pub-lic [pŭb'lĭk] *adjective*, 1) open
to everyone, *noun*, 2) the people
constituting a community

pub-li-ca-tion [pŭb'lĭ-kā'shən]
noun, published material

pub-lic-i-ty [pŭ-blĭs′ĭ-tē] *noun*, public notice and attention

pub-lish [pŭb′lĭsh] *verb*, to print and sell a manuscript

pub-lish-er [pŭb′lĭ-shər] *noun*, one who creates and offers for sale books, magazines, or newspapers

pud-ding [pŏŏd′ĭng] *noun*, a custard made of milk, sugar and eggs boiled together

pud-dle [pŭd′l] *noun*, a small pool of water on the ground

puff [pŭf] *noun*, 1) a quick burst of air, smoke, etc. *verb*, 2) to make short quick breaths of air

pu-is-sant [pwĭs′ənt] *adjective*, powerful, strong, potent

pul-chri-tude [pŭl′krĭ-tōōd′] *noun*, beauty, loveliness, charm

pull [pŏŏl] *noun*, 1) a force, a power, strength, *verb*, 2) to move something toward oneself through grasping and drawing with force 3) to put on

pul-ley [pŏŏl′ē] *noun*, a wheel used to transmit power by means of a band, rope, or chain

pul-mo-nar-y [pŏŏl′mə-nĕr′ē] *adjective*, affecting the lungs

pulp [pŭlp] *noun*, 1) a soft, moist, sticky mass of fibers made up of wood, straw, etc. and used to make paper and paperboard 2) the soft moist part of fruit that is left after the seeds and peels are removed

pul-pit [pŏŏl′pĭt] *noun*, an elevated desk or platform in a church for a preacher or orator

pul-sate [pŭl′sāt′] *verb*, to throb

pulse [pŭls] *noun*, a heart beat

pul-ver-ize [pŭl′və-rīz′] *verb*, to reduce to a fine powder or dust

pum-mel [pŭm′əl] *verb*, to beat

pump [pŭmp] *noun*, 1) a machine for making liquid or gas move by putting air into something or to extract water or other liquid from the ground using a special lever, *verb*, 2) to draw up using up and down action

pump-kin [pŭmp′kĭn] *noun*, a large orange gourd-like fruit, used in cooking

punch [pŭnch] *noun*, 1) a hit aimed squarely with a fist 2) a fruit drink, *verb*, 3) to hit someone very hard with a fist

punc-tu-al [pŭngk′chōō-əl] *adjective*, on time

punc-tu-al-i-ty [pŭngk′chōō-ăl′ĭ-tē] *adverb*, being prompt, making a point of being on time

punc-tu-ate [pŭngk′chōō-āt′] *verb*, to add proper signs and marks of grammar to writing

punc-tu-a-tion [pŭngk′chōō-ā′shən] *noun, no plural*, signs like „ : , ? and ! used in writing so that the reader emphasizes certain ideas expressed on paper

punc-ture [pŭngk′chər] *noun*, 1) perforation with something pointed, *verb*, 2) to penetrate, to pierce, to make a hole

pun-dit [pŭn′dĭt] *noun*, a critic, a commentator, an expert

pun-gent [pŭn′jənt] *adjective*, powerful or strong odor or flavor

pun-ish [pŭn′ĭsh] *verb*, to discipline, to reprove, to correct

pun·ish·ment [pŭn′ĭsh-mənt] *noun*, 1) a penalty 2) rough or injurious treatment inflicted for an offense

pu·ny [pyōō′nē] *adjective*, insignificant, tiny, weak

pu·pa [pyōō′pə] *noun*, the inactive stage of development of an insect before reaching the adult, often in a cocoon or a cell

pu·pil [pyōō′pəl] *noun*, 1) a boy or girl being taught by a teacher 2) a black dot, the expanding and contracting opening in the iris of the eye, in the middle of the eye that receives light

pup·pet [pŭp′ĭt] *noun*, a small doll that looks like a person or an animal that is moved by wires or with hands on a small stage

pup·py [pəpē] *noun*, a young dog

pur·chase [pûr′chĭs] *noun*, 1) something that has been paid for, an acquisition, *verb*, 2) to acquire for a price, to buy

pure [pyōōr] *adjective*, without anything mixed into it, clean, unpolluted, spotless

pur·ga·to·ry [pûr′gə-tôr′ē] *noun*, 1) a place where a sinner makes amends 2) limbo

purge [pûrj] *verb*, 1) to clean by removing impurities 2) to remove all criminal charges from public record

pu·ri·fy [pyōōr′ə-fī′] *verb*, to make clean, to free from imperfection

pu·ri·ty [pyōōr′ĭ-tē] *noun*, cleanliness, pureness, innocence

pur·loin [pər-loin′] *verb*, to steal

pur·ple [pûr′pəl] *adjective*, 1) the color of violets, *noun*, 2) a color made by mixing red and blue pigments

pur·pose [pûr′pəs] *noun*, the reason for which something exists or has been done

purse [pûrs] *noun*, 1) a small hand bag to hold money and personal items 2) prize money from a race

pur·sue [pər-sōō′] *verb*, to follow with a view to overtake, to persecute, to capture

pur·suit [pər-sōōt′] *noun*, an aim, an objective, a search

pur·vey·or [pər-vā′ər] *noun*, someone who provides food

push [pŏŏsh] *noun*, 1) a shove, a rigorous effort, *verb*, 2) to press or lean against, so as to move

put [pŏŏt] *verb*, 1) to move, to place 2) to impose, to assign

pu·trid [pyōō′trĭd] *adjective*, foul, rotten, decayed, disgusting

put·ty [pŭt′ē] *noun*, an adhesive, glue, compound to seal something

puz·zle [pŭz′əl] *noun*, 1) a picture cut into many pieces to be put back together 2) a game designed to make it difficult for you to finish, *verb*, 3) to exercise one's mind 4) to perplex

pyg·my [pĭg′mē] *noun*, a dwarf, a very small person in Africa

pyr·a·mid [pĭr′ə-mĭd] *noun*, 1) a solid shape with a square base and pointed top, whose faces are triangles with a common corner 2) huge monuments in the desert of Egypt, built as burial vaults for ancient Egyptian kings

py·ro·ma·ni·ac [pī′rō-mā′nē-ăk′] *noun*, a person who likes to set fires and watch them burn

Q

quack [kwăk] *noun*, 1) an imposter, a fake, *verb*, 2) to make noises like a duck

quad-ri-lat-er-al [kwŏd′rə-lăt′ər-əl] *noun*, a four-sided polygon

quad-ru-ped [kwŏd′rə-pĕd′] *noun*, a four-footed animal

quad-ru-ple [kwŏ-drōō′pəl] *verb*, to multiply by four

quaff [kwŏf] *verb*, to drink with enthusiasm

quag-mire [kwăg′mīr′] *noun*, a bog, a marsh, muddy soil

quail [kwāl] *noun*, 1) a game bird that lives in the woods, *verb*, 2) to cower in fear

quaint [kwānt] *adjective*, pleasing in an unusual way, peculiar, odd

quake [kwāk] *verb*, to shake, to shudder, to tremble, to quiver

qual-i-fi-ca-tion [kwŏl′ə-fĭ-kā′shən] *noun*, special training or knowledge

qual-i-fy [kwŏl′ə-fī′] *verb*, 1) to meet the set criteria 2) to make less harsh or strict

qual-i-ty [kwŏl′ĭ-tē] *noun*, 1) a characteristic that distinguishes one person or thing from another 2) a degree of excellence

qualm [kwäm] *noun*, misgiving, remorse, uneasiness

quan-ti-ty [kwŏn′tĭ-tē] *noun*, an amount of something

quar-an-tine [kwôr′ən-tēn′] *noun*, 1) separation of people with a contagious disease, *verb*, 2) to be separated from society while suffering from an infectious disease in order to prevent the disease from spreading

quar-rel [kwôr′əl] *noun*, 1) an angry dispute, *verb*, 2) to argue or disagree, to fight with words

quar-ry [kwôr′ē] *noun*, 1) a deep pit where stone is cut out of the gound to use for buildings 2) prey, an animal that is being hunted

quart [kwôrt] *noun*, a liquid measurement equal to 32 ounces, 1.13 liters, ¼ of a gallon, 2 pints

quar-ter [kwôr′tər] *noun*, 1) a coin worth 25 cents 2) one of four equal parts, one fourth

quar-ters [kwôr′tərz] *noun*, *plural*, a place where people live, especially if they live where they work

quar-tet [kwôr-tĕt′] *noun*, a group of musicians or singers, also, a piece of music for four instruments or voices

quartz [kwôrts] *noun*, a mineral compound of silica

queen [kwēn] *noun*, 1) the wife of the king 2) any woman who rules a country born to that title, a female monarch 4) any woman who is very important

queer [kwîr] *adjective*, odd, different in a strange way

quell [kwĕl] *verb*, to put down, to quiet, to suppress, to pacify

quench [kwĕnch] *verb*, 1) to satisfy one's thirst or appetite 2) to extinguish, to put out

que-ry [kwîr′ē] *noun*, 1) a question, *verb*, 2) to doubt or question, to interrogate

quest [kwĕst] *noun*, a search, a journey, an exploration

ques-tion [kwĕs'chən] *noun*, 1) an inquiry, *verb*, 2) to ask, to inquire 3) to doubt, to quiz

ques-tion mark [kwĕs'chən-märk] *noun*, the sign ? used at the end of a sentence which asks a question

ques-tion-naire [kwĕs'chə-nâr'] *noun*, a survey submitted to obtain information, a printed set of questions

queue [kyōō] *noun*, a line of people waiting for something

quib-ble [kwĭb'əl] *verb*, to equivocate, to play with words

quick [kwĭk] *adjective*, very fast,

quick-en [kwĭk'ən] *verb*, to hurry, to rush, to hasten, to accelerate

quick-ly [kwĭk'lē] *adverb*, in a hurry, fast, promptly

qui-et [kwī'ĭt] *adjective*, 1) making little or no sound, not moving , still, *noun*, 2) silence

qui-et-ly [kwī'ĭt- lē] *adverb*, calmly, peacefully, silently

quill [kwĭl] *noun*, 1) a large feather from a bird's tail or wing 2) a sharp spine of a porcupine

quilt [kwĭlt] *noun*, a blanket made by sewing small pieces of fabric together in a design

quince [kwĭns] *noun*, a bitter fruit resembling a large yellow apple

quin-tes-sence [kwĭn-tĕs'əns] *noun*, the purest and highest essence of something

quip [kwĭp] *noun*, a taunting remark, a witty remark

quit [kwĭt] *verb*, to give up, to surrender, to depart, to leave

quite [kwīt] *adverb*, completely, entirely, actually, really

quiv-er [kwĭv'ər] *noun*, 1) a case for holding arrows 2) a trembling, *verb*, 2) to shake a little, to tremble nervously

quix-ot-ic [kwĭk-sŏt'ĭk] *adjective*, romantically idealistic

quiz [kwĭz] *noun*, 1) a small test, *verb*, 2) to ask a few questions

quo-rum [kwôr'əm] *noun*, the number of members of any body legally necessary to transact business

quo-ta [kwō'tə] *noun*, the amount, quantity, proportion

quo-ta-tion [kwō-tā'shən] *noun*, 1) words spoken or written lifted from their context 2) the naming or publishing of the current prices of stocks, bonds, etc., the price named

quote [kwōt] *verb*, to say or write something that has been said before by another person

quo-tient [kwō'shənt] *noun*, the product from the division of one number by another

R

rab-bi [răb'ī] *noun*, the leader or teacher of the Jewish religion

rab-bit [răb'ĭt] *noun*, a small furry animal with long ears and long back legs that burrows holes and lives underground

rab-ies [rā'bēz] *noun*, disease caused by organisms called a virus that attacks the nervous system

rac-coon [rǎ-kōōn'] *noun*, a large black and white nocturnal rodent that lives in trees

race [rās] *noun*, 1) a competitive event, a contest 2) a part of humanity, an ethnic background, *verb*, 3) to run in a dash or compete in a contest

ra-cial [rā'shəl] *adjective*, pertaining to ethnic differences

rac-ism [rā'sĭz'əm] *noun*, narrow-mindedness, intolerance, bigotry

rack [rǎk] *noun*, 1) a frame on which things are hung or kept, *verb*, 2) to cause pain

rack-et [rǎk'ĭt] *noun*, 1) a light bat with a netted hoop 2) a loud or confused noise

ra-dar [rā'där] *noun*, *no plural*, the ability to find the position of something using radio waves

ra-di-ant [rā'dē-ənt] *adjective*, shining, luminous, bright

ra-di-ate [rā'dē-āt'] *verb*, to give off energy or rays of light

ra-di-a-tor [rā'dē-ā'tər] *noun*, a device used to cool a liquid

rad-i-cal [rǎd'ĭ-kəl] *adjective*, unconventional, odd

ra-di-o [rā'dē-ō] *noun*, 1) the use of electromagnetic waves to carry messages without the use of wires, *verb*, 2) to send or transmit messages in this manner

ra-di-o-act-ive waste [rā'dē-ōǎk'tĭvwǎst] *noun*, waste left over from creating nuclear power or atomic energy which gives off radiant energy in rays

or particles by the disintegration of the atomic nuclei such as found in radium and uranium

ra-di-um [rā'dē-əm] *noun*, a metallic element found in minute quantities, capable of spontaneously emitting rays

ra-di-us [rā'dē-əs] *noun*, the distance from the center of a circle to its perimeter

raft [rǎft] *noun*, large pieces of wood joined to make a flat floating surface

raft-er [rǎf'tər] *noun*, a crossbeam

rag [rǎg] *noun*, 1) an old torn garment 2) a piece of cloth used to clean, polish or dust

rage [rāj] *noun*, fierce anger that continues with great force

raid [rād] *noun*, 1) a surprise attack, *verb*, 2) to invade, to attack, to assault

rail [rāl] *noun*, 1) a bar extending horizontally between two posts as in a fence or railroad ties, *verb*, 2) to scold, to rant and rave

rail-ing [rā'lĭng] *noun*, a fence or partition to keep people or animals from straying into places they should not go

rail-road [rāl'rōd'] *noun*, a road or track built with parallel steel rails and used by trains

rain [rān] *noun*, 1) part of a weather pattern when water droplets fall from the sky *verb*, 2) to fall as water in drops from the sky 3) to fall like rain

rain-bow [rān'bō'] *noun*, an arc-shaped spectrum of colors seen in the sky, caused by sunlight refracted by raindrops

rain-coat [rān′kōt′] *noun*, a coat
that protects a person from
getting wet

rain-fall [rān′fôl′] *noun*, the
amount of water falling within a
given time or area

raise [rāz] *noun*, 1) an increase in
salary, *verb*, 2) to lift up, to
elevate 3) to grow or cultivate

rai-sin [rā′zĭn] *noun*, a dried grape

rake [rāk] *noun*, 1) a long-handled
tool with teeth used to gather
leaves on the ground, *verb*,
2) to use a tool to gather fallen
leaves or to smooth soil

ral-ly [răl′ē] *noun*, 1) a mass
meeting, an assembly, a
gathering, *verb*, 2) to restore, to
recover, to renew 3) to counter
attack

ram [răm] *noun*, a male sheep

ram-ble [răm′bəl] *verb*, 1) to
wander in a leisurely manner
2) to chatter, to babble

ram-i-fi-ca-tion [răm′ə-fĭ-kā′shən]
noun, 1) branching out, a
diversion 2) a complicated
outcome of a statement or plan

ramp [rămp] *noun*, a slanted walk
or roadway that connects a lower
to a higher place

ram-page [răm′pāj′] *noun*, a rage,
violent behavior, a tantrum

ramp-ant [răm′pənt] *adjective*,
unruly, furious, unrestrained

ram-part [răm′pärt′, -pərt] *noun*, a
defensive mound of earth

ran [răn] *past tense* of **run**

ranch [rănch] *noun*, a rural estate
where livestock are raised

ran-cid [răn′sĭd] *adjective*, rotten,
an unpleasant smell or taste

ran-cor [răng′kər] *noun*, bitterness,
hatred, malice, ill will

ran-dom [răn′dəm] *adjective*,
without order or logic

range [rānj] *noun*, 1) a line of
mountains or hills 2) an
extensive variety of different
things 3) the distance something
can travel on its own

rang-er [rān′jər] *noun*, an
enforcement agent assigned to
patrol a certain region

rank [răngk] *adjective*, 1) bad
smell, *noun*, 2) a class thought of
as higher or lower than other
groups, *verb*, 3) to classify in a
group 4) to arrange in a row

ran-kle [răng′kəl] *verb*, to make
someone annoyed and mad

ran-sack [răn′săk′] *verb*, to search
a house thoroughly

ran-som [răn′səm] *noun*, the
release of a person in return for
payment

rant [rănt] *verb*, to talk wildly, to
speak bombastically

rap [răp] *verb*, 1) to knock, to tap,
to thump 2) to talk, to chat

rap-id [răp′ĭd] *adjective*, very fast,
quick, swift, often, speedy

rap-id-ly [răp′ĭd-lē] *adverb*,
quickly, hurriedly, swiftly

rap-ids [răp′ĭdz] *noun*, *plural*, a
place in the river where the river
descends creating swift currents
producing white water

rap-ture [răp′chər] *noun*, joy,
delight, bliss, exhilaration

rare [râr] *adjective*, unusual, uncommon, hard to come by

rare-ly [râr′lē] *adverb*, once in a great while, very seldom

ras-cal [răs′kəl] *noun*, a deceitful person, a mischievous child

rash [răsh] *adjective*, 1) acting quickly without thinking about the results, reckless, *noun*, 2) a patch of red spots on the skin, indicating an allergy

rasp-ber-ry [răz′bĕr′ē] *noun*, a red or purple berry that grows on a bush

rat [răt] *noun*, 1) a small rodent, larger than a mouse, that lives on grain 2) an informer

rate [rāt] *noun*, 1) a relative amount 2) the speed of something, *verb*, 3) to evaluate, to rank, to grade, to consider

rath-er [răth′ər, rä′thər] *adjective*, preferable, instead, somewhat

rat-i-fy [răt′ə-fī′] *verb*, to approve formally, to endorse, to confirm

ra-tio [rā′shē-ō′] *noun*, the relation between quantity, degree or rate between one thing and another

rati-oci-na-tion [răsh′ē-ŏs′ə-nā′shən] *noun*, reasoning logically, the process of exact thinking

ra-tion [răsh′ən, rā′shən] *noun*, 1) a fixed amount of something that is given or allowed, *verb*, 2) to limit the amount of resources to each person or group

ra-tion-al [răsh′ə-nəl] *adjective*, reasonable, intelligent, sensible

ra-tion-al-i-za-tion [răsh′ə-nə-lĭ-zā′shən] *noun*, bringing into conformity with reason, logical thinking

ra-tion-al-ize [răsh′ə-nə-līz′] *verb*, to explain away, to make excuses, to justify

rat-tle [răt′l] *noun*, 1) a toy with beads inside that makes a noise when shaken, *verb*, 2) to shake, to clatter, to move noisily

rau-cous [rô′kəs] *adjective*, rough in sound, harsh, strident

rav-age [(răv′ĭj] *verb*, to plunder, to destroy or wreck, to ruin

rave [rāv] *verb*, to speak enthusiastically about something

rav-en-ous [răv′ə-nəs] *adjective*, extremely hungry

ra-vine [rə-vēn′] *noun*, a long, deep ditch or valley in the earth with steep sides, a gorge

raw [rô] *adjective*, 1) uncooked, bloody 2) something in its natural, unadulterated state 3) cold and damp

ray [rā] *noun*, 1) a thin beacon of light 2) part of a line that has an endpoint and goes on and on in one direction 3) a trace, hint

ray-on [rā′ŏn] *noun*, a fine cloth that looks like glossy silk

raze [rāz] *verb*, to level to the ground, to destroy completely

ra-zor [rā′zər] *noun*, a keen edged cutting instrument especially used for shaving

reach [rēch] *noun*, *no plural*, 1) the distance that can be attained 2) grasp, *verb*, 3) to stretch out to touch or take something 4) to get to a place, to arrive at

re-act [rē-ăkt′] *verb*, to act because of something that has happened

re-ac-tion [rē-ăk′shən] *noun*, a direct response

re-ac-tion-ar-y [rē-ăk′shə-nĕr′ē] *adjective*, someone who is against progress

read [rĕd] *verb*, to look at words and understand them

read-i-ly [rĕd′ə-lē] *adverb*, easily, willingly, without hesitating

read-i-ness [rĕd′ē-nĭs] *adjective*, willingness, eagerness, ease

re-ad-just [rē′ə-jŭst′] *verb*, to adapt, to fine tune

read-y [rĕd′ē] *adjective*, 1) willing, able, prepared 2) at hand

re-al [rē′əl, rēl] *adjective*, factual, actually existing, true

re-al-ism [rē′ə-lĭz′əm] *noun*, a philosophy that regards external objects as the most fundamentally real things, and perceptions or ideas as secondary

re-al-ist [rē′ə-lĭst] *noun*, 1) someone with a practical outlook 2) an attempt to make art and literature resemble life

re-al-i-ty [rē-ăl′ĭ-tē] *noun*, fact, truth, real things, actuality

re-al-i-za-tion [rē′ə-lĭ-zā′shən] *noun*, a perception, an awareness

re-al-ize [rē′ə-līz′] *verb*, 1) to understand, to make real, to become aware of 2) to achieve

re-al-ly [rē′ə-lē′] *adverb*, truly, very, genuinely, indeed

realm [rĕlm] *noun*, region, sphere, domain or province, kingdom

ream [rēm] *noun*, in the paper business, 500 sheets of paper

reap [rēp] *verb*, to gather the harvest by cutting

rear [rîr] *adjective*, 1) at the back, behind, *noun*, 2) the back part, *verb*, 3) to raise children or livestock until they grow up

rea-son [rē′zən] *noun*, 1) the answer or explanation given as part of a belief 2) a cause for acting, thinking, or feeling a special way, *verb*, 3)) to think things through to present an opinion, to conclude, to infer

rea-son-a-ble [rē′zə-nə-bəl] *adjective*, governed by reason, not excessive, logical

rea-son-a-bly [rē′zə-nə-blē] *adverb*, adequately, fairly, sufficiently, logically

rea-son-ing [rē′zə-nĭng] *noun*, the thought process, logic

re-as-sure [rē′ə-shŏŏr′] *verb*, to help feel safe and comfortable

re-as-sur-ance [rē′ə-shŏŏr′əns] *noun*, restoring confidence

re-bate [rē′bāt′] *noun*, a deduction offered on a purchase in the form of a refund

re-bel [rĕb′əl] *noun*, 1) someone who fights against the leadership of a government, *verb*, 2) to defy any authority, to revolt

re-bel-lion [rĭ-bĕl′yən] *noun*, a revolt, resistance, mutiny

re-bel-lious [rĭ-bĕl′yəs] *adjective*, someone who resists the law, defying authority

re-bound [rē′bound′] *verb*, 1) to bounce back, to move like a boomerang 2) to recover

re-buff [rǐ-bŭf′] *noun*, 1) a sharp rejection, a snub, *verb*, 2) to drive back or repel 3) to refuse bluntly, to reject, to snub

re-build [rē-bǐld′] *verb*, to repair, to put in working order

re-but-tal [rǐ-bŭt′l] *noun*, an argument introduced by the other side in a debate

re-cal-ci-trant [rǐ-kǎl′sǐ-trənt] *adjective*, obstinately stubborn

re-call [rǐ-kôl′] *verb*, 1) to summon or call back, 2) to remember

re-cant [rǐ-kǎnt′] *verb*, to withdraw a previous statement

re-ca-pit-u-late [rē′kə-pǐch′ə-lāt′] *verb*, to repeat the principal points in a program

re-cede [rǐ-sēd′] *verb*, to move back, to retreat, to withdraw

re-ceipt [rǐ-sēt′] *noun*, a written acknowledgment that something was received

re-ceive [rǐ-sēv′] *verb*, to take in, to come into possession of, accept

re-cent [rē′sənt] *adjective*, something that happened a little while ago, prior to the present

re-cent-ly [rē′sənt-lē] *adverb*, a short while ago

re-cep-ta-cle [rǐ-sĕp′tə-kəl] *noun*, 1) a container 2) a contact device in an electrical outlet

re-cep-tion [rǐ-sĕp′shən] *noun*, 1) a greeting, a welcome 2) a social gathering 3) receiving television or radio wave frequencies

re-cess [rē′sĕs′] *noun*, 1) a short intermission 2) a space formed by indentation, a niche

re-ces-sion [rǐ-sĕsh′ən] *noun*, an economic slow down

characterized by high unemployment

re-cid-i-vism [rǐ-sǐd′ə-vǐz′əm] *noun*, habitual return to crime

rec-i-pe [rĕs′ə-pē′] *noun*, a list of ingredients and directions for preparing something

re-cip-i-ent [rǐ-sǐp′ē-ənt] *noun*, the person who accepts something

re-cip-ro-cal [rǐ-sǐp′rə-kəl] *adjective*, mutual, exchangeable, interacting, given in return

re-cip-ro-cate [rǐ-sǐp′rə-kāt′] *verb*, to return the favor for something done or given, to respond in kind

re-cit-al [rǐ-sīt′l] *noun*, a program of music or dance before an audience

re-cite [rǐ-sīt′] *verb*, to repeat from memory, to narrate in detail

reck-less [rĕk′lǐs] *adjective*, careless and dangerous

reck-on [rĕk′ən] *verb*, to come to terms with, to settle matters

re-claim [rǐ-klām′] *verb*, to bring back, to restore, to recover

re-cline [rǐ-klīn′] *verb*, to lean back

rec-luse [rĕk′lōōs′] *noun*, a hermit

rec-og-ni-tion [rĕk′əg-nǐsh′ən] *noun*, 1) acknowledgment of something done or given 2) acknowledgment of acquaintance

rec-og-nize [rĕk′əg-nīz′] *verb*, 1) to acknowledge formally 2) to identify as previously seen or known 3) to greet, to accept

re-coil [rǐ-koil′] *verb*, to retreat, or drawback, to shrink back

rec-ol-lect [rĕk′ə-lĕkt′] *verb*, to call to mind, to remember

rec·ol·lec·tion [rĕk′ə-lĕk′shən] *noun*, something remembered

re·com·mend [rĕk′ə-mĕnd′] *verb*, to present favorably, to urge or suggest, to advise

re·com·men·da·tion [rĕk′ə-mĕn-dā′shən] *noun*, a suggestion for what is most suitable

rec·om·pense [rĕk′əm-pĕns′] *noun*, 1) compensation, payment, *verb*, 2) to pay for service received

rec·on·cil·able [rĕk′ən-sī′lə-bəl, rĕk′ən-sĭ′-] *adjective*, capable of being restored to friendship

rec·on·cile [rĕk′ən-sīl′] *verb*, 1) to restore to friendship 2) to settle a quarrel 3) to be resigned to accept things as they are

rec·on·dite [rĕk′ən-dīt′] *adjective*, hard to understand, profound, hidden from view, abstruse

re·con·nais·sance [rĭ-kŏn′ə-səns] *noun*, a survey of the enemy by soldiers, reconnoitering

re·con·struct [rē′kən-strŭkt′] *verb*, to put something together or make it again, to rebuild

re·cord [rĭ-kôrd′,rĕk′ərd] *noun*, 1) a written account of what happened 2) a plastic disk on which sounds are recorded, *verb*, 3) to write down something that has happened 4) to register on a disk in a permanent form

re·cord·er [rĭ-kôr′dər] *noun*, 1) an office worker, a secretary 2) a wooden or plastic flute played like a whistle

re·count [rĭ-kount′] *verb*, 1) to tell a story, to narrate 2) to count again

re·course [rē′kôrs′] *noun*, resorting to help when in trouble

re·cov·er [rĭ-kŭv′ər] *verb*, 1) to get back to a normal state, to get well, to recuperate 2) to get back what was lost or stolen

re·cov·er·y [rĭ-kŭv′ə-rē] *noun*, a return to good health

rec·re·a·tion [rĕk′rē-ā′shən] *noun*, sports or play as a form of enjoyable relaxation

re·crim·i·na·tion [rĭ-krĭm′ə-nā′shən] *noun*, an accusation in response to another, a countercharge

re·cruit [rĭ-krōōt′] *noun*, 1) a new member who agrees to join a group, *verb*, 2) to enlist, to attract new members

rec·tan·gle [rĕk′tăng′gəl] *noun*, a flat shape with four straight sides and four right angles, the opposite sides of which are parallel

rec·ti·fy [rĕk′tə-fī′] *verb*, to correct, to make better

rec·ti·tude [rĕk′tĭ-tōōd′] *noun*, uprightness, moral virtue

re·cu·per·ate [rĭ-kōō′pə-rāt′] *verb*, to recover, to restore to health

re·cur [rĭ-kûr′] *verb*, to return again and again as an event

re·cur·rent [rĭ-kûr′ənt] *adjective*, returning repeatedly or coming again to mind

re·cy·cle [rē-sī′kəl] *verb*, to process materials so they can be made into new products

red [rĕd] *adjective*, 1) crimson, cherry, ruby, *noun*, 2) one of the three primary colors: blue, yellow and red

re-deem [rĭ-dēm′] *verb*, 1) to recover, 2) to buy back 3) to exchange 4) to fulfill

red-o-lent [rĕd′l-ənt] *adjective*, 1) fragrant, a pleasant odor 2) strongly reminiscent

re-dress [rĭ-drĕs′] *verb*, to set right, to make amends for, to remedy

re-duce [rĭ-dōōs′] *verb*, to diminish, to make smaller or use less

re-duc-tion [rĭ-dŭk′shən] *noun*, 1) the act of making less 2) the amount by which something is diminished

re-dun-dant [rĭ-dŭn′dənt] *adjective*, superfluous, excessively wordy, repetitious

reed [rēd] *noun*, 1) a small, thin, flat piece of wood, metal, or plastic used in the mouthpiece of some musical instruments, such as the clarinet 2) a tall sturdy grass that grows in the water

reef [rēf] *noun*, a ridge of sand, rocks, or coral, off shore but close to the water's surface

reek [rēk] *verb*, to give off an odor

reel [rēl] *noun*, 1) a spool that turns, winding or unwinding a fishing line 2) a Scottish dance *verb*, 3) to feel happy, elated with excitement

re-enact [rē′ĕn-ăkt′] *verb*, to perform as if for the first time

re-fer [rĭ-fûr′] *verb*, 1)to recommend someone go to see someone for help 2) to use a resource for help

ref-er-ee [rĕf′ə-rē′] *noun*, a sports official with final authority for conducting a game

ref-er-ence [rĕf′ər-əns] *noun*, 1) a specific direction of the attention 2) a written statement of the qualifications of another 3) a resource such as a, dictionary, encyclopedia, thesaurus

re-fine [rĭ- fīnd] *verb*, to reduce to a fine, unmixed, or pure state by removing all unwanted matter

re-fined [rĭ-fīn′d] *adjective*, 1) in a pure state 2) having good manners and good taste

re-fin-er-y [rĭ-fĭ′nə-rē] *noun*, a factory where resources like oil or sugar are made ready for use

re-flect [rĭ-flĕkt′] *verb*, 1) to throw or cast back the sun's rays 2) to think, to ponder, to meditate

re-flec-tion [rĭ-flĕk′shən] *noun*, 1) an image bounced back from a shiny surface 2) an object seen in a mirror 3) serious thought, thinking back on something

re-flec-tor [rĭ-flĕk′tər] *noun*, a polished surface that reflects light, heat or sound

re-flex [rē′flĕks′] *adjective*, mechanical, unthinking, habitual

re-form [rĭ-fôrm′] *noun*, 1) a change, an improvement, *verb*, 2) to correct, to revise, to renew

re-form-a-to-ry [rĭ-fôr′mə-tôr′ē] *noun*, a jail or prison

re-frac-tion [rĭ-frăk′shən] *noun*, the bending of a ray of light

re-frain [rĭ-frān′] *noun*, 1) a chorus, a verse, *verb*, 2) to abstain, to keep oneself from doing something

re-fresh [rĭ-frĕsh′] *verb*, to revive with rest or food, to renew

re-fresh-ment [rĭ-frĕsh′mənt]
noun, food and drink

re-frig-er-a-tor [rĭ-frĭj′ə-rā′tər]
noun, an appliance that preserves
food in cool temperatures

ref-uge [rĕf′yoōj] *noun*, a safe
place, a place to hide

ref-u-gee [rĕf′yoō-jē′] *noun*, an
exile, someone forced to leave
his country

re-fund [rĭ-fŭnd′] *noun*,
1) reimbursement, repayment,
verb, 2) to give back, to
reimburse, to repay

re-fur-bish [rē-fûr′bĭsh] *verb*, to
repair and make useful

re-fus-al [rĭ-fyoō′zəl] *noun*, a
denial, a refusal

ref-use [rĭ-fyoōz′] *noun*, 1) a
general term for solid waste
materials, also called garbage or
trash, *verb*, 2) to deny, not to
allow or agree, to decline to do

re-fute [rĭ-fyoōt′] *verb*, to prove
false, to disprove, to rebut

re-gain[rĭ- gān/ *verb,* to get back

re-gal [rē′gəl] *adjective*, royal,
majestic, stately, splendid

re-gale [rĭ-gāl′] *verb*, to entertain

re-gard [rĭ-gärd′] *noun*, *plural*,
1) respect, concern, *verb*, 2) to
think of or see 3) to relate to

re-gard-ing [rĭ-gär′dĭng]
preposition, about, concerning

re-gard-less [rĭ-gärd′lĭs] *adverb*,
without concern for problems or
objections, despite, although

re-gat-ta [rĭ-gä′tə] *noun*, a boat or
yacht race

re-gen-er-ate [rĭ-jĕn′ə-rāt′] *verb*, to
renew, to reconstitute

re-gen-er-a-tion [rĭ-jĕn′ə-rā′shən]
noun, a spiritual rebirth

re-gime [rā-zhēm′] *noun*, character
of government or of a prevailing
social system

reg-i-men [rĕj′ə-mən] *noun*, a
systematic plan usually to
improve health

reg-i-ment [rĕj′ə-mənt] *noun*, 1) an
infantry unit 2) to systematize

re-gion [rē′jən] *noun*, an indefinite
area, a territory, a district

reg-is-ter [rĕj′ĭ-stər] *noun*, 1) an
entry in a book or record, *verb*,
2) to enter one's name on the
list, catalogue, record

reg-is-trar [rĕj′ĭ-strär′] *noun*, a
keeper of records

reg-is-tra-tion [rĕj′ĭ-strā′shən]
noun, enrollment

re-gret [rĭ-grĕt′] *noun*, 1) an
apology, a feeling of remorse,
verb, 2) to feel sadness for doing
something wrong

re-gret-ta-ble [rĭ-grĕt′ə-bəl]
adjective, unfortunate, sad

reg-u-lar [rĕg′yə-lər] *adjective*,
habitual, customary, routine

reg-u-lar-i-ty [rĕg′yə-lăr′ĭ-tē]
noun, *no plural*, uniformity,
constancy, on a regular basis

reg-u-late [rĕg′yə-lāt′] *verb*, to
control or direct by a rule

reg-u-la-tion [rĕg′yə-lā′shən]
noun, a rule of conduct, a law

re-gur-gi-tate [rē-gûr′jĭ-tāt′] *verb*,
to spit out food that has been
swallowed, to vomit

re-ha-bil-i-tate [rē′hə-bĭl′ĭ-tāt′]
verb, to restore to proper
condition, i.e., good health

re-hears-al [rĭ-hûr′səl] *noun*, a practice performance

re-hearse [rĭ-hûrs′] *verb*, to practice in preparation for a performance, to try out

reign [rān] *noun*, 1) the period of time when a king or queen rules, *verb*, 2) to rule or govern

re-im-burse [rē′ĭm-bûrs′] *verb*, to make restoration, to repay

rein [rān] *noun*, 1) a strap of a bridle for governing a horse or other animal, *verb*, 2) to pull back or control, to check

rein-deer [rān′dîr′] *noun*, a large deer of arctic regions used to pull a cart or sled

re-in-force [rē′ĭn-fôrs′] *verb*, to strengthen, to stiffen, to fortify

re-it-er-ate [rē-ĭt′ə-rāt′] *verb*, to repeat, to say or do again

re-ject [rĭ-jĕkt′] *verb*, to refuse to accept or use

re-joice [rĭ-jois′] *verb*, to celebrate, to be joyful, to delight

re-ju-ve-nate [rĭ-jōō′və-nāt′] *verb*, to make more lively or youthful

re-lapse [rĭ-lăps′] *noun*, 1) a set back 2) to return to a former state, to regress, to revert

re-late [rĭ-lāt′] *verb*, to tell, to connect in thought and meaning

re-la-tion [rĭ-lā′shən] *noun*, 1) a member of the same family 2) an association or friendship

re-la-tion-ship [rĭ-lā′shən-shĭp′] *noun*, 1) the association between things, the affect one thing has on another 2) kinship

rel-a-tive [rĕl′ə-tĭv] *adjective*, 1) in relation to or compared to, *noun*,

2) a person connected by blood or marriage, kin

re-lax [rĭ-lăks′] *verb*, 1) to rest, to be at ease 2) to make less rigid

re-lax-a-tion [rē′lăk-sā′shən] *noun*, *no plural*, a time of rest or ease, leisure, recreation

re-lay [rē′lā] *verb*, 1) to send, to give out, to beam 2) to convey

re-lease [rĭ-lēs′] *noun*, 1) freedom from confinement, *verb*, 2) to let go, to set free, to liberate, to untie 3) to relieve

rel-e-gate [rĕl′ĭ-gāt′] *verb*, to banish, to consign to an inferior position, place or condition

re-lent [rĭ-lĕnt′] *verb*, to become less strict, to become more forgiving

rel-e-van-cy [rĕl′ə-vəns-ē] *noun*, reference to the case at hand

re-li-a-ble [rĭ-lī′ə-bəl] *adjective*, dependable, trustworthy

re-li-ance [rĭ-lī′əns] *adjective*, confidence, something dependent on something else

rel-ic [rĕl′ĭk] *noun*, that which remains or is left, anything valued as a memento

re-lief [rĭ-lēf′] *noun*, *no plural*, 1) assistance, aid, help 2) comfort, ease, cheer 3) the contour of the front of a building

re-lieve [rĭ-lēv′] *verb*, to free from worry, to alleviate, to ease

re-lig-ion [rĭ-lĭj′ən] *noun*, possessing beliefs concerning the nature and purpose of the universe and the supernatural

re-li-gious [rĭ-lĭj′əs] *adjective*, 1) possessing pious beliefs in a God or gods 2) dedicated, devout

re-lin-quish [rǐ-lǐng′kwǐsh] *verb*, to abandon, to leave alone

rel-ish [rěl′ǐsh] *noun*, 1) a condiment made of vegetables and herbs 2) great enjoyment, *verb*, 3) to savor, to appreciate, to like, to enjoy, to delight in

re-luc-tance [rǐ-lŭk′təns] *noun*, unwillingness or aversion marked by hesitation

re-luc-tant [rǐ-lŭk′tənt] *adjective*, unwilling to participate

re-ly [rǐ-lī′] *verb*, to depend confidently, to trust

re-main [rǐ-mān′] *verb*, to stay

re-main-der [rǐ-mān′dər] *noun*, 1) the fraction left over after a number has been divided 2) what is left over, remnant, residue

re-mains [rǐ-mānz′] *noun*, *plural*, parts which are left behind exposed to air, a corpse

re-make [rē-māk′] *noun*, something that has been redone or given a different form

re-mark [rǐ-märk′] *noun*, 1) something spoken, *verb*, 2) to take notice of, to observe

re-mark-a-ble [rǐ-mär′kə-bəl] *adjective*, uncommon, extraordinary, unusual

re-mark-a-bly [rǐ-mär′kə-blē] *adverb*, notably, especially

re-me-di-al [rǐ-mē′dē-əl] *adjective*, intended to improve poor skills

rem-e-dy [rěm′ǐ-dē] *noun*, 1) a cure 2) a solution to the problem

re-mem-ber [rǐ-měm′bər] *verb*, to keep in your mind, to not forget

re-mem-brance [rǐ-měm′brəns] *noun*, a recollection, a memory

re-mind [rǐ-mīnd′] *verb*, to help someone remember

rem-i-nisce [rěm′ə-nǐs′] *verb*, to think back or talk about past experiences, to recollect

rem-i-nis-cence [rěm′ə-nǐs′əns] *noun*, the act of recalling the past, recollection, memory

rem-i-nis-cent [rěm′ə-nǐs′ənt] *adjective*, suggestive, remindful

re-miss [rǐ-mǐs′] *adjective*, negligent, careless, neglectful

re-mit [rǐ-mǐt′] *verb*, to send money usually as a payment

re-mit-tance [rǐ-mǐt′ns] *noun*, the sending of money, bills, etc, money transmitted

rem-nant [rěm′nənt] *noun*, something left over

re-mon-strate [rǐ-mŏn′strāt′] *verb*, to plead in protest, to argue repeatedly in opposition

re-morse [rǐ-môrs′] *noun*, deep regret for wrongdoing

re-mote [rǐ-mōt′] *adjective*, far away, a great distance from everyone, slight, aloof

re-mov-al [rǐ-mōō′vəl] *noun*, to take something away

re-move [rǐ-mōōv′] *verb*, to take and move somewhere else

re-mu-ner-a-tion [rǐ-myōō′nə-rā′shən] *noun*, compensation, reward

ren-ais-sance [rěn′ǐ-säns′] *noun*, a rebirth, revival, renewal

rend [rěnd] *verb*, 1) to split, to tear apart with violence 2) to disturb

rend-er [rěn′dər] *verb*, to cause to become, to provide, to perform

ren-dez-vous [rän'dā-vōō'] *noun*, a meeting place, an appointment

ren-di-tion [rĕn-dĭsh'ən] *noun*, the artistic interpretation of a song or dance and how it is performed

ren-e-gade [rĕn'ĭ-gād'] *noun*, a person who deserts a cause to join another, an outlaw

re-new [rĭ-nōō'] *verb*, 1) to bring up to date 2) to start again

re-new-al [rĭ-nōō'əl] *noun*, a new start, recurrence, a revival

re-nounce [rĭ-nouns'] *verb*, to abandon, to discontinue, to disown, to repudiate

ren-o-vate [rĕn'ə-vāt'] *verb*, to make repairs, to redecorate

re-nown [rĭ-noun'] *adjective*, fame, widespread recognition

rent [rĕnt] *noun*, 1) money paid to use property, *verb*, 2) to pay someone to use something that they own, to lease

re-pair [rĭ-pâr'] *noun*, 1) the result of renewing or reconditioning something, *verb*, 2) to fix or mend, to make new again

re-par-a-tion [rĕp'ə-rā'shən] *noun*, compensation demanded by a victorious nation from a defeated nation, amends

rep-ar-tee [rĕp'ər-tē',] *noun*, a clever reply

re-pay [rĭ-pā'] *verb*, to give someone money owed to them

re-peal [rĭ-pēl'] *noun*, 1) retraction, withdrawal, *verb*, 2) to cancel

re-peat [rĭ-pēt'] *verb*, to cause to happen again, to duplicate

re-peat-ed-ly [rĭ-pē'tĭd-lē] *adverb*, said or done again and again

re-pel [rĭ-pĕl'] *verb*, to drive back, to oppose, to ward off

re-pel-lent [rĭ-pĕl'ənt] *adjective*, offensive, disgusting, repugnant

re-pent [rĭ-pĕnt'] *verb*, to express regret, to feel contrite

re-pent-ance [rĭ-pĕn'təns] *noun*, sorrow for what one has done or omitted to do, regret

re-per-cus-sion [rē'pər-kŭsh'ən, rĕp'ər-] *noun*, the effect of something said or done

rep-er-toire [rĕp'ər-twär'] *noun*, a list of musical scores which a company or a person has thoroughly rehearsed and is prepared to perform

rep-e-ti-tion [rĕp'ĭ-tĭsh'ən] *noun*, doing or saying something over and over

re-place [rĭ-plās'] *verb*, 1) to take the place of 2) to make good

re-place-ment [rĭ-plās'mənt] *noun*, a person or thing that takes the place of another, a substitute

re-plen-ish [rĭ-plĕn'ĭsh] *verb*, to fill again, to supply anew

re-plete [rĭ-plēt'] *adjective*, filled to capacity, abundantly supplied

rep-li-ca [rĕp'lĭ-kə] *noun*, a copy

re-ply [rĭ-plī'] *noun*, 1) an answer, a response, *verb*, 2) to give an answer, to reply

re-port [rĭ-pôrt'] *noun*, 1) an oral or written account containing information in an organized form, *verb*, 2) to give an account of something

re-port-er [rĭ-pôr'tər] *noun*, 1) a person who reports legal proceedings or legislative

271

debates 2) a person who gathers information and writes reports for a newspaper, magazine, etc.

re-pose [rĭ-pōz/] *noun*, 1) a state of calm *verb*, 2) to be in a state of rest or relaxation

re-pos-i-to-ry [rĭ-pŏz/ĭ-tôr/ē] *noun*, a storehouse, a vault where things are kept

rep-re-hen-si-ble [rĕp/rĭ-hĕn/sə-bəl] *adjective*, wicked, deceitful

rep-re-sent [rĕp/rĭ-zĕnt/] *verb*, 1) to portray, to stand for something 2) to act or speak for someone or something

rep-re-sen-ta-tion [rĕp/rĭ-zĕn-tā/shən] *noun*, something that expresses a point of view or represents something else, a portrayal

rep-re-sen-ta-tive [rĕp/rĭ-zĕn/tə-tĭv] *adjective*, 1) typically characteristic, *noun*, 2) one acting in the place of another

re-press [rĭ-prĕs/] *verb*, to check, to overpower, to keep down

re-prieve [rĭ-prēv/] *noun*, 1) a delay, *verb*, 2) to postpone, to delay the punishment of

rep-ri-mand [rĕp/rə-mănd/] *verb*, to censure, to scold

re-pris-al [rĭ-prī/zəl] *noun*, retaliation, an act short of war by which a nation seeks to redress a wrong committed against it by another nation

re-proach [rĭ-prōch/] *noun*, 1) blame, censure, *verb*, 2) to blame someone, to criticize

re-pro-duce [rē/prə-dōōs/] *verb*, 1) to produce offspring or create babies 2) to make a copy of something, to duplicate

re-prove [rĭ-prōōv/] *verb*, to reprimand someone for doing something wrong, to correct someone, to rebuke, to chide

rep-tile [rĕp/tĭl] *noun*, a cold-blooded, egg laying vertebrate, such as a lizard or snake

re-pub-lic [rĭ-pŭb/lĭk] *noun*, government in which people elect their representatives

re-pub-li-can [rĭ-pŭb/lĭ-kən] *noun*, a person who advocates a republican government, one who believes in government elected by representatives

Re-pub-li-can Par-ty *noun*, A U.S. political party that developed in the 1850s from those who opposed slavery and that elected Abraham Lincoln as the first Republican president

re-pu-di-ate [rĭ-pyōō/dē-āt/] *verb*, to disown, to renounce, to reject

re-pug-nant [rĭ-pŭg/nənt] *adjective*, disgusting, vulgar, offensive, objectionable

re-pulse [rĭ-pŭls/] *verb*, 1) to offend 2) to drive back, to repel

re-pul-sion [rĭ-pŭl/shən] *noun*, a feeling of aversion or disgust

rep-u-ta-tion [rĕp/yə-tā/shən] *noun*, what people generally think about a person or thing

re-quest [rĭ-kwĕst/] *noun*, 1) something that is asked for, *verb*, 2) to ask for something

re-quire [rĭ-kwīr/] *verb*, to have need of, to insist or compel

re-quire-ment [rĭ-kwīr/mənt] *noun*, a condition that must be met to satisfy a need, something needed, a necessity

req-ui-site [rĕk′wĭ-zĭt] *noun*, a demand, something required

req-ui-si-tion [rĕk′wĭ-zĭsh′ən] *noun*, an official written request for something

re-quite [rĭ-kwīt′] *verb*, to repay, to seek revenge, to get even

re-scind [rĭ-sĭnd′] *verb*, to cut off or remove, to cancel or abolish

res-cue [rĕs′kyōō] *noun*, 1) help or aid offered to save someone, *verb*, 2) to aid or save from danger or confinement

re-search [rĭ-sûrch′] *noun*, *no plural*, 1) a careful study conducted to learn and discover something new, *verb*, 2) to investigate or study something, to make a careful inquiry

re-sem-blance [rĭ-zĕm′bləns] *noun*, similarity, likeness

re-sem-ble [rĭ-zĕm′bəl] *verb*, to have similar qualities and likeness, to be like

re-sent [rĭ-zĕnt′] *verb*, to feel angry with someone because they were unfair, to feel indignant

re-sent-ment [rĭ-zĕnt′mənt] *noun*, exasperation, ill will, anger

res-er-va-tion [rĕz′ər-vā′shən] *noun*, 1) a commitment to hold a space for someone 2) hesitation, qualm 3) land the government gave Indians to live on

re-serve [rĭ-zûrv′] *noun*, 1) restraint, *verb*, 2) to hold something for someone

res-er-voir [rĕz′ər-vwär′] *noun*, 1) a place where a large supply of anything is built up over time 2) a body of water stored for use

re-side [rĭ-zīd′] *verb*, to live, to inhabit, to remain, to dwell in

res-i-dence [rĕz′ĭ-dəns] *noun*, living or dwelling in a place for some time, a home

res-i-dent [rĕz′ĭ-dənt] *noun*, someone who resides in a place

res-i-den-tial [rĕz′ĭ-dĕn′shəl] *adjective*, where people live

re-sid-u-al [rĭ-zĭj′ōō-əl] *adjective*, effects remaining after the principal action

res-i-due [rĕz′ĭ-dōō′] *noun*, the remainder of something

re-sign [rĭ-zīn′] *verb*, 1) to relinquish, to withdraw from an office or position 2) to calmly accept an unpleasant outcome

res-ig-na-tion [rĕz′ĭg-nā′shən] *noun*, *no plural*, 1) reluctant acceptance of something 2) a formal statement to leave a job

re-signed [rĭ-zīnd′] *adjective*, unresisting, patiently submissive

re-sil-ient [rĭ-zĭl′yənt] *adjective*, the ability to return to the same size and condition after excessive stress and strain

res-in [rĕz′ĭn] *noun*, in plastics manufacturing, the different compounds used to create the different forms of plastic

re-sist [rĭ-zĭst′] *verb*, to oppose, to refuse to comply, to withstand

re-sist-ance [rĭ-zĭs′təns] *noun*, the act of resisting or opposing, any opposing force

res-o-lute [rĕz′ə-lōōt′] *adjective*, determined, purposeful, earnest

res-o-lu-tion [rĕz′ə-lōō′shən] *noun*, a firm decision, a course of action decided on

resolve

re-solve [rĭ-zŏlv'] *noun,*
1) determination, conviction,
verb, 2) to decide to take action

res-o-nant [rĕz'ə-nənt] *adjective,*
echoing, , vibrant, deep in sound

re-sort [rĭ-zôrt'] *noun,* 1) a
luxurious place visited for rest
and relaxation, *verb,* 2) to turn to
for help or support

re-source [rē'sôrs'] *noun,* 1) a
source of help or supply 2) a
supply or form of wealth

re-source-ful [rē'sôrs' fəl]
adjective, able to deal promptly
and effectively with difficulties

re-spect [rĭ-spĕkt'] *noun,* 1) an
aspect 2) consideration for
someone, *verb,* 3) to have a good
opinion of someone, look up to

re-spect-a-ble [rĭ-spĕk'tə-bəl]
adjective, honorable, well
thought of, worthy of respect

re-spect-ful [rĭ-spĕkt'fəl] *adjective,*
polite, well-mannered, civil

re-spect-ful-ly [rĭ-spĕkt'fə-lē]
adverb, in a civil manner,
courteously, politely

re-spec-tive [rĭ-spĕk'tĭv] *adjective,*
in relation to one another

re-spec-tive-ly [rĭ-spĕk'tĭv-lē]
adverb, for each separately in
precisely the order given

res-pi-ra-tion [rĕs'pə-rā'shən]
noun, the emission of vapor,
exhaling, breathing

res-pite [rĕs'pĭt] *verb,*
postponement or delay

re-splend-ent [rĭ-splĕn'dənt]
adjective, brilliant, lustrous

re-spond [rĭ-spŏnd'] *verb,* to
answer, to act in return

re-spond-ent [rĭ-spŏn'dənt] *noun,*
one who answers actions in a
civil case, the defendant

re-sponse [rĭ-spŏns'] *noun,* an
answer, a reaction, something
said or done in reply

re-spon-si-bil-i-ty
[rĭ-spŏn'sə-bĭl'ĭ-tē] *noun,* a
person or thing that one is
obligated to

re-spon-si-ble [rĭ-spŏn'sə-bəl]
adjective, 1) the willingness to
respond appropriately
2) trustworthy, reliable

rest [rĕst] *noun,* 1) a quiet time for
relaxation 2) those remaining or
left behind, *verb,* 3) to relax and
feel at ease 4) to lean or put
something on something else

res-tau-rant [rĕs'tər-ənt] *noun,* a
public eating establishment

res-ti-tu-tion [rĕs'tĭ-tōō'shən]
noun, an agreement as terms of a
settlement to pay for loss or
damage

rest-less [rĕst'lĭs] *adjective,* not
relaxed, eager to do something
else, uneasy, fitful, anxious

re-store [rĭ-stôr'] *verb,* 1) to revive
2) to bring back to a former
condition, to reinstate 3) to give
back, to return

re-strain [rĭ-strān'] *verb,* 1) to hold
back with force 2) to limit

re-straint [rĭ-strānt'] *noun,* a
means of holding back from
action, a check, a control

re-strict [rĭ-strĭkt'] *verb,* to
confine, to keep within limits

re-stric-tive [rĭ-strĭk'tĭv] *adjective,*
confining, limiting, inhibiting

274

re-sult [rĭ-zŭlt′] *noun*, 1) the product or consequence of an action, the outcome, *verb*, 2) to have been the cause of something, to end up

re-su-me [′rez-ə-mā] *noun*, 1) a summary of personal or work history [rĭ-zoom′] *verb*, 2) to start again after stopping for a period of time

res-ur-rec-tion [rĕz′ə-rĕk′shən] *noun*, 1) the act of rising from the dead 2) renewal, revival

re-sus-ci-tate [rĭ-sŭs′ĭ-tāt′] *verb*, to revive, especially from apparent death, to restore to consciousness

re-tail [rē′tāl′] *verb*, to sell goods directly to the consumer

re-tain [rĭ-tān′] *verb*, 1) to keep in place or hold 2) to have

re-tal-i-a-tion [rĭ-tăl′ē-ā′shən] *noun*, to repay like for like, especially evil for evil

re-tard [rĭ-tärd′] *verb*, to delay the progress of, to hinder

re-ten-tion [rĭ-tĕn′shən] *noun*, 1) holding something 2) having a good memory, retaining

ret-i-cence [rĕt′ĭ-səns] *noun*, not revealing your thoughts or feelings openly

ret-i-na [rĕt′n-ə] *noun*, the light-sensitive colored circle of the eye

re-tire [rĭ-tīr′] *verb*, 1) to withdraw to a place of privacy 2) to go to bed 3) to remove from active service, to stop working

re-tire-ment [rĭ-tīr′mənt] *noun*, the period at the end of a career when a person stops working to relax and enjoy their savings

re-tort [rĭ-tôrt′] *verb*, a response

re-tract [rĭ-trăkt′] *verb*, 1) to draw back as claws or fangs 2) to disavow, to recant, to disclaim

re-treat [rĭ-trēt′] *noun*, 1) a place to rest and relax, a place of safety *verb*, 2) to turn around and go back, to withdraw

re-trench [rĭ-trĕnch′] *verb*, to cut down, to economize, to reduce

ret-ri-bu-tion [rĕt′rə-byoo′shən] *noun*, vengeance, compensation, punishment for offenses

re-trieve [rĭ-trēv′] *verb*, 1) to get and bring back 2) to recover

ret-ro-ac-tive [rĕt′rō-ăk′tĭv] *adjective*, that which takes effect from a time in the past

ret-ro-grade [rĕt′rə-grād′] *verb*, going or inclined to go backward

ret-ro-spect [rĕt′rə-spĕkt′] *noun*, looking back on things past

re-turn [rĭ-tûrn′] *noun*, 1) to go to a place and come back from it 2) the income from an investment 3) a response, in kind, *verb*, 4) to give or put something back 5) to go back to the same place 6) to reply

re-use [rē-yooz′] *verb*, to use again, to recycle

re-veal [rĭ-vēl′] *verb*, 1) to make known, to tell, to disclose, to admit 2) to bring into view

rev-e-la-tion [rĕv′ə-lā′shən] *noun*, a sign from heaven, a disclosure

rev-el-ry [rĕv′əl-rē] *noun*, boisterous merrymaking

re-venge [rĭ-vĕnj′] *noun*, 1) a form of getting even, retribution, *verb*, 2) to administer punishment

rev-e-nue [rĕv′ə-noō] *noun*, 1) the return from an investment 2) an item of income 3) the taxes collected by a government

re-ver-ber-ate [rĭ-vûr′bə-rāt′] *verb*, to echo, to resound

re-vere [rĭ-vîr′] *verb*, to regard with respect, to admire, to look upon with awe, to honor

rev-er-ence [rĕv′ər-əns] *adjective*, honor or respect given to someone because of position

rev-er-end [rĕv′ər-ənd] *noun*, the title of the leader of a church

rev-er-ie [rĕv′ə-rē] *noun*, a daydream, a musing

re-verse [rĭ-vûrs′] *adjective*, 1) opposite, backward, *noun*, 2) a lever that makes a machine go backward, *verb*, 3) to turn around 4) to evoke or annul

re-vert [rĭ-vûrt′] *verb*, to return to a former habit, practice or belief

re-view [rĭ-vyoō′] *noun*, 1) an examination or inspection 2) a critical essay 3) a periodical, *verb*, 4) to go over or examine critically 5) to look back upon

re-vile [rĭ-vīl′] *verb*, to denounce using abusive language

re-vise [rĭ-vīz′] *verb*, to review and make changes in order to bring up to date, to correct, to improve

re-vi-sion [rĭ-vĭzh′ən] *noun*, the changed version of writing, or work, or doctrine

re-vive [rĭ-vīv′] *verb*, to bring back to life or to regain strength

re-volt [rĭ-vōlt′] *noun*, 1) a planned attack on a government or its leaders, *verb*, 2) to fight as a mass of rebels against leaders

3) to offend someone by making them feel ill or disgusted

rev-o-lu-tion [rĕv′ə-loō′shən] *noun*, 1) the motion of a body about a center or axis 2) the overthrow of a government or system by those who are governed, usually by force, a rebellion 3) a radical change

rev-o-lu-tion-ar-y [rĕv′ə-loō′shə-nĕr′ē] *noun*, an activist who promotes change

re-volve [rĭ-vŏlv′] *verb*, 1) to orbit the way the Earth moves slowly around the sun 2) to rotate 3) to recur in cycles

re-volv-er [rĭ-vŏl′vər] *noun*, a firearm with a cylinder of several chambers to be discharged repeatedly

re-vul-sion [rĭ-vŭl′shən] *noun*, a sickening feeling, a negative reaction

re-ward [rĭ-wôrd′] *noun*, 1) something given to praise good work, kindness or bravery, *verb*, 2) to give a prize in appreciation of good work

rhap-so-dize [răp′sə-dīz′] *verb*, to speak or write enthusiastically

rhap-so-dy [răp′sə-dē] *noun*, delight, happiness, enthusiasm

rhet-o-ric [rĕt′ər-ĭk] *noun*, the art of using language effectively

rheu-ma-tism [roō′mə-tĭz′əm] *noun*, a painful disease of the muscles and joints, accompanied by swelling and stiffness

rhi-noc-er-os [rī-nŏs′ər-əs] *noun*, a large wild animal with tough skin and two large horns on its nose, which lives in Africa

Rhode Island [rōd-ī′lənd] *noun*, one of 50 United States located on the East Coast, the capital is Providence. The state flower of **Rhode Island** is the violet and the motto is "Hope."

rhom-bus [rŏm′bəs] *noun*, a form with four equal sides in the shape of a diamond

rhu-barb [rōō′bärb′] *noun*, acid leafstalks of a coarse herb, used for sauce or pie

rhyme [rīm] *noun*, 1) correspondence of sounds in words, *verb*, 2) to end the line of a verse with the same sound

rhythm [rĭth′əm] *noun*, 1) repetition of a beat played throughout a musical piece 2) the recurrence of stress or a sound at regular intervals

rib [rĭb] *noun*, one of many narrow bones surrounding the chest

rib-ald [rĭb′əld] *adjective*, wanton, indecent, offensive, vulgar

rib-bon [rĭb′ən] *noun*, a narrow fabric usually of silk, used to trim a dress or a present

rice [rīs] *noun*, a grain in the cereal family from a grass-like plant grown in warm climates

rich [rĭch] *adjective*, 1) wealthy, having many resources 2) plentiful, abundant

rich-es [rĭch′ĭz] *noun*, *plural*, treasures, things worth a lot of money, valuable possessions

rid [rĭd] *verb*, to dispose of

rid-dance [rĭd′ns] *noun*, a cleaning up or out, relief or deliverance

rid-dle [rĭd′l] *noun*, 1) a clever question with an unexpected answer, a puzzling thing, *verb*, 2) to pierce with many holes

ride [rīd] *verb*, to travel on an animal or in a vehicle to go somewhere

rid-er [rī′dər] *noun*, 1) modification of a law, by adding a clause to legislation 2) a passenger on a journey

ridge [rĭj] *noun*, a long narrow raised part of the hill, a crest

rid-i-cule [rĭd′ĭ-kyōōl′] *noun*, 1) taunts, jeers, *verb*, 2) to sneer at, to mock, to make fun of

ri-dic-u-lous [rĭ-dĭk′yə-ləs] *adjective*, unworthy of serious consideration, absurd

rife [rīf] *adjective*, prevalent, common or frequent in occurrence, widespread

ri-fle [rī′fəl] *noun*, 1) a firearm with a long barrel, *verb*, 2) to search hurriedly through

rift [rĭft] *noun*, an opening, a break

rig [rĭg] *noun*, 1) heavy equipment, 2) a vehicle used to haul large shipments, *verb*, 3) to equip

rig-ging [rĭg′ĭng] *noun*, the ropes used to raise and hold the sails of a ship

right [rīt] *adjective*, 1) of or located to the opposite of left 2) what is fair and good 3) correct, *adverb*, 4) correctly 5) straight to, without stopping, *noun*, 6) a freedom allowed by law

right triangle [rīt-trī′ăng′gəl] *noun*, a triangle that has one right angle in it, the opposite side is the hypotenuse

right-eous [rī′chəs] *adjective*, just, virtuous, acting in a moral way

rig-id [rĭj/ĭd] *adjective*, hard, fixed, immovable, not flexible

rig-or [rĭg/ər] *noun*, severity

rig-or-ous [rĭg/ər-əs] *adjective*, strict, severe discipline

rim [rĭm] *noun*, the often-curved edge of something, a border

rind [rīnd] *noun*, the tough outer skin of fruit or cheese that protects it from air

ring [rĭng] *noun*, 1) small circular band worn on the finger 2) a circular object with a vacant center 3) an area where fights are held 4) a small group 5) the sound created by a bell, *verb*, 6) to make a sound like a bell

rink [rĭngk] *noun*, a playing field, a ring, a smooth floor

rinse [rĭns] *verb*, to wash lightly, to cleanse with water after washing

ri-ot [rī/ət] *noun*, 1) unrestrained fighting, vandalism, *verb*, 2) to fight in an angry crowd

ri-ot-ous [rī/ət-əs] *adjective*, unrestrained, tumultuous

rip [rĭp] *verb*, to tear away

ripe [rīp] *adjective*, completely mature, ready to be eaten, aged

rip-en [rī/pən] *verb*, to grow to maturity or when ready to eat

rip-ple [rĭp/əl] *noun*, 1) a little wave, *verb*, 2) to cause little waves of sound or movement

rise [rīz] *noun*, 1) an increase, *verb*, 2) to stand up 3) to move upward 4) to move in a positive direction 5) to get up out of bed

risk [rĭsk] *noun*, 1) a sense or chance of danger involved with doing something, *verb*, 2) to take a chance of injury or loss

rite [rīt] *noun*, a formal ceremony, a ritual, a custom

rit-u-al [rĭch/ōō-əl] *noun*, a set action or series of actions

ri-val [rī/vəl] *noun*, a competitor

ri-val-ry [rī/vəl-rē] *noun*, a contention, a contest, jealousy

riv-er [rĭv/ər] *noun*, a large natural stream of fresh water that flows into the ocean

riv-et [rĭv/ĭt] *noun*, 1) a metal pin for passing through holes in two or more pieces to hold them together, *verb*, 2) to hold attention firmly, to engross

roach [rōch] *noun*, a cockroach or small insect that eats food

road [rōd] *noun*, a route used for vehicles to travel, an avenue or thoroughfare, a street

roam [rōm] *verb*, to wander about, to travel aimlessly

roar [rôr, rōr] *noun*, 1) a deep angry sound, a loud growl or snarl, *verb*, 2) to growl in a loud voice 3) to laugh loudly

roast [rōst] *noun*, 1) a piece of meat that has been cooked over fire or will be, *verb*, 2) to cook without water over a fire or in an oven, to expose to great heat

rob [rŏb] *verb*, to deprive, to steal from someone, to defraud

rob-ber [rŏb/ər] *noun*, a thief, someone who steals

robe [rōb] *noun*, a piece of clothing that covers most of the body worn for lounging or relaxing

rob-in [rŏb′ĭn] *noun*, a brown song bird with a red breast found in North America

ro-bot [rō′bət] *noun*, a programmed machine designed to do the work of a person

ro-bust [rō-bŭst′] *adjective*, vigorous, strong, healthy

rock [rŏk] *noun*, 1) a hard, natural material of mineral origin, a stone 2) the mineral matter that makes up a large portion of the earth's crust *verb*, 3) to move back and forth or from side to side

rock-et [rŏk′ĭt] *noun*, 1) a missile, a spacecraft, *verb*, 2) to shoot into space 3) to move swiftly

rod [rŏd] *noun*, a thin pole

rode [rōd] *past tense* of **ride**

ro-dent [rōd′nt] *noun*, a small furry four-legged animal, like a squirrel, mouse or rat

ro-de-o [rō′dē-ō′] *noun*, a contest for people like cowhands to show their skill

rogue [rōg] *noun*, a scoundrel

role [rōl] *noun*, a part played by an actor, a characterization

roll [rōl] *noun*, 1) a long, drawn out noise 2) a small rounded portion of bread, *verb*, 3) to move along by turning over and over 4) to make something flat by pressing it with a roller

ro-man nu-mer-als [rō-mäṇ′nōō′mər-əlz] *noun*, *plural*, the number symbols used by Romans 2,500 years ago

ro-mance [rō-măns′] *noun*, 1) the state of being in love 2) a story

about love between people, *verb*, 3) to court romantically

ro-man-tic [rō-măn′tĭk] *adjective*, 1) not based in fact, fanciful 2) thoughts of love and adventure

roof [rōōf] *noun*, 1) the external cover of a building 2) the top of the mouth

rook-ie [rōōk′ē] *noun*, a beginner

room [rōōm] *noun*, 1) one of the parts of a house separated by walls and doors 2) space

roost [rōōst] *noun*, 1) a perch, *verb*, 2) to sleep or rest on a perch

roost-er [rōō′stər] *noun*, a domestic male fowl, a cock

root [rōōt] *noun*, 1) the part of the plant that grows in the ground, it's purpose is to hold it in the earth, to absorb moisture and to store nourishment, *verb*, 2) to dig up with the nose or snout 3) to cheer for a team

rope [rōp] *noun*, a heavy cord

rose [rōz] *noun*, 1) a shrub with a beautiful sweet smelling flower, *verb*, 2) the *past tense* of **rise**

ros-ter [rŏs′tər] *noun*, a list

ros-trum [rŏs′trəm] *noun*, a stage for public speaking, a pulpit

rot [rŏt] *verb*, to decompose, to decay gradually, to spoil

ro-ta-ry [rō′tə-rē] *noun*, rotating parts turning on an axis

ro-tate [rō′tāt] *verb*, to turn around a center or axis, to revolve

ro-ta-tion [rō-tā′shən] *noun*, 1) a revolution, turning in a full circle 2) the growing of different crops in the same place usually in regular order

rote [rōt] *noun*, learning through repetition and memorization

rot-ten [rŏt'n] *adjective*, no good

ro-tun-da [rō-tŭn'də] *noun*, a round building or rooms

rough [rŭf] *adjective*, 1) not smooth, uneven, difficult 2) not calm or gentle 3) unfinished, crude, coarse to the touch

rough-ly [rŭf-lē] *adverb*, approximately

round [round] *adjective*, 1) in the shape of a circle or a ball, circular, *noun*, 2) a circle, an orb, a globe 3) a course, a cycle, a series, *verb*, 4) to turn a bend or corner 5) to express a number to the nearest whole number

round-a-bout [round'ə-bout'] *adjective*, 1) not direct or straightforward, *noun*, 2) a traffic circle

rouse [rouz] *verb*, to stimulate, to spur, to wake up, to excite

rout [rout] noun1) a disorganized retreat from an attack, *verb*, 2) to drive out, to defeat decisively

route [rōōt, rout] *noun*, a road for traveling ffrom one place to another

rou-tine [rōō-tēn'] *adjective*, 1) habitual, 2) unoriginal, *noun*, 3) a regular course or procedure

row [rō] *noun*, 1) an arrangement set in a line 2) a quarrel or fight, *verb*, 3) to propel with oars

roy-al [roi'əl] *adjective*, 1) of a king or queen 2) fit for a king or queen

roy-al-ty [roi'əl-tē] *noun*, 1) a member of the monarchy 2) a

payment to the owner for the use of copyright material

rub [rŭb] *verb*, 1) to create friction by moving something back and forth 2) to massage

rub-ber [rŭb'ər] *noun*, *no plural*, a mix of chemicals to form a pliable substance

rub-bish [rŭb'ĭsh] *noun*, *no plural*, refuse, garbage, trash

rub-ble [rŭb'əl] *noun*, pieces or fragments of large objects

ru-by [rōō'bē] *noun*, a precious stone of deep red

rud-der [rŭd'ər] *noun*, a steering mechanism at the back of a boat

rud-dy [rŭd'ē] *adjective*, reddish, healthy-looking

rude [rōōd] *adjective*, not polite, unkind, discourteous behavior

ru-di-men-ta-ry [rōō'də-měn'tə-rē] *adjective*, the first undeveloped state, unrefined, imperfect

ruf-fi-an [rŭf'ē-ən] *noun*, a person who is rough and dirty

ruf-fle [rŭf'əl] *noun*, 1) trimming, frill, *verb*, 2) to make uneasy

rug [rŭg] *noun*, a floor covering made of woven fibers or wool

rug-ged [rŭg'ĭd] *adjective*, 1) rough and wild 2) full of rocks with a jagged surface

ruin [rōō'ĭn] *noun*, 1) destruction, damage, *verb*, 2) to destroy

ru-in-ous [rōō'ə-nəs] *adjective*, destructive, harmful

rule [rōōl] *noun*, 1) a law, a guide for conduct or procedure, *verb*, 2) to have power and control over someone 3) to decree

rul-er [rōō'lər] *noun*, 1) someone who governs a body of people 2) an instrument used to measure and draw straight lines

rum-ble [rŭm'bəl] *noun*, 1) a loud sound, like earth moving, *verb*, 2) to make a continuous loud, low noise, like thunder

ru-mi-nate [rōō'mə-nāt'] *verb*, 1) to chew the cud 2) to ponder, to think about, to meditate

rum-mage [rŭm'ĭj] *noun*, 1) odds and ends, miscellany, *verb*, 2) to look for, to rifle through

ru-mor [rōō'mər] *noun*, gossip, common talk, hearsay

run [rŭn] *noun*, 1) the distance traveled, a journey 2) a score in a baseball game 3) movement that is quicker than walking, *verb*, 4) to move quickly or swiftly 5) to work or make work 6) to go over 7) to escape, to retreat 8) to ooze or pour out

rung [rŭng] *noun*, 1) a rod on a ladder that is a step, *verb*, 2) *past participle* of **ring**

run-ner [rŭn'ər] *noun*, 1) a long ski on the bottom of a sled used on snow and ice 2) a person who runs errands 3) a carpet or decorative cover placed on a walk way

rup-ture [rŭp'chər] *noun*, 1) a fracture, a breach, *verb*, 2) to break, to fracture, to split

ru-ral [rōōr'əl] *adjective*, having to do with, in, or like farmland, country people, or life in the country, rustic

ruse [rōōs] *noun*, a trick, gimmick

rush [rŭsh] *noun*, *no plural*, 1) rapid movement, haste 2) reeds that grow near the water, *verb*, 3) to move quickly or fast as in a hurry 4) to charge, to attack

rust [rŭst] *noun*, *no plural*, 1) a corrosive red-brown substance that forms on iron when it has been exposed to water and oxygen, *verb*, 2) to become covered with the corrosive element 3) to deteriorate

rus-tic [rŭs'tĭk] *adjective*, referring to country people, rural

rus-tle [rŭs'əl] *verb*, to make a soft rubbing sound, to crackle

rut [rŭt] *noun*, a deep narrow groove in the ground

ruth-less [rōōth'lĭs] *adjective*, cruel, harsh, merciless

rye [rī] *noun*, a grass with seeds that are made into flour

S

Sab-bath [săb'əth] *noun*, a holy day, a day of rest

sack [săk] *noun*, a large bag made of woven material strong enough to carry things

sa-cred [sā'krĭd] *adjective*, holy, worshipped, religious

sac-ri-fice [săk'rə-fīs'] *noun*, 1) an offering, the surrender of anything, *verb*, 2) to give up something, to forfeit

sac-ro-sanct [săk'rō-săngkt'] *adjective*, most sacred, holy

sad [săd] *adjective*, unhappy, sorrowful, affected by grief

sad-dle [săd′l] *noun*, 1) a seat made of leather that fits over the back of a horse, *verb*, 2) to burden

sa-dis-tic [sə-dĭs′tĭk] *adjective*, pleasure in being cruel

sa-fa-ri [sə-fär′ē] *noun*, a journey into the jungles and plains of Africa, to look at wild animals

safe [sāf] *adjective*, 1) not in danger, protected, secure, *noun*, 2) a vault, a chest, a repository

safe-ty [sāf′tē] *noun*, a way of preventing damage to workers or a machine, protection

sag [săg] *verb*, 1) to hang down, to droop 2) to decline in value

sa-ga [sä′gə] *noun*, any legend, a long drawn out story

sa-ga-cious [sə-gā′shəs] *adjective*, shrewd, wise, knowledgeable

sage [sāj] *noun*, a wise man

said [sĕd] *past tense* of **say**

sail [sāl] *noun*, 1) a large cloth used to catch the wind and move a boat, *verb*, 2) to travel on water 3) to navigate a ship equipped with sails 4) to glide

sail-or [sā′lər] *noun*, a seaman, someone who works on a ship

saint [sānt] *noun*, a person who has lived a religious life and has been recognized by a group of religious leaders for good work

sake [sāk] *noun*, used with **for**, to show purpose or reason

sal-ad [săl′əd] *noun*, a preparation of vegetables, fruit, fish, or meat, dressed with vinegar, oil, etc.

sal-a-ry [săl′ə-rē] *noun*, fixed wages for a year, quarter, or month, for services

sale [sāl] *noun*, 1) the exchange of goods for money 2) when merchandise is sold at a reduced price 3) an auction

sales-man [sālz′mən] *noun*, one whose occupation is to sell goods or merchandise

sales tax [sālz-tăks] *noun*, a tax people pay for purchases

sa-li-ent [sā′lē-ənt] *adjective*, standing out, conspicuous, striking, noticeable, obvious

sa-line [sā′lēn′] *adjective*, salty

sa-li-va [sə-lī′və] *noun*, a fluid secreted in the mouth to help digest food, spit

salm-on [săm′ən] *noun*, a species of fresh-water fish

sa-loon [sə-loon′] *noun*, a bar or tavern, a large public room

salt [sôlt] *noun*, a white crystal compound of chlorine and sodium found in sea water, used to flavor food and preserve meat

sa-lu-bri-ous [sə-loo′brē-əs] *adjective*, promoting health

sal-u-ta-tion [săl′yə-tā′shən] *noun*, a greeting, something uttered

sal-u-tar-y [săl′yə-tĕr′ē] *adjective*, beneficial, wholesome

sa-lute [sə-loot′] *noun*, 1) a sign of respect for a superior officer *verb*, 2) to hold up the right hand to the forehead

sal-vage [săl′vĭj] *noun*, 1) property saved from destruction, *verb*, 2) to save, to recycle, to recover, to rescue

sal-va-tion [săl-vā′shən] *noun*, saved from harm or loss, deliverance from evil

salve [săv] *noun*, a medicinal ointment, a lotion, a balm

same [sām] *adjective*, 1) having the same identifying characteristics, identical 2) not different

sam-ple [săm/pəl] *noun*, 1) a part of anything shown as quality, a specimen, *verb*, 2) to try or test

san-a-to-ri-um [săn/ə-tôr/ē-əm] *noun*, a place where chronically ill patients receive the care they need until they recover

sanc-ti-fy [săngk/tə-fī/] *verb*, to bless, to make holy, to anoint

sanc-tion [săngk/shən] *noun*, 1) coercive measures directed against a nation that has violated international law, *verb*, 2) to approve, to support, to confirm

sanc-tu-ar-y [săngk/chōō-ĕr/ē] *noun*, church, a place of refuge

sand [sănd] *noun*, fine grains made of rocks usually found near the sea or in deserts

san-dal [săn/dl] *noun*, a kind of shoe consisting of a sole strapped to the foot

sand-bar [sănd' bär] *noun*, a ridge of sand formed in a river or along a shore by the action of currents or tides

sand-wich [sănd/wĭch, săn/-] *noun*, slices of bread with something between them such as meat, cheese or jelly

sane [sān] *adjective*, reasonable, sensible, sound in mind

sang [săng] *past tense* of **sing**

san-i-tar-y [săn/ĭ-tĕr/ē] *adjective*, very clean, antiseptic

san-i-ta-tion [săn/ĭ-tā/shən] *noun*, the job of keeping things clean

san-i-ty [săn/ĭ-tē] *noun*, a healthy outlook, a state of being sane

sank [săngk] *verb*, *past tense* of **sink**

sap [săp] *noun*, 1) the fluid in a plant that gives nourishment, *verb*, 2) to take strength from something, to weaken gradually

sapi-ent [sā/pē-ənt] *adjective*, wise, shrewd, showing great wisdom

sap-ling [săp/lĭng] *noun*, a young tree, a sprout of a tree

sap-phire [săf/īr/] *noun*, a precious gemstone of a blue color

sar-casm [sär/kăz/əm] *noun*, scornful remarks, ridicule

sar-cas-tic [sär-kăs/tĭk] *adjective*, taunting, satirical, scornful

sar-coph-a-gus [sär-kŏf/ə-gəs] *noun*, a coffin, a casket

sar-dine [sär-dēn/] *noun*, a small herring preserved in olive oil

sash [săsh] *noun*, a wide ribbon worn around the waist or over the shoulder

sat-el-lite [săt/l-īt/] *noun*, 1) a small, heavenly body orbiting around a planet 2) an artificial object launched to orbit the earth or other celestial bodies

sa-ti-ate [sā/shē-āt/] *verb*, to satisfy fully, to supply to excess

sat-ire [săt/īr/] *noun*, a form of writing that makes fun of habits, ideas or customs

sat-is-fac-tion [săt/ĭs-făk/shən] *noun*, the feeling of having what you need and want

sat-is-fac-to-ry [săt/ĭs-făk/tə-rē] *adjective*, sufficient, fulfilling all requirements

283

sat-is-fy [săt′ĭs-fī′] *verb*, 1) to be enough or good enough, to meet a need 2) to make sure

sat-u-rate [săch′ə-rāt′] *verb*, to become thoroughly soaked

Sat-ur-day [săt′ər-dē′]*noun*, the seventh day of the week

sauce [sôs] *noun*, a liquid of a variety of ingredients eaten with food to improve its taste

sauc-er [sô′sər] *noun*, a small plate put under a cup to catch the drips

saun-ter [sôn′tər] *verb*, to stroll, to walk slowly, to meander

sau-sage [sô′sĭj] *noun*, minced and seasoned meat, commonly enclosed in an edible casing

sav-age [săv′ĭj] *adjective*, wild, untamed, uncivilized

sa-van-nah [sə-văn′ə] *noun*, a grassy plain that has very few trees in a tropical climate

sa-vant [să-vänt′] *noun*, a scholar, a learned person, a sage

save [sāv] *verb*, 1) to rescue from harm or danger 2) to set aside for future use, to store 4) to copy on a disk or the hard drive of a computer

sav-ings [sā′vĭngz] *noun, plural, no singular*, money set aside to use in the future, reserve funds

sav-ior [sāv′yər] *noun*, someone who saves others from danger

savoir faire [săv′wär-fâr′] *noun*, ease and dexterity in social and practical affairs from the French, meaning "to know how to act"

sa-vor [sā′vər] *noun*, 1) a taste or aroma, *verb*, 2) to taste or enjoy a distinctive flavor, smell or quality, to appreciate

saw [sô] *noun*, 1) a tool with a jagged blade and handle used to cut wood, *verb*, 2) *past tense of* **see** 3) to cut with a serrated blade or saw

say [sā] *verb*, to speak something

say-ing [sā′ĭng] *noun*, a wise statement that is full of truth like a proverb or aphorism

scab [skăb] *noun*, 1) a hard covering of coagulated blood on a wound 2) a worker who stays on the job while others go on strike

scaf-fold [skăf′əld] *noun*, 1) a temporary platform built to hold laborers while working on a building 2) the gallows

scald [skôld] *verb*, 1) to injure or burn with hot liquid 2) to heat to just below the boiling point

scale [skāl] *noun*, 1) a balance to measure weight 2) a range, degree, or progression 3) a protective crust or plate, *verb*, 4) to adjust to a standard of measure 5) to climb or ascend

sca-lene tri-an-gle [skā′lēn′ trī′ăng′gəl] *noun*, a triangle with unequal sides

scalp [skălp] *noun*, 1) the skin and hair of the head, *verb*, 2) to remove the hair from someone

scam-per [skăm′pər] *verb*, to scurry, to run quickly

scan [skăn] *verb*, to look over, to review, to examine closely

scan-dal [skăn′dl] *noun*, an incident which offends conscience or moral feelings

scape-goat [skāp′gōt′] *noun*, someone who bears the blame for others

scar [skär] *noun*, 1) a mark left where a wound has been, *verb*, 2) to make a deep cut or to cause deep unforgettable pain

scar-ab [skăr′əb] *noun*, a large beetle with a dark shell

scarce [skârs] *adjective*, hard to find or come by, not plentiful

scarce-ly [skârs′lē] *adverb*, with difficulty, barely, hardly

scare [skâr] *noun*, 1) a sudden fright, *verb*, 2) to make someone afraid, to frighten, to terrify

scarf [skärf] *noun, plural,* **scarves,** a triangular or rectangular piece of cloth worn around the head, neck, or shoulders

scat-ter [skăt′ər] *verb*, to throw things everywhere, to disperse

scav-en-ger [skăv′ən-jər] *noun*, collector of garbage, an animal that devours refuse

scene [sēn] *noun*, 1) a setting, a view 2) a specific place 3) a shocking incident, a fight

scen-er-y [sē′nə-rē] *noun, no plural,* the setting of a place or landscape, the panorama

sce-nic [sē′nĭk] *adjective*, full of natural splendors, breathtaking

scent [sĕnt] *noun*, a smell that characterizes things

sched-ule [skĕj′ool] *noun*, 1) a written formal list or inventory, *verb*, 2) to arrange tasks to fit into a specific time frame

scheme [skēm] *noun*, 1) a plan, purpose, arrangement, *verb*, 2) to make secret plans, to plot

schol-ar [skŏl′ər] *noun*, a person who studies a particular subject at great length, a student

schol-ar-ship [skŏl′ər-shĭp′] *noun*, money granted to a worthy student to pay for their education

scho-las-tic [skə-lăs′tĭk] *noun*, pertaining to a school or education

school [skool] *noun*, 1) a place of instruction 2) a style, manner, method 3) a company of fish

schoon-er [skoo′nər] *noun*, a sailing ship with fore and aft rigging

sci-ence [sī′əns] *noun*, classified knowledge of the physical world

sci-en-tif-ic [sī′ən-tĭf′ĭk] *adjective*, in conformity to the principles of the physical world

sci-en-tist [sī′ən-tĭst] *noun*, a person who studies information about natural phenomena

scis-sors [sĭz′ərz] *noun*, a tool with two blades and handles, pivoted together that is used for cutting

scoff [skŏf] *verb*, to ridicule or make fun of, to deride, to jeer

scold [skōld] *verb*, to angrily tell someone they did something wrong, to reprimand

scoop [skoop] *noun*, 1) a large spoonful, *verb*, 2) to take some from the whole thing with a spoon or the hands

scoot [skoot] *verb*, to move quickly

scope [skōp] *noun*, 1) a range of one's mind 2) opportunity to function, capacity

scorch [skôrch] *verb*, to burn lightly leaving a brown stain

score [skôr] *noun*, 1) the points you get on a test or in a game 2) the reason, the purpose 3) a copy of a musical composition showing all parts for instruments and voices 4) a period of twenty years, *verb*, 5) to make points in a game 6) to achieve success

scorn [skôrn] *noun*, 1) disrespect, disdain, hate, *verb*, 2) to act like someone is worthless

scor-pi-on [skôr′pē-ən] *noun*, an arachnid that kills with its poisonous tail

scoun-drel [skoun′drəl] *noun*, a villain, a rascal, a mean person

scour [skour] *verb*, 1) to search an area thoroughly 2) to clean and polish by rubbing

scout [skout] *noun*, a person sent out ahead to obtain information

scowl [skoul] *noun*, 1) an angry, disgusted expression on the face, *verb*, 2) to make an angry expression showing disapproval

scram-ble [skrăm′bəl] *verb*, to move quickly on hands and knees, to hurry along

scrap [skrăp] *noun*, waste with some value, particularly material left over from construction or manufacturing suitable for reprocessing

scrap-book [skrăp′bŏŏk′] *noun*, a blank book filled with mementos, cards and souvenirs

scrape [skrāp] *noun* 1) a predicament, *verb*, 2) to take something off by rubbing it with a sharp edge 3) to make a grating sound

scratch [skrăch] *noun*, 1) a mark or wound made with something

pointed, *verb*, 2) to make marks with something pointed 3) to remove a layer by rubbing

scream [skrēm] *noun*, 1) a loud, shrill cry made in pain or fear, *verb*, 2) to make a sharp cry

screech [skrēch] *noun*, 1) a loud high-pitched noise, *verb*, 2) to make a harsh shrill noise

screen [skrēn] *noun*, 1) a portable covered framework in the nature of a partition 2) a grating of fine wire 3) protection, *verb*, 4) to eliminate applicants

screw [skrŏŏ] *noun*, 1) a cylinder, grooved in an advancing spiral on its surface, *verb*, 2) to turn around and around

scrib-ble [skrĭb′əl] *verb*, to write or draw in an unintelligible manner

scrim-mage [skrĭm′ĭj] *noun*, a football or soccer practice

script [skrĭpt] *noun*, 1) a form of handwriting that connects the letters in a word 2) written words to be spoken from a text

scrip-ture [skrĭp′chər] *noun*, any sacred writing or book

scroll [skrōl] *noun*, 1) a rod with a roll of paper attached, *verb*, 2) to move a cursor smoothly, vertically or sideways, to review data on a computer screen

scrub [skrŭb] *verb*, to rub with a hard brush to clean off a spot

scru-pu-lous [skrŏŏ′pyə-ləs] *adjective*, careful, cautious, exact, punctilious, precise

scru-ti-nize [skrŏŏt′n-īz′] *verb*, to examine closely and critically

scru-ti-ny [skrŏŏt′n-ē] *noun*, close examination, minute inspection

sculpt [skŭlpt] *verb*, to make or shape a statue

sculp-tor [skŭlp′tər] *noun*, one who carves or cuts wood or stone into statues

sculp-ture [skŭlp′chər] *noun*, works of art made from stone, clay or metal, a statue

scythe [sīth] *noun*, a long, curved-blade instrument for cutting grass by hand

sea [sē] *noun*, part of an ocean that is partially surrounded by land

sea-gull [sē′gŭl′] *noun*, a bird that lives near the ocean and scavenges for food

seal [sēl] *noun*, 1) an emblem affixed to a document to prove authenticity 2) a mammal with sleek skin, and limbs in the form of flippers, that lives in the water and on land, *verb*, 3) to close tightly so that something cannot be opened

seam [sēm] *noun*, 1) the stitches joining two pieces of cloth together 2) a fissure

search [sûrch] *noun*, 1) a careful investigation, a hunt, *verb*, 2) to look for someone or something

sea-son [sē′zən] *noun*, 1) a special time of year devoted to a certain activity 2) one of the four divisions of the year: fall, winter, spring, summer, *verb*, 3) to flavor or add zest to food

seat [sēt] *noun*, 1) a place to sit or a cushion to sit on 2) a location

se-cede [sĭ-sēd′] *verb*, to withdraw from association or fellowship

se-ces-sion [sĭ-sĕsh′ən] *noun*, withdrawal

se-clu-sion [sĭ-kloo′zhən] *noun*, 1) hiding, isolation 2) privacy

sec-ond [sĕk′ənd] *adjective*, 1) coming directly after the first, *noun*, 2) a very short period of time, a moment, *verb*, 3) to endorse, to support a motion

se-cre-cy [sē′krĭ-sē] *noun*, keeping things quiet without the knowledge of others

se-cret [sē′krĭt] *adjective*, 1) kept from the knowledge of most people, *noun*, 2) something known only to oneself or a few

sec-re-tar-y [sĕk′rĭ-tĕr′ē] *noun*, 1) a person who writes letters, files documents 2) a government officer who is a member of the President's Cabinet

se-crete [sĭ-krēt′] *verb*, to produce a liquid discharge

se-cre-tion [sĭ-krē′shən] *noun*, a fluid emitted from a gland

sec-tion [sĕk′shən] *noun*, a piece of the whole, a portion, a part

sec-tor [sĕk′tər, -tôr′] *noun*, region, area, zone, a distinct part

sec-u-lar [sĕk′yə-lər] *adjective*, worldly, not pertaining to church matters, temporal, earthly

se-cure [sĭ-kyoor′] *adjective*, 1) safe 2) strong and fixed firmly, *verb*, 3) to assure

se-cu-ri-ty [sĭ-kyoor′ĭ-tē] *noun*, 1) safety, assurance, freedom from danger 2) stocks and bonds

se-date [sĭ-dāt′] *adjective*, dignified, composed, calm

sed-a-tive [sĕd′ə-tĭv] *noun*, a tranquilizer, a narcotic

sed-en-tar-y [sĕd′n-tĕr′ē] *adjective*, accustomed to sitting most of the day

sed-i-ment [sĕd′ə-mənt] *noun*, the solid matter that settles to the bottom in a liquid

se-di-tion [sĭ-dĭsh′ən] *noun*, resistance to authority, insubordination

sed-ul-ous-ly [sĕj′ə-ləs-lē] *adjective*, diligently, loyally

see [sē] *verb*, 1) to perceive or acknowledge with the eye 2) to understand, to comprehend 3) to go and look 4) to meet 5) to think about something and to act if necessary

seed [sēd] *noun*, a small grain from which a plant grows

seek [sēk] *verb*, to look for

seem [sēm] *verb*, to appear to be true or probable

seen [sēn] past perfect of **see**

seep [sēp] *verb*, a liquid flowing slowly or leaking through

seg-ment [sĕg′mənt] *noun*, any part which separates from the whole

seg-re-ga-tion [sĕg′rĭ-gā′shən] *noun*, the act of setting groups apart, separation, racial discrimination

seis-mic [sīz′mĭk] *adjective*, pertaining to an earthquake

seize [sēz] *verb*, 1) to grasp suddenly, to take possession of by force

sel-dom [sĕl′dəm] *adjective*, rarely, not often, infrequently

se-lect [sĭ-lĕkt′] *adjective*, 1) first rate, *verb*, 2) to choose, to pick

se-lec-tion [sĭ-lĕk′shən] *noun*, a collection of things chosen

self [sĕlf] *noun*, your person, what you are, individuality

self-ish [sĕl′fĭsh] *adjective*, to think only of oneself and not others

self-ish-ness [sĕl′fĭsh- nĭs] *noun*, pettiness, being small-minded

sell [sĕl] *verb*, 1) to offer in exchange for money 2) to deal in

sem-blance [sĕm′bləns] *noun*, likeness, resemblance, similarity

se-mes-ter [sə-mĕs′tər] *noun*, either of two terms into which the period of instruction is divided in an academic year

sem-i-cir-cle [sĕm′ĭ-sûr′kəl] *noun*, the shape of half a circle

sem-i-co-lon [sĕm′ĭ-kō′lən] *noun*, a mark of punctuation (;) used to separate two independent clauses when there is no conjunction, as *When you make a mistake, admit it; learn from it and don't repeat it. Bear Bryant*

sen-ate [sĕn′ĭt] *noun*, one of the groups that constitutes the legislative government

sen-a-tor [sĕn′ə-tər] *noun*, an elected representative who serves in the Senate

send [sĕnd] *verb*, to cause something or someone to go somewhere, to dispatch

se-nil-i-ty [sĭ-nĭl′ĭ-tē] *noun*, feeble mindedness of old age

sen-ior [sēn′yər] *adjective*, 1) older, *noun*, 2) a student in the final year of a four year course

sen-sa-tion [sĕn-sā′shən] *noun*, 1) a feeling 2) excited interest

sense [sĕns] *noun*, 1) reason or understanding 2) a reasonable meaning 3) a feeling, *verb*, 4) to recognize, to detect

sense-less [sĕns/lĭs] *adjective*, stupid, unconscious

sen-si-bil-i-ty [sĕn/sə-bĭl/ĭ-tē] *noun*, awareness, understanding

sen-si-ble [sĕn/sə-bəl] *adjective*, reasonable, possessing good sense, smart, perceptive

sen-si-tive [sĕn/sĭ-tĭv] *adjective*, quick to notice or feel, easily affected by slight change

sen-su-ous [sĕn/shoo-əs] *adjective*, exuding sexuality, appealing

sen-tence [sĕn/təns] *noun*, 1) a combination of words which express a thought, beginning with a capital letter, ending with a period 2) the judgment of a court, *verb*, 3) to pass judgment on, to inflict punishment

sen-ti-ment [sĕn/tə-mənt] *noun*, feelings, thoughts, emotions

sen-ti-men-tal [sĕn/tə-mĕn/tl] *adjective*, expressing feelings of love, pity, etc. to excess

sen-ti-nel [sĕn/tə-nəl] *noun*, a soldier set to guard an army from surprise, a sentry

sen-try [sĕn/trē] *noun*, a lookout, guard, watchman, a soldier

sep-a-rate [sĕp/ə-rāt/] *adjective*, 1) not together, not joined, *verb*, 2) to divide, to disconnect, to keep apart

sep-a-ra-tion [sĕp/ə-rā/shən] *noun*, time apart from each other

Sep-tem-ber [sĕp-tĕm/bər] *noun*, the ninth month of the year, having 30 days

sep-tic [sĕp/tĭk] *adjective*, decaying, putrid or toxic, rotten

sep-ul-cher [sĕp/əl-kər] *noun*, a tomb, a burial place

se-quel [sē/kwəl] *noun*, the continuation of a story

se-quence [sē/kwəns] *noun*, a succession, a following or coming after in a series

se-ques-ter [sĭ-kwĕs/tər] *verb*, to go into isolation, to remove, to seclude, to set apart

sere-nade [sĕr/ə-nād/] *verb*, to court, to play music under a window to charm a sweetheart

ser-en-dip-i-tous [sĕr/ən-dĭp/ĭ-t əs] *adjective*, fortunate discoveries made by accident

se-rene [sə-rēn/] *adjective*, calm, tranquil, placid, clear, peaceful

se-ren-i-ty [sə-rĕn/ĭ-tē] *noun*, tranquility, peace

ser-geant [sär/jənt] *noun*, a noncommissioned officer ranking next above a corporal

se-ries [sîr/ēz] *noun*, a number of things connected in a sequence

se-ri-ous [sîr/ē-əs] *adjective*, 1) grave in manner or disposition, earnest, solemn, sincere 2) weighty

se-ri-ous-ly [sîr/ē-əs-lē] *adverb*, thoughtfully, sincerely

ser-mon [sûr/mən] *noun*, a talk given by a religious leader

ser-pent [sûr/pənt] *noun*, a snake

ser-pen-tine [sûr/pən-tēn/] *adjective*, winding, twisting

ser-rat-ed [sĕr/ā/tĭd] *adjective*, having a saw-toothed or jagged edge

se-rum [sîr′əm] *noun*, the watery protein of an animal fluid

serv-ant [sûr′vənt] *noun*, one who works for another especially in a home to perform chores

serve [sûrv] *verb*, 1) to do work for other people, to give assistance or help 2) to offer food to someone 3) to answer a purpose

serv-ice [sûr′vĭs] *noun*, 1) a job that one performs for others 2) a church ceremony

ser-vile [sûr′vəl] *adjective*, subservient, groveling, cringing

ses-sion [sĕsh′ən] *noun*, the sitting of an organized body for the transaction of business, a term

set [sĕt] *noun*, 1) a group of things thought of together, *verb*, 2) to put in a place 3) to make something happen, to begin 4) to sink in the sky below the horizon

set-tle [sĕt′l] *verb*, 1) to go and live in a place 2) to make comfortable 3) to come to a mutual agreement, to decide

set-tle-ment [sĕt′l-mənt] *noun*, 1) an agreement 2) a new place for people to live, a colony

set-tler [sĕt′lər] *noun*, one who travels to a new place to build a community

sev-en [sĕv′ən] *adjective, noun*, the number 7, one plus six

sev-en-teen [sĕv′ən-tēn′] *adjective, noun*, the number 17, 7 + 10

sev-en-teenth [sĕv′ən-tēnth′] *adjective, noun*, number 17 in order, written 17th

sev-enth [sĕv′ənth] *adjective, noun*, number 7 in order, 7th

sev-en-ti-eth [sĕv′ən-tē-ĭth] *adjective, noun*, number 70 in order, written 70th

sev-en-ty [sĕv′ən-tē] *noun, adjective*, the number 70, 7 X 10

sev-er [sĕv′ər] *verb*, to cut off, to separate, to disconnect

sev-er-al [sĕv′ər-əl] *adjective*, 1) separate or distinct, *pronoun*, 2) being more than two but fewer than many

sev-er-ance [sĕv′ər-əns] *noun*, the act of dividing or separating

se-vere [sə-vîr′] *adjective*, 1) strict in judgment or discipline, harsh 2) hard to deal with

se-ver-i-ty [sə-vĕr′ĭ-tē] *noun*, harshness, intensity

sew [sō] *verb*, to bind two pieces of cloth together with a needle and thread, to repair with a thread

sew-age [sōō′ĭj] *noun*, liquid waste from homes and businesses transported away by sewers

sew-er [sōō′ər] *noun*, 1) a gutter, a disposal system 2) one who works with needle and thread

sex [sĕks] *noun*, to be male or female

shab-by [shăb′ē] *adjective*, old, cheap, dirty, falling apart

shack-le [shăk′əl] *noun*, 1) chains, handcuffs, *verb*, 2) to confine, to restrain with chains

shade [shād] *noun, no plural*, 1) shelter from the sun or light, *verb*, 2) to cut off from a bright light 3) to dim or darken

shad-ow [shăd′ō] *noun*, 1) the image cast on a surface when something blocks the light 2) a faint indication

shaft [shăft] *noun*, 1) a long thin pole 2) a long hole leading to a mine 3) a ray or beam

shake [shāk] *verb*, 1) to move quickly up and down or from side to side 2) to tremble

shall [shăl] *verb*, 1) a word used instead of will with *I* and *we*, to indicate something is going to happen 2) used with *I* and *we* in questions when asking or offering to do something

shal-low [shăl'ō] *adjective*, 1) not deep 2) superficial, with little seriousness or deep thought

sham [shăm] *noun*, a farce

sham-bles [shăm'bəlz] *noun*, *plural*, a scene of disorder

shame [shām] *noun*, 1) dishonor, guilt, humiliation, *verb*, 2) to humiliate or embarrass

sham-poo [shăm-pōo'] *noun*, 1) a liquid soap used to wash hair, *verb*, 2) to wash hair with soap

shape [shāp] *noun*, 1) the outline or form of something, 2) a form or condition in which something exists, *verb*, 3) to mold or give form to something

share [shâr] *noun*, 1) a portion of something after it has been divided, *verb*, 2) to use, experience, or enjoy with others 3) to participate in 4) to disclose to others

shark [shärk] *noun*, a large predatory fish with a huge jaw

sharp [shärp] *adjective*, 1) having a thin cutting edge 2) the ability to see things far away that are small 3) severe or intense, having a strong odor or flavor 4) alert 5) quick and forceful

sharp-en [shär'pən] *verb*, 1) to make a pointed edge 2) to make acute or more sensitive to feeling

sharp-ly [sharp-lē] *adverb*, clearly, suddenly, distinctly

shat-ter [shăt'ər] *verb*, to break into many pieces or fragments

shave [shāv] *verb*, to cut very close

shawl [shôl] *noun*, a long piece of cloth worn over the shoulders and head

she [shē] *pronoun*, the female person or subject in the sentence

sheaf [shēf] *noun*, 1) a quantity of grain bound together 2) any bundle of things such as papers

shear [shîr] *verb*, to cut wool from a sheep, to cut hair, to clip

shears [shîrz] *noun*, *plural*, large-sized scissors

sheathe [shēth] *noun*, a case that holds a sword or dagger

shed [shĕd] *noun*, 1) a hut for tools, *verb*, 2) to let something fall or come off naturally 3) to pour forth, to stream out

sheep [shēp] *noun*, *no plural*, an animal that is kept for meat and wool from its thick, fleecy coat

sheep-ish-ly [shē'pĭsh-lē] *adverb*, with a feeling of embarrassment

sheer [shîr] *adjective*, 1) totally, entirely 2) transparent

sheet [shēt] *noun*, 1) a large flat piece of something such as paper or cloth 2) a rope used to secure a ship's sail

shelf [shĕlf] *noun*, *plural*, **shelves,** a board attached to a wall to put things on it

shell [shĕl] *noun*, the hard protective covering of fruit, eggs or some fish and reptiles

shel-lac [shə-lăk′] *noun*, a wood finishing such as varnish, lacquer or polyurethane

shell-fish [shĕl′fĭsh′] *noun*, any fish with a hard armor such as crabs, lobsters, or oysters

shel-ter [shĕl′tər] *noun*, 1) a safe place, a place to hide, *verb*, 2) to hide someone, to guard or keep safe from harm, to protect

shep-herd [shĕp′ərd] *noun*, a person who herds and tends sheep, goats, or livestock

sher-bet [shûr′bĭt] *noun*, flavored dessert ice

sher-iff [shĕr′ĭf] *noun*, a law enforcement officer of a county

shib-bo-leth [shĭb′ə-lĭth] *noun*, a slogan, a catchword

shield [shēld] *noun*, 1) a piece of defensive armor carried on the arm, protection, *verb*, 2) to defend or protect 3) to conceal

shift [shĭft] *noun*, 1) a set of people who work together at one time 2) the length of time that one group works, *verb*, 3) to move, to transfer, to displace

shil-ling [shĭl′ĭng] *noun*, English currency equal to a dollar

shim-mer [shĭm′ər] *verb*, to glimmer intermittently, to glisten

shin [shĭn] *noun*, the part of the leg between the knee and the ankle

shine [shīn] *verb*, to give off light

shin-gle [shĭng′gəl] *noun*, a tile made of wood or tar used to protect a roof or wall

ship [shĭp] *noun*, a boat which travels in deep water like the sea

ship-ment [shĭp′mənt] *noun*, the act of sending goods by air, water, or, on the road

shirt [shûrt] *noun*, a garment which covers the upper part of the body consisting of a collar, sleeves, and buttons, similar to a blouse

shiv-er [shĭv′ər] *verb*, to shake from feeling cold or afraid

shoal [shōl] *noun*, 1) a sandbar or piece of rising ground in shallow water 2) a school of fish

shock [shŏk] *noun*, 1) the feeling caused by a horrible unpleasant surprise 2) a pain caused by electricity going through the body, *verb*, 3) to startle, to disturb or surprise

shod-dy [shŏd′ē] *adjective*, not genuine, inferior quality

shoe [shoō] *noun*, an outer covering for the foot consisting of a sole and a covering made of leather, vinyl, or cloth

shoot [shoōt] *noun*, 1) new growth, a sprout, *verb*, 2) to discharge a weapon 3) to move quickly, to dart 4) to photograph or film

shop [shŏp] *noun*, 1) a place to go and buy different things that you need, *verb*, 2) to visit stores to make purchases

shore [shôr] *noun*, 1) land bordering water such as a lake or the sea, *verb*, 2) to prop up

short [shôrt] *adjective*, 1) the opposite of tall, small 2) not long, brief, concise

short-age [shôr′tĭj] *noun*, not enough, a deficiency

short-en [shôr′tn] *verb*, to cut down

short-en-ing [shôr′tn-ĭng] *noun*, lard or oil used in cooking

shorts [shôrts] *noun*, *plural*, pants worn above the knee

shot [shŏt] *noun*, 1) injection with a hypodermic needle 2) a try, an attempt 3) lead for a shotgun

should [shŏŏd] *verb*, obligated to do something, ought to

shoul-der [shōl′dər] *noun*, 1) part of the body between the neck and the chest, connected to the arms, *verb*, 2) to carry a burden

shouldn't [shŏŏd′nt] the *contraction* of the words **should** and **not**

shout [shout] *noun*, 1) a sudden cry in a loud voice, *verb*, 2) to speak or yell in a loud voice

shove [shŭv] *noun*, 1) a push, *verb*, 2) to push roughly or rudely

shov-el [shŭv′əl] *noun*, 1) a broad scoop with a long handle for lifting dirt, *verb*, 2) to lift things from the ground or another surface with a shovel

show [shō] *noun*, 1) a display in which things are gathered for people to see 2) a performance for an audience, *verb*, 3) to let someone see something

show-er [shou′ər] *noun*, 1) precipitation from clouds in the form of rain 2) a bath in which water is sprayed from a nozzle, 3) a party to honor someone, *verb*, 4) to pour down or fall on 5) to bestow

shred [shrĕd] *noun*, 1) a small amount, *verb*, 2) to tear or cut into small pieces

shrew [shrōō] *noun*, 1) a scolding woman, a nag 2) a mouse-like rodent with a long snout

shrewd [shrōōd] *adjective*, clever in business, sharp-witted

shriek [shrēk] *noun*, 1) a high-pitched scream, *verb*, 2) to utter a shrill cry, to scream

shrill [shrĭl] *adjective*, having a high-pitched sound, piercing

shrimp [shrĭmp] *noun*, a shellfish that is considered a delicacy

shrine [shrīn] *noun*, 1) a place for worship and keeping sacred relics 2) a saint's tomb

shrink [shrĭngk] *verb*, 1) to withdraw, to pull away from 2) to lessen, to make smaller

shriv-el [shrĭv′əl] *verb*, to contract in size, to shrink and wrinkle

shrub [shrŭb] *noun*, a bush

shrug [shrŭg] *verb*, 1) to raise the shoulders to show uncertainty about something 2) to disregard

shud-der [shŭd′ər] *noun*, a quick, light shaking motion, a shiver

shuf-fle [shŭf′əl] *verb*, 1) to walk dragging your feet 2) to rearrange the order, to mix together like a deck of cards

shun [shŭn] *verb*, to deliberately avoid, to stay away from

shut [shŭt] *verb*, to close, to adjust something so that it is not open

shut-ter [shŭt′ər] *noun*, 1) a wooden cover for a window 2) a mechanical device for opening and closing the aperture of a camera lens to expose the film

shy [shī] *adjective*, bashful, wary

sib-ling [sĭb′lĭng] *noun*, a brother or sister

sick [sĭk] *adjective*, not feeling well

sick-ness [sĭk′nĭs] *noun*, illness, nausea, ailment, disease

side [sīd] *noun*, 1) the right or left portion of the body 2) the surface of something flat that has a top and bottom 3) the boundaries of a surface or image 4) a perspective, a way of looking at something, an aspect

siege [sēj] *noun*, a continued attempt of an army to gain possession around a fortified place to compel its surrender

sieve [sĭv] *noun*, a bowl with holes in the bottom to let water drain out, or used to separate fine matter from the coarser materials

sigh [sī] *noun*, 1) a deep breath indicating that you are tired, sad, etc, *verb*, 2) to let out a deep breath, to exhale

sight [sīt] *noun*, 1) the ability or power to see 2) something worth seeing 3) a scene, an incident

sign [sīn] *noun*, 1) a gesture which sends a message to the person who sees it 2) an indication 3) a plaque with a message, *verb*, 4) to write your name 5) to hire or transfer by written agreement

sig-nal [sĭg′nəl] *noun*, 1) a movement or thing which tells you what to do, *verb*, 2) to warn someone, or send another message through motions

sig-na-ture [sĭg′nə-chər] *noun*, 1) a name written in ink, in cursive, as an endorsement to an agreement 2) the printed sheet so marked as to leave an imprint

sig-net [sĭg′nĭt] *noun*, an emblem

sig-nif-i-cance [sĭg-nĭf′ĭ-kəns] *noun*, *no plural*, the meaning

sig-nif-i-cant [sĭg-nĭf′ĭ-kənt] *adjective*, expressive or suggestive, important

sig-ni-fy [sĭg′nə-fī′] *verb*, to indicate, to signal, to denote

si-lence [sī′ləns] *noun*, *no plural*, 1) complete quiet 2) secrecy

si-lent [sī′lənt] *adjective*, making no sound, quiet, speechless

si-lent-ly [sī′lənt-lē] *adverb*, quietly, without a sound

sil-hou-ette [sĭl′ōō-ĕt′] *noun*, an outline of something against a light background

sil-i-con [sĭl′ĭ-kən, -kŏn′] *noun*, a nonmetallic element, very abundant in nature

silk [sĭlk] *noun*, a fine fabric made by the threads from a silkworm

sill [sĭl] *noun*, the rim on a window

sil-ly [sĭl′ē] *adjective*, not serious, or reasonable, playful, foolish

silt [sĭlt] *noun*, sediment of fine particles that hangs in stagnant water or forms on the bottom left there by running water

sil-ver [sĭl′vər] *adjective*, 1) the color of this metal, *noun*, *no plural*, 2) a soft shiny gray metal used for jewelry, coins, and ornaments

sim-i-an [sĭm′ē-ən] *adjective*, like a monkey or an ape

sim-i-lar [sĭm′ə-lər] *adjective*, somewhat like, nearly, corresponding, resembling

sim-i-lar-i-ty [sĭm′ə-lăr′ĭ-tē] *noun*, likeness, resemblance

sim-i-le [sĭm′ə-lē] *noun*, a phrase comparing unlike subjects using like or as, i.e., *he is as slow as a turtle; she sings like a bird*

sim-ple [sĭm′pəl] *adjective*, 1) easy to understand 2) plain, common, not complicated, not showy

sim-plic-i-ty [sĭm-plĭs′ĭ-tē] *adjective*, freedom from duplicity, absence of excess

sim-pli-fy [sĭm′plə-fī′] *verb*, to make easier to understand

sim-ply [sĭm′plē] *adverb*, only, or merely, absolutely

sim-u-late [sĭm′yə-lāt′] *verb*, to feign, to pretend, to fake

si-mul-ta-ne-ous [sī′məl-tā′nē-əs] *adjective*, occurring at the same time, concurrent

si-mul-ta-ne-ous-ly [sī′məl-tā′nē-əs-lē] *adverb*, happening at the same time

sin [sĭn] *noun*, 1) an act that harms someone, *verb*, 2) to err, to violate a moral principle

since [sĭns] *adverb*, 1) between a moment in the past and now, *conjunction*, 2) because, *preposition*, 3) after, from a certain time until the present

sin-cere [sĭn-sîr′] *adjective*, being in reality what it appears to be, genuine 3) free of hypocrisy

sin-cere-ly [sĭn-sîr′lē] *adverb*, honestly, truly

sin-cer-i-ty [sĭn-sĕr′ĭ-tē] *adverb*, honesty, truthfulness

sin-ew [sĭn′yo͞o] *noun*, 1) a tendon, a muscle 2) strength, power

sing [sĭng] *verb*, to make music with the voice

singe [sĭnj] *verb*, to scorch or burn the surface or ends

sin-gle [sĭng′gəl] *adjective*, 1) consisting of only one person or thing, alone, *noun*, 2) in baseball, a hit in which the batter reaches first base

sin-gu-lar [sĭng′gyə-lər] *adjective*, 1) when speaking of one, *noun*, 2) unique, particular, individual

sin-is-ter [sĭn′ĭ-stər] *adjective*, dishonest, indicative of lurking evil, wicked, menacing

sink [sĭngk] *noun*, 1) a basin that holds water, *verb*, 2) to go down 3) to become submerged in water 4) to permeate the mind

sin-ner [sĭn′ər] *noun*, one who does wrong to others

sip [sĭp] *noun*, 1) a small amount of liquid swallowed, *verb*, 2) to drink a very small amount

si-phon [sī′fən] *noun*, a bent tube, having one end longer than the other, used for drawing liquids from a higher to a lower level

sir [sûr] *noun*, 1) the title of a knight 2) a respectful term to use when talking to a man

si-ren [sī′rən] *noun*, a device that makes a loud warning sound

sis-ter [sĭs′tər] *noun*, 1) a girl who shares the same parents with a sibling 2) the title for a nun

sit [sĭt] *verb*, to rest in a chair or on the floor, with the back straight

site [sīt] *noun*, a particular piece of land or a building whose location has a specific use

sit-u-ate [sĭch′o͞o-āt′] *verb*, to put in place, to position, to locate

sit-u-a-tion [sĭch′ōō-ā′shən] *noun*, 1) circumstances, the state of affairs, a case 2) a position

six [sĭks] *adjective, noun*, the number 6, five plus one

six-teen [sĭk-stēn′] *adjective, noun*, the number 16, six plus ten

six-teenth [sĭk-stēnth′] *adjective, noun*, the number 16 in order, written 16th

sixth [sĭksth] *adjective, noun*, the number 6 in order, written 6th

six-ti-eth [sĭk′stē-ĭth] *adjective, noun*, number 60 in order, 60th

six-ty [sĭks′tē] *adjective*, the number 60, six times ten

size [sīz] *noun*,) the amount of space that something takes up

skate [skāt] *noun*, 1) a runner or set of wheels with a frame to fit the shoe, *verb*, 2) to glide over the ground with shoes fit with a runner or wheels

skel-e-ton [skĕl′ĭ-tn] *noun*, the bones of a person or animal that show its size and structure

skep-tic [skĕp′tĭk] *noun*, a doubter, a cynic, a questioner

skep-ti-cal [skĕp′tĭ-kəl] *adjective*, distrustful, not convinced

sketch [skĕch] *noun*, 1) a drawing that lacks detail, an image, *verb*, 2) to draw in a rough manner

ski [skē] *noun, plural*, **skis**, one of a pair of long, flat, wood or metal runners used for gliding on snow or water *verb*, 2) to slide over snow on blades using poles for balance

skid [skĭd] *noun*, 1) a pallet to elevate something off the ground, *verb*, 2) to slip on a wet surface 3) to lose traction

skill [skĭl] *noun*, having the ability and knowledge to do something

skil-let [skĭl′ĭt] *noun*, a small metal pan with a handle, a frying pan

skill-ful [skĭl′fəl] *adjective*, very good at doing something

skill-ful-ly [skĭl′fə-lē] *adverb*, intelligently, sensibly

skim [skĭm] *verb*, 1) to read over quickly 2) to take off the floating layer 3) to glide in a sail boat

skimp [skĭmp] *verb*, to hold back, to offer small portions

skin [skĭn] *noun*, 1) the covering of the body, or fruit, often covered with hairs, *verb*, 2) to remove the outer layer of something

skip [skĭp] *verb*, 1) to move by alternately hopping or stepping, to move with light steps 2) to omit, to pass over, to miss

skirt [skûrt] *noun*, 1) woman's clothing that hangs from the waist, *verb*, 2) to avoid

skit [skĭt] *noun*, a short play or theatrical performance

skit-tish [skĭt′ĭsh] *adjective*, lively, frisky, nervous

skulk [skŭlk] *verb*, 1) to move slyly and secretly in a sneaking manner 2) to lie in hiding

skull [skŭl] *noun*, the bony structure of the head enclosing the brain

skunk [skŭngk] *noun*, a rodent with black fur distinguished by a white strip that goes down its back and tail and sprays a horrible smell when scared

sky [skī] *noun*, the space above the Earth which can be seen when you look up

sky-scrap-er [skī′skrā′pər] *noun*, a very tall building

slack [slăk] *adjective*, 1) lazy, slow, barely moving 2) a mixture of small pieces of coal, dust, and dirt left from screening coal

slacks [slăks] *noun*, *plural*, pants or trousers for casual wear

slain [slān] *past tense* of **slay**

slam [slăm] *verb*, 1) to fling or put down violently, with a loud noise 2) to shut, to close

slan-der [slăn′dər] *noun*, 1) a malicious, false statement made about a person, *verb*, 2) to lie about someone, to malign

slang [slăng] *noun*, *no plural*, language used in conversation, that is crude, disrespectful, or inappropriate, an expletive

slant [slănt] *verb*, to lean or slope

slap [slăp] *noun*, 1) the blow or hit caused by something flat, *verb*, 2) to hit with something flat, such as an open hand

slate [slāt] *adjective*, 1) a gray color, *noun*, 2) a fine-grained metamorphic rock used as a roof tile or writing tablet with chalk

slaugh-ter [slô′tər] *noun*, *no plural*, 1) a killing of a large number of people or animals, *verb*, 2) to kill animals or people as a group

slave [slāv] *noun*, a person owned as property, subject to the will of his owner

slav-er-y [slā′və-rē] *noun*, *no plural*, the concept of owning people as property

slay [slā] *verb*, to kill or murder

slea-zy [slē′zē] *adjective*, low class

sled [slĕd] *noun*, a wooden device with runners on the bottom used to carry people or things on snow or ice as a vehicle

sleek [slēk] *adjective*, smooth, neat, glossy, well-groomed

sleep [slēp] *noun*, *no plural*, 1) a time of rest, when not awake and alert, *verb*, 2) to rest semi-unconscious with eyes closed

sleet [slēt] *noun*, 1) a combination of ice and rain falling from the sky, *verb*, 2) to shower frozen rain

sleeve [slēv] *noun*, the part of clothing that covers the arm

sleigh [slā] *noun*, a horse drawn vehicle with runners used for carrying people over snow

sleight [slīt] *noun*, skill, cunning

slen-der [slĕn′dər] *adjective*, 1) slim, thin, narrow 2) meager

slice [slīs] *noun*, 1) a piece cut from the whole, a share, *verb*, 2) to cut a flat piece off from a whole

slick [slĭk] *adjective*, 1) a surface that is slippery when wet 2) clever, deceitful

slide [slīd] *noun*, 1) a piece of playground equipment you can sit on and slip on the surface, an image for projection on a screen, *verb*, 3) to move smoothly along a surface, to slip

slight [slīt] *adjective*, of a small degree, of little importance

slim [slĭm] *adjective*, slender, thin, lean, narrow, meager

slime [slīm] *noun*, mire, mud

sling [slĭng] *noun*, 1) a piece of cloth holding something in place, *verb*, 2) to throw

slip [slĭp] *noun*, 1) a piece of paper 2) a garment worn under a dress 3) a mistake 4) a stall to keep something 5) a cutting from a plant, *verb*, 6) to lose your balance and fall off your feet 7) to enter or exit unobserved 8) to escape, to pass from

slip-per [slĭp′ər] *noun*, a light shoe which may be slipped on easily

slip-per-y [slĭp′ə-rē] *adjective*, 1) smooth and slick or slimy, hard to grip 2) devious, not trustworthy, shifty

slit [slĭt] *noun*, 1) a long narrow opening, *verb*, 2) to make a small cut as an opening

slith-er [slĭth′ər] *verb*, to move, walk or slide like a snake

slo-gan [slō′gən] *noun*, a catch phrase, a motto

slope [slōp] *noun*, 1) a surface which is higher on one side than the other, *verb*, 2) to lean on an incline, to slant up or downward

slot [slŏt] *noun*, a long, narrow opening for receiving something

sloth [slôth] *noun*, a lazy person

slov-en-ly [slŭv′ən-lē] *adjective*, untidy, careless in work habits

slow [slō] *adjective*, 1) to move without much speed, not fast 2) behind others, sluggish

slow-ly [slō′lē] *adverb*, to crawl or move sluggishly, not quickly

slug [slŭg] *noun*, 1) a soft, slimy creature without bones or legs that eats plants 2) a false coin, *verb*, 3) to hit with a fist

sluice [slōōs] *noun*, an artificial channel for water with a gate

slum [slŭm] *noun*, a poor neighborhood, a ghetto

slum-ber [slŭm′bər] *noun*, 1) a deep sleep, a long nap, *verb*, 2) to sleep, to doze

slump [slŭmp] *noun*, 1) a drop, a decline, a decrease, *verb*, 2) to sink, to slouch, to stoop

slur [slûr] *noun*, 1) a disrespectful remark, *verb*, 2) to insult, to offend 3) to garble, to speak unintelligibly

slush [slŭsh] *noun*, partially melting snow or ice

sly [slī] *adjective*, clever and deceiving, calculating

smack [smăk] *noun*, 1) a kiss, a pat, a peck, *verb*, 2) to hit with a hand as a slap 3) to put the lips together after a tasty meal

small [smôl] *adjective*, 1) little, not big 2) mean-spirited, petty

small-er [smôl′ər] *adjective*, not as big, shorter, less, fewer

small-est [smôl′ĕst] *adjective*, tiny, very little

smart [smärt] *adjective*, 1) intelligent, bright 2) well dressed, attractive, fashionable, *verb*, 3) to feel a painful, stinging sensation

smash [smăsh] *verb*, to break violently into many pieces

smat-ter-ing [smăt′ər-ĭng] *noun*, slight knowledge, a few facts

smell [smĕl] *noun*, 1) the identifying scent that distinguishes something, *verb*, 2) to use one of the five senses to recognize a scent

smelt [smĕlt] *verb*, to melt in a furnace, to purify an ore

smile [smīl] *noun*, 1) an expression formed by the upward curve of the mouth, to show pleasure, amusement, etc., 2) to look pleased, happy, or amused

smirk [smûrk] *noun*, a twisted smile, a smug expression

smog [smŏg] *noun*, a mixture of pollutants, principally ground-level ozone, produced by chemical reactions in the air

smoke [smōk] *noun*, 1) visible fumes, *verb*, 2) to puff or inhale the vapor from a lit cigarette, 3) to make fumes come from fire

smol-der [smōl′dər] *verb*, 1) to burn without flame 2) anger, liable to break out at any moment

smooth [smōōth] *adjective*, 1) even, level, without bumps 2) calm, composed, unruffled

smoth-er [smŭth′ər] *verb*, to hide or suppress, to suffocate

smudge [smŭj] *noun*, 1) a dirty mark, a smear, *verb*, 2) to make dirty with streaks, to blur

smug-gle [smŭg′əl] *verb*, to bring things across borders secretly or illegally to avoid tariffs or border police

snack [snăk] *noun*, 1) food eaten between meals, *verb*, 2) to eat when you are hungry

snail [snāl] *noun*, a slug in a spirally coiled shell

snake [snāk] *noun*, 1) a cold-blooded reptile without legs that slithers on the ground, *verb*,

2) to move like a snake by slithering and sliding

snap [snăp] *noun*, 1) the sharp sound of something breaking 2) a clothing fastener like a button that locks together, *verb*, 3) to make a crackling sound 4) to try to nip or bite someone

snare [snâr] *noun*, 1) a trap that entangles its victim, *verb*, 2) to trap with a noose, to entangle

snarl [snärl] *verb*, 1) to make an angry noise with bared teeth 2) to talk in a vicious tone

snatch [snăch] *verb*, to take something directly from someone in a rough manner

sneak [snēk] *noun*, 1) a cheater, a rascal, *verb*, 2) to creep, to slink

sneer [snîr] *verb*, to laugh at someone to make fun of them

sneeze [snēz] *noun*, to blow air through the nose and mouth without warning

sniff [snĭf] *verb*, to breathe in, to inhale, to get a whiff, to smell

snif-fle [snĭf′əl] *verb*, to gasp as if to hold back tears

snob [snŏb] *noun*, someone who thinks he is better than others

snore [snôr] *verb*, to make a loud guttural noise in the nose or throat when sleeping

snout [snout] *noun*, the part of an animal's head projecting forward containing the nose and jaws

snow [snō] *noun*, *no plural*, 1) precipitation that forms in the cloud as ice and falls to the ground as white crystals, *verb*, 2) a flurry of crystals or snowflakes falling to the earth from the sky

snug [snŭg] *adjective*, close fitting, comfortable, cozy, compact

so [sō] *adverb*, 1) in such a way, to such a point 2) also 3) very, very much 4) therefore 5) in order that 6) the same thing 7) used to show agreement 8) factual

soak [sōk] *verb*, to leave something immersed in liquid for awhile

soap [sōp] *noun*, a substance used for cleaning, manufactured as bars, flakes, or liquid, mixed with water

soar [sôr] *verb*, to fly or ascend quickly and easily upward

sob [sŏb] *verb*, to cry aloud in large gasps, to weep convulsively

so-ber [sō'bər] *adjective*, serious, somber, not gay or laughing

so-bri-e-ty [sə-brĭ'ĭ-tē] *noun*, self-restraint, recovery from abusing alcohol, not drunk

so-bri-quet [sō'brĭ-kā'] *noun*, a nickname, a pet name

soc-cer [sŏk'ər] *noun*, *no plural*, a game of football as played in South America and Europe

so-cial [sō'shəl] *adjective*, pertaining to men as living in society, friendly, gregarious

so-cial-ism [sō'shə-lĭz'əm] *adjective*, a communist philosophy of equal sharing

so-ci-e-ty [sə-sī'ĭ-tē] *noun*, a group of persons associated together

so-ci-ol-o-gy [sō'sē-ŏl'ə-jē] *noun*, the science of the development of human society

sock [sŏk] *noun*, 1) a short stocking made of soft material to cover the foot, *verb*, 2) to punch

sock-et [sŏk'ĭt] *noun*, any device that holds a complementary part

sod [sŏd] *noun*, earth, dirt, soil

so-da [sō'də] *noun*, a carbonated beverage flavored with syrup

so-di-um [sō'dē-əm] *noun*, a silver-white, metallic element of the alkali group, also known as salt

so-fa [sō'fə] *noun*, a long couch, usually upholstered and having a back and arms

soft [sôft] *adjective*, 1) tender 2) not hard 3) low volume, not loud 4) smooth or silky

soft-en [sô'fən] *verb*, 1) to ease, to relax, 2) to melt

soft-ly [sôft'lē] *adverb*, 1) quietly 2) mildly, gently

soft-ware [sôft'wâr'] *noun*, instructions that allow a computer to do specific tasks

soil [soil] *noun*, 1) the earth where plants and organisms grow, the top layer of the earth, *verb*, 2) to make something dirty

so-journ [sō'jûrn'] *noun*, 1) a brief stay, *verb*, 2) to stay temporarily

sol-ace [sŏl'ĭs] *noun*, comfort in trouble, or sorrow, or distress

so-lar [sō'lər] *adjective*, referring to or relating to the sun

sold [sōld] *verb*, *past tense* of **sell**

sol-der [sŏd'ər] *noun*, a metal or a metallic alloy used to join by fusing metallic surfaces together

sol-dier [sōl'jər] *noun*, a man or woman enlisted in the army

sole [sōl] *adjective*, 1) single, one *noun*, 2) the part of the shoe on which the bottom of the foot rests 3) an edible flat fish

sole-ly [sōl′lē] *adverb*, the only one

sol-emn [sŏl′əm] *adjective*, serious, grave, formal and dignified

so-le-noid [sō′lə-noid′] *noun*, a cylindrical round coil

so-lic-it [sə-lĭs′ĭt] *verb*, to ask someone to buy something or commit themselves

so-lic-i-tor [sə-lĭs′ĭ-tər] *noun*, the law officer of a city or government

sol-id [sŏl′ĭd] *adjective*, 1) not hollow, hard, firm, *noun*, 2) any substance that cannot flow, takes up space and has a definite shape and volume

sol-id-ify [sə-lĭd′ə-fī′] *verb*, to change from liquid to solid

so-lil-o-quy [sə-lĭl′ə-kwē] *noun*, the act of talking to one's self, a monologue in a drama

sol-i-taire [sŏl′ĭ-târ′] *noun*, 1) a gem set by itself 2) a card game involving one person

sol-i-tar-y [sŏl′ĭ-tĕr′ē] *adjective*, living or being by one's self, single, alone, secluded

sol-i-tude [sŏl′ĭ-tōōd′] *noun*, state of being alone, a lonely place

so-lo [sō′lō] *noun*, 1) something accomplished by one person 2) an unaccompanied performance

sol-stice [sŏl′stĭs] *noun*, either of two times a year when the sun is farthest from the equator, June 21 and December 22

sol-u-ble [sŏl′yə-bəl] *adjective*, something that dissolves

so-lu-tion [sə-lōō′shən] *noun*, 1) a liquid containing another

substance dissolved in it 2) the answer found

solve [sŏlv] *verb*, to find the answer by studying the problem

sol-vent [sŏl′vənt] *adjective*, 1) debt free, able to pay one's debts, *noun*, 2) a usually liquid substance in which another substance can dissolve

som-ber [sŏm′bər] *adjective*, gloomy, serious, melancholy

some [sŭm] *adjective*, 1) an amount of, not all, *pronoun*, 2) referring to people or things without saying exactly who or what

some-bod-y [sŭm′bŏd′ē] *pronoun*, any person, an unknown or unnamed individual

some-how [sŭm′hou′] *adverb*, in some unknown way

some-one [sŭm′wŭn′] *pronoun*, a person or an individual

some-place [sŭm′plās′] *noun*, a location, somewhere

som-er-sault [sŭm′ər-sôlt′] *noun*, the act of turning upside down with your head between your legs as though rolling in a ball

some-thing [sŭm′thĭng] *pronoun*, an unknown amount or identity

some-time [sŭm′tīm′] *adverb*, a time in the past or future

some-times [sŭm′tīmz′] *adverb*, occasionally, not always

some-what [sŭm′hwŏt′] *adverb*, slightly, to some degree

some-where [sŭm′hwâr′] *adverb*, in or to, in an unspecified space

som-nam-bu-list [sŏm-năm′byə-lĭst] *noun*, a sleepwalker

son [sŭn] *noun*, a male child, or offspring

song [sông] *noun*, 1) a piece of music with words 2) poetry

son-net [sŏn′ĭt] *noun*, a poem, fourteen lines in iambic pentameter

soon [sōōn] *adverb*, something that will happen in a short time from now, promptly, quickly

soon-er [sōō′nər] *adjective*, before long, earlier than

soot [sŏŏt] *noun*, *no plural*, a black powder separated from fuel during combustion, smoke residue, cinders, ash

soothe [sōōth] *verb*, 1) to ease or comfort 2) to quiet, to pacify

so-phis-ti-ca-tion [sə-fĭs′tĭ-kā′shən] *noun*, intelligence, social grace, elegance, style, good taste

soph-o-more [sŏf′ə-môr] *noun*, a student in his second year of a four-year course

so-pran-o [sə-prăn′ō] *noun*, a singer who can reach the high notes in music

sor-cer-ess [sôr′sər-ĭs] *noun*, a magician, a witch

sor-did [sôr′dĭd] *adjective*, filthy, base, vile, squalid, selfish

sore [sôr, sōr] *adjective*, 1) hurting, painful, *noun*, 2) a painful, injured place on the body

so-ror-i-ty [sə-rôr′ĭ-tē] *noun*, a society or club of girls or women

sor-row [sŏr′ō] *noun*, sadness

sor-ry [sŏr′ē] *adjective*, feeling pity or sorrow, regret

sort [sôrt] *noun*, 1) a special kind, *verb*, 2) to separate into different categories, to classify

sought [sôt] *past tense* of **seek**

soul [sōl] *noun*, the part of you that is not body, and that some people think does not die with your body, the spirit of a person

sound [sound] *adjective*, 1) healthy, strong 2) deep or undisturbed, *noun*, 3) something heard from transmitted vibrations, *verb*, 4) to make a noise

soup [sōōp] *noun*, a liquid food made by boiling meat, fish, or vegetables and seasonings

sour [sour] *adjective*, 1) having a sharp, biting, acid taste, tart, tangy, 2) soured, curdled

source [sôrs] *noun*, 1) the thing or place from which something comes 2) someone who provides information 3) a reference book

south [south] *adjective*, just opposite the north

South America [south-ə-mĕr′ĭ-kə] *noun*, one of the seven continents, containing 5.9% of the world's population

South Carolina [south-kăr′ə-lī′nə] *noun*, one of the 50 United States, located in the Southeast, the capital is Columbia. The state flower of **South Carolina** is the yellow jasmine, and the motto is "Prepared in mind and resources." or "While I breathe I hope."

South Dakota [south-də-kō′tə] *noun*, one of the 50 United States, located in the Northwest, the capital is Pierre. The state flower of **South Dakota** is the pasque flower and the motto is "Under God the people rule."

sou·ve·nir [sōō′və-nîr′] *noun*, that which serves as a reminder, a keepsake, a memento

sov·er·eign [sŏv′ər-ĭn] *adjective*, highest in power, chief

sow *noun*, 1) [sou] *noun*, an adult female pig, [sō]*verb*, 2) to scatter seeds over the ground where they will grow to produce a crop

soy·bean [soi′bēn′] *noun*, a bean plant grown for its edible, nutritious seeds

space [spās] *noun*, 1) the limitless three-dimensional area all around 2) an empty place or unoccupied time

space·ship [spās′shĭp′] *noun*, a rocket orbiting the earth

spa·cious [spā′shəs] *adjective*, vast in extent, roomy, sizeable

spade [spād] *noun*, 1) a small hand-held shovel 2) one of the four suits in a deck of cards

spa·ghet·ti [spə-gĕt′ē] *noun*, a variety of macaroni made in cords of small diameter

span [spăn] *noun*, a stretch, the measure, the distance

spare [spâr] *adjective*, 1) kept in addition to what is already available, *verb*, 2) to give something extra, to share 3) to omit, to leave out

spark [spärk] *noun*, 1) a tiny burning particle, *verb*, 2) to stimulate, to set in motion

spar·kle [spär′kəl] *verb*, to shine with bright points of light

spark plug [spärk′plŭg′] *noun*, a device with two electrodes between which an electric jump spark is made

spar·row [spăr′ō] *noun*, a small brownish-gray bird

sparse [spärs] *adjective*, not thick, thinly scattered or distributed

spasm [spăz′əm] *noun*, a seizure, a fit, a convulsive movement

spa·tial [spā′shəl] *adjective*, relating to space

spat·u·la [spăch′ə-lə] *noun*, an instrument with a broad blade used for spreading or mixing

spawn [spôn] *verb*, to lay eggs

spay [spā] *verb*, to remove the reproductive organs, to castrate

speak [spēk] *verb*, to say words aloud, to talk, to express ideas

speak·er [spē′kər] *noun*, 1) the one talking 2) an orator, a lecturer

spear [spîr] *noun*, a long thin weapon with a pointed end

spe·cial [spĕsh′əl] *adjective*, 1) individual, peculiar, particular, different from the others 2) extra nice, distinct 3) having a particular function or purpose

spe·cial·ist [spĕsh′ə-lĭst] *noun*, one who studies a subject at great length, an expert, an authority

spe·cial·ize [spĕsh′ə-līz′] *verb*, to focus on a particular profession

spe·cies [spē′shēz] *noun*, *plural*, 1) a distinct sort, kind, or variety 2) a group of living things that can breed together

spe·cif·ic [spĭ-sĭf′ĭk] *noun*, special, explicit, precise, peculiar

spec-i-fy [spĕs′ə-fī′] *verb*, to designate, to name explicitly, to indicate, to point out in detail

spec-i-men [spĕs′ə-mən] *noun*, a part of anything to show the quality of the whole

spe-cious [spē′shəs] *adjective*, something false that appears to be true, deceptively attractive

speck [spĕk] *noun*, a small piece or particle of something

spec-ta-cle [spĕk′tə-kəl] *noun*, 1) a public scene that attracts a lot of attention   2) a display, a show

spec-ta-cles [spĕk′tə-kəlz] *noun*, *plural*, glasses for the eyes to improve vision set in a frame that rests on the nose and ears

spec-tac-u-lar [spĕk-tăk′yə-lər] *adjective*, impressive, amazing, breath-taking, dramatic

spec-ta-tor [spĕk′tā′tər] *noun*, one who looks on and watches, an observer, a bystander

spec-ter [spĕk′tər] *noun*, any object of dread, a ghost, an apparition

spec-tro-scope [spĕk′trə-skōp′] *noun*, an optical instrument

spec-trum [spĕk′trəm] *noun*, a broad range or sequence, especially of light separated into its constituent colors

spec-u-late [spĕk′yə-lāt′] *verb*, 1) to gamble, to buy and sell on the market 2) to consider, to contemplate, to hypothesize

spec-u-la-tion [spĕk′yə-lā′shən] *noun*, theory, hypothesis, conjecture, contemplation

speech [spēch] *noun*, 1) a long talk usually prepared in advance 2) language, words, dialect

speed [spēd] *noun*, 1) the rate at which something moves, *verb*, 2) to quicken, to hasten

speed-om-e-ter [spǐ-dŏm′ǐ-tər] *noun*, an instrument for indicating speed or velocity

spell [spĕl] *noun*, 1) a charm that captivates some and holds their attention 2) a short period of time, *verb*, 3) to indicate the correct arrangement of letters in a word 4) to be a sign of

spell-ing [spĕl′ĭng] *noun*, the correct use of the proper letters composing a word

spend [spĕnd] *verb*, 1) to give money for something 2) to use or consume energy or time

spent [spĕnt] *past tense* of **spend**

sphere [sfîr] *noun*, 1) a ball or globe 2) a social position

spher-i-cal [sfîr′ĭ-kəl] *adjective*, globular, round like a ball

spice [spīs] *noun*, a seed or root of a plant used to add flavor to food

spi-der [spī′dər] *noun*, a bug that uses threads from its body to make a web to capture insects

spill [spĭl] *noun*, 1) a fall, *verb*, 2) to accidentally allow a liquid to flow out of a container

spin [spĭn] *verb*, 1) to whirl quickly 2) to twist into a thread from wool or other natural fibers

spin-ach [spĭn′ĭch] *noun*, an edible green leafy vegetable

spine [spīn] *noun*, the long row of bones in your back

spi-ral [spī′rəl] *noun*, helical, like the thread of a screw

spire [spīr] *noun*, the pointed top of a tower or steeple

spir-it [spĭr′ĭt] *noun*, 1) a mysterious part of you which is not the body, the soul 2) an eager state of mind, enthusiasm 3) a supernatural being such as a ghost or a fairy 4) an alcoholic drink such as liquor

spir-it-u-al [spĭr′ĭ-choo-əl] *adjective*, airy, motivated by the soul, holy, religious

spit [spĭt] *noun*, 1) saliva, *verb*, 2) to spray water or saliva out of the mouth

spite [spīt] *noun*, *no plural*, hatred, feelings of malice, wanting to hurt or annoy another person

splash [splăsh] *noun*, 1) the sound made when a solid falls into liquid, *verb*, 2) to make a wave of water in a pool

splen-did [splĕn′dĭd] *adjective*, excellent, magnificent, very fine

splice [splīs] *verb*, to insert a shoot, to join, to graft, to unite

splint [splĭnt] *noun*, a board used to keep a thing from bending, a way of protecting a broken limb

splin-ter [splĭn′tər] *noun*, a thin piece of wood split off from a larger piece, a sliver

split [splĭt] *noun*, 1) a tear or a break, *verb*, 2) to break lengthwise 3) to share

spoil [spoil] *verb*, 1) to damage or injure, to ruin 2) to become rotten or decayed 3) to overindulge or corrupt by coddling someone

spoke [spōk] *noun*, 1) one of the bars joining the rim of a wheel to the center to support weight, *verb*, 2) *past tense* of **speak**

sponge [spŭnj] *noun*, 1) a substance that absorbs water, used to clean and wipe 2) a primitive marine animal having a soft elastic skeleton with many pores, *verb*, 3) to live off the generosity of others

spon-sor [spŏn′sər] *noun*, 1) a patron, supporter, mentor, backer, *verb*, 2) to offer financial support for a program

spon-ta-ne-i-ty [spŏn′tə-nē′ĭ-tē] *noun*, acting at the spur of the moment, impromptu

spon-ta-ne-ous [spŏn-tā′nē-əs] *adjective*, 1) acting on impulse, a result of natural law 2) energy

spool [spo͞ol] *noun*, a round object used to wind thread or wire

spoon [spo͞on] *noun*, a utensil consisting of a shallow bowl with a long handle attached to it, used for eating or stirring

spoon-ful [spo͞on′fo͞ol′] *noun*, the quantity that sits in a spoon

spo-rad-ic [spə-răd′ĭk] *adjective*, occurring irregularly

spore [spōr] *noun*, a cell that reproduces itself such as fungus, fern, or bacterium

sport [spōrt] *noun*, 1) games and exercises done for pleasure 2) a rascal 3) a teammate

spot [spŏt] *noun*, 1) a small mark, drop, or blemish 2) a location, *verb*, 3) to see something

spouse [spous] *noun*, a partner in marriage, husband or wife

spout [spout] *noun*, 1) the lip of a container from which liquid is poured, *verb*, 2) to spurt, to discharge, to gush forth

sprain [sprān] *verb*, to strain a body part by twisting it, causing it to swell

sprawl [sprôl] *verb*, to stretch, to spread out, to extend

spray [sprā] *noun*, 1) liquid dispersed in very fine drops 2) a small branch of flowers or foliage, *verb*, 3) to blow or scatter liquid in very fine drops

spread [sprĕd] *verb*, 1) to cover a surface with a thin layer 2) to unfold or stretch out 3) to circulate, to disseminate 4) to extend over a large area

spring [sprĭng] *noun*, 1) a river or pond coming up from the ground 2) a piece of wire twisted into a spiral 3) the season following winter, when plants begin to grow, *verb*, 4) to jump 5) to come from 6) to grow

sprin-kle [sprĭng′kəl] *verb*, 1) to spatter, to disperse in drops 2) to drizzle 3) to dust lightly

sprint [sprĭnt] *noun*, 1) a short fast run, *verb*, 2) to dash at top speed

sprock-et [sprŏk′ĭt] *noun*, a projection shaped so as to engage with a chain

sprout [sprout] *verb*, to grow out of a seed, to start to grow

spruce [sprōōs] *noun*, one of the softwood evergreen trees

sprung [sprŭng] *past tense* of **spring**

spry [sprī] *adjective*, vigorous, quick, energetic, agile

spun [spŭn] *past tense* of **spin**

spur [spûr] *noun*, 1) a short side track connecting with the main railroad 2) a sharp blade on the back of a boot used to urge horses to run faster 3) an impetus, a cause to action, *verb*, 4) to urge, to goad, to stimulate

spu-ri-ous [spyŏŏr′ē-əs] *adjective*, counterfeit, false, not genuine

spurn [spûrn] *verb*, to reject, to scorn, to despise, to shun

spy [spī] *noun*, *plural*, **spies** 1) an agent hired to secretly obtain information *verb*, 2) to see or catch sight of 3) to keep secret watch

squad [skwŏd] *noun*, a small group of soldiers who work, train, and fight together 2) a sports team

squad-ron [skwŏd′rən] *noun*, armed military aircraft

squal-id [skwŏl′ĭd] *adjective*, dirty through neglect, foul, filthy

squan-der [skwŏn′dər] *verb*, to spend money or resources without being accountable

square [skwâr] *noun*, 1) a surface with four equal sides and right angle corners 2) an open area in the center of town surrounded by stores and businesses

square number [skwâr- nŭm′bər] *noun*, the product of two equal factors, e.g., 12 x 12 = 144

squash [skwŏsh] *noun*, 1) a vegetable that grows on a vine 2) a game similar to racquetball, *verb*, 3) to repress 4) to flatten by pressing or from pressure

squat [skwŏt] *adjective*, 1) wide, stocky, *verb*, 2) to stoop

squeak [skwēk] *noun*, 1) a very sharp, high-pitched sound, *verb*, 2) to make a short high-pitched sound

squeal [skwēl] *noun*, 1) a shrill cry, *verb*, 2) to make such a sound 3) to turn informer

squeeze [skwēz] *verb*, to press hard upon 2) to put pressure on to extract liquid 3) to force one's way 4) to crowd

squid [skwĭd] *noun*, a sea creature similar to an octopus

squint [skwĭnt] *verb*, to scrunch one's eyes to strain to see

squirm [skwûrm] *verb*, to wiggle

squir-rel [skwûr′əl, skwŭr′-] *noun*, a small rodent with a bushy tail that lives in trees

sta-bil-i-ty [stə-bĭl′ĭ-tē] *adjective*, firmness, constancy, strength

sta-ble [stā′bəl] *adjective*, 1) firm, steady, *noun*, 2) a building in which horses are kept

stac-ca-to [stə-kä′tō] *adjective*,—— music played in a quick, abrupt manner

stack [stăk] *noun*, 1) a large pile arranged in some sort of order 2) a chimney, *verb*, 3) to pile up

sta-di-um [stā′dē-əm] *noun*, an open arena where concerts, sports, or events are held

staff [stăf] *noun*, a long stick carried in the hand for support

stag [stăg] *noun*, a male deer

stage [stāj] *noun*, 1) a raised floor or platform in a theatre or auditorium 2) a period of time or a step in development

stage-coach [stāj′kōch′] *noun*, a carriage on wheels pulled by horses on a fixed route

stag-ger [stăg′ər] *verb*, to stand or walk unsteadily, to reel

stag-nant [stăg′nənt] *adjective*, not moving or flowing, motionless

stag-nate [stăg′nāt′] *verb*, to be or become stagnant, motionless, to become impure or foul

staid [stād] *adjective*, sober, sedate

stain [stān] *noun*, 1) a mark that changes the appearance of something, a blot, *verb*, 2) to make a strong mark that cannot be removed 3) to disgrace

stair [stâr] *noun*, a step connecting different levels, *plural*, **stairs,** a series or flight of steps

stake [stāk] *noun*, 1) a pointed post in the ground 2) a gamble, a risk

stale [stāl] *adjective*, 1) old, dry, not fresh, tasteless 2) dull, bored

stale-mate [stāl′māt′] *noun*, a deadlock, a situation in which no side may take action

stalk [stôk] *noun*, 1) the long part of a plant that supports leaves or flowers, *verb*, 2) to follow someone with the intention of harming them 3) to pursue prey

stall [stôl] *noun*, 1) a small open room, as in a barn or a market place, *verb*, 2) to stop due to a mechanical failure

stal-lion [stăl′yən] *noun*, a male horse used for breeding

stal-wart [stôl′wərt] *adjective*, 1) sturdy, strong 2) brave 3) firm, steadfast

stam-i-na [stăm′ə-nə] *noun*, the power to endure, strength

stamp [stămp] *noun*, 1) a small piece of paper purchased to put on an envelope or package to send it through the mail 2) the imprint left from a device used to leave a mark, *verb*, 3) to mark with a special sign or design 4) to set the foot down heavily

stam-pede [stăm-pēd'] *verb*, to rush wildly in a group

stance [stăns] *noun*, 1) posture 2) political or social position

stand [stănd] *verb*, 1) to be your feet 2) to take a position 3) to mean or signify, to represent

stand-ard [stăn'dərd] *adjective*, 1) regular, official, basic, *noun*, 2) a fixed quality by which things are compared

stands [stăndz] *noun*, *plural*, the bleachers where spectators watch sports events

stan-za [stăn'zə] *noun*, a verse in a poem, a song, or a refrain

sta-ple [stā'pəl] *noun*, 1) principle commodity, an essential element 2) a wire used to attach things together such as cloth or paper, *verb*, 3) to bond two things together with staples applied by a stapler

star [stär] *noun*, 1) a point of light that can be seen in the sky at night 2) a five-pointed shape, *verb*, 3) to stand out in a positive way, to be outstanding

star-board [stär'bərd] *adjective*, the right-hand side of a ship as one faces the bow

starch [stärch] *noun*, a carbohydrate, that turns into sugar in the body

stare [stâr] *verb*, to look at someone or something with a steady gaze, to glare

stark [stärk] *adjective*, extremely

simple or severe, bleak

start [stärt] *noun*, 1) an act of beginning, *verb*, 2) to begin

star-tle [stär'tl] *verb*, to surprise, to greet unexpectedly, to shock

star-va-tion [stär-vā'shən] *noun*, to suffer from hunger and lack of food or nourishment

starve [stärv] *verb*, to be hungry, unfed, deprived of food

state [stāt] *noun*, 1) the condition of something 2) a mood 3) a group of people living under an independent government, that is part of a country or federal union which governs itself as a republic or commonwealth, *verb*, 5) to say, to declare

state-ment [stāt'mənt] *noun*, 1) a spoken or written commitment 2) a summary of an account showing the balance

states-man [stāts'mən] *noun*, a leader of government

stat-ic [stăt'ĭk] *noun*, applicable to electricity not in motion

sta-tion [stā'shən] *noun*, a stopping place for the convenience of passengers, a depot

sta-tion-ar-y [stā'shə-nĕr'ē] *adjective*, fixed in a place, not changing, not moving

sta-tion-er-y [stā'shə-nĕr'ē] *noun*, writing paper, pens, and ink

sta-tis-tic [stə-tĭs'tĭk] *noun*, a fact stated as a number or in tables

stat-ue [stăch'ōō] *noun*, replica of a figure, real or imaginary made of plaster, bronze, marble, etc.

stat-ure [stăch'ər] *noun*, 1) a person's status or place in the community 2) the height of a person or animal

sta-tus [stā′təs] *noun*, position, high standing, prestige

stat-ute [stăch′oot] *noun*, a law enacted by a legislature

stat-u-to-ry [stăch′ə-tôr′ē] *adjective*, authorized, made as a law, regulated by statute

stay [stā] *verb*, to continue to be

stead-fast [stĕd′făst′] *adjective*, firmly fixed or established, unwavering, constant

stead-i-ly [stĕd′ə-lē] *adverb*, constantly, continually, regularly

stead-y [stĕd′ē] *adjective*, 1) regular 2) firm, immovable

steak [stāk] *noun*, a thick flat piece of meat or fish

steal [stēl] *verb*, 1) to take or carry away without permission 2) to move or pass very quietly without making noise

stealth [stĕlth] *noun*, secrecy, sneakiness, surreptitious

steam [stēm] *noun*, 1) the vapor created by boiling water, *verb*, 2) to cook using the intense heat of steam 3) to expose to steam

steel [stēl] *noun*, a hard metal made of specially treated iron, used in machines, knives, hardware, etc.

steep [stēp] *adjective*, 1) having a sharp incline or slope, *verb*, 2) to soak in liquid, to saturate

stee-ple [stē′pəl] *noun*, a free-standing column or tower

steer [stîr] *noun*, 1) a form of long horned cattle, *verb*, 2) to drive, to fly, to conduct, or navigate

stel-lar [stĕl′ər] *adjective*, 1) star-like, referring to heavenly bodies, celestial 2) outstanding

stem [stĕm] *noun*, 1) the stalk of a plant that supports the flower and leaves, *verb*, 2) to develop from 3) to stop the flow of

sten-cil [stĕn′səl] *noun*, a thin sheet with perforations in the form of a design through which ink or paint may be forced

step [stĕp] *noun*, 1) a stair 2) the movement with a foot 3) one of a series of events in a program, *verb*, 4) to walk or go on foot

ster-e-o [stĕr′ē-ō′] *noun*, an electronic device that makes sounds or plays music

ster-e-o-type [stĕr′ē-ə-tīp′] *noun*, 1) a typical example or pattern 2) a conventional, oversimplified conception

ste-rile [stĕr′əl, -īl′] *adjective*, clean, antiseptic, germ free

ster-i-lize [stĕr′ə-līz′] *verb*, 1) to render incapable of germination, 2) to disinfect, to clean

ster-ling [stûr′lĭng] *adjective*, 1) genuine, superior, excellent, of the highest quality *noun*, 2) made of pure silver

stern [stûrn] *adjective*, 1) unsympathetic, firm, strict 2) the back of the boat

steth-o-scope [stĕth′ə-skōp′] *noun*, an instrument used to convey to the ear, sounds produced by the body, especially in the chest

stew [stoo] *noun*, 1) a thick soup, *verb*, 2) to cook in liquid

stew-ard [stoo′ərd] *noun*, 1) one who watches over something

2) one in charge of provisions and dinner arrangements

stew-ard-ess [stoo′ər-dĭs] *noun*, a woman who serves passengers on an airplane, a flight attendant

stick [stĭk] *noun*, 1) a short branch or twig, *verb*, 2) to make something adhere to a surface 3) to push in or out 4) to stay with something, to persevere

stiff [stĭf] *adjective*, 1) something that cannot be moved or bent easily 2) difficult or hard to accomplish 3) sore

stiff-en [stĭf′ən] *verb*, to harden, to brace, to become rigid

stig-ma [stĭg′mə] *noun*, 1) the apex of a flower pistil where pollen is placed 2) a disgraceful blot

stig-ma-tize [stĭg′mə-tīz′] *verb*, to brand, to mark as wicked

still [stĭl] *adjective*, 1) without movement, *adverb*, 2) up to this or that time 3) even so

still-ness [stĭl′nĭs] *noun*, the state of being still or without sound

stim-u-late [stĭm′yə-lāt′] *verb*, to rouse to action, to excite

stim-u-lus [stĭm′yə-ləs] *noun*, an incentive, an impetus

sting [stĭng] *noun*, 1) the pain or wound of an insect or bug bite 2) a sharp or smarting pain 3) the attack of an insect on its prey or an enemy

stin-gy [stĭn′jē] *adjective*, not generous, miserly

stink [stĭngk] *noun*, 1) an unpleasant or awful odor, *verb*, 2) to give off a bad smell

sti-pend [stī′pĕnd′] *noun*, payment

given for services

stip-u-late [stĭp′yə-lāt′] *verb*, to bargain, to specify as a condition of an agreement

stir [stûr] *verb*, to mix or blend

stir-ring [stûr′ĭng] *adjective*, moving, touching, exciting

stitch [stĭch] *noun*, 1) a needle and thread, piercing cloth and then coming out again to form a link of thread in a fabric 2) a loop added to a knitting needle, *verb*, 3) to sew, to mend

stock [stŏk] *noun*, goods kept on hand, merchandise

stock-ade [stŏ-kād′] *noun*, a barricade, a fence, an obstacle

stock-hold-er [stŏk′hōl′dər] *noun*, an owner of stocks or shares in a corporation

stock-ing [stŏk′ĭng] *noun*, a sheer leg covering made of nylon

sto-ic [stō′ĭk] *noun*, one who is indifferent to pleasure or pain, stern, impassive, cold

stoke [stōk] *verb*, to provide with fuel, to feed abundantly

stole [stōl] *past tense* of **steal**

stom-ach [stŭm′ək] *noun*, a large muscular bag in the body that receives food, digests some of it, and passes it on, to the intestines

stone [stōn] *noun*, 1) a small piece of rock, *verb*, 2) to pelt or kill

stool [stool] *noun*, 1) a seat with three or four legs and no back 2) a bowel movement

stoop [stoop] *verb*, 1) to bend or squat the body over forwards 2) to lower oneself, to condescend

stop [stŏp] *verb*, 1) to end, to make something come to an end 2) to prevent something from happening, moving, etc. 3) to complete the action

stor-age [stôr′ĭj] *noun*, the act of keeping things for future use

store [stōr] *noun*, 1) supply of things saved for use later 2) a shop, a place to buy things, *verb*, 3) to put away for later use

stork [stôrk] *noun*, a large bird with long legs and a large beak

storm [stôrm] *noun*, a period of high winds, possibly thunder and lightning, rain, or snow

sto-ry [stōr′ē] *noun*, 1) telling of events with words spoken or written in a book 2) an account of something either true or fictitious, a report, a description 3) the floor of a building

stout [stout] *adjective*, 1) short, heavy-set 2) resolute, firm

stove [stōv] *noun*, an appliance which is heated and used for cooking or heating

stow [stō] *verb*, to store, to put away for future use

straight [strāt] *adjective*, 1) not curved, *adverb*, 2) without going anywhere else or doing anything else 3) moving on a path that is unchanged 4) without delay, *noun*, 5) a poker hand of 5 cards in consecutive order

straight-en [strāt′n] *verb*, to fix or arrange in order

strain [strān] *noun*, 1) the change of a body produced by an external force 2) a new species or breed 3) a suggestion, a hint, a

trace, *verb*, 4) to cause stress, to stretch to reach

strait [strāt] *noun*, a narrow channel between two bodies of water, an inlet

strand [strănd] *noun*, 1) seashore, waterfront 2) a piece of something long and thin like a rope or a piece of hair

strange [strănj] *adjective*, unusual, odd, unfamiliar, exotic

strange-ly [strănj′lē] *adverb*, unusually, seldom, rarely

stran-ger [strān′jər] *noun*, a person you do not know, an outsider

stran-gle [străng′gəl] *verb*, 1) to kill by suffocating 2) to suppress

strap [străp] *noun*, 1) a narrow piece of leather, plastic, cloth, etc, used for fastening something, *verb*, 2) to fasten something on with a tie

strat-e-gy [străt′ə-jē] *noun*, a plan

strat-o-sphere [străt′ə-sfîr′] *noun*, the upper atmosphere extending 12-30 miles above the earth's surface, the ozone

stra-tum [strā′təm] *noun*, *plural* **strata**, a layer of the earth's surface, sedimentary rock

straw [strô] *noun*, 1) a dry stem 2) something that people use to suck a drink out of a glass

straw-ber-ry [strô′bĕr′ē] *noun*, the edible red berry of a small plant, that grows on the ground

stray [strā] *adjective*, 1) anything that does not seem to be part of a group, *verb*, 2) to wander away from home or the rest of the group 3) to digress

streak [strēk] *noun*, 1) a long mark like a band or a stripe, *verb*, 2) to draw stripes on something in a wild pattern

stream [strēm] *noun*, 1) a small river, a small body of running water 2) a steady or continuous flow, *verb*, 3) to flow, to pour

street [strēt] *noun*, a public paved road as in a town or city

strength [strĕngkth] *noun*, power, capacity for exertion or endurance, force in numbers

strength-en [strĕngk'thən] *verb*, to reinforce, to make stronger

stren-u-ous [strĕn'yoō-əs] *adjective*, characterized by vigorous exertion, difficult

stress [strĕs] *noun*, 1) anxiety, tension, *verb*, 2) to apply force to cause strain 3) to emphasize

stretch [strĕch] *noun*, 1) an expanse, a range or reach, *verb*, 2) to exaggerate 3) to extend across a given space 4) to extend one's body or limbs to feel limber

stretch-er [strĕch'ər] *noun*, a framework consisting of two poles with a cloth fastened between them used to carry an injured person

strict [strĭkt] *adjective*, close adherence to the rules

stride [strīd] *noun*, 1) a large step or movement, *verb*, 2) to step

stri-dent [strīd'nt] *adjective*, loud and harsh in sound, grating

strike [strīk] *noun*, 1) an unsuccessful attempt by a batter to hit a ball 2) a protest of wages that stops work in order to bargain for better working conditions, *verb*, 3) to hit or attack 4) to refuse to work because of a pay dispute 5) to discover, to uncover 7) to eliminate 8) to find, reach

string [strĭng] *noun*, 1) a thin twine made by twisting several threads together 2) a fine piece of wire used in some musical instruments, *verb*, 3) to thread a bead onto a string

strin-gent [strĭn'jənt] *adjective*, strict, inflexible, exacting

stripe [strīp] *noun*, a long thin line

strive [strīv] *verb*, to exert oneself, to try, to make a strenuous effort

stroke [strōk] *noun*, 1) a mild heart attack, *verb*, 2) to caress, to rub

stroll [strōl] *verb*, to walk idly or leisurely, to saunter

strong [st rŏng] *adjective*, 1) able to take stress or strain 2) intense in degree 3) physically powerful

struck [strŭk] *verb*, *past tense* of **strike**

struc-ture [strŭk'chər] *noun*, 1) the framework of something, 2) a complete building

strug-gle [strŭg'əl] *noun*, 1) the fight, the effort, *verb*, 2) to continue to endure set backs

strut [strŭt] *verb*, to swagger, to walk with bravado

stub-born [stŭb'ərn] *adjective*, unwilling to give in to, obstinate

stuck [stŭk] *verb*, *past tense* of **stick**

stu-dent [stoōd'nt] *noun*, a person enrolled at a school or college

stud-y [stŭd'ē] *noun*, 1) a room to read and write 2) a subject or area of learning, *verb*, 3) to apply the mind to the effort to learn by reading and thinking

stuff [stŭf] *noun*, 1) substance, material, *verb*, 2) to fill by crowding, to push or crowd in

stum-ble [stŭm'bəl] *verb*, to walk unsteadily so that you appear to be about to fall, to trip while running 2) to come upon unexpectedly 3) to stammer

stump [stŭmp] *noun*, 1) what is left of a tree after it has been cut down, *verb*, 2) to perplex

stun [stŭn] *verb*, to astonish, to amaze, to confound, to shock

stunt [stŭnt] *noun*, 1) a performance, a skit, an act, *verb*, 2) to suppress, to retard the growth of something

stu-pen-dous [stoo-pĕn'dəs] *adjective*, astonishing, wonderful, amazing

stu-por [stoo'pər] *noun*, a state of apathy, a daze, lack of awareness

stur-dy [stûr'dē] *adjective*, very strongly built, durable

stut-ter [stə'tər] *verb*, to speak with difficulty, to stammer

sty [stī] *noun*, 1) a painful red swelling on the end of an eyelid 2) a pig pen

style [stīl] *noun*, 1) the fashion or design 2) elegance or flair

sty-mie [stī'mē] *verb*, to present an obstacle, to stump, to hinder

suave [swäv] *adjective*, smoothly, agreeable, courteous, polite

sub-due [səb-doo'] *verb*, to calm down, to bring under control

sub-ject [sŭb'jĭkt] *adjective*, 1) owing obedience to 2) likely to

be affected by, *noun*, 3) an area of study 4) a person under the authority or control of another 5) in a sentence the word about which the predicate makes a statement, *verb*, 6) to expose to

sub-jec-tive [səb-jĕk'tĭv] *adjective*, based on personal feelings or opinions

sub-lime [sə-blīm'] *adjective*, great, noble, majestic, superior

sub-lim-i-nal [sŭb-lĭm'ə-nəl] *adjective*, affecting the subconscious, below the threshold of consciousness

sub-ma-rine [sŭb'mə-rēn'] *noun*, a naval warship armed with torpedoes and missiles that operates under water

sub-merge [səb-mûrj'] *verb*, to put or sink under water

sub-mis-sive [səb-mĭs'ĭv] *adjective*, to take orders without questioning authority

sub-mit [səb-mĭt'] *verb*, 1) to put something forward for someone else's reaction 2) to give in to someone or something

sub-or-di-nate [sə-bôr'dn-ĭt] *adjective*, 1) being in a lower rank or class, subject to another's authority *noun*, 2) one lower in rank, inferior

sub-poe-na [sə-pē'nə] *noun*, a writ commanding a person listed in it to appear in court

sub-scribe [səb-scrīb'] *verb*, 1) to sign up for something 2) to give permission or consent

sub-scrip-tion [səb-skrĭp'shən] *noun*, money paid as a form of membership for service

subsequent

sub-se-quent [sŭb′sĭ-kwĕnt′]
adjective, following, after

sub-ser-vi-ent [səb-sûr′vē-ənt]
adjective, useful in an inferior
capacity, subordinate

sub-side [səb-sīd′] *verb*, to settle
down to become calm

sub-si-dize [sŭb′sĭ-dīz′] *verb*, 1) to
add to 2) to provide financial aid

sub-si-dy [sŭb′sĭ-dē] *noun*, direct
financial aid by the government

sub-stance [sŭb′stəns] *noun*, the
material something is made of

sub-stan-tial [səb-stăn′shəl]
adjective, 1) strong 2) having
considerable property

sub-stan-ti-ate [səb-stăn′shē-āt′]
verb, to verify, to support

sub-stan-tive [sŭb′stən-tĭv]
adjective, essential, pertaining to

sub-sti-tute [sŭb′stĭ-tōōt′] *noun*,
1) a person or thing put in place
of another, a replacement *verb*,
2) to act in place of another

sub-ter-fuge [sŭb′tər-fyōōj′] *noun*,
a device or plan by which one
can hide an objective

sub-ter-ra-ne-an [sŭb′tə-rā′nē-ən]
adjective, 1) under the earth's
surface 2) hidden, secret

sub-tle [sŭt′l] *adjective*, difficult to
to detect, not obvious, slight

sub-tle-ty [sŭt′l-tē] *noun*, cunning,
delicately understated

sub-tract [səb-trăkt′] *verb*, to take
away one number from another

sub-trac-tion [səb-trăk′shən] *noun*,
an equation in which one
number is taken from another

sub-urb [sŭb′ûrb′] *noun*, a
residential community on the
outskirts of a city or town

sub-ur-ban [sə-bûr′bən] *adjective*,
pertaining to a smaller district
next to a city or town

sub-ver-sive [səb-vûr′sĭv]
adjective, tending to overthrow
or ruin, corruption

sub-way [sŭb′wā′] *noun*, an
underground electric train or
urban transit system

suc-ceed [sək-sēd′] *adjective*, to
follow in order, to be successful

suc-cess [sək-sĕs′] *noun*,
accomplishment, achievement

suc-cess-ful [sək-sĕs′fəl] *adjective*,
prosperous, enjoying abundance

suc-ces-sion [sək-sĕsh′ən] *noun*,
sequence, order, descent

suc-ces-sive [sək-sĕs′ĭv] *adjective*,
following one after another in a
line or series, consecutive

suc-ces-sor [sək-sĕs′ər] *noun*,
someone appointed to a title or
position to replace another

suc-cinct [sək-sĭngkt′] *adjective*,
concise, terse, brief, to the point

suc-cor [sŭk′ər] *noun*, aid, help,
assistance, relief

suc-cul-ent [sŭk′yə-lənt] *adjective*,
1) juicy 2) delicious, tasty

suc-cumb [sə-kŭm′] *verb*, 1) to
yield, to give way 2) to die

such [sŭch] *pronoun*, 1) of this or
that kind 2) used in some
phrases to mean like

suck [sŭk] *verb*, to draw into the
mouth by action of the lips and
tongue when inhaling

suc-tion [sŭk′shən] *noun*, the
pulling force of a partial vacuum

sud-den [sŭd′n] *adjective*, when something happens unexpectedly

sud-den-ly [sŭd′n-lē] *adverb*, happening at once without warning, unexpectedly

sue [soo] *verb*, to ask a court to solve a problem legally

suede [swād] *noun*, a leather made when the flesh, or underside of the leather is buffed or sandpapered to produce a nap

suf-fer [sŭf′ər] *verb*, 1) to feel pain or hurt 2) to tolerate, to allow

suf-fer-ing [sŭf′ər-ĭng] *noun*, enduring pain and sadness

suf-fice [sə-fīs′] *verb*, 1) to be enough for 2) to satisfy a need

suf-fi-cient [sə-fĭsh′ənt] *adjective*, adequate for the purpose

suf-fix [sŭf′ĭks] *noun*, letters added to the end of a word, to change the meaning

suf-fo-cate [sŭf′ə-kāt′] *verb*, 1) to smother, to choke 2) to kill by depriving of oxygen

suf-frage [sŭf′rĭj] *noun*, the right to vote, a vote

sug-ar [shoog′ər] *noun*, a sweet substance made from sugar cane or sugar beets and used in food products

sug-gest [səg-jĕst′] *verb*, 1) to bring up or call to mind, to propose for consideration 2) to imply, to indicate

sug-ges-tion [səg-jĕs′chən] *noun*, 1) a proposal, an idea, advice, a recommendation 2) a slight trace, a hint, an implication

su-i-cide [soo′ĭ-sīd′] *noun*, the act of taking one's own life

suit [soot] *noun*, 1) a set of clothes made from the same fabric, *verb*, 2) to adapt, to perform

suit-a-ble [soo′tə-bəl] *adjective*, fitting, proper, appropriate

suit-case [soot′kās′] *noun*, a rectangular piece of luggage

suite [swēt] *noun*, a number of rooms connected together

suit-or [soo′tər] *noun*, a beau, a man who courts a woman

sul-fur [sŭl′fər] *noun*, a nonmetallic yellow element used in gunpowder, matches and medicine

sulk [sŭlk] *verb*, to pout and feel angry or sorry for yourself

sul-len [sŭl′ən] *adjective*, silent from anger or hurt, moody, indulging in self-pity

sul-ly [sŭl′ē] *verb*, 1) to disgrace, to tarnish someone's character 2) to soil, to stain

sul-try [sŭl′trē] *adjective*, sweltering, hot and humid

sum [sŭm] *noun*, the total of different amounts added together

sum-ma-ry [sŭm′ə-rē] *noun*, a shorter version of an article

sum-ma-tion [sə-mā′shən] *noun*, the act of finding the total

sum-mer [sŭm′ər] *noun*, the third season of the year, between spring and autumn

sum-mit [sŭm′ĭt] *noun*, 1) the top leaders, the highest point 2) the top of a mountain

sum-mon [sŭm′ən] *verb*, 1) to call, to send for by command 2) to gather, to call forth

sum-mons [sŭm′ənz] *noun*, a formal letter requesting a response in court

sump-tu-ous [sŭmp′choo-əs] *adjective*, luxurious, splendid

sun [sŭn] *noun*, the central star of the solar system around which the planets revolve which gives off light, heat and energy

sun-dae [sŭn′dē] *noun*, a dish of ice cream covered with toppings such as chocolate syrup, nuts and whipped cream

Sun-day [sŭn′dē] *noun*, the first day of the week, observed as the Sabbath by most Christians

sun-der [sŭn′dər] *verb*, to separate, to cut apart

sun-dry [sŭn′drē] *adjective*, various, more than one or two

sun-flow-er [sŭn′flou′ər] *noun*, a plant with large yellow-rayed flower heads that produces edible seeds that yield an oil

sun-rise [sŭn′rīz′] *noun*, the early morning, the crack of dawn

sun-set [sŭn′sĕt′] *noun*, close of the day, twilight

sun-shine [sŭn′shīn′] *noun*, *no plural*, bright light, day light

su-perb [soo-pûrb′] *adjective*, exceptional, extraordinary

su-per-cil-i-ous [soo′pər-sĭl′ē-əs] *adjective*, lofty with pride, haughty, contemptuous

su-per-fi-cial [soo′pər-fĭsh′əl] *adjective*, 1) relating to the surface or appearance only 2) actions thoughtlessly done, insincere, shallow

su-per-flu-ous [soo-pûr′floo-əs] *adjective*, excessive, more than needed, unnecessary

su-per-in-tend-ent [soo′pər-ĭn-tĕn′dənt] *noun*, one having oversight and direction of some place or agency, a supervisor

su-pe-ri-or [soo-pîr′ē-ər] *adjective*, 1) beyond power or influence of another 2) greater in number or amount 3) above average

su-per-mar-ket [soo′pər-mär′kĭt] *noun*, a large store with shelves that have a variety of household products, a large grocery store

su-per-nat-u-ral [soo′pər-năch′ər-əl] *adjective*, beyond what is explainable by natural law, metaphysical

su-per-sede [soo′pər-sēd′] *adjective*, to come to take the place of something that is not as modern, efficient,or appropriate

su-per-son-ic [soo′pər-sŏn′ĭk] *adjective*, faster than sound

su-per-sti-tion [soo′pər-stĭsh′ən] *noun*, reverence for that which is unknown or mysterious

su-per-sti-tious [soo′pər-stĭsh′əs] *adjective*, awe of that which is unknown or mysterious based on an irrational belief

su-per-vise [soo′pər-vīz′] *verb*, to direct, to watch over

su-per-vi-sion [soo′pər-vĭzh′ən] *noun*, *no plural*, overseeing someone's work

su-per-vi-sor [soo′pər-vī′zər] *noun*, a person who oversees a department of workers

sup-per [sŭp'ər] *noun*, dinner, a meal eaten in the evening

sup-ple [sŭp'əl] *adjective*, 1) flexible, pliant, bending easily without breaking 2) limber

sup-ple-ment [sŭp'lə-mənt] *noun*, a part added, an appendix

sup-ply [sə-plī'] *noun*, 1) a resource, *verb*, 2) to provide something needed

sup-port [sə-pôrt', -pōrt'] *noun*, 1) aid, assistance 2) an abutment, a brace, a stay, *verb*, 3) to advocate, to back, to uphold 4) to provide food and clothing and other necessities

sup-pose [sə-pōz'] *verb*, to hold that something is true, to assume that something is true

sup-posi-tion [sŭp'ə-zĭsh'ən] *noun*, presumption, theory, principle

sup-press [sə-prĕs'] *verb*, to subdue, to quell, to inhibit, to keep from being known

su-preme [soŏ-prēm'] *adjective*, highest in authority, highest in degree, above all

Su-preme Court [soŏ-prēm'kôrt] *noun*, the most important court in the United States, consisting of 9 members appointed by the President

sure [shoŏr] *adjective*, best, convinced, certain beyond question, having no doubt

sure-ly [shoŏr'lē] *adverb*, certainly

surf [sûrf] *noun*, *no plural*, 1) the waves of the sea as they hit the shores, *verb*, 2) to ride a surfboard 3) to search the Internet on a computer

sur-face [sûr'fəs] *noun*, 1) a plane, the outside or topside face 2) the top of the water, *verb*, 3) to rise to the top of a body of water

sur-feit [sûr'fĭt] *noun*, 1) excess *verb*, 2) to over-indulge, to give to use in excess

surge [sûrj] *noun*, 1) a rush, a flow, *verb*, 2) to swell, to pour out

sur-geon [sûr'jən] *noun*, one whose profession it is to cure diseases or injuries by performing an operation

sur-ger-y [sûr'jə-rē] *noun*, an operation when a doctor cuts into a person's body with instruments to help cure them

sur-ly [sûr'lē] *adjective*, rude, cross, bad-tempered, grumpy

sur-mise [sər-mīz'] *noun*, 1) a guess, speculation, likelihood, *verb*, 2) to assume, to guess, to suppose, to conjecture

sur-mount [sər-mount'] *verb*, to be overcome, to prevail over

sur-name [sûr'nām'] *noun*, a person's last name

sur-pass [sər-păs'] *verb*, to exceed, to do better, to go beyond

sur-plus [sûr'pləs, -plŭs'] *noun*, the amount that is more than needed

sur-prise [sər-prīz'] *noun*, 1) something sudden and unexpected, *verb*, 2) to cause to feel astonished

sur-ren-der [sə-rĕn'dər] *verb*, to give up, to yield

sur-rep-ti-tious [sûr'əp-tĭsh'əs] *adjective*, kept secret (especially if it would not be approved of), hidden, unauthorized

317

sur-ro-gate [sûr′ə-gĭt] *adjective*, substitute, replacement

sur-round [sə-round′] *verb*, to encircle, to enclose on all sides

sur-round-ings [sə-roundĭngs] *noun, plural,* the conditions or things around a person or place

sur-veil-lance [sər-vā′ləns] *noun*, a close watch over something

sur-vey [sər-vā′] *noun*, 1) a study, a poll, a review, *verb*, 2) to examine, to inspect, to watch

sur-vey-or [sər-vā′ər] *noun*, 1) one who collects data 2) one whose business it is to measure land

sur-viv-al [sər-vī′vəl] *noun, no plural*, what one needs to stay alive or exist

sur-vive [sər-vīv′] *verb*, to continue to exist, to live longer

sur-vi-vor [sər-vīv′ər] *noun*, the one to make it when the others died or failed

sus-cep-ti-ble [sə-sĕp′tə-bəl] *adjective*, capable of being affected emotionally

sus-pect [sə-spĕkt′] *noun*, 1) a person who is believed to have done something wrong, *verb*, 2) to suppose that something is true

sus-pend [sə-spĕnd′] *verb*, 1) to hang 2) to stop temporarily

sus-pend-ers [sə-spĕn′dərz] *noun*, two adjustable bands or straps arranged to pass over the shoulders to hold the trousers up

sus-pense [sə-spĕns′] *noun, no plural*, a pause or delay which leaves people anxious

sus-pen-sion [sə-spĕn′shən] *noun*, a delay, postponement

sus-pi-cion [sə-spĭsh′ən] *noun*, 1) distrust, doubt, 2) trace

sus-pi-cious [sə-spĭsh′əs] *adjective*, having a feeling that something is wrong, distrustful

sus-tain [sə-stān′] *verb*, to bear, to hold up, to keep going

sus-te-nance [sŭs′tə-nəns] *noun*, that which supports life, a food or nourishment

su-ture [soo′chər] *noun*, stitches sewn to hold the cut edges of a wound or incision

swag-ger [swăg′ər] *verb*, to walk in a boastful and proud manner

swal-low [swŏl′ō] *noun*, 1) a small bird with a tail split into two parts, *verb*, 2) to consume food through the mouth into the stomach, to take in

swamp [swŏmp, swômp] *noun*, 1) wet, boggy ground which is always soft, wet land, *verb*, 2) to overwhelm with too much of something, to inundate

swan [swŏn] *noun*, a large white bird with a long neck that lives in the water and on land

swap [swŏp] *verb*, to exchange

swarm [swôrm] *noun*, 1) a large group of insects or bugs, *verb*, 2) to travel in a large group, to teem, to crowd

swarth-y [swôr′thē] *adjective*, having a dark complexion

swat [swŏt] *verb*, to hit, to whack

sway [swā] *verb*, to swing gently

swear [swâr] *verb*, 1) to curse 2) to promise, to make a solemn declaration

sweat [swĕt] *noun*, 1) moisture which comes out of the skin due to exertion, *verb*, 2) to give off water through the skin

sweat-er [swĕt/ər] *noun*, a wool or woven garment worn over the top of the body

sweep [swēp] *noun*, 1) a continuous search, *verb*, 2) to clean or clear with a brush or broom

sweet [swēt] *adjective*, 1) tasting like sugar or honey 2) a pleasant or kind personality 3) having a pleasant smell

swell [swĕl] *verb*, to become larger

swel-ter [swĕl/tər] *verb*, to be tired by the heat, to perspire, to sweat

swept [swĕpt] *past tense* of **sweep**

swerve [swûrv] *verb*, to move suddenly to one side when you are moving along

swift [swĭft] *adjective*, very fast or quick to act, nimble, prompt

swim [swĭm] *noun*, 1) the act of staying afloat or traveling in water, *verb*, 2) to move through the water by using movements of the body

swin-dle [swĭn/dl] *verb*, to cheat, to deceive or defraud

swing [swĭng] *noun*, 1) a seat hanging on ropes or chains, *verb*, 2) to move freely from a fixed point, to pivot, to turn

swirl [swûrl] *noun*, 1) a spiral, *verb*, 2) to turn in a circular motion

switch [swĭch] *noun*, 1) a device that can be used to turn something on or off, *verb*, 2) to exchange 3) to turn something

on or off 4) to change as in direction, to shift, to transfer

swiv-el [swĭv/əl] *verb*, to turn or pivot on a bolt or pin

swol-len [swō/lən] *adjective*, inflamed, puffed up, enlarged

swoon [swoon] *noun*, 1) a fainting spell, *verb*, 2) to faint

swoop [swoop] *verb*, to fly down very quickly like a bird

sword [sôrd] *noun*, a weapon similar to a knife with a very long blade

swore [swôr] *past tense* of **swear**

syl-la-ble [sĭl/ə-bəl] *noun*, a combination of letters indicating a sound in a word

syl-lo-gism [sĭl/ə-jĭz/əm] *noun*, logical formula using a major premise, a minor premise leading to a conclusion

sym-bol [sĭm/bəl] *noun*, anything that stands for or represents something else

sym-met-ri-cal [sĭ-mĕt/rĭ-kəl] *adjective*, alike on both sides, proportional

sym-me-try [sĭm/ĭ-trē] *noun*, the balance of things on opposite sides, excellence of proportion

sym-pa-thet-ic [sĭm/pə-thĕt/ĭk] *adjective*, kind, warm, compassionate, tender

sym-pa-thize [sĭm/pə-thīz/] *verb*, to feel sorry for another, to share in a feeling

sym-pa-thy [sĭm/pə-thē] *noun*, the understanding of another person's feelings, pity

sym-pho-ny [sĭm/fə-nē] *noun*, a musical composition performed by an orchestra

symp-tom [sĭm′təm] *noun*, a sign indicating the presence of an illness or something

syn-a-gogue [sĭn′ə-gŏg′] *noun*, a house of worship, a temple

syn-chro-nic-i-ty [sĭng′krə-nĭs′ĭ-tē] *noun*, the simultaneous occurrence of unrelated events

syn-chro-nize [sĭng′krə-nīz′] *verb*, to agree in time, to move together simultaneously

syn-di-cate [sĭn′dĭ-kĭt] *verb*, a network of businesses that sell news to the media

syn-er-gy: [sĭn′/ər-jē] *noun*, the interaction or cooperation of two or more organizations, substances, or other agents to produce a combined effect greater than the sum of their separate effects

syn-o-nym [sĭn′ə-nĭm′] *noun*, a word that means the same as another

syn-op-sis [sĭ-nŏp′sĭs] *noun*, a general view of a whole, an abstract or summary

syn-the-sis [sĭn′thĭ-sĭs] *noun*, combining parts into a whole

syn-thet-ic [sĭn-thĕt′ĭk] *adjective*, produced by artificial processes, not real or genuine

sy-ringe [sə-rĭnj′] *noun*, an instrument with a needle used to inject a fluid into the blood stream through a vein

syr-up [sĭr′əp] *noun*, a sugar liquid

sys-tem [sĭs′təm] *noun*, a network of things working together as a functioning unit

sys-tem-at-ic [sĭs′tə-măt′ĭk] *adjective*, methodical, proceeding according to method

T

ta-ble [tā′bəl] *noun*, 1) a piece of furniture with a flat top and legs used for placing things on or using as a work space 2) a list, often using lines or boxes to arrange facts

tab-leau [tăb′lō′] *noun*, a scene of persons in costume remaining silent and motionless

ta-ble-cloth [tā′bəl-klôth′] *noun*, a decorative cloth spread over the top of a table

tab-let [tăb′lĭt] *noun*, a hard flat piece or block of something

ta-boo [tə-bōō′] *adjective*, 1) forbidden, banned by religion or custom, *noun*, 2) a restriction, a ban or prohibition

tac-it [tăs′ĭt] *adjective*, implied or indicated, but not expressed

tac-i-tur-n [tăs′ĭ-tûr′n] *adjective*, habitually silent, not given to conversation, saying little

tack [tăk] *noun*, a small nail with a broad, flat head on it

tack-le [tăk′əl] *noun*, 1) ropes and pulleys arranged for hoisting or pulling 2) equipment used for fishing 3) downing a player 4) a football player, *verb*, 5) to undertake, to get down to business to solve a problem 6) to bring to the ground

tact [tăkt] *noun*, *no plural*, to do or say the right thing at the right time, diplomacy

tac-tic [tăk′tĭk] *noun*, a maneuver

tac-tile [tăk′təl, -tīl′] *adjective*, referring to the sense of touch

tag [tăg] *noun*, 1) a small identifying card or label 2) a game children play in which one player chases the others, *verb*, 3) to follow along

tail [tāl] *adjective*, 1) coming from behind, *noun*, 2) the part of an animal which sticks out at the end of its back

tai-lor [tā'lər] *noun*, one who cuts out and makes clothes

taint [tānt] *noun*, 1) a trace of something bad, *verb*, 2) to contaminate, to corrupt

take [tāk] *verb*, 1) to get hold of something 2) to carry something or go with someone to another place 3) to swallow something 4) to travel 5) to accept

tak-en [tā'kən] *adjective*, 1)held, seized 2) accepted

tale [tāl] *noun*, a story

tal-ent [tăl'ənt] *noun*, a natural ability, aptitude

tal-is-man [tăl'ĭs-mən] *noun*, an amulet or lucky charm

talk [tôk] *verb*, to speak or be able to speak, to consult or confer

talk-a-tive [tô'kə-tĭv] *adjective*, inclined to talk a lot, loquacious

tall [tôl] *adjective*, above average height, high

tal-ly [tăl'ē] *noun*, 1) a mark made to keep score or to count, *verb*, 2) to keep track of an account

tal-on [tăl'ən] *noun*, a claw, especially of a bird of prey

tame [tām] *adjective*, 1) trained to live with man, not wild, *verb*, 2) to make a wild animal obedient and friendly to people

tan [tăn] *adjective*, 1) light brown, tawny 2) to convert into leather

tan-gent [tăn'jənt] *noun*, a straight line that touches a curve without intersecting it

tan-ge-rine [tăn'jə-rēn'] *noun*, a variety of a small-sized orange

tan-gi-ble [tăn'jə-bəl] *adjective*, capable of being touched, real

tank [tăngk] *noun*, 1) a large container to hold liquid or gas fuel 2) an armed vehicle used in combat 3) the metal box in the back of the car to hold gasoline

tan-ta-lize [tăn'tə-līz'] *verb*, to tease, to tempt, to promise by offering a reward

tan-ta-mount [tăn'tə-mount'] *adjective*, equal in value

tan-trum [tăn'trəm] *noun*, an exaggerated display of anger, a fit of irritability, bad temper

tap [tăp] *noun*, 1) the faucet attached to a pipe to release liquid or gas, *verb*, 2) to hit or touch lightly 3) to drain by making a hole 4) to draw from

tape [tāp] *noun*, 1) a narrow piece of cloth or other material, *verb*, 2) to bind or secure with an adhesive material 3) to record

tap-er [tā'pər] *verb*, to decrease in amount or size by becoming narrower or smaller

tap-es-try [tăp'ĭ-strē] *noun*, a heavy, decorative, woven fabric used especially for wall hangings

tap-i-o-ca [tăp'ē-ō'kə] *noun*, cassava, a coarse granular preparation used in puddings

tar [tär] *noun*, a thick black liquid made from heating coal or wood

ta-ran-tu-la [tə-răn′chə-lə] *noun*, a hairy, large, poisonous spider

tar-get [tär′gĭt] *noun*, 1) somebody or something aimed at 2) an objective, a goal, *verb*, 3) to make the focus or object of something

tar-iff [tăr′ĭf] *noun*, a tax on goods exported or imported

tar-nish [tär′nĭsh] *verb*, 1) to dull the luster of metal 2) to taint

tart [tärt] *adjective*, 1) sour to the taste, *noun*, 2) a piece of pastry with fruit cooked on top of it

task [tăsk] *noun*, a chore or small job that must be done

tas-sel [tăs′əl] *noun*, a pendant ornament ending in a tuft of loose threads

taste [tāst] *noun*, 1) a preference, liking 2) the sense that distinguishes the flavor of things placed in the mouth 3) a flavor itself, *verb*, 4) to sample a food by taking it in the mouth

tat-too [tă-tōō′] *noun*, a design on the skin made using a needle filled with dye

taught [tôt] *past tense* of **teach**

taunt [tônt] *verb*, to tease or make fun of someone, to ridicule

taut [tôt] *adjective*, stretched tight

tau-tol-o-gy [tô-tŏl′ə-jē] *noun*, needless repetition of an idea in different words

taw-dry [tô′drē] *adjective*, cheap and showy or flashy, gaudy

tax [tăks] *noun*, 1) a sum of money to be paid to the government imposed by the elected leaders, *verb*, 2) to put a fee on service or goods that are given to the government

tax-i [tăk′sē] *noun*, a car for hire to take people from one place to another

tax-i-der-mist [tăk′sĭ-dûr′mĭst] *noun*, one who prepares dead animals to look life-like

tea [tē] *noun*, 1) prepared dried leaves of a plant 2) a small meal or snack in the afternoon 3) a cup of a beverage made with leaves steeped in hot water

teach [tēch] *verb*, to show someone how to do something

teach-er [tē′chər] *noun*, a person whose occupation is to instruct

team [tēm] *noun*, 1) persons working or playing together as a group 2) two or more animals harnessed together to pull a vehicle

tear [tîr] *noun*, 1) a drop of water that comes out of the eye 2) a hole or cut in material, *verb*, 3) to make a hole in material by pulling on it

tease [tēz] *verb*, to make fun of someone, to annoy

tea-spoon [tē′spōōn′] *noun*, a small spoon used to measure

tech-ni-cal [tĕk′nĭ-kəl] *adjective*, having knowledge of a mechanical or scientific subject

tech-ni-cian [tĕk-nĭsh′ən] *noun*, a person who works with machines or instruments

tech-nique [tĕk-nēk′] *noun*, a procedure used to perform a task

tech-nol-o-gy [tĕk-nŏl′ə-jē] *noun*, the use of science to produce useful things for practical purposes that improve life

te-di-ous [tē′dē-əs] *adjective*, tiresome, dull, boring

tel-e-phone [tĕl′ə-fōn′] *noun*, 1) an instrument for reproducing sounds at a distance, *verb*, 2) to call to someone on an instrument that communicates voices through airwaves or wires

tel-e-scope [tĕl′ĭ-skōp′] *noun*, an optical instrument for viewing distant objects

tel-e-vi-sion [tĕl′ə-vĭzh′ən] *noun*, a set used to transmit visual images onto a screen

tell [tĕl] *verb*, 1) to speak to someone, to inform, to narrate 2) to deduce, to differentiate

tell-er [tĕl′ər] *noun*, a bank employee who handles customer transactions

tem-per [tĕm′pər] *noun*, 1) a person's state of mind, mood, *verb*, 2) to tone down, to moderate, to bring under control

tem-per-a-ment [tĕm′prə-mənt] *noun*, the physical and mental character of an individual

tem-per-ance [tĕm′pər-əns] *noun*, habitual moderation of the appetites and passions

tem-per-a-ture [tĕm′pər-ə-cho͞or′] *noun*, the degree of heat or cold

tem-ple [tĕm′pəl] *noun*, 1) a holy building 2) the part of the head above and in front of the ear

tem-po [tĕm′pō] *noun*, the rate of speed of a musical passage

tem-po-ral [tĕm′pər-əl] *adjective*, not lasting forever, limited by time, the present

tem-po-rar-y [tĕm′pə-rĕr′ē] *adjective*, intended to last a short period of time, not permanent

tempt [tĕmpt] *verb*, to entice someone to do something unwise or wrong, to provoke

temp-ta-tion [tĕmp-tā′shən] *noun*, an instant urge or desire

ten [tĕn] *noun*, the number 10, one plus nine

te-na-cious [tə-nā′shəs] *adjective*, holding fast, stubborn

te-nac-i-ty [tə-năs′ĭ-tē] *noun*, firmness, persistency

ten-ant [tĕn′ənt] *noun*, someone who lives in a house or apartment, an occupant

tend [tĕnd] *verb*, 1) to be likely to, to do something habitually 2) to take care of something

ten-den-cy [tĕn′dən-sē] *noun*, inclination, propensity, leaning

tend-er [tĕn′dər] *adjective*, 1) kind and gentle 2) sensitive to touch or movement 3) cooked so that the food is easy to eat, *noun*, 4) a car attached to a train for carrying food and water

ten-der-ness [tĕn′dər-nĭs] *noun*, affection, gentleness

ten-e-ment [tĕn′ə-mənt] *noun*, a building divided into flats

ten-et [tĕn′ĭt] *noun*, doctrine, dogma, belief, conviction

Ten-nes-see [tĕn′ĭ-sē′] *noun*, one of the 50 United States located in the South, the capital city is Nashville. The state flower of **Tennessee** is the iris, and the motto is "Agriculture and commerce."

ten-nis [tĕn′ĭs] *noun*, a game played on a court by two or four people in which a ball is hit over a net with rackets

ten-or [tĕn′ər] *noun*, 1) a tendency, a trend 2) the highest range of the male singing voice

tense [tĕns] *adjective*, 1) full of excitement 2) tightly stretched, taut, *noun*, 3) the form of a verb that shows when the action happens – past, present, or future

ten-sion [tĕn′shən] *noun*, 1) act of stretching, 2) mental or nervous state or condition of strain

tent [tĕnt] *noun*, a shelter of canvas or nylon, supported by poles

ten-ta-cle [tĕn′tə-kəl] *noun*, a long, arm-like structure growing out of an animal's body, used for movement or catching food

ten-ta-tive [tĕn′tə-tĭv] *adjective*, experimental, based on trials

tenth [tĕnth] *adjective*, the number after nine in series, written 10th

ten-u-ous [tĕn′yōō-əs] *adjective*, weak and unconvincing

ten-ure [tĕn′yər] *noun*, the period of time during which an office is held, occupation

te-pee [tē′pē] *noun*, a cone shaped tent in which some Indians live

tep-id [tĕp′ĭd] *adjective*, 1) not hot or cold, lukewarm 2) showing little enthusiasm

term [tûrm] *noun*, 1) a fixed amount of time 2) an expression or form of language 3) a condition or stipulation

ter-min-al [tûr′mə-nəl] *adjective*, 1) end, final, concluding, *noun*, 2) a depot where buses or trains make connections 3) a conducting circuit arranged for connection in an electrical circuit

ter-mi-nate [tûr′mə-nāt′] *verb*, to end, to finish, to cancel

ter-mi-na-tion [tûr′mə-nā′shən] *noun*, the ending

ter-mi-nol-o-gy [tûr′mə-nŏl′ə-jē] *noun*, vocabulary used in a specialized subject

ter-race [tĕr′ĭs] *noun*, 1) a level area cut out from the side of a hill 2) a flat area outside a house

ter-rain [tə-rān′] *noun*, the features of an area of land

ter-rar-i-um [tə-râr′ē-əm] *noun*, a transparent container in which land animals or plants are kept

ter-res-tri-al [tə-rĕs′trē-əl] *adjective*, of the earth

ter-ri-ble [tĕr′ə-bəl] *adjective*, appalling, dreadful

ter-rif-ic [tə-rĭf′ĭk] *adjective*, awesome, wonderful

ter-ri-fy [tĕr′ə-fī′] *verb*, to fill or overwhelm with fear or horror

ter-ri-to-ry [tĕr′ĭ-tôr′ē] *noun*, a portion of land belonging to someone else, a region

ter-ror [tĕr′ər] *adjective*, fear

terse [tûrs] *adjective*, concise, abrupt, a short answer

ter-ror-ist [tĕr′ər] *noun*, one who uses violence to intimidate

tes-sel-la-tion [tĕs′ə-lā′shən] *adjective*, a mosaic pattern

test [tĕst] *noun*, 1) a trial 2) an examination to assess the level of knowledge of a subject, *verb*,

3) to try something out to see how it works 4) to examine

tes-ta-ment [tĕs′tə-mənt] *noun*,1) a will 2) scripture from the Bible

tes-ti-fy [tĕs′tə-fī′] *verb*, to establish some fact, under oath to serve as evidence

tes-ti-mo-ny [tĕs′tə-mō′nē] *noun*, the statement of a witness under oath, evidence or proof

tet-a-nus [tĕt′n-əs] *noun*, a painful, often fatal, infectious disease, marked by painful spasms of the muscles, frequently of the jaw

teth-er [tĕth′ər] *noun*, 1) a cord, a link, a supply line, *verb*, 2) to tie up with a rope, to hitch, to fasten

Tex-as [tĕk′səs] *noun*, one of the 50 United States located in the Southwest, the capital city is Austin. The state flower of **Texas**, is the bluebonnet, and the motto is "Friendship."

text [tĕkst] *noun*, the words used in a book, a written work

text-book [tĕkst′bŏŏk′] *noun*, an instruction book, a manual

tex-tile [tĕks′tĭl′] *noun*, fabrics such as cotton, wool, linen, silk, rayon, and polyester

tex-ture [tĕks′chər] *noun*, the feel or appearance of something – soft, smooth, rough, etc.

than [thăn] used as a *conjunction*, to introduce the second part of a comparison

thank [thăngk] *verb*, to express gratitude and appreciation

thank-ful [thăngk′fəl] *adjective*, to be grateful, to show appreciation or praise

thanks [thăngks] *noun*, an acknowledgment of a kindness, favor, or the like

Thanks-giv-ing [thăngks-gĭv′ĭng] *noun*, a national holiday celebrated the fourth Thursday in November

that [thăt] *pronoun, plural*, **those**, 1) the one singled out 2) used instead of which 3) used to join two parts of a sentence 4) so 5) used to show the result

that's [thăts] the *contraction* of the words **that** and **is**

thatch [thăch] *noun*, bundles of straw woven together

thaw [thô] *verb*, to pass from a frozen to a liquid state

the [thē before a vowel, thə before a consonant] *article*, a word that usually goes before a noun

the-a-ter [thē′ə-tər] *noun*, a building where dramatic performances are presented

the-at-ri-cal [thē-ăt′rĭ-kəl] *adjective*, dramatic, eloquent

thee [thē] *pronoun*, the objective case of thou

theft [thĕft] *noun*, the act or instance of stealing

their [thâr] *pronoun*, belonging to them, a form of the possessive case of **they** used attributively, e.g., *their school*

theirs [thârz] *pronoun*, showing personal ownership

them [thĕm] *pronoun*, the objective case of **they** , *plural*, **them-selves**, the same people, animals, or things mentioned in the sentence as the subject

theme [thēm] *noun*, the subject on which one writes or speaks

then [thĕn] *adverb*, 1) at another time, not now 2) afterwards, next 3) if that is true

the-ol-o-gy [thē-ŏl′ə-jē] *noun*, the science of God or of religion

the-o-rem [thē′ər-əm] *noun*, a principle, a theoretical proposition, a thesis

the-o-ret-i-cal [thē′ə-rĕt′ĭ-kəl] *adjective*, depending on conjecture, speculative

the-o-ry [thē′ə-rē] *noun*, an abstract plan, an hypothesis

ther-a-peu-tic [thĕr′ə-pyōō′tĭk] *adjective*, curative, healing, good for personal growth and health

ther-a-py [thĕr′ə-pē] *noun*, exercises and treatment designed to rehabilitate a person

there [thâr] *adverb*, 1) referring to that place 2) referring to at that moment

there's [thârz] the *contraction* of the words **there** and **is**

there-fore [thâr′fôr′] *adverb*, for that reason, as a result

ther-mal [thûr′məl] *adjective*, pertaining to heat

ther-mo-dy-nam-ics [thûr′mō-dī-năm′ĭks] *noun*, the relationship between heat and other forms of energy

ther-mom-e-ter [thər-mŏm′ĭ-tər] *noun*, a device for measuring how hot or cold something is

ther-mos [thûr′məs] *noun*, a vacuum bottle for keeping liquids hot or cold for hours

ther-mos-tat [thûr′mə-stăt′] *noun*, a mechanical device used to control temperature

the-sau-rus [thĭ-sôr′əs] *noun*, a dictionary of synonyms and antonyms

these [thēz] *pronoun, adjective,* the *plural* of **this**

the-sis [thē′sĭs] *noun*, an essay presented by a candidate for a degree or diploma

they [thā] *pronoun*, the *plural* of **he**, **she** or **it**

they'll [thāl] the *contraction* of the words **they** and **will**

they're [thâr] the *contraction* of the words **they** and **are**

thick [thĭk] *adjective*, 1)occupying much space in depth or from side to side, measuring between sides 2) a dense fluid

thief [thēf] *noun, plural,* **thieves,** one who steals

thigh [thī] *noun*, the thick part of the leg above the knee and below the hip

thim-ble [thĭm′bəl] *noun*, a metal cap put over the finger to keep it from being pricked by a needle

thin [thĭn] *adjective*, 1) not thick, slender 2) lacking volume or density 3) sparse, insufficient, inadequate, flimsy

thing [thĭng] *noun*, 1) an object, a creature, a form of life 2) conditions 3) an act or event

think [thĭngk] *verb*, 1) to form or have on the mind 2) to use the power of reason 3) to believe, to suppose

third [thûrd] *adjective, noun*, the number three in order after second, written 3rd

thirst [thûrst] *noun, no plural*, the need to drink something

thirst-y [thûr′stē] *adjective*, needing something to drink

thir-teen [thûr-tēn′] *adjective, noun*, the number 13, 3+10

thir-teenth [thûr-tēnth′] *adjective, noun*, number 13 in order, 13th

thir-ti-eth [thûr′tē-ĭth] *adjective, noun*, number 30 in order, 30th

thir-ty [thûr′tē] *adjective, noun*, the number 30, ten times three

this [thĭs] *adjective*, 1) being the one present, near or just mentioned 2) being the one nearer or last mentioned, *adverb* 3) the thing the speaker is talking about, *pronoun*, 4) the one nearer than that one in time or space 5) what is in the present or being talked about

this-tle [thĭs′əl] *noun*, a plant with prickly leaves and stems and a purple flower

thorn [thôrn] *noun*, a sharp projection on a stem

thor-ough [thûr′ō] *adjective*, complete, perfect, absolute

thor-ough-ly [thûr′ō-lē] *adverb*, completely done

those [thōz] *pronoun*, the ones over there, the ones further away than this one

thou [thou] you

though [thō] *adverb*, 1) however, *conjunction*, 2) although

thought [thôt] *noun*, 1) an idea, the process of thinking 2) intention

3) attention, *verb*, 4) the *past tense* of **think**

thought-ful [thôt′fəl] *adjective*, showing consideration for others, kind, discreet

thou-sand [thou′zənd] *noun, adjective*, the number 1,000, ten times 100

thread [thrĕd] *noun*, 1) a fine string made of cotton, linen, or silk, *verb*, 2) to wind through openings 3) to pass a thread through the eye of a needle

threat [thrĕt] *noun*, a warning, that one may do harm

threat-en [thrĕt′n] *verb*, to torment, to scare, to pose a danger

three [thrē] *adjective, noun*, the number 3, one plus two, 3

thresh-old [thrĕsh′ōld′] *noun*, a piece of timber which lies under a door, an entrance

thrift [thrĭft] *noun*, the practice of being careful with money

thrift-y [thrĭf′tē] *adjective*, careful about money, economical

thrill [thrĭl] *noun*, 1) an excited feeling, *verb*, 2) to fill with excitement

thrive [thrīv] *verb*, 1) to grow abundantly 2) to prosper, to flourish, to succeed

throat [thrōt] *noun*, the passage inside of the neck attached to the mouth, lungs, and stomach

throb [thrŏb] *verb*, to beat strongly, to pulsate steadily

throne [thrōn] *noun*, a special chair for a king or queen

throng [thrông, thrŏng] *noun*, a crowd of people, a multitude

throt-tle [thrŏt′l] *noun*, 1) a lever
or pedal that helps an engine
start and run, *verb*, 2) to strangle,
to suppress 3) to regulate the
speed of an engine

through [thrōō] *adjective*,
1) finished, *adverb*, 2) from one
side or end of something to the
other, *preposition*, 3) in one side
and out the other 4) by the
agency of, by way of, via 5)
among or between 6) past or
beyond, in the midst of

through-out [thrōō-out′]
preposition, all over, from end to
end, from inside and out

throw [thrō] *noun*, 1) the distance
something is tossed, cast or
projected, *verb*, 2) to toss, fling,
or hurl something through the air

thrust [thrŭst] *verb*, to push
suddenly and hard

thud [thŭd] *noun*, a dull sound
made when something heavy
falls and lands

thumb [thŭm] *noun*, the short,
thick finger on the hand which is
separate from the others

thun-der [thŭn′dər] *noun*, *no
plural*, the loud sound heard in
the sky during a storm

Thurs-day [thûrz′dē] *noun*, the
fifth day of the week

thus [thŭs] *adverb*, in this way

thy [thī] *pronoun*, your

thyme [tīm] *noun*, a pungent
aromatic plant used in seasoning

tick [tĭk] *noun*, 1) the sound made
by a clock 2) a check mark 3) a
small parasite that sucks blood
from animals and people, *verb*,
4) to make a soft, steady,
rhythmic sound 5) to check off

things on a list that fit a certain
criteria

tick-et [tĭk′ĭt] *noun*, 1) the ballot of
a political party 2) an admission
3) a fine 4) a price tag

tick-le [tĭk′əl] *noun,* 1) a tingling
sensation, *verb*, 2) to touch a
person lightly to make them
laugh 3) to delight, to amuse

tid-al [tīd′l] *adjective*, action of the
sea caused by the tides

tide [tīd] *noun*, regular rise and fall
of the ocean caused by the pull
of the sun and the moon

ti-dy [tī′dē] *adjective*, 1) neat and
orderly 2) substantial, more than
enough, *verb*, 3) to make neat
and well organized

tie [tī] *noun*, 1) a narrow piece of
cloth used as a man's dress
accessory worn around the neck
with a shirt, a necktie 2) a bond
3) a standoff, *verb*, 4) to fasten
something with string or rope
5) to come out equal in a contest

tier [tûr] *noun*, one of several levels

ti-ger [tī′gər] *noun*, a fierce animal,
one of the big cats, which has
yellow fur with black stripes

tight [tīt] *adjective*, 1) so close in
structure that air cannot pass
through 2) drawn close together

tight-en [tīt′n] *verb*, to draw closer
together, to squeeze

tight-ly [tīt′lē] *adverb*, pulled
closely together, tautly

tile [tīl] *noun*, a flat piece of baked
clay that has been painted with
glaze or linoleum used as a floor
covering

till [tĭl] *adverb*, 1) before, up to,
noun, 2) a cash box, a place to
store money, *verb*, 3) to plow

till-er [tĭl′ər] *noun*, a handle used to steer a boat

tilt [tĭlt] *verb*, to make something lean so that it is not straight

tim-ber [tĭm′bər] *noun*, a piece of wood, lumber, hardwood

tim-bre [tăm′bər] *noun*, quality of a musical tone produced by a musical instrument

time [tīm] *noun*, 1) minutes, hours, days, weeks, months, years, etc. 2) a limited duration 3) a period or event 4) a certain hour or part of the day 5) a moment, *verb*, 6) to measure how long it takes to do something 7) a number multiplied by another

time-ly [tīm′lē] *adjective*, prompt, well-timed, favorable

tim-id [tĭm′ĭd] *adjective*, shy, afraid, uncertain, hesitant

tim-id-i-ty [tĭm′ĭ-dĭ-tē] *noun*, lack of self-confidence or courage

tim-or-ous [tĭm′ər-əs] *adjective*, fearful and hesitant

tin [tĭn] *noun*, *no plural*, a soft white metal mixed with alloys

tinc-ture [tĭngk′chər] *noun*, a slight quality added to anything, a trace or tinge, pigment

tin-gle [tĭng′gəl] *verb*, 1) to get excited 2) to have a prickling sensation

tin-ker [tĭng′kər] *verb*, to fool around, to dabble, to putter

tin-sel [tĭn′səl] *noun*, showy or flashy decorations

tint [tĭnt] *noun*, 1) the color, shade, hue, *verb*, 2) to color, to shade

ti-ny [tī′nē] *adjective*, very small

tip [tĭp] *noun*, 1) a point, a peak 2) a hint, advice, a pointer 3) compensation for a service, gratuity, *verb*, 4) to tilt, to list, to lean to one side, to slant 5) to give someone money for a small service 6) to cause something to fall over and become unbalanced

tip-toe [tĭp′tō′] *noun*, 1) the tip of the toe, *verb*, 2) to walk on toes so that the step is not heard

ti-rade [tī′rād′] *noun*, extended scolding, denunciation

tire [tīr] *noun*, 1) a rubber, air-filled part of a car or bicycle wheel that cushions the ride, *verb*, 2) to make someone feel that he needs to rest 3) to become bored, fatigued

tir-ed [tīrd] *adjective*, feeling sleepy, needing rest

tis-sue [tĭsh′ōō] *noun*, very thin paper that feels soft

ti-tan-ic [tī-tăn′ĭk] *adjective*, gigantic, of great size or power

tithe [tīth] *noun*, a payment of one tenth of an income given to the church

ti-tle [tīt′l] *noun*, 1) the name of a story, a book, a film, etc. 2) a word used in front of a person's name 3) a sport's championship

to [tōō; tə when unstressed] *adverb*, 1) in a direction toward *preposition*, 2) in the direction of 3) as far as 4) on or against 5) until 6) used to show why 7) used with the infinitive of a verb 8) for the purpose of, in order to 9) belonging to

toad [tōd] *noun*, an animal that looks like a frog but does not live in water

toast [tōst] *noun*, 1) sliced bread broiled to make it crisp, *verb*, 2) to praise someone when raising your drink in a glass 3) to brown by the heat of a fire

to-bac-co [tə-băk′ō] *noun*, the dried leaves of a plant used for smoking and chewing

to-bog-gan [tə-bŏg′ən] *noun*, a very long, flat-bottomed sled used to glide on snow

to-day [tə-dā′] *adverb*, 1) on or for this very day, *noun*, 2) modern times, present day, contemporary

toe [tō] *noun*, 1) one of the five digits at the end of the foot 2) the part of a shoe or sock that covers this end of the foot

tof-fee [tô′fē] *noun*, a hard, chewy candy made of caramel candy, brown sugar and butter

to-ga [tō′gə] *noun*, Roman outer robe of free born men

to-geth-er [tə-gĕth′ər] *adverb*, 1) at the same time 2) in conjunction, with each other, in cooperation

toil [toil] *verb*, to work very hard

toi-let [toi′lĭt] *noun*, an instrument to flush human waste into a sewage system, a bathroom

to-ken [tō′kən] *noun*, 1) a sign or symbol or gesture to indicate something 2) a souvenir 3) a coin used as a currency for travel

told [tōld] *past tense* of **tell**

tol-er-ance [tŏl′ər-əns] *noun*, compassion, open-mindedness

tol-er-ant [tŏl′ər-ənt] *noun*, broad minded, liberal, accepting

toll [tōl] *noun*, 1) the charge for a privilege or a service 2) the number lost or taken, *verb*, 3) to ring a bell slowly

to-ma-to [tə-mā′tō] *noun*, a red berry with seeds and a pulpy fruit eaten as a vegetable

tomb [tōōm] *noun*, a sacred place where a dead person is buried

tomb-stone [tōōm′stōn′] *noun*, the carved stone with an inscription identifying the person on a grave

tome [tōm] *noun*, a large volume

to-mor-row [tə-môr′ō] *adverb* 1) the day after today, *noun*, 2) sometime in the future

ton [tŭn] *noun*, the weight of measure of a very heavy quantity, e.g., 2,000 pounds

tone [tōn] *noun*, 1) the sound of a voice or a musical instrument 2) manner, expression

tongs [tôngz] *noun*, *plural*, (pair of) a tool with two legs used to pick up or to hold something

tongue [tŭng] *noun*, 1) the organ of speech covered by taste buds 2) the flap of leather or cloth under the lacing of a shoe 3) the language spoken 4) dialect

ton-ic [tŏn′ĭk] *noun*, a refreshing drink that invigorates

to-night [tə-nīt′] *adverb*, 1) the night following today, *noun*, 2) on or during the night of today

too [tōō] *adverb*, 1) also 2) more than is needed or wanted

took [tōōk] *past tense* of **take**

tool [tōōl] *noun*, an instrument used to do work or make repairs

tooth [tōōth] *noun*, *plural* **teeth**, 1) a white bony material that grows in the mouth, set in sockets around the jaw, used for chewing 2) the projections on a blade

tooth-ache [tŏŏth′āk′] *noun*, a pain in the mouth due to an infection from a tooth

top [tŏp] *adjective*, 1) best, first, chief, *noun*, 2) the highest part of something 3) the lid or cover of something 4) a toy that spins quickly on a point

to-paz [tō′păz′] *noun*, a precious stone, usually yellow or tan

top-ic [tŏp′ĭk] *noun*, the subject of a conversation, speech, or essay

to-pog-ra-phy [tə-pŏg′rə-fē] *noun*, detailed representation of the features of a region

top-ple [tŏp′əl] *verb*, to make or become unsteady and fall over

torch [tôrch] *noun*, an instrument that gives off a very hot flame

tor-ment [tôr′mĕnt′] *noun*, 1) abuse, agony, chaos, *verb*, 2) to torture, to annoy, to abuse, to afflict with mental suffering

tor-na-do [tôr-nā′dō] *noun*, a storm with a strong wind which spins very fast, a violent windstorm

tor-pe-do [tôr-pē′dō] *noun*, a self-propelled underwater missile containing explosives fired through the water from a ship to destroy another ship

tor-pid [tôr′pĭd] *adjective*, dormant, dull, lethargic

tor-rent [tôr′ənt] *noun*, a fast flow of water or energy

tor-ren-tial [tô-rĕn′shəl] *adjective*, a rushing, violent stream of water

tor-rid [tôr′ĭd] *adjective*, 1) passionate 2) tropical, sweltering, oppressively hot

tor-so [tôr′sō] *noun*, the trunk of a statue with head and limbs missing, the upper part of the human body

tort [tôrt] *noun*, a civil wrong, independent of a contract

tor-toise [tôr′tĭs] *noun*, a very large turtle that lives on the land

tor-tu-ous [tôr′chōō-əs] *adjective*, winding, full of curves

tor-ture [tôr′chər] noun1) 1) the act of causing great pain or suffering 2) distress of mind or body, *verb*, 3) to inflict pain on someone mercilessly, in order to dominate and control

toss [tôs, tŏs] *verb*, 1) to throw 2) to move about up and down

to-tal [tōt′l] *adjective*, 1) making up the whole, *noun*, 2) the sum of numbers in addition, entire

to-tal-ly [tōt′l-ē] *adverb*, wholly, entirely, completely

touch [tŭch] *noun*, 1) one of the five senses that communicates texture, temperature, and density to the brain through feeling, *verb*, 2) to make physical contact with something or someone

tough [tŭf] *adjective*, 1) yielding to force without breaking, strong 2) brave 3) difficult to chew

tour [tŏŏr] *noun*, 1) a journey that ends where it starts, *verb*, 2) to make a trip through a place

tour-ist [tŏŏr′ĭst] *noun*, a person traveling for pleasure

tour-na-ment [tŏŏr′nə-mənt] *noun*, a set number of contests that includes many teams or players and produces one winner

tow [tō] *verb*, to pull or haul with a rope or a chain

to-ward [tôrd] *preposition*, 1) in the direction of, in relation to 2) facing from the opposite direction

tow-el [tou′əl] *noun*, a piece of cloth or paper used for wiping or drying

tow-er [tou′ər] *noun*, 1) a tall building or structure that stands above the rest 2) a turret or spire attached to a building or fortress, *verb*, 3) to stand or reach above everything

town [toun] *noun*, buildings creating a hamlet, larger than a village and smaller than a city

tox-ic [tŏk′sĭk] *adjective*, poisonous, harmful

toy [toi] *noun*, an ornament or article that children play with

trace [trās] *noun*, 1) a small amount of something, *verb*, 2) to draw an image using paper to duplicate it 3) to try to follow the path of someone by using evidence that indicates what they are doing or have done

track [trăk] *noun*, 1) a mark left by something that has passed 2) a path or roadway, *verb*, 3) to follow or trace, to trail

trac-ta-ble [trăk′tə-bəl] *adjective*, 1) obedient, willing, compliant 2) easily shaped, malleable

trac-tion [trăk′shən] *noun*, adhesive friction of being drawn or pulled, the grip of the wheels on the ground

trac-tor [trăk′tər] *noun*, a machine used to plow and to till soil

trade [trād] *noun, no plural*, 1) the exchange of goods 2) a form of business or enterprise 3) a specific type of work that requires training, *verb*, 4) to buy and sell goods, to barter

tra-di-tion [trə-dĭsh′ən] *noun*, a custom or belief passed within families or other groups from one generation to another

tra-di-tion-al [trə-dĭsh′ə-nəl] *adjective*, following a custom that has been handed down through generations

traf-fic [trăf′ĭk] *noun*, 1) commerce, trade, the interchange of goods 2) the movement of cars and people on streets or of planes and ships

trag-e-dy [trăj′ĭ-dē] *noun*, 1) drama that deals with sad or terrible themes, as opposed to comedy 2) a fatal or unfortunate event

trag-ic [trăj′ĭk] *adjective*, very sad, dreadful, unfortunate

trail [trāl] *noun*, 1) the marks or evidence left behind in a place by someone or something that was there earlier, *verb*, 2) to follow someone, to track

train [trān] *noun*, 1) railroad cars for people or freight pulled by an engine, *verb*, 2) to teach someone or yourself to do something difficult that is considered a skill

train-ing [trā′nĭng] *noun*, preparation, instruction

trait [trāt] *noun*, a distinguishing characteristic or feature

trai-tor [trā′tər] *noun*, one who betrays a cause or any trust

tra-jec-to-ry [trə-jĕk′tə-rē] *noun*, a path taken by a projectile, or a flying object in its flight

tramp [trămp] *noun*, a person with no home or job who travels from place to place begging, a vagrant

tram-ple [trăm′pəl] *verb*, to crush into the ground, to walk heavily on something

trance [trăns] *noun*, a dream-like state of deep relaxation

tran-quil [trăng′kwəl] *adjective*, calm, peaceful, quiet, serene

tran-quil-li-ty [trăng-kwĭl′ĭ-tē] *noun*, calmness, composure

tran-quil-iz-er [trăng′kwə-līz′ər] *noun*, a sleeping pill, sedative

trans-ac-tion [trăn-săk′shən] *noun*, the doing of any business

tran-scend [trăn-sĕnd′] *verb*, to rise above and beyond the ordinary limits, to surpass, to exist independent of

tran-scen-den-tal [trăn′sĕn-dĕn′tl] *adjective*, beyond understanding, supernatural, spiritual

trans-con-ti-nen-tal [trăns′kŏn-tə-nĕn′tl] *adjective*, spanning a continent

tran-scribe [trăn-skrīb′] *verb*, to write out a spoken message

tran-script [trăn′skrĭpt′] *noun*, a written copy of a broadcast or a student's academic record

trans-fer [trăns-fûr′] *noun*, 1) the act of moving something from one place to another, *verb*, 2) to convey from one place to another

trans-fer-a-ble [trăns-fûr′ə-bəl] *noun*, capable of being moved from one place to another

trans-form [trăns-fôrm′] *verb*, 1) to change the form or appearance of something 2) to glorify

trans-form-er [trăns-fôr′mər] *noun*, a device that transfers electrical energy from one circuit to another

trans-fu-sion [trăns-fyōō′zhən] *noun*, the introduction into a person's bloodstream of blood from another

trans-gres-sion [trăns-grĕsh′ən] *noun*, violation of law, a going beyond some limit

tran-sient [trăn′shənt] *adjective*, of short duration, not permanent

tran-sis-tor [trăn-zĭs′tər] *noun*, an electrical device like a radio that sends and amplifies voices

tran-sit [trăn′sĭt] *noun*, 1) a passage 2) the state of being carried 3) a surveyor's instrument for measuring angles

tran-si-tion [trăn-zĭsh′ən] *noun*, going from one state of action or position to another

tran-si-tive [trăn′sĭ-tĭv] *adjective*, when the action of a verb is done to something or somebody

tran-si-to-ry [trăn′sĭ-tôr′ē] *adjective*, temporary, brief, lasting only a short time

trans-late [trăns-lāt′] *verb*, to pass along the message from one language to another

trans-la-tion [trăns-lā′shən] *noun*, the conversion of a message into another language

trans-lu-cent [trăns-lōō′sənt] *adjective*, a material that lets

light in diffusively, but cannot
be seen through

trans-mis-sion [trăns-mĭsh′ən]
noun, 1) the apparatus by which
the power is transmitted from
high-speed motor power to the
wheels 2) the sending of
pictures or sounds through the
air, as by radio or television

trans-mit [trăns-mĭt′] *verb*, 1) to
send information over wires or
air waves 2) to convey or send
from one person or place to
another 3) to pass along

trans-mit-ter [trăns-mĭt′ər] *noun*,
the part of a telegraph or
telephone instrument used to
send a message

trans-mute [trăns-myo͞ot′] *verb*, to
change, to convert from one
substance to another

trans-par-ent [trăns-pâr′ənt]
adjective, easily or capable of
being seen through, clear

tran-spire [trăn-spīr′] *verb*, 1) to
exhale 2) to happen, to occur

trans-plant [trăns-plănt′] *verb*,
1) to take out of the ground to
plant somewhere else 2) a
medical operation that
exchanges a body organ

trans-port [trăns-pôrt′] *noun*,
1) a means of carrying people or
goods, *verb*, 2) to carry

trans-por-ta-tion
[trăns′pər-tā′shən] *noun*, the act
of carrying things from one
place to another

trans-pose [trăns-pōz′] *verb*, 1) to
switch 2) to put into a different
key in music

trap [trăp] *noun*, 1) a device used

to capture someone or something
2) a strategy used to capture or
entrap someone, *verb*, 3) to
capture something or someone
so that they cannot escape

tra-peze [tră-pēz′] *noun*, a short
horizontal bar suspended by two
parallel ropes, one at each end

trap-e-zoid [trăp′ĭ-zoid′] *noun*, a
closed figure of four sides, two
sides are parallel but unequal in
length

trash [trăsh] *noun*, something
considered worthless that is
thrown away, garbage

trau-mat-ic [trou′măt′ĭk] *adjective*,
upsetting, stressful, disturbing

tra-vail [trə-vāl′] *noun*, 1) painful
labor, hard work 2) anguish

trav-el [trăv′əl] *noun*, 1) taking a
trip, seeing the world, *verb*, 2) to
go from one place to another 3)
to be transmitted

trav-erse [trə-vûrs′] *verb*, to cross
in traveling, to pass over

trawl-er [trô′lər] *noun*, a fishing
boat which drags a cone-shaped
net across the floor of the ocean

tray [trā] *noun*, a flat shallow
receptacle with raised edges on
which things can be carried

treach-er-ous [trĕch′ər-əs]
adjective, dangerous

trea-cle [trē′kəl] *noun*, syrup
obtained in refining sugar

tread [trĕd] *noun*, 1) the part
touching the ground, *verb*, 2) to
stand on, to walk on, to step

trea-son [trē′zən] *noun*, *no plural*,
the crime of helping the enemy
of a group or country one is
loyal to

treas-ure [trĕzh′ər] *noun*, 1) an accumulation of wealth such as money or jewels, *verb*, 2) to regard as precious, to cherish

treas-ur-er [trĕzh′ər-ər] *noun*, one who is in charge of the disbursement of money

treas-ur-y [trĕzh′ə-rē] *noun*, a place to store and control money such as a bank

treat [trēt] *noun*, 1) something that gives pleasure or enjoyment, *verb*, 2) to behave toward 3) to deal with, to serve 4) to give medicine as a doctor

trea-tise [trē′tĭs] *noun*, an article treating a subject systematically and thoroughly

treat-ment [trēt′mənt] *noun*, 1) handling, processing 2) therapeutic care

trea-ty [trē′tē] *noun*, a formal agreement between two or more nations

tree [trē] *noun*, a large plant with a trunk, branches, and leaves

trek [trĕk] *verb*, to travel with hardship, a difficult journey

trem-ble [trĕm′bəl] *verb*, to shake, to feel nervous, to quiver

tre-men-dous [trĭ-mĕn′dəs] *adjective*, 1) marvelously great 2) very large 3) extraordinary

trem-or [trĕm′ər] *noun*, trembling, involuntary shaking of the body

trem-u-lous [trĕm′yə-ləs] *adjective*, quivering, wavering

trench [trĕnch] *noun*, a long narrow ditch, dug into the earth

trench-ant [trĕn′chənt] *adjective*, cutting, sarcastic, critical, keen

trench-er-man [trĕn′chər-mən] *noun*, a person with a hearty appetite

trend [trĕnd] *noun*, the popular course, direction, tendency

trep-i-da-tion [trĕp′ĭ-dā′shən] *noun*, fear, apprehension

tres-pass [trĕs′pəs] *noun*, 1) an intrusion, *verb*, 2) to encroach on another's right 3) to unlawfully enter land of another

tri-al [trī′əl] *noun*, 1) when people in a court of law decide whether a person is guilty of a crime 2) a test to see if something is good or bad 3) suffering or adversity

tri-an-gle [trī′ăng′gəl] *noun*, a plane figure with three sides and three angles

tri-an-gu-lar [trī-ăng′gyə-lər] *adjective*, having three points like a triangle

tribe [trīb] *noun*, a group of people descended from a common ancestor who practice the same traditions, customs, and rituals, lead by a chief, a clan

trib-u-la-tion [trĭb′yə-lā′shən] *noun*, distress, suffering

tri-bu-nal [trī-byōō′nəl] *noun*, a court or forum for justice

trib-u-tar-y [trĭb′yə-tĕr′ē] *adjective*, 1) under another's control, *noun*, 2) a stream or river that flows into a larger river

trib-ute [trĭb′yōōt] *noun*, a gesture to show respect for someone, an accolade, a statement of praise

trick [trĭk] *noun*, 1) a clever prank, a sly or ingenious feat, *verb*, 2) to deceive by cunning

trick-le [trĭk′əl] *verb*, to flow in a small, gentle stream

tri-cy-cle [trī′sĭk′əl] *noun*, a light three-wheeled vehicle

tried [trīd] *verb, past tense* of **try**

tri-fle [trī′fəl] *noun*, 1) a small amount, *verb*, 2) to play or toy

tri-fling [trī′flĭng] *adjective*, of little value or importance

trig-ger [trĭg′ər] *noun*, 1) the firing mechanism on a gun, *verb*, 2) to set something off

trig-o-nom-e-try [trĭg′ə-nŏm′ĭ-trē] *noun*, measuring the sides and angles of triangles

tril-lion [trĭl′yən] *noun*, a million million: 1,000,000,000,000

tril-o-gy [trĭl′ə-jē] *noun*, a novel or play in three separate works that are closely related

trim [trĭm] *noun*, 1) ornaments or decorations, *verb*, 2) to decorate 3) to cut back to make tidy

tri-o [trē′ō] *noun*, a set of three

trip [trĭp] *noun*, 1) a journey, a jaunt, an excursion, a pilgrimage, *verb*, 2) to skip, to stumble 3) to release a lever suddenly

tri-ple [trĭp′əl] *adjective*, three-way, multiplied three times

tri-pod [trī′pŏd′] *noun*, a stand or support with three legs

trite [trīt] *adjective*, 1) trivial, insignificant 2) hackneyed

tri-umph [trī′əmf] *noun*, 1) exultation over success, *verb*, 2) to gain victory

tri-um-phant [trī-ŭm′fənt] *adjective*, feeling thrilled and happy about winning

triv-i-a [trĭv′ē-ə] *noun*, small, unimportant matters, trifles

triv-i-al [trĭv′ē-əl] *adjective*, of little worth or importance

trol-ley [trŏl′ē] *noun*, a public bus that travels on a train track

troop [trōōp] *noun*, 1) a collection of people 2) a body of cavalry

tro-phy [trō′fē] *noun*, something as a prize or award, given to show success in an activity

trop-ic [trŏp′ĭk] *noun*, a place between the two imaginary lines that run parallel above and below the equator, where it is hot and humid all year long

trop-i-cal [trŏp′ĭ-kəl] *adjective*, temperatures of a jungle or warm climate near the equator

trot [trŏt] *verb*, to keep at a steady pace, to canter, to gallop

trou-ble [trŭb′əl] *noun*, 1) a problem, a difficulty 2) a dangerous or difficult situation 3) extra work or effort, *verb*, 4) to impose on 5) to distress

troupe [trōōp] *noun*, a troop or company of musicians or performers

trou-sers [trou′zərz] *noun*, *plural*, loose fitting pants

trout [trout] *noun*, a small fresh-water fish in the salmon family

trow-el [trou′əl] *noun*, 1) a hand-held spade 2) a tool with a flat blade used to apply mortar or cement

tru-ant [trōō′ənt] *noun*, 1) absent from school without good reason or permission 2) neglectful

truce [trōōs] *noun*, a peace agreement, cease-fire

truck [trŭk] *noun*, a large, heavy vehicle with at least four wheels used to transport heavy materials or large quantities

true [trōō] *adjective*, 1) honest, real, based on fact 2) loyal

tru-ism [trōō′ĭz′əm] *noun*, a self-evident obvious truth

tru-ly [(trōō′lē] *adverb*, in fact or in truth, sincerely, really

trump [trŭmp] *verb*, 1) to fabricate or invent 2) to beat someone's hand in bridge

trum-pet [trŭm′pĭt] *noun*, a horn, one of the brass instruments with several keys and a flaring bell

trunk [trŭngk] *noun*, 1) a chest, a storage box 2) the main stem of a tree 3) a compartment in a car 4) the snout of an elephant

trust [trŭst] *noun*, 1) belief in someone's goodness and integrity, *verb*, 2) to believe that someone will do as they say

trust-ee [trŭ-stē′] *noun*, one entrusted to administer the property of another

truth [trōōth] *noun*, the facts, what actually happened, reality

try [trī] *verb*, 1) to attempt, to make an effort 2) to sample

try-ing [trī′ĭng] *adjective*, difficult, straining one's patience

tryst [trĭst] *noun*, an appointed meeting, a rendezvous

tsu-nami [tsōō-nä′mē] *noun*, a giant, fast-moving ocean wave produced by an earthquake

tub [tŭb] *noun*, a container used to take a bath or wash clothes

tu-ba [tōō′bə] *noun*, the lowest-pitched of the brass instruments

tube [tōōb] *noun*, 1) a pipe made of metal, plastic 2) a glass usually used to conduct liquids or gases

tu-ber-cu-lo-sis [tōō-bûr′kyə-lō′sĭs] *noun*, a disease due to the presence of small granular tumors in an organ marked by coughing

tuck [tŭk] *noun*, 1) a pleat, fold or gather, *verb*, 2) to wrap snugly

Tues-day [tōōz′dē] *noun*, the third day of the week

tuft [tŭft] *noun*, a cluster, a clump

tug [tŭg] *verb*, to jerk, to yank, to pull with force or effort

tu-i-tion [tōō-ĭsh′ən] *noun*, the price paid for instruction

tu-lip [tōō′lĭp] *noun*, a flower grown from a bulb in a variety of colors, in the shape of a cup

tulle [tōōl] *noun*, a thin, fine net of acetate, silk or rayon, often used for veils or a tutu

tum-ble [tŭm′bəl] *verb*, 1) to turn over and over 2) to fall down

tu-mult [tōō′mŭlt′] *noun*, commotion, a riot, noise

tu-mul-tu-ous [tōō-mŭl′chōō-əs] *adjective*, wild, chaotic

tu-na [tōō′nə] *noun*, a large fish that lives in the ocean

tun-dra [tŭn′drə] *noun*, the vast treeless plains in Siberia and arctic North America

tune [tōōn] *noun*, 1) musical notes arranged as a medley, *verb*, 2) to put the right amount of tension on a stringed instrument so that it gives a certain sound 3) to be in sync, to put in harmony

tung-sten [tŭng′stən] *noun*, a hard, brittle, whitish-gray metal with an extremely high melting point at 6100 degrees Fahrenheit

tun-nel [tŭn′əl] *noun*, 1) a long passageway dug underground, *verb*, 2) to dig, to burrow

tur-ban [tûr′bən] *noun*, a man's headdress having no brim

tur-bid [tûr′bĭd] *adjective*, muddy, having the sediment disturbed

tur-bine [tûr′bĭn] *noun*, a machine that uses the gas from liquid to turn a shaft

tur-bu-lence [tûr′byə-ləns] *noun*, 1) wind currents that move very quickly 2) confusion, excited activity

tur-bu-lent [tûr′byə-lənt] *adjective*, 1) agitated, violent, stormy 2) full of disorder

tu-reen [tŏŏ-rēn′] *noun*, a deep table dish for holding soup

tur-key [tûr′kē] *noun*, a bird of the pheasant family, originally found wild in North America

tur-moil [tûr′moil′] *adjective*, a state of great confusion, agitation or disturbance

turn [tûrn] *noun*, 1) a chance or opportunity to participate 2) a bend, a twist or a bend in the road 3) a reversal, *verb*, 4) to cause to rotate or revolve 5) to change or to transform

tur-nip [tûr′nĭp] *noun*, a thick, edible, roundish root in the mustard family, eaten when boiled

tur-pen-tine [tûr′pən-tīn′] *noun*, a thin flammable oil made from certain pine trees used as a solvent to thin paint

tur-quoise [tûr′kwoiz′] *noun*, a bluish green or greenish gray mineral used as a gem

tur-tle [tûr′tl] *noun*, a reptile with a soft body protected by a hard shell covering, into which it can draw its head, legs and tail

tusk [tŭsk] *noun*, a long pointed tooth, as of an elephant or walrus

tu-tor [tŏŏ′tər] *noun*, a private teacher

tweed [twēd] *noun*, a soft, flexible fabric made of wool

tweez-ers [twē′zərz] *noun*, *plural*, a pincer-like tool for picking up small things

twelfth [twĕlfth] *adjective*, the ordinal of twelve, written 12th

twelve [twĕlv] *noun*, *adjective*, the number 12, the number after eleven

twen-ti-eth [twĕn′tē-ĭth] *adjective*, the number 20 in order, 20th, the ordinal of 20

twen-ty [twĕn′tē] *adjective*, *noun*, the number 20, ten times two

twice [twīs] *adverb*, two times, in double amount or degree

twig [twĭg] *noun*, a small outshoot of a branch or stem

twi-light [twī′līt′] *noun*, the light as the sun sets from the west

twin [twĭn] *adjective*, 1) two of the same, matched 2) an identical couple 3) either of two offspring brought forth at birth

twine [twīn] *noun*, 1) a strong cord of threads twisted together, *verb*, 2) to wrap around

twinge [twĭnj] *verb*, to feel a sharp physical or mental pain

twin-kle [twĭng′kəl] *verb*, to sparkle, to shimmer, to flicker

twins [twĭnz] *noun*, *plural*, two children born at the same time from the same mother

twirl [twûrl] *verb*, to whirl or spin in a circle, to rotate rapidly

twitch [twĭch] *verb*, to jerk or make spastic movements

two [tōō] *adjective, noun*, the number that follow one, written 2, one plus one

type [tīp] *noun*, 1) the letters in different fonts used to print words 2) category, sort, *verb*, 3) to use a keyboard to write words printed on a computer

type-writ-er [tīp′rī′tər] *noun*, an instrument for writing by means of type

ty-phoid [tī′foid′] *noun*, a fever which eats up the walls of the intestines causing internal bleeding

ty-phoon [tī-fōōn′] *noun*, a cyclone in the region of the Philippines or the China Sea

typ-i-cal [tĭp′ĭ-kəl] *adjective*, representing something by a model, characteristic

ty-rant [tī′rənt] *noun*, an absolute ruler who is usually cruel

U

u-biq-ui-tous [yōō-bĭk′wĭ-təs] *adjective*, being everywhere at the same time, omnipresent

UFO [yōō′ĕf-ō′] *noun*, Unidentified Flying Object, a thing in the sky that seems too strange to be real

ug-ly [ŭg′lē] *adjective*, revolting to look at, unattractive

ul-cer [ŭl′sər] *noun*, a sore on the skin filled with pus and infection that is painful, a lesion

ul-te-ri-or [ŭl-tîr′ē-ər] *adjective*, intentionally concealed

ul-ti-mate [ŭl′tə-mĭt] *adjective*, 1) final, supreme 2) farthest

ul-ti-mate-ly [ŭl′tə-mĭt-lē] *adverb*, finally, at the end, totally

ul-ti-ma-tum [ŭl′tə-mā′təm] *noun*, a final proposition or condition offered to end a dispute

um-brage [ŭm′brĭj] *noun*, resentment, anger, a sense of injury or insult, offense

um-brel-la [ŭm-brĕl′ə] *noun*, a covering held in the hand as protection from the rain

um-pire [ŭm′pīr′] *noun*, the judge or referee in a baseball game

un-a-ble [ŭn-ā′bəl] *adjective*, not capable of doing something

un-alienable [ŭn-āl′yə-nə-bəl] *adjective*, cannot be taken away

u-nan-i-mous [yōō-năn′ə-məs] *adjective*, being of one mind, showing complete agreement

u-nan-i-mous-ly [yōō-năn′ə-məs-lē] *adverb*, with all voters voting the same way

un-as-sum-ing [ŭn′ə-sōō′mĭng] *adjective*, modest, simple, plain

un-a-void-a-ble [ŭn′ə-voi′də-bəl] *adjective*, inevitable

un-bear-a-ble [ŭn-bâr′ə-bəl]
adjective, not capable of being
endured, intolerable

un-bri-dled [ŭn-brīd′ld] *adjective*,
not restrained, uninhibited

un-can-ny [ŭn-kăn′ē] *adjective*,
strange, mysterious, seeming to
have a supernatural basis

un-cer-tain-ty [ŭn-sûr′tn-tē] *noun*,
doubtfulness, not reliable

un-cle [ŭng′kəl] *noun*, the brother
of one of your parents, or the
husband of the sister of one of
your parents

un-com-fort-a-ble
[ŭn-kŭm′fər-tə-bəl] *adjective*,
uneasy, causing discomfort

un-com-mon [ŭn-kŏm′ən]
adjective, not usual, unique, odd

un-con-cern [ŭn′kən-sûrn′]
adjective, indifference, lack of
interest, freedom from anxiety

un-con-scion-a-ble
[ŭn-kŏn′shə-nə-bəl] *adjective*,
unscrupulous, something done
that shows no conscience,
excessive, unreasonable

un-con-scious [ŭn-kŏn′shəs]
adjective, not knowing what is
happening, lacking awareness

un-cov-er [ŭn-kŭv′ər] *verb*, 1) to
take something from on top of
something 2) to find out

un-der [ŭn′dər] *adjective*, 1) less
than 2) working for or obeying,
preposition, 3) below, beneath
the surface

un-der-es-ti-mate
[ŭn′dər-ĕs′tə-māt′] *verb*, to
undervalue, to minimize

un-der-go [ŭn′dər-gō′] *verb*, 1) to
experience 2) to suffer or endure

un-der-grad-u-ate
[ŭn′dər-grăj′ōō-ĭt] *noun*, a
student at a college who has not
received a bachelor's degree

un-der-ground [ŭn′dər-ground′]
adjective, 1) in the recesses of
the earth 2) in secrecy

un-der-growth [ŭn′dər-grōth′]
noun, no plural, thickly growing
plants underneath trees

un-der-line [ŭn′dər-līn′] *verb*, to
put a line under a word or words

un-der-neath [ŭn′dər-nēth′]
adverb, down from below

un-der-rate [ŭn′dər-rāt′] *verb*, to
evaluate too low, to undervalue

un-der-stand [ŭn′dər-stănd′] *verb*,
1) to grasp the nature of 2) to
know what something means

un-der-take [ŭn′dər-tāk′] *verb*, to
take upon oneself as a task

un-der-tak-er [ŭn′dər-tā′kər]
noun, one whose business is to
take charge of funerals

un-der-tak-ing [ŭn′dər-tā′kĭng]
noun, a duty, an obligation, a
job, a task

un-der-wear [ŭn′dər-wâr′] *noun*,
the garments worn next to the
skin under outside clothing

un-der-write [ŭn′dər-rīt′] *verb*, to
insure, to guarantee

un-do [ŭn-dōō′] *verb*, 1) to untie or
unfasten 2) to reverse or annul

un-doubt-ed-ly [ŭn-dou′tĭd-lē]
adverb, certainly, indeed

un-du-late [ŭn′jə-lāt′] *verb*, to
move with a wave-like motion

un-earth [ŭn-ûrth′] *verb*, to dig up

un-eas-y [ŭn-ē′zē] *adjective*, a little
afraid, worried, uncomfortable

un-em-ployed [ŭn′ĕm-ploid′]
adjective, having no paid work

un-equal [ŭn-ē′kwəl] *adjective*,
unmatched, not even

un-e-quiv-o-cal [ŭn′ĭ-kwĭv′ə-kəl]
adjective, plain, obvious

un-err-ing-ly [ŭn-ûr′ĭng-lē]
adverb, correctly, precisely

un-e-ven [ŭn-ē′vən] *adjective*, not
level or flat, not uniform

un-ex-pect-ed [ŭn′ĭk-spĕk′tĭd]
adjective, sudden, coming
without warning, unforeseen

un-fair [ŭn-fâr′] *adjective*, unjust,
unethical, biased, dishonest

un-faltering [ŭn-fôl′tər-ĭng]
adjective, steadfast

un-fas-ten [ŭn-făs′ən] *verb*, to
undo, to detach, to separate

un-fit [ŭn-fĭt′] *adjective*, made
unsuitable, not prepared or
adapted, not competent

un-fold [ŭn-fōld′] *verb*, 1) to open
out 2) to explain, to evolve

un-for-tu-nate [ŭn-fôr′chə-nĭt]
adjective, unlucky, regrettable

un-for-tu-nate-ly
[ŭn-fôr′chə-nĭt-lē] *adverb*,
having bad luck, unsuccessful

un-friend-ly [ŭn-frĕnd′lē]
adjective, not nice, unkind

un-grate-ful [ŭn-grāt′fəl]
adjective, thankless

un-hap-py [ŭn-hăp′ē] *adjective*,
dissatisfied, not content, sad

un-health-y [ŭn-hĕl′thē] *adjective*,
1) ill, sickly 2) food or habits
that make a person feel tired

u-ni-corn [yōō′nə- kôrn] *noun,* a
mythical horse-like animal with
a horn growing from the center
of its forehead

u-ni-form [yōō′nə-fôrm′]
adjective, 1) unvaried, regular,
constant, *noun*, 2) clothes worn
for a special job or school

u-ni-form-i-ty [yōō′nə-fôrm′ĭ-tē]
noun, sameness, consistency,
identical with others

u-ni-fy [yōō′nə-fī′] *verb*, to
combine, to integrate, to connect

u-ni-lat-er-al [yōō′nə-lăt′ər-əl]
adjective, one sided

un-im-ag-i-na-tive
[ŭn-ĭ-măj′ə-nə-tĭv] *adjective*,
uninspired, common, everyday

un-impeach-a-ble
[ŭn′ĭm-pē′chə-bəl] *adjective*,
blameless and exemplary

un-im-por-tant [ŭn′ĭm-pôr′tnt]
adjective, 1) trivial, trite
2) extraneous, irrelevant

un-in-hib-it-ed [ŭn′ĭn-hĭb′ĭ-tĭd]
adjective, not restrained by
convention, free and easy

un-in-tel-li-gi-ble
[ŭn′ĭn-tĕl′ĭ-jə-bəl] *adjective*, not
capable of being understood

un-ion [yōōn′yən] *noun*, 1) a
coming together 2) a group of
people who join to support a
cause 3) a combination, **Union**
4) the army fighting for northern
states during the Civil War

u-nique [yōō-nēk′] *adjective*,
single in kind or excellence

u-ni-son [yōō′nĭ-sən] *noun*, unity
of pitch, complete accord

u-nit [yōō′nĭt] *noun*, 1) a complete
set 2) a specific amount, a form of
measurement 3) a single entity

u-nite [yōō-nīt′] *verb*, to join, to
connect, to put together as one

United States
[yōō-nī′tĭd-stāts-ŭv-ə-měr′ĭ-kə]
noun, the country made up of the
land between the Pacific Ocean
and the Atlantic Ocean, between
Canada and Mexico, including
Alaska and Hawaii, the capital,
is the District of Columbia

u-ni-ver-sal [yōō′nə-vûr′səl]
adjective, mutual understanding,
totally accepted everywhere

u-ni-verse [yōō′nə-vûrs′] *noun*, all
existing things including the
earth and all of space

u-ni-ver-si-ty [yōō′nə-vûr′sĭ-tē]
noun, an institution of higher
learning empowered to confer
degrees in the arts and sciences

un-just [ŭn-jŭst′] *adjective*, unfair,
biased, unworthy

un-kempt [ŭn-kĕmpt′] *adjective*, a
messy, uncared for appearance

un-kind [ŭn-kīnd′] *adjective*,
indifferent, unfeeling, rude

un-known [ŭn-nōn′] *adjective*,
1) unfamiliar 2) unexplored

un-less [ŭn-lĕs′] *adjective*, under
different conditions, or else

un-like [ŭn-līk′] *adjective*,
dissimilar, different

un-load [ŭn-lōd′] *verb*, 1) to take
something off a vehicle such as
cargo 2) to dispose of

un-luck-y [ŭn-lŭk′ē] *adjective*,
unsuccessful, unfortunate

un-mis-tak-a-ble [ŭn′mĭ-stā′kə-
bəl] *adjective*, clear, obvious

un-mis-take-a-bly
[ŭn′mĭ-stā′kə-blē] *adverb*,
undoubtedly, obviously

un-nec-es-sar-y [ŭn-nĕs′ĭ-sĕr′ē]
adjective, uncalled for, needless

un-ob-tru-sive [ŭn′əb-trōō′sĭv]
adjective, not noticeable

un-pleas-ant [ŭn-plĕz′ənt]
adjective, disagreeable

un-prec-e-dent-ed
[ŭn-prĕs′ĭ-dĕn′tĭd] *adjective*,
novel, unparalleled

un-pre-pare-ed [ŭn′prĭ-pârd′]
adjective, unaware, not ready

un-rav-el [ŭn-răv′əl] *verb*, to
untangle, to unwind, to unfold

un-rea-son-a-ble [ŭn-rē′zə-nə-bəl]
adjective, absurd, illogical, not
guided by sound judgment

un-re-li-a-ble [ŭn′rĭ-lī′ə-bəl]
adjective, not dependable

un-ru-ly [ŭn-rōō′lē] *adjective*,
disobedient, lawless

un-safe [ŭn-sāf′] *adjective*,
hazardous, dangerous, risky

un-sat-is-fac-to-ry
[ŭn-săt′ĭs-făk′tə-rē] *adjective*,
not good enough, inadequate

un-scru-pu-lous
[ŭn-skrōō′pyə-ləs] *adjective*,
unprincipled, devious, deceptive

un-seem-ly [ŭn-sēm′lē] *adjective*,
unbecoming, inappropriate

un-so-phis-ti-cat-ed
[ŭn′sə-fĭs′tĭ-kā′tĭd] *adjective*,
innocent, showing inexperience

un-stead-y [ŭn-stĕd′ē] *adjective*,
1) not safe 2) unsure, wavering

un-suit-a-ble [ŭn-sōō′tə-bəl]
adjective, inappropriate,
unsatisfactory, something that
does not fit in, irregular

un-sure [ŭn-sho͞or] *adjective*,
uncertain, doubtful

un-ten-a-ble [ŭn-tĕn′ə-bəl]
adjective, not able to be
maintained or supported against
criticism

un-ti-dy [ŭn-tī′dē] *adjective*,
messy, cluttered, not orderly

un-tie [ŭn-tī′] *verb*, to unfasten a
string or rope, to loosen

un-til [ŭn-tĭl′] *conjunction*, 1) up to
the time that, or when,
preposition, 2) up to the time of,
as far as, before

un-to-ward [ŭn-tôrd′] *adjective*,
unfortunate, annoying

un-true [ŭn-tro͞o′] *adjective*, false,
incorrect, deceptive, disloyal

un-u-su-al [ŭn-yo͞o′zho͞o-əl]
adjective, strange, odd

un-veil [ŭn-vāl′] *verb*, to remove
the covering from something

un-well [ŭn-wĕl′] *adjective*, ill,
sick, upset, ailing

un-wield-y [ŭn-wēl′dē] *adjective*,
not easily carried, bulky

un-will-ing [ŭn-wĭl′ĭng] *adjective*,
reluctant, resistant, averse

un-wind [ŭn-wīnd′] *verb*, 1) to
undo something, to loosen 2) to
relax, to get rid of tension

un-wise [ŭn-wīz′] *adjective*, not
reasonable or sensible

un-wit-ting [ŭn-wĭt′ĭng] *adjective*,
unintentional, not knowing

un-wrap [ŭn-răp′] *verb*, to remove
the protective covering on a
package or product

up [ŭp] *adverb*, 1) to put in a
higher place, *preposition*, 2) to

position something so that it is
standing 3) toward

up-braid [ŭp-brād′] *verb*, to scold,
to reproach, to criticize

up-heav-al [ŭp-hē′vəl] *noun*, an
eruption, a change, a quake

up-hold [ŭp-hōld′] *verb*, to
maintain, to support, to sustain

up-hol-ster-y [ŭp-hōl′stə-rē] *noun*,
furniture with stuffing, springs,
covers or trim

up-lift [ŭp-lĭft′] *noun*, 1) upgrade,
restoration, *verb*, 2) to raise up
3) to improve socially or morally

up-on [ə-pŏn′] *adverb*, 1) on,
preposition, 2) on top of
something

up-per [ŭp′ər] *adjective*, 1) in a
higher position, further up 2) the
part of the shoe above the sole

up-right [ŭp′rīt′] *adjective*,
1) straight up and down
2) righteous, honest, honorable

up-set [ŭp-sĕt′] *verb*, 1) to knock
over 2) to make unhappy or
worried 3) to spoil something
that was planned, to disrupt

up-stairs [ŭp′stârz′] *adverb*, 1) up
the steps, *noun*, *plural*, 2) the
next level, the upper level

up-ward [ŭp′wərd] *adjective*, to a
higher place or position

u-ra-ni-um [yo͞o-rā′nē-əm] *noun*, a
rare radioactive, metallic element
found in certain minerals used in
nuclear weapons

ur-ban [ûr′bən] *noun*,
characteristic of a city

ur-bane [ûr-bān′] *adjective*, suave,
refined, elegant, sophisticated,
charming, courteous, polite

ur-chin [ûr′chĭn] *noun*, a needy or mischievous child

urge [ûrj] *noun*, 1) a strong wish, *verb*, 2) to try to make someone do something, to push along

ur-gen-cy [ûr′jən-sē] *adverb*, calling for quick action

ur-gent [ûr′jənt] *adjective*, pressing, calling for immediate attention or action, insistent

us [ŭs] *pronoun*, the person who is speaking, and the others mentioned in the sentence, the objective case of **we**

us-age [yōō′sĭj] *noun*, a long continued practice, a custom

use [yōōz] *noun*, 1) a purpose, *verb*, 2) to find a purpose for something, to utilize

used to [yōōst-tōō] *adjective*, 1) knowing what something or someone is like, familiar 2) something done in the past but is not done now

use-ful [yōōs′fəl] *adjective*, helpful, serviceable

use-less [yōōs′lĭs] *adjective*, hopeless, valueless, vain

u-su-al [yōō′zhōō-əl] *adjective*, something that happens regularly, not uncommon

u-su-al-ly [yōō′zhōō-əl-lē] *adverb*, almost always, frequently

u-surp [yōō-sûrp′] *verb*, to seize, to take possession of by force or without right or authority

u-sur-pa-tion [yōō′sər-pā′shən] *noun*, the act of seizing power and rank of another

u-su-ry [yōō′zhə-rē] *noun*, interest in excess of a legal rate

Utah [yōō′tô′] *noun*, one of the 50 United States, the capital is Salt Lake City. The state flower of **Utah** is the sego lily; and the motto is "Industry."

u-ten-sil [yōō-tĕn′səl] *noun*, a tool, a dining instrument

u-til-i-ty [yōō-tĭl′ĭ-tē] *noun*, a company that provides products such as gas, water, or electricity

u-ti-lize [yōōt′l-īz′] *verb*, to make use of something

ut-most [ŭt′mōst′] *adjective*, the furthest or most extreme

u-to-pi-a [yōō-tō′pē-ə] *noun*, heaven on earth, an ideal place

ut-ter [ŭt′ər] *verb*, to speak

ut-ter-ance [ŭt′ər-əns] *noun*, vocal expression, articulation, speech

ut-ter-ly [ŭt′ər-lē] *adverb*, completely, entirely, without doubt, totally, absolutely

V

va-can-cy [vā′kən-sē] *noun*, 1) emptiness 2) an unfilled place, job or position 3) an opening

va-cant [vā′kənt] *adjective*, empty

va-cate [vā′kāt′] *verb*, 1) to leave, to empty, to abandon 2) to annul

va-ca-tion [vā-kā′shən] *noun*, time away from work for rest, amusement, and relaxation

vac-ci-nate [văk′sə-nāt′] *verb*, to inoculate with a vaccine as a form of protection

vac-ci-na-tion [văk′sə-nā′shən] *noun*, a remedy which is the injection of the virus itself

vac-cine [văk-sēn] *noun*, a weakened germ of a virus used to protect a person from the often fatal or debilitating effects of a virus

vac-il-late [văs′ə-lāt′] *verb*, to change one's mind often, to waffle, to waver

vac-u-um [văk′yōō-əm] *noun*, 1) the absolute absence of matter 2) an appliance used to rid carpets of dust and litter 3) a kind of bottle that keeps liquids hot or cold, such as a thermos

vag-a-bond [văg′ə-bŏnd′] *noun*, a tramp, an outcast, a vagrant

va-grant [vā′grənt] *adjective*, 1) wandering, begging, *noun*, 2) one who wanders around outside from place to place

vague [vāg] *adjective*, not clearly understood, indistinct

vain [vān] *adjective*, 1) conceited, too proud of the way one looks 2) fruitless, unsuccessful

vale [vāl] *noun*, valley, lowland

val-en-tine [văl′ən-tīn′] *noun*, a love poem, a card expressing love and affection

val-et [văl′ĭt] *noun*, a manservant who takes care of the clothes and personal needs of his employer

val-iant [văl′yənt] *adjective*, full of courage, brave

val-iant-ly [văl′yənt-lē] *adverb*, unafraid of danger, boldly, courageously, bravely

val-id [văl′ĭd] *adjective*, 1) significant, sound 2) legally effective or binding

val-i-date [văl′ĭ-dāt′] *verb*, to confirm, to ratify, to substantiate

va-lid-i-ty [vă-lĭ′dĭ-tē] *noun*, confirmation that proves something is what it says it is

val-ley [văl′ē] *noun*, a region of low land situated between hills or mountains, a dale

val-or [văl′ər] *noun*, bravery, courage, boldness

val-u-a-ble [văl′yōō-ə-bəl] *adjective*, precious, of considerable worth, costly

val-ue [văl′yōō] *noun*, 1) what something is worth, what someone will pay, *verb*, 2) to think that something is important and meaningful 3) to say how much something is worth on the market, to rate

valve [vălv] *noun*, any device by which the flow of liquid, air, or gas may be regulated

van [văn] *noun*, a small, covered truck carrying goods or passengers or animals

van-dal-ism [văn′dl-ĭz′əm] *noun*, willful destruction of property

vane [vān] *noun*, a piece of metal or cloth set up high to swing with the wind to tell which way the wind is blowing

va-nil-la [və-nĭl′ə] *noun*, a flavoring made from the beans of the vanilla plant

van-ish [văn′ĭsh] *verb*, to disappear, to dissolve, to end

van-i-ty [văn′ĭ-tē] *noun*, 1) too proud of oneself 2) a small table for putting on makeup

van-quish [văng′kwĭsh] *verb*, to conquer, to overcome, to overpower in battle

van-tage [văn′tĭj] *noun*, the position giving an advantage

va-por [vā′pər] *noun*, 1) condensation 2) a substance in the gaseous state such as fog

var-i-a-ble [vâr′ē-ə-bəl] *noun*, changeable, liable to vary

var-i-ance [vâr′ē-əns] *noun*, degree of difference, disagreement

var-i-a-tion [vâr′ē-ā′shən] *noun*, a change in form, position, or condition

vari-ed [vâr′ēd] *adjective*, changing, various, mixed

va-ri-e-ty [və-rī′ĭ-tē] *noun*, many different things, assortment

var-i-ous [vâr′ē-əs] *adjective*, , several different kinds, diverse

var-nish [vär′nĭsh] *noun*, 1) a liquid that hardens into a clear, tough film, *verb*, 2) to paint with a clear protective sealant

var-si-ty [vär′sĭ-tē] *adjective*, a principal team in sports representing a school or college

var-y [vâr′ē] *verb*, to change

vase [vās] *noun*, a decorated pot for cut flowers or for decoration

vas-sal [văs′əl] *noun*, 1) in feudalism, one who held land of a superior lord 2) one subordinate to another

vast [văst] *adjective*, limitless, expansive, very great in size, quantity or degree, widespread

vat [văt] *noun*, a large container for liquids, a barrel

vaude-ville [vôd′vĭl′] *noun*, a theatrical performance with music, songs, and dance

vault [vôlt] *noun*, 1) a room constructed to protect its contents, *verb*, 2) to jump or leap over with hands holding a pole

veer [vîr] *verb*, to change in direction or position

veg-e-ta-ble [věj′tə-bəl] *noun*, an edible plant grown for food

veg-e-tate [věj′ĭ-tāt′] *verb*, to live in a monotonous, inactive way

veg-e-ta-tion [věj′ĭ-tā′shən] *noun*, plant growth, foliage, flowers

ve-he-ment [vē′ə-mənt] *adjective*, fervent, impassioned, fiery

ve-hi-cle [vē′ĭ-kəl] *noun*, something used to carry or transport people or things

veil [vāl] *noun*, a piece of transparent material worn to hide or protect the face, a cover

vein [vān] *noun*, 1) one of the vessels which conveys blood back to the heart 2) a crack in a rock filled with mineral matter

vel-lum [věl′əm] *noun*, a form of paper made out of animal skin that will hold ink

ve-loc-i-ty [və-lŏs′ĭ-tē] *noun*, speed, rapidity of motion

vel-vet [věl′vĭt] *noun*, *no plural*, a type of cloth with a soft surface

ven-det-ta [věn-dět′ə] *noun*, revenge for a dispute

ven-dor [věn′dər] *noun*, someone who sells something, a merchant

ve-neer [və-nîr′] *noun*, 1) a thin layer of finished wood, glued on a cheap surface 2) a superficial show of quality

ven-er-a-ble [věn′ər-ə-bəl] *adjective*, deserving high respect

ven·er·a·tion [vĕn′ə-rā′shən] *noun*, deep, respectful love, adoration

venge·ance [vĕn′jəns] *noun*, retribution, revenge, reprisal

ve·ni·al [vē′nē-əl] *adjective*, capable of being forgiven

ven·i·son [vĕn′ĭ-sən] *noun*, deer meat used for food

venn di·a·gram [vĕn-dī′ə-grăm′] *noun*, a tool to indicate the relationship of elements

ven·om [vĕn′əm] *noun*, poison stored in some animals

ven·om·ous [vĕn′ə-məs] *adjective*, poisonous, deadly, toxic

vent [vĕnt] *noun*, a small opening or outlet to let out air

ven·ti·late [vĕn′tl-āt′] *verb*, to cause fresh air to circulate through, to admit air into

ven·tril·o·quist [vĕn-trĭl′ə-kwĭst] *noun*, one who can make a voice seem to come from another person or thing

ven·ture [vĕn′chər] *noun*, 1) doing something that involves the risk of a loss *verb*, 2) to dare to go, to do or to say

ven·tur·ous [vĕn′chər-əs] *adjective*, daring, bold

ven·ue [vĕn′yōō] *noun*, an agreed place where something will take place, the scene of an event

ve·rac·i·ty [və-răs′ĭ-tē] *noun*, habitual observance of truth, truthfulness, accuracy

ve·ran·da [və-răn′də] *noun*, a long porch with a roof along the outside of a building

verb [vûrb] *noun*, the word expressing action in a sentence

ver·ba·tim [vər-bā′tĭm] *adverb*, word for word, literal

ver·bi·age [vûr′bē-ĭj] *noun*, a pompous array of words

ver·dant [vûr′dnt] *adjective*, green with vegetation

ver·dict [vûr′dĭkt] *noun*, the decision of a jury given to the court, a judgment or decision

verge [vûrj] *noun*, 1) the point beyond which something begins, the brink, *verb*, 2) to tend to incline

ver·i·fy [vĕr′ə-fī′] *verb*, to prove to be true, to confirm, to ascertain

ver·i·si·mil·i·tude [vĕr′ə-sĭ-mĭl′ĭ-tōōd′] *noun*, appearance of truth, likelihood

ver·i·ty [vĕr′ĭ-tē] *noun*, truth, reality, a statement that is true

ver·mi·cel·li [vûr′mĭ-chĕl′ē] *noun*, a pasta dried in a slender tube form, like spaghetti

ver·mic·u·lite [vər-mĭk′yə-līt′] *noun*, slivers of mica that expand when heated, used for insulation, and water absorption

Ver·mont [vər-mŏnt′] *noun*, one of the 50 United States located in New England, the capital is Montpelier. The state flower of **Vermont** is the red clover, the motto is "Freedom and unity."

ver·nal [vûr′nəl] *adjective*, of or occurring in spring

ver·sa·tile [vûr′sə-təl′] *adjective*, having the capability of doing many different things

verse [vûrs] *noun*, 1) lines of writing which have a rhythm often a rhyme 2) a few lines of this from a longer piece called a poem 3) a

small part of the Bible, the
Jewish Koran, or Talmud

ver-sion [vûr′zhən] *noun*, a
description from a particular
point of view, a standpoint

ver-sus [vûr′səs] *preposition*,
against, in contrast with

ver-te-brate [vûr′tə-brĭt] *noun*, an
animal with a backbone

ver-tex [vûr′tĕks′] *noun*, 1) the point
where two planes intersect to form
an angle 2) the highest point

ver-ti-cal [vûr′tĭ-kəl] *adjective*,
straight up and down,
perpendicular to the horizon

verve [vûrv] *noun*, enthusiasm,
liveliness, energy, vitality

ver-y [vĕr′ē] *adjective*, 1) the same
2) the right or appropriate one,
adverb, 3) to emphasize the
word to make it stronger

ves-sel [vĕs′əl] *noun*, 1) a hollow
utensil for holding anything 2) a
boat 3) a channel for delivering
blood, a vein or capillary

vest [vĕst] *noun*, a sleeveless top
worn over a shirt with pockets

ves-ti-bule [vĕs′tə-byōōl′] *noun*,
foyer, lobby, entry way

ves-tige [vĕs′tĭj] *noun*, a trace or
visible sign of something gone

vet-er-an [vĕt′ər-ən] *noun*,
experienced, old in service,
especially in military life

vet-er-i-nar-i-an
[vĕt′ər-ə-nâr′ē-ən] *noun*, a
doctor who specializes in animal
health and surgery

ve-to [vē′tō] *noun*, 1) a denial, a
rejection, *verb*, 2) to deny, to
reject, to turn down

vex [vĕks] *verb*, to annoy, to
distress, to irritate

vex-a-tion [vĕk-sā′shən] *noun*,
irritation, a cause of trouble or
disquiet, annoyance

vi-a [vī′ə, vē′ə] *preposition*,
traveling through, by way of

vi-a-ble [vī′ə-bəl] *adjective*,
capable of maintaining life

vi-a-duct [vī′ə-dŭkt′] *noun*, a
bridge for carrying a road,
resting on masonry arches, often
in water to allow boats through

vi-and [vī′ənd] *noun*, food, provision

vi-brate [vī′brāt′] *verb*, 1) to shake,
to quiver or tremble 2) to echo, to
reverberate, to move to and fro

vi-bra-tion [vī-brā′shən] *noun*,
shaking, quivering, pulsation

vi-car-i-ous-ly [vī-kâr′ē-əs-lē]
adverb, acting as a substitute,
done in place of another

vice [vīs] *noun*, corruption,
weakness, faults, foible

vice president [vīs′prĕz′ĭ-dənt]
noun, the person chosen to
replace the president

vi-cin-i-ty [vĭ-sĭn′ĭ-tē] *noun*, the
region near or close by

vi-cious [vĭsh′əs] *adjective*,
corrupt, wicked, mean

vi-cis-si-tude [vĭ-sĭs′ĭ-tōōd′] *noun*,
change or succession from one
thing to another

vic-tim [vĭk′tĭm] *noun*, 1) someone
who suffers from an illness or
action 2) a person who is
deceived or cheated

vic-to-ri-ous [vĭk-tôr′ē-əs]
adjective, conquering,
triumphant in war, winning

vic-to-ry [vĭk′tə-rē] *noun*, winning a contest or a battle, a triumph

vict-uals [vĭt′lz] *noun*, *plural*, food for human beings, provisions

vid-e-o [vĭd′ē-ō′] *noun*, film for showing on a television set

vie [vī] *verb*, to contend, to strive in competition, to struggle

view [vyo͞o] *noun*, 1) ideas, beliefs, opinion 2) something you see

vig-i-lance [vĭj′ə-ləns] *noun*, watchfulness, caution

vi-gnette [vĭn-yĕt′] *noun*, a scene depicting a moment in time

vig-or [vĭg′ər] *noun*, physical or mental strength, stamina, energy

vig-or-ous [vĭg′ər-əs] *adjective*, exhibiting strength, powerful, energetic, intense, robust

vig-or-ous-ly [vĭg′ər-əs-lē] *adverb*, very fast, with a great deal of energy, healthy, spirited

vile [vīl] *adjective*, disgusting

vil-i-fy [vĭl′ə-fī′] *verb*, to slander, to malign, to slur, to revile

vil-lage [vĭl′ĭj] *noun*, a group of houses that form a community, usually smaller than a town

vil-lain [vĭl′ən] *noun*, one capable or guilty of great crimes, a rascal

vin-di-cate [vĭn′dĭ-kāt′] *verb*, to support or maintain as true, to free from suspicion

vin-dic-tive [vĭn-dĭk′tĭv] *adjective*, revengeful, retaliatory, mean

vine [vīn] *noun*, a name given to some plants with climbing stems

vin-e-gar [vĭn′ĭ-gər] *noun*, a sour liquid made by fermentation of wine or cider

vi-o-late [vī′ə-lāt′] *verb*, 1) to break the rules, to disregard as a promise 2) to break in on

vi-o-lence [vī′ə-ləns] *noun*, commotion, disorder, outrage, brutality, intense force

vi-o-lent [vī′ə-lənt] *adjective*, brutal, vicious, forceful, rough

vi-o-lent-ly [vī′ə-lənt-lē] *adverb*, harshly, with destructive force

vi-o-let [vī′ə-lĭt]) *adjective*, 1) the color that is a combination of blue and red, *noun*, 2) a purple flower that grows wild as a ground cover

vi-o-lin [vī′ə-lĭn′] *noun*, one of the small stringed instruments played with a bow

vi-o-lin-ist [vī′ə-lĭn′ĭst] *noun*, a person who plays a violin

vi-per [vī′pər] *noun*, a poisonous snake, a serpent

vir-gin [vûr′jĭn] *adjective*, pure, chaste, unused

Vir-gin-i-a [vər-jĭn′yə] *noun*, one of the 50 United States located in the Southeast, the capital is Richmond. The state flower of **Virginia** is the flowering dogwood and the motto is "Ever thus to tyrants."

vir-ile [vîr′əl] *adjective*, manly, showing masculine strength

vir-tue [vûr′cho͞o] *noun*, uprightness, moral excellence

vir-tu-ous [vûr′cho͞o-əs] *adjective*, righteous, honest, moral

vi-rus [vī′rəs] *noun*, an group of infectious submicroscopic agents that cause various diseases in plants or animals

vis-age [vĭz′ĭj] *noun*, the face, countenance, or look of a person

vis-cer-al [vĭs′ər-əl] *adjective*, felt in one's inner organs

vis-cous [vĭs′kəs] *adjective*, thick and sticky in consistency

vise [vīs] *noun*, a clamp, clasp, grip

vis-i-ble [vĭz′ə-bəl] *adjective*, as seen by the eye, in view

vi-sion [vĭzh′ən] *noun*, 1) something imagined, as in a dream, 2) a look into the future, unusual foresight 3) eyesight

vi-sion-ar-y [vĭzh′ə-nĕr′ē] *noun*, existing only in the imagination

vis-it [vĭz′ĭt] *noun*, 1) a social call, an appointment, *verb*, 2) to go to see someone 3) to chat casually

vis-i-tor [vĭz′ĭ-tər] *noun*, one who comes and goes to see another

vi-sor [vī′zər] *noun*, a sunshade or shield to block out glare

vi-su-al [vĭzh′ōō-ə-l] *adjective,* of or used in seeing, pleasing to the sight

vi-su-al-ize [vĭzh′ōō-ə-līz′] *verb*, 1) to imagine 2) to see something in the mind

vi-tal [vīt′l] *adjective*, necessary to life, important, indispensable

vi-tal-i-ty [vī-tăl′ĭ-tē] *noun*, spirit, energy, the capacity to develop

vi-ta-min [vī′tə-mĭn] *noun*, nutrients that give the body nourishment to help balance the metabolism

vi-ti-ate [vĭsh′ē-āt′] *verb*, to spoil the effect of something, to make less effective or imperfect

vit-ri-ol-ic [vĭt′rē-ŏl′ĭk] *adjective*, mean, causing bad feelings

vi-tu-per-a-tive [vī-tōō′pər-ə-tĭv] *adjective*, abusive, scolding

vi-va-cious [vĭ-vā′shəs] *adjective*, lively, sprightly, energetic

viv-id [vĭv′ĭd] *adjective*, easy to see, life-like, bright, colorful

vo-cab-u-lar-y [vō-kăb′yə-lĕr′ē] *noun*, the words of a language

vo-cal [vō′kəl] *adjective*, 1) voiced, outspoken 2) expressed by singing 3) relating to the sound made by the voice

vo-ca-tion [vō-kā′shən] *noun*, a job, employment or occupation, a profession

vogue [vōg] *adjective*, fashionable, in style, popular

voice [vois] *noun*, 1) the sounds coming through a person's mouth made by the respiratory system 2) to express an idea in writing or speaking

void [void] *noun*, an empty space

vol-ca-no [vŏl-kā′nō] *noun*, an opening in the earth's crust where melted rock, lava, gases, and ash are thrust onto the land

vo-li-tion [və-lĭsh′ən] *noun*, the act of making a conscious choice

vol-ley [vŏl′ē] *noun*, 1) a barrage, a bombardment, *verb*, 2) to return a ball before it hits the ground

vol-ley-ball [vŏl′ē-bôl′] *noun, no plural*, a game in which a large ball is knocked back and forth across a net, by hand

volt [vōlt] *noun*, 1) a measure of electricity, the unit of electromotive force 2) a turning movement of a horse, sideways around a center

vol-ume [vŏl′yoōm] *noun*, 1) a single book, *no plural*, 2) the space something occupies 3) the amount of sound something makes, loudness of a sound

vo-lu-mi-nous [və-loō′mə-nəs] *adjective*, bulky, large

vol-un-tar-y [vŏl′ən-tĕr′ē] *adjective*, proceeding from the will, to give or perform freely

vol-un-teer [vŏl′ən-tîr′] *noun*, 1) one who offers his services of his own free will, *verb*, 2) to give freely of your time

vo-lup-tu-ous [və-lŭp′choō-əs] *adjective*, shapely, full in form

vom-it [vŏm′ĭt] *noun*, 1) something from the stomach that comes out of the mouth, *verb*, 2) to throw up, to regurgitate food

vo-ra-cious [vô-rā′shəs] *adjective*, 1) someone who has a large appetite 2) exceedingly eager

vor-tex [vôr′tĕks′] *noun*, whirlpool, a whirling mass of water

vote [vōt] *noun*, 1) a choice made during an election, *verb*, 2) to decide by general consent

vot-er [vōt′ər] *noun*, a person registered to participate in an election

vouch [vouch] *verb*, to certify, to assure, to provide proof

vouch-er [vou′chər] *noun*, a document exchangeable for certain goods or services

vow [vou] *noun*, 1) an oath, a solemn pledge, *verb*, 2) to promise something important

vow-el [vou′əl] *noun*, the letters – **a,e,i,o,u**, and sometimes **y**

voy-age [voi′ĭj] *noun*, a long journey by sea, water, air, or space from one place to another

vul-gar [vŭl′gər] *adjective*, rude or rough in behavior, taste, etc, lacking good taste or refinement

vul-ner-a-ble [vŭl′nər-ə-bəl] *adjective*, capable of being easily harmed or injured

vul-ture [vŭl′chər] *noun*, a large bird that eats as a scavenger of dead animals

W

wade [wād] *verb*, to walk through a substance such as water or snow

waf-fle [wŏf′əl] *noun*, 1) a cake baked in an iron skillet, *verb*, 2) to change one's mind often

waft [wäft] *verb*, to move or be moved lightly over water or air, to drift

wag [wăg] *verb*, to move or cause to move briskly from side to side or up and down

wage [wāj] *noun*, 1) money paid for work or services, salary, *verb*, 2) to carry on (a war or campaign)

wa-ger [wā′jər] *noun*, the amount of money or property risked in an uncertain event, a bet

wag-on [wăg′ən] *noun*, a vehicle with four wheels that is pulled

waif [wāf] *noun*, a homeless child or animal, an orphan

wail [wāl] *verb*, to make a long cry showing sadness or pain, to sob

waist [wāst] *noun*, the part of the body between the ribs and the hips where the stomach is

wait [wāt] *noun*, 1) a time of anticipation, *verb*, 2) to expect, to anticipate, to be ready

wait-er [wā′tər] *noun*, a person hired to serve meals

waive [wāv] *verb*, to give up temporarily, to yield

waiv-er [wā′vər] *noun*, the act of giving up a right or claim

wake [wāk] *noun*, 1) the ripple of waves created by the motion of a boat, *verb*, 2) to stop sleeping, to be aroused from sleep

walk [wôk] *noun*, 1) a journey on foot, *verb*, 2) to take one step at a time, to proceed along

wall [wôl] *noun*, 1) a partition made of bricks or stone that separates space 2) one of the sides of a room or building, *verb*, 3) to divide, to partition

wal-let [wŏl′ĭt] *noun*, a small flat case for money, cards, etc., usually carried in a pocket

wal-low [wŏl′ō] *verb*, to tumble in the mud 2) move slowly

wall-pa-per [wôl′pā′pər] *noun*, *no plural*, decorative paper used to cover walls of a room

wal-nut [wôl′nŭt′] *noun*, 1) a hard, dark-colored wood, valuable for furniture 2) the nut or seed from the walnut tree in a hard shell

wal-rus [wôl′rəs] *noun*, a sea animal like the sea lion and the seal, with outside ears and tusks

waltz [wôlts] *verb*, a dance with four beats per measure

wand [wŏnd] *noun*, a baton, a rod, a fairy's magic stick

wan-der [wŏn′dər] *verb*, 1) to travel from place to place with no plan or purpose in mind 2) to slip easily off the subject

wane [wān] *verb*, to grow gradually smaller or decrease in strength, to decline, to dwindle

wan-gle [wăng′gəl] *verb*, to bring about by manipulation or trickery, to scheme

want [wŏnt] *noun*, 1) a need, lack, not having something necessary, *verb*, 2) to desire or feel a need for something, to require

wan-ton [wŏn′tən] *adjective*, unruly, wild, excessive

war [wôr] *noun*, a deadly conflict

war-ble [wôr′bəl] *verb*, to sing melodiously, to trill as a bird

ward [wôrd] *noun*, 1) a child in the care of someone other than his family 2) a district, a division, *verb*, 3) to fend off, to avert

war-den [wôrd′n] *noun*, the chief administrator of a prison

ward-robe [wôr′drōb′] *noun*, a closet of clothes

ware [wâr] *noun*, a product, merchandise, stock

ware-house [wâr′hous′] *noun*, a storehouse for wares or goods

war-i-ly [wâr′ĭ-lē] *adverb*, in a watchful and suspicious way

warm [wôrm] *adjective*, 1) a temperature between hot and cold, moderately hot, *verb*, 2) to add heat to make something no longer cold

warmth [wôrmth] *noun*,
1) friendship, 2) bodily heat

warn [wôrn] *verb*, 1) to inform
someone that something bad will
happen or of impending danger
2) to admonish, to exhort

warn-ing [wôr′nĭng] *noun*, the act
of telling someone to proceed
with caution, notification

warp [wôrp] *verb*, 1) to twist out of
shape, to distort 2) to pervert

war-rant [wôr′ənt] *noun*, 1) a
written request by a judge to
make an arrest, *verb*, 2) to be
sure of something, to justify

war-ran-ty [wôr′ən-tē] *noun*, a
guarantee that a contract will be
carried out by a manufacturer

war-ri-or [wôr′ē-ər] *noun*, a man
engaged in war, a soldier

wart [wôrt] *noun*, a small growth
or lump that looks like a blemish
on skin often caused by a virus

war-y [wâr′ē] *adjective*, very
cautious, on guard, watchful

was [wŭz] *verb*, *past tense* of **be**,
used in the 3rd singular, **he**, **she**
or **it**

wash [wŏsh] *noun*, 1) a batch of
clothes that are to be or have
been washed, *verb*, 2) to clean
with soap and water

wash-er [wŏsh′ər] *noun*, 1) a
machine for washing clothes or
dishes 2) a round piece of metal
put under a bolt to reinforce it

Wash-ing-ton [wŏsh′ĭng-tən]
noun, one of the 50 United
States located in the Northwest,
the capital is Olympia. The state
flower of **Washington** is the
coast rhododendron and the
motto is "By and by."

wasn't [wŭz′ənt] the *contraction* of
the words **was** and **not**

wasp [wŏsp] *noun*, a flying insect
like a bee, the female inflicts the
sting

waste [wāst] *adjective*, 1) useless,
noun, 2) anything discarded that
is not considered useful, *verb*,
3) to throw away, to spend or
use carelessly, to squander

waste-ful [wāst′fəl] *adjective*,
squandering or spending in a
needless manner

waste-wa-ter [wāst′wô′tər] *noun*,
water that has been used either
to manufacture a product or in
the home, which requires
treatment and purification before
it can be used again

watch [wŏch] *noun*, 1) a small
clock worn on the wrist or hung
from a chain, *verb*, 2) to observe,
to look at attentively 3) to guard,
to be on the lookout

wa-ter [wô′tər] *noun*, 1) a
molecule made of one oxygen
atom and two hydrogen atoms,
that covers 75% of the earth's
surface, *verb*, 2) to moisten or
soak with water 3) to dilute

wa-ter-fall [wô′tər-fôl′] *noun*, a
cascade of water coming down
from a precipice or cliff

wa-ter-way [wô′tər-wā′] *noun*, a
body of water that ships can use

wa-ter-wheel [wô′tər-hwēl] *noun*,
a wheel turned by water running
against or falling on paddles or
steps, used as a source of power

watt [wŏt] *noun*, a measure of electrical power equivalent to one joule per second

wave [wāv] *noun*, 1) a breaker, a surge, surf, *verb*, 2) to greet

wa-ver [wā′vər] *verb*, 1) to be unsteady and show signs of falling or giving way 2) to be indecisive

wax [wăks] *noun*, a substance made from fats and oils that melts when heated

way [wā] *noun*, 1) the manner, custom, style 2) means, process 3) a passage, a road

way-ward [wā′wərd] *adjective*, unruly, disobedient, willful

we [wē] *plural*, *pronoun*, referring to one or more people, including oneself and others

weak [wēk] *adjective*, not strong, fragile, feeble, lacking strength

weak-ness [wēk′nĭs] *noun*, inclination, tendency

wealth [wĕlth] *noun*, riches, abundance of property

wealth-y [wĕl′thē] *adjective*, rich

wean [wēn] *verb*, 1) to accustom a baby not to nurse 2) to give up a cherished activity or habit

weap-on [wĕp′ən] *noun*, an instrument used for attack or defense in combat

wear [wâr] *noun*, 1) the diminished value caused by use, *verb*, 2) to put on, to carry on the body or over the body 3) to exhibit a certain expression on the face 4) to diminish through friction 5) to cause someone to be weary or exhausted 6) to impair by use

wear-y [wîr′ē] *adjective*, 1) tired, fatigued, physically or mentally exhausted 2) bored or resigned

weath-er [wĕth′ər] *noun*, 1) the condition of the atmosphere with regard to temperature, moisture, wind, etc., *verb*, 2) to endure the elements and the conditions outside in the atmosphere

weath-er-ing [wĕth′ər-ĭng] *noun*, the effects of the elements such as wind and water on rock surfaces, and homes or buildings

weave [wēv] *verb*, 1) to interlace threads into cloth 2) to wind or curve in and out

web [wĕb] *noun*, an entanglement, a mesh, an intricate network

web-li-og-ra-phy [wĕb′lē-ŏg′rə-fē] a bibliography of Internet sites used for research

we'd [wēd] the *contraction* of the words **we** and **would**

wed [wĕd] *verb*, to join in marriage

wed-ding [wĕd′ĭng] *noun*, the ceremony when two people are married

wedge [wĕj] *noun*, 1) a flat edge used to split objects or to lift them up 2) a chunk or a block of something, *verb*, 3) to force something to split apart 4) to press in closely, to cramp

Wednes-day [wĕnz′dē, -dā′] *noun*, the fourth day of the week

weed [wēd] *noun*, 1) an unwanted or unwelcome plant, *verb*, 2) to remove as being undesirable

week [wēk] *noun*, the seven days between Sunday and Saturday

weep [wēp] *verb*, 1) to cry, to shed tears 2) to grieve, to mourn 3) to exude liquid

weigh [wāt] *verb*, to measure the heaviness of someone or something by use of a scale

weight [wāt] *noun*, 1) the force with which a body is attracted to the earth or other celestial bodies 2) how heavy a thing is

weight-less [wāt′lĭs] *adjective*, having little or no weight

weird [wîrd] *adjective*, strange

wel-come [wĕl′kəm] *verb*, to receive gladly or hospitably

weld [wĕld] *verb*, 1) to join or connect by fusing two metals with heat 2) to unite

wel-fare [wĕl′fâr′] *noun*, assistance from the government, a state of well-being, health, happiness

well [wĕl] *adjective*, 1) healthy, strong, fit, *adverb*, 2) in a thoroughly manner, *noun*, 3) a narrow, deep hole made in the ground to reach water, oil, or natural gas, *verb*, 4) to rise to the surface and flow out

welt [wĕlt] *noun*, a large blister caused by heat or a beating

went [wĕnt] past participle of **go**

were [wûr] *past tense* of the *verb* **to be** used with **we, you** and **they**

west [wĕst] *noun*, the direction in which the sunsets

west-ern [wĕs′tərn] *adjective*, of or typical of Europe and the Americas

West Vir-gin-i-a [wĕst- vər-jĭn′yə] *noun*, one of the 50 United States, located in the Southeast, the capital is Charleston. The state flower of **West Virginia** is the great rhododendron, and the motto is "Mountaineers are always free."

wet [wĕt] *adjective*, 1) covered with water, moistened, *verb*, 2) to soak with water

wet-lands [wĕt′lăndz′] *noun*, *plural*, a habitat of swamps and marshes for wildlife

whale [hwāl] *noun*, the largest animal in the sea, it is not a fish, it feeds its young with milk

wharf [hwôrf] *noun*, a dock along a shore where boats moor and wait to load or unload

what [hwŏt, hwŭt, wôt] *adjective*, 1) used to ask about the identity of a person, object, or matter 2) how remarkable, *adverb*, 3) how, in what way? *pronoun*, 4) which thing, an event, etc. 5) used to ask a question

what-ev-er [hwŏt-ĕv′ər] *pronoun*, anything that, so what

what's [hwŏts] the *contraction* of the words **what** and **is**

wheat [hwēt] *noun*, one of the cereal plants used to make flour

whee-dle [hwēd′l] *verb*, to coax

wheel [hwēl] *noun*, 1) a circular frame or disk attached to an axis that rolls, *verb*, 2) to move on rollers or wheels, to turn

wheel-bar-row [hwēl′băr′ō] *noun*, a hand cart with three wheels

when [hwĕn] *adverb*, at what time

when-ev-er [hwĕn-ĕv′ər] *adjective*, any time, no matter when

where [hwâr] *adverb*, 1) to or toward a certain place 2) at or in what place

wher-ev-er [hwâr-ĕv′ər] *adjective*, at or any place at all

whet [hwĕt] *verb*, 1) to sharpen by friction 2) to stimulate, to stir

wheth-er [hwĕth′ər] *conjunction*, used to express two choices

which [hwĭch] *noun*, the one, that

while [hwīl] *noun*, a period of time

whim-per [hwĭm′pər] *verb*, to pout, to whine, to cry

whim-si-cal [hwĭm′zĭ-kəl] *adjective*, capricious, given to fanciful notions, playful

whine [hwīn] *verb*, 1) to make a high and sad sound 2) to complain in an annoying manner

whip [hwĭp] *noun*, 1) an instrument made with a handle attached to a long lash made of leather or rope used to hit animals, *verb*, 2) to strike with a lash, to flog 3) to beat to a froth 4) to defeat

whirl [hwûrl] *verb*, to move or make something move around and around very fast

whisk [hwĭsk] *noun*, 1) a cooking utensil used to mix ingredients, *verb*, 2) to move quickly

whisk-er [hwĭs′kər] *noun*, 1) hair growing on the sides of a man's face 2) one of the long stiff hairs that grow near the mouth of dogs, cats, rats, etc.

whis-per [hwĭs′pər] *noun*, 1) a murmur, *verb*, 2) to speak very softly, to murmur quietly

whis-tle [hwĭs′əl] *noun*, 1) an instrument that makes a shrill, musical sound, *verb*, 2) to make a clear, high-pitched sound by forcing breath through the teeth or by pursing the lips, or using a device

white [hwīt] *adjective*, the color of snow, the absence of pigment

whit-tle [hwĭt′l] *verb*, 1) to shape by carving or sculpting 2) to reduce gradually, to diminish

who [hoō] *pronoun*, 1) the person or people 2) that 3) what persons?

who-ev-er [hoō-ĕv′ər] *pronoun*, any one that

whole [hōl] *adjective*, 1) total, complete, *noun*, 2) the total amount of a thing or entity

whole-some [hōl′səm] *adjective*, sound, healthy, beneficial

whom [hoōm] *pronoun*, objective case for **who**

whose [hoōz] *pronoun*, of whom, possessive case of who

why [hwī] *adverb*, 1) for what reason or purpose, *noun*, 2) the cause or interaction

wick [wĭk] *noun*, the string in a candle or oil lamp that burns the flame, the fuse

wick-ed [wĭk′ĭd] *adjective*, evil, very bad, sinful, vicious

wide [wīd] *adjective*, 1) broad, expansive 2) to be fully open or accessible, of great scope

wid-ow [wĭd′ō] *noun*, a woman whose husband is dead

wid-ow-er [wĭd′ō-ər] *noun*, a man whose wife is dead

width [wĭdth] *noun*, the distance across, the breadth

wield [wēld] *verb*, to control, to sway, to manage, to use

wife [wīf] *noun*, *plural*, **wives**, a married woman

wig [wĭg] *noun*, a covering for the head, hair from other people or animals, a toupee, a hairpiece

wig-gle [wĭg′əl] *verb*, to move around, to squirm

wild [wīld] *adjective*, untamed or domesticated, a natural state

wil-der-ness [wĭl′dər-nĭs] *noun*, a tract of land undeveloped and uninhabited by people

wild-life [wīld′līf′] *noun*, any species of animals living in their natural habitat

will [wĭl] *noun*, 1) power in the mind or character, what we want to do 2) a declaration by a person concerning their estate, *verb*, 3) used with other verbs to show something is going to happen 4) used in questions when asking to do something

will-ful [wĭl′fəl] *adjective*, 1) determined, intentional, deliberate 2) obstinate, stubborn

will-ing [wĭl′ĭng] *adjective*, ready, given or done gladly

wil-low [wĭl′ō] *noun*, a tree with long drooping branches

wilt [wĭlt] *verb*, 1) to wither, to lose freshness 2) to lose strength

wil-y [wī′lē] *adjective*, sly, clever

win [wĭn] *verb*, 1) to achieve victory over others 2) to receive an award for a performance

wind [wĭnd] *noun*, 1) currents of air moving quickly with force, [wīnd] *verb*, 2) to wrap around

over and over 3) to turn round and round, to move in curves and bends, not a straight line

win-dow [wĭn′dō] *noun*, 1) a pane of glass 2) an opening with a transparent cover that allows light in

win-dow-sill [wĭn′dō-sĭl′] *noun*, a shelf or ledge below a window

wine [wīn] *noun*, an alcoholic drink made from fermented grapes

wing [wĭng] *noun*, 1) one of the two limbs of a bird or insect with which it flies 2) a side projection of an airplane 3) an extension of a building

wink [wĭngk] *noun*, 1) one eye closed and one eye open as a gesture of approval, *verb*, 2) to close and open one eye quickly

win-now [wĭn′ō] *verb*, to sift, to separate good parts from bad

win-some [wĭn′səm] *adjective*, agreeable, gracious, engaging

win-ter [wĭn′tər] *noun*, the first season of the year, between autumn and spring, when the earth is farthest from the sun

wipe [wīp] *verb*, to make dry or clean with a cloth

wire [wīr] *noun*, 1) a metal thread or strand, *verb*, 2) to connect

Wis-con-sin [wĭs-kŏn′sĭn] *noun*, one of the 50 United States located in the North,, the capital city is Madison. The state flower of **Wisconsin** is the violet, and the motto is "Forward."

wis-dom [wĭz′dəm] *noun*, learning and the capacity to use it

wise [wīz] *adjective*, showing good judgment, intelligent

357

wish [wĭsh] *noun*, 1) a desire that is beyond your reach, *verb*, 2) to desire, to long for, to want

wist-ful [wĭst′fəl] *adjective*, hopeful, yearning, longing

wit [wĭt] *noun*, cleverness, quickness of mind, ingenuity

witch [wĭch] *noun*, a woman who is believed to have magic powers

with [wĭth] *preposition*, 1) in the company of 2) using it 3) having 4) because of 5) next to

with-draw [wĭth-drô′] *verb*, to retreat, to go back on or to take back an offer, to remove

with-drawn [wĭth-drôn′] *adjective*, 1) taken back, removed 2) shy, introverted, retiring

with-er [wĭth′ər] *verb*, to shrivel, to fade, to lose freshness

with-hold [wĭth-hōld′] *verb*, to restrain, to hold back

with-in [wĭth-ĭn′] *adverb*, 1)inside, *preposition*, 2) in, not beyond

with-out [wĭth-out′] *adverb*, 1) outside, *preposition*, 2) lacking, wanting, not having

wit-ness [wĭt′nĭs] *noun*, 1) a spectator, an onlooker, a person who sees something, *verb*, 2) to see, to observe, to hear

wiz-ard [wĭz′ərd] *noun*, someone who performs sorcery and magic

woe [wō] *noun*, self-pity, sadness

wolf [wŏŏlf] *noun*, *plural*, **wolves**, a wild dog similar to a coyote

wom-an [wŏŏm′ən] *noun*, *plural*, **women**, a female, a lady

won-der [wŭn′dər] *noun*, 1) a feeling of surprise and admiration, amazement, *verb*, 2) to express a wish to know

won-der-ful [wŭn′dər-fəl] *adjective*, extra special, very good, remarkable, marvelous

won-drous [wŭn′drəs] *adjective*, amazing, astonishing, marvelous

wont [wônt] *adjective,* 1) apt or likely, *noun,* 2) custom, habitual procedure

won't [wōnt] the *contraction* of the words **will** and **not**

wood [wŏŏd] *noun*, 1) the material which comes from trees 2) a small forest

wood-land [wŏŏd′lənd] *noun*, the forest, land covered with trees

wood-peck-er [wŏŏd′pĕk′ər] *noun*, a black and white bird with a red head that pecks at wood with its bill to find bugs to eat

wood-winds [wŏŏd′wĭndz′] *noun*, musical instruments, usually made of wood or metal, that you blow into or use a reed to play such as a clarinet

wool [wŏŏl] *noun*, the, soft, thick fur on a sheep often used for yarn or fabric

word [wûrd] *noun*, 1) speech, sound or symbol used to communicate 2) a message 3) a promise 4) news, information

work [wûrk] *noun*, 1) a job, a form of labor, a duty 2) a deed, an accomplishment, *verb*, 3) to do a job, to labor, to perform a duty

world [wûrld] *noun*, the Earth and its people and all living things

world-ly [wûrld′lē] *adjective*, known all over, cosmopolitan

worm [wûrm] *noun*, a thin creature without bones or legs that lives in the earth

wor-ry [wûr′ē] *verb*, to fret, to show concern, to feel anxious

worse [wûrs] *adjective*, to go from bad to a more damaged condition, inferior, less good or well

wor-ship [wûr′shĭp] *noun*, 1) ardent devotion, *verb*, 2) to pray to or show reverence or respect for someone

worth [wûrth] *noun*, the value or importance of something

worth-less [wûrth′lĭs] *adjective*, without worth, valueless

wor-thy [wûr′thē] *adjective*, admirable, deserving, noble

would [wŏŏd] future perfect tense of the *verb* **be**

wouldn't [wŏŏd′nt] the *contraction* of the words **would** and **not**

wound [wōŏnd] *noun*, 1) an injury, *verb*, 2) to injure, to hurt 3) to disturb or upset someone

wrap [răp] *noun*, 1) a removable covering, *verb*, 2) to cover with a substance usually to protect something, to enclose

wrath [răth] *noun*, anger, fury

wreak [rēk] *verb*, to produce rage, to perpetuate, to inflict

wreath [rēth] *noun*, a ring of flowers and leaves

wreck [rĕk] *noun*, 1) the remains of something partially destroyed like a car, building, or ship, *verb*, 2) to destroy or damage something, to tear down

wren [rĕn] *noun*, a brown-grey bird that lives in the south

wrench [rĕnch] *noun*, 1) a tool used to grip a bolt or a nut 2) a forcible twist, *verb*, 3) to pull or turn suddenly with a violent twist, to damage or injure

wrest [rĕst] *verb*, 1) to pull away 2) to take away by force, to seize, to wrench away

wres-tle [rĕs′əl] *verb*, to struggle hand to hand, to grapple

wretch [rĕch] *noun*, a vile, disgusting person

wrig-gle [rĭg′əl] *verb*, to squirm or twist like a worm, to writhe

wring [rĭng] *verb*, to twist, to squeeze by twisting tightly

wrin-kle [rĭng′kəl] *noun*, 1) a crease or fold, *verb*, 2) to pucker into folds and creases

wrist [rĭst] *noun*, the joint between the hand and the forearm

write [rīt] *verb*, to use an instrument such as a pencil to draw symbols to communicate

writer [rī′tər] *noun*, author, journalist, newsperson

writhe [rīth] *verb*, to twist or move as if in pain, to thrash

wrong [rông*adjective*, 1) in error, mistaken, incorrect 2) contrary to morality or laws

wrote [rĭt′n] *past tense* of **write**

Wy-o-ming [wī-ō′mĭng] *noun*, one of the 50 United States, located in the Northwest, the capital is Cheyenne. The state flower of **Wyoming** is the indian paintbrush and the motto is "Equal rights."

X

xeno-phile [zĕn′ə-fĭl′] *noun*, one attracted to foreign people, manners, and styles

xeno-pho-bi-a [zĕn′ə-fō′bē-ə] *noun*, fear and hatred of foreigners or strangers

xy-lem [zī′ləm] *noun*, the system of a woody tissue that transports water in a plant

xy-log-ra-phy [zī-lŏg′rə-fē] *noun*, the art of engraving on wood

x-ray [ĕks′rā′] *noun*, 1) a photograph of a part of the inside of your body, *verb*, 2) to take a picture of the inside of the body

xy-lo-phone [zailəfoun] *noun*, a percussion instrument played by striking horizontal wooden bars with small wooden hammers

Y

yacht [yät] *noun*, a boat used either for private pleasure or as a vessel of state to convey people from one place to another

yam [yăm] *noun*, a sweet potato, the root of a potato plant

yard [yärd] *noun*, 1) a length of measure equal to 36 inches 2) the lawn or property around someone's house

yarn [yärn] *noun*, 1) knitting thread, spun wool 2) a story, a fictional account

yawn [yôn] *verb*, to open the mouth wide and breathe deeply as if tired or bored

year [yîr] *noun*, 365 days, or each cycle that the earth rotates around the sun divided, into 12 months, 52 weeks, and 4 seasons

yearn [yûrn] *verb*, to strive, to long for, to desire something

yeast [yēst] *noun*, a fungus which produces fermentation, used as a leavening agent in dough

yell [yĕl] *verb*, to shout, to scream

yel-low [yĕl′ō] *noun*, one of the primary colors, the color of corn

yen [yĕn] *noun*, a longing, an urge

yes [yĕs] *interjection*, a word of agreement, an affirmative reply

yes-ter-day [yĕs′tər-dā′] *adverb*, 1) on the day before today, *noun*, 2) the day before today

yet [yĕt] *conjunctionb*, until now

yield [yēld] *verb*, 1) to give way to, to submit, to surrender 2) to produce, to supply, to give forth

yo-gurt [yō′gərt] *noun*, a dairy product that is thicker than milk and has a slightly sour taste

yolk [yōk] *noun*, the yellow and principal part inside an egg

yon-der [yŏn′dər] *adjective*, distant, far-off, faraway

you [yo͞o] *pronoun*, the person or persons spoken to

you'd [yo͞od] the *contraction* of the words **you** and **would**

you'll [yo͞ol] the *contraction* of the words **you** and **will**

young [yŭng] *adjective*, the early stage of life, not old

young-ster [yŭng′stər] *noun*, a very young person, a child

your [yo͝or] *pronoun*, belonging to
you

you're [yo͝or] the *contraction* of
the words, **you** and **are**

yours [yo͝orz, yôrz, yōrz] *pronoun*,
belonging to you

your-self [yo͝or-sĕlf′] *pronoun*, a
form of you for emphasis when
the object of a verb and the
subject are the same

youth [yo͞oth] *noun*, a young
person, an adolescent

Z

za-ny [zā′nē] *adjective*, crazy,
whimsically comical

zeal [zēl] *noun*, ardor in pursuit of
anything, fervor, eagerness

zeal-ous [zĕl′əs] *adjective*,
enthusiastic, eager, devoted

ze-bra [zē′brə] *noun*, an African
wild animal like a horse which
has brown and white stripes all
over its body

Zen Buddhism [zĕn- bo͞o′dĭz′əm]
noun, a faith in which Buddha is
the embodiment of the superior
man and followers seek
enlightenment through
meditation and intuition instead
of the traditional scriptures

ze-nith [zē′nĭth] *noun*, the point
directly overhead in the sky, the
summit, the highest point

zeph-yr [zĕf′ər] *noun*, a gentle
breeze, the west wind

ze-ro [zîr′ō] *noun, plural*, **zeros** *or*
zeroes, 1) the numberical symbol 0
2) nothing 3) the temperature on
a scale indicated by this symbol

zest [zĕst] *noun*, feeling of
enjoyment, excitement

zinc [zĭngk] *noun*, a bluish white
metallic element often mixed
with other metals

zip-per [zĭp′ər] *noun*, a slide
fastener for clothing, briefcases,
and other objects, with two sets
of interlocking teeth

zone [zōn] *noun*, a division of an
area that differs or is
distinguished in some respect
from adjoining areas

zoo [zo͞o] *noun*, a place where
different animals are kept for
people to look at

zo-ol-o-gy [zō-ŏl′ə-jē] *noun*, the
science of animals

zy-gote [zī′gōt′] *noun*, the single
cell resulting from the union of
an egg and a sperm

Weights and Measures

ENGLISH

Length
12 inches	=	1 foot
3 feet	=	1 yard
5280 feet	=	1 mile
1760 yards	=	1 mile

Area
144 square inches	=	1 square foot
9 square feet	=	1 square yard
4840 square yards	=	1 acre
640 acres	=	1 square mile

Volume
1728 cubic inches	=	1 cubic foot
27 cubic feet	=	1 cubic yard

Capacity (Dry)
2 pints	=	1 quart
8 quarts	=	1 peck
4 pecks	=	1 bushel

Capacity (Liquid)
16 fluid ounces	=	1 pint
2 pints	=	1 quart
4 quarts	=	1 gallon (8 pints)

Mass
16 ounces	=	1 pound (7000 grains)
20 cwt	=	1 ton
2000 lbs	=	1 ton

Troy Weights
12 ounces	=	1 pound (5760 grains)

Measures
16 fluid ounces	=	1 pint

inches

metric

METRIC

Length

1 millimeter	=	1000 micrometers
1 centimeter	=	10 millimeters
1 meter	=	1000 millimeters
1 meter	=	100 centimeters
1 kilometer	=	1000 meters

Area

1 square centimeter	=	100 square millimeters
1 square meter	=	10,000 square centimeters
1 square meter	=	1,000,000 square millimeters
1 square kilometer	=	1,000,000 square meters

Volume

1 milliliter	=	1 cubic centimeter
1 liter	=	1000 milliliters
1 liter	=	0.001 cubic meter

Mass

1 gram	=	1000 milligrams
1 kilogram	=	1000 grams
1 metric ton	=	1000 kilograms

Volume

1 cubic inch (in^3) = 16.39 cubic centimeters (cm^3)

1,728 cubic inches = 1 cubic foot = 0.02832 cubic meters (m^3)

27 cubic feet (ft^3) = 1 cubic yard (y^3) = 0.7646 cubic meters

Temperature Conversions

°C Celsius = 5/9 (°F −32)   °F Fahrenheit = (9/5 °C) +32

Square Measure

100 sq. meters	= 1 sq. decameter	= 119.6 sq. yards
100 sq. decameters	= 1 sq. hectometer	= 2.471 acres
100 sq. hectometers	= 1 sq. kilometer	= 0.386 sq. miles or
		247.1 acres

Words for Large Numbers

Term	Calculation	# of Zeros
Ten	10 Ones	1
1 Hundred	100 Ones	2
1 Thousand	1,000 Ones	3
1 Million	1,000 Thousands	6
1 Billion	1,000 Millions	9
1 Trillion	1,000 Billions	12
1 Quadrillion	1,000 Trillions	15
1 Quintillion	1,000 Quadrillions	18
1 Sextillion	1,000 Quintillions	21
1 Septillion	1,000 Sextillions	24
1 Octillion	1,000 Septillions	27
1 Nonillion	1,000 Octillions	30
1 Decillion	1,000 Nonillions	33
1 Googol	1036 Decillions	100

Five Big Words

PNEUMONOULTRAMICROSCOPICSILICOVOLCANOCONIOSIS
This word is forty-five letters and its definition is a lung disease caused by breathing in certain particles.

ANTITRANSUBSTANTIATIONALIST
one who doubts the validity of transubstantiation.

FLOCCINAUCINIHILIPILIFICATION
the estimation of a thing as worthless.

ANTIDISESTABLISHMENTARIANISM
opposition of those who oppose the link between Church and State.

SUPERCALIFRAGILISTICEXPIALIDOCIOUS
The word to say when you don't know what to say.

Common Standard International Unit Symbols

km   = kilometer
m   = meter
dm   = decimeter
cm   = centimeter
mm   = millimeter

m^3 = L   = cubic decimeter = Liter
m^3 = cubic meter
mg = T = megagram
kg   = kilogram
g   = gram
mg   = milligram

ks   = kilosecond
s   = second
ms   = millisecond
us   = microsecond

Standard International Factor Prefixes

10^{24}	yotta	Y	septillion
10^{21}	zeta	Z	sextillion
10^{18}	exa	E	quintillion
10^{15}	peta	P	quadrillion
10^{12}	tera	T	trillion
10^{9}	giga	G	billion
10^{6}	mega	M	million
10^{3}	kilo	k	thousand
10^{2}	hecto	h	hundred
10^{1}	deka	da	ten
10^{-1}	deci	d	tenth
10^{-2}	centi	c	hundredth
10^{-3}	milli	m	thousandth
10^{-6}	micro	u	millionth
10^{-9}	nano	n	billionth
10^{-12}	pico	p	trillionth
10^{-15}	femto	f	quadrillionth
10^{-18}	atto	a	quintillionth
10^{-21}	zepto	z	sextillionth
10^{-24}	yocto	y	septillionth

Periodic Table of the Elements

Multiplication Table

x	1	2	3	4	5	6	7	8	9	10	11	12
1	1	2	3	4	5	6	7	8	9	10	11	12
2	2	4	6	8	10	12	14	16	18	20	22	24
3	3	6	9	12	15	18	21	24	27	30	33	36
4	4	8	12	16	20	24	28	32	36	40	44	48
5	5	10	15	20	25	30	35	40	45	50	55	60
6	6	12	18	24	30	36	42	48	54	60	66	72
7	7	14	21	28	35	42	49	56	63	70	77	84
8	8	16	24	32	40	48	56	64	72	80	88	96
9	9	18	27	36	45	54	63	72	81	90	99	108
10	10	20	30	40	50	60	70	80	90	100	110	120
11	11	22	33	44	55	66	77	88	99	110	121	132
12	12	24	36	48	60	72	84	96	108	120	132	144

Prefix Table

one	uni	unicorn, unilateral, universal
	mon, mono	monarchy, monotone, monochromatic
two	bi	bicycle, binoculars, bifocals, biceps
three	tri	triangle, trilogy, triplet, triceratops
four	quad	quadruped, quadrant, quadruplet
	tetra	tetrahedron, tetrad, tetragon, tetrapod
five	penta	pentagon, pentathlon, pentagram,
	quint	quintuplet, quintet, quintile
six	hexa	hexagon, hexagram, hexahedron
	sext	sextuplet, sextet, sextuple
seven	hepta	heptathlon, heptad, heptahedron,
	sept	septuplet, septuple, septennial
eight	oct	octave, octopus, octagon, octogenarian
nine	nona *before a vowel*	nonagon, nonet
ten	deca *before a vowel*	decade, decibel, decimal, decimeter
hundred	cent	century, centennial, centigrade
thousand	kilo	kilogram, kilometer, kilobyte, kilobit,
million	mega	megabyte, megahertz, megaton
billion	giga	gigabyte, gigawatt, gigaflop, gigaton

367

Roman Numerals

I	1
II	2
III	3
IV	4
V	5
VI	6
VII	7
VIII	8
IX	9
X	10
XX	20
XXX	30
XL	40
L	50
LX	60
LXX	70
LXXX	80
XC	90
C	100
CC	200
CD	400
D	500
DC	600
DCC	700
DCCC	800
CM	900
M	1,000
$\overline{V}$	5,000
$\overline{X}$	10,000
$\overline{L}$	50,000
$\overline{C}$	100,000
$\overline{D}$	500,000
$\overline{M}$	1,000,000

The United States Flag

I pledge allegiance to the flag of the United States of America and to the republic for which it stands, one nation under God, indivisible, with liberty and justice for all.

Betsy Ross sewed the first American flag in May, 1776. No one wrote down the story of why the United States decided to use stars and stripes when the design was chosen for our new nation. The only record indicates that the United States flag was first authorized by the Second Continental Congress on June 14, 1777. It said: "Resolved That the Flag of the United States be 13 stripes alternate red and white, that the Union be 13 stars white in a blue field representing a new constellation." This date is now observed as "Flag Day" throughout America.

The flag has changed 26 times since 1777 – each time a new state joined the Union. The flag was first modified in 1795 to have 15 stars and 15 stripes to include Vermont and Kentucky. In 1818, when Tennessee, Ohio, Louisiana, Indiana and Mississippi became states, the flag was designed to have 20 stars and 13 stripes. At that time, Congress decreed that the flag should have 13 stripes, and that new admissions to the Union should be recognized with a new star. Since then, the only change has been the adding of new stars. The 50-star flag was raised for the first time at 12:01 a.m. on July 4, 1960, at Fort McHenry National Monument, when Hawaii joined the Union.

Civics Lesson

Civics is the study of government and laws. Laws are made to maintain order, mutual respect and peace by defining people's rights and responsibilities toward society. They are agreed upon by people and enforced by governments. Most of the laws we use are based on English law, which was based on a judge's ruling in a dispute. This decision set a precedent and so any similar violation was considered with the most recent ruling used as a guide or common practice. These decisions created what is known as common law. Since these laws do not cover all situations, the elected officials created their own laws, known as statute law. There are three branches of government: legislative, executive and judicial.

The **Legislative** branch: Congress, which is made up of the Senate and the House of Representatives. Congress has the power to make laws, declare war, charge taxes, borrow money, and regulate business between states. To become law, a bill must be approved by both houses, the Senate and the House of Representatives and signed by the President. The **Senate** consists of 100 members, two members from each state, who serve six year terms. The Senate approves or rejects appointees made by the President to his Cabinet and to the Supreme Court and treaties made by the President.

The **House of Representatives** consists of 435 members. Members are elected from their home state based on population. Each member is elected to serve a two year term. The House of Representatives has the responsibility of initiating all bills for raising taxes and determines how money collected in taxes is spent.

The **Executive** branch consists of the President, the Vice President and the members of the President's Cabinet. The President and Vice President are elected together to a four year term by the Electoral College after a popular vote. Duties of the President: chief executive, head of state, and commander in chief of the armed forces.

The Vice President acts in the President's capacity if the President is unable to perform the duties of the office; presides over the Senate and only votes to break a tie. The Cabinet members are advisors to the President, appointed by the President. All members must be approved by the Senate. The Cabinet consists of the secretaries of the departments of State, Agriculture, Commerce, Defense, Education, Energy, Health and Human Services, Homeland Security, Housing and Urban Development, the Interior, Labor, Transportation, the Treasury, Veterans Affairs, and the Justice Department headed by the Attorney General.

The **Judicial** Branch is made up of the Supreme Court which consists of the Chief Justice and eight Associate Justices, who make the law of the land. Each justice is appointed for life terms after the President's nomination to the court is approved by the Senate. The Supreme Court is responsible for hearing appeals of cases from the federal and state court systems and all matters between individual states. It presides over all cases affecting ambassadors to the United States and public ministers including the president.

Courts handle two kinds of disputes: civil and criminal. A civil dispute is between two private citizens; a criminal case is a dispute between an individual and the whole community. Our justice system is comprised of courts within 94 Federal Judicial Districts in the 50 states, Washington D.C. and Puerto Rico. Some of the other courts within the Federal Judicial Branch include the U.S. Tax Court, the U.S. Court of International Trade, 13 circuit Courts of Appeals: one Federal Circuit Court, 11 for the districts throughout the states, and one in the District of Columbia.

THE DECLARATION OF INDEPENDENCE
of the Thirteen Colonies In CONGRESS, July 4, 1776

The unanimous Declaration of the thirteen united States of America,

When in the Course of human events, it becomes necessary for one people to dissolve the political bands which have connected them with another, and to assume among the powers of the earth, the separate and equal station to which the Laws of Nature and of Nature's God entitle them, a decent respect to the opinions of mankind requires that they should declare the causes which impel them to the separation.

We hold these truths to be self-evident, that all men are created equal, that they are endowed by their Creator with certain unalienable Rights, that among these are Life, Liberty and the pursuit of Happiness. – That to secure these rights, Governments are instituted among Men, deriving their just powers from the consent of the governed, – That whenever any Form of Government becomes destructive of these ends, it is the Right of the People to alter or to abolish it, and to institute new Government, laying its foundation on such principles and organizing its powers in such form, as to them shall seem most likely to effect their Safety and Happiness. Prudence, indeed, will dictate that Governments long established should not be changed for light and transient causes; and accordingly all experience hath shown, that mankind are more disposed to suffer, while evils are sufferable, than to right themselves by abolishing the forms to which they are accustomed. But when a long train of abuses and usurpations, pursuing invariably the same Object evinces a design to reduce them under absolute Despotism, it is their right, it is their duty, to throw off such Government, and to provide new Guards for their future security. --Such has been the patient sufferance of these Colonies; and such is now the necessity which constrains them to alter their former Systems of Government. The history of the present King of Great Britain [George III] is a history of repeated injuries and usurpations, all having in direct object the establishment of an absolute Tyranny over these States. To prove this, let Facts be submitted to a candid world.

He has refused his Assent to Laws, the most wholesome and necessary for the public good.

He has forbidden his Governors to pass Laws of immediate and pressing importance, unless suspended in their operation till his Assent should be obtained; and when so suspended, he has utterly neglected to attend to them.

He has refused to pass other Laws for the accommodation of large districts of people, unless those people would relinquish the right of Representation in the Legislature, a right inestimable to them and formidable to tyrants only.

He has called together legislative bodies at places unusual, uncomfortable, and distant from the depository of their public Records, for the sole purpose of fatiguing them into compliance with his measures.

He has dissolved Representative Houses repeatedly, for opposing with manly firmness his invasions on the rights of the people.

He has refused for a long time, after such dissolutions, to cause others to be elected; whereby the Legislative powers, incapable of Annihilation, have returned to the People at large for their exercise; the State remaining in the mean time exposed to all the dangers of invasion from without, and convulsions within.

He has endeavored to prevent the population of these States; for that purpose obstructing the Laws for Naturalization of Foreigners; refusing to pass others to encourage their migrations hither, and raising the conditions of new Appropriations of Lands.

He has obstructed the Administration of Justice, by refusing his Assent to Laws for establishing Judiciary powers.

He has made Judges dependent on his Will alone, for the tenure of their offices, and the amount and payment of their salaries.

He has erected a multitude of New Offices, and sent hither swarms of Officers to harass our people, and eat out their substance.

He has kept among us, in times of peace, Standing Armies without the consent of our legislatures.

He has affected to render the Military independent of and superior to the Civil power.

He has combined with others to subject us to a jurisdiction foreign to our constitution and unacknowledged by our laws; giving his Assent to their Acts of pretended Legislation:

For Quartering large bodies of armed troops among us:

For protecting them, by a mock Trial, from punishment for any Murders which they should commit on the Inhabitants of these States:

For cutting off our Trade with all parts of the world:

For imposing Taxes on us without our Consent:

For depriving us, in many cases, of the benefits of Trial by Jury:

For transporting us beyond Seas to be tried for pretended offences:

For abolishing the free System of English Laws in a neighbouring Province, establishing therein an Arbitrary government, and enlarging its Boundaries so as to render it at once an example and fit instrument for introducing the same absolute rule into these Colonies:

For taking away our Charters, abolishing our most valuable Laws, and altering fundamentally the Forms of our Governments:

For suspending our own Legislatures, and declaring themselves invested with power to legislate for us in all cases whatsoever.

He has abdicated Government here, by declaring us out of his Protection and waging War against us.

He has plundered our seas, ravaged our Coasts, burnt our towns, and destroyed the lives of our people.

He is at this time transporting large Armies of foreign Mercenaries to complete the works of death, desolation and tyranny, already begun with circumstances of Cruelty and perfidy scarcely paralleled in the most barbarous ages, and totally unworthy the Head of a civilized nation.

He has constrained our fellow Citizens taken Captive on the high Seas to bear Arms against their Country, to become the executioners of their friends and Brethren, or to fall themselves by their Hands.

He has excited domestic insurrections amongst us, and has endeavored to bring on the inhabitants of our frontiers, the merciless Indian Savages, whose known rule of warfare, is an undistinguished destruction of all ages, sexes and conditions.

In every stage of these Oppressions We have Petitioned for Redress in the most humble terms: Our repeated Petitions have been answered only by repeated injury. A Prince whose character is thus marked by every act which may define a Tyrant, is unfit to be the ruler of a free people.

Nor have We been wanting in attentions to our British brethren. We have warned them from time to time of attempts by their legislature to extend an unwarrantable jurisdiction over us. We have reminded them of the circumstances of our emigration and settlement here. We have appealed to their native justice and magnanimity, and we have conjured them by the ties of our common kindred to disavow these usurpations, which, would inevitably interrupt our connections and correspondence. They too have been deaf to the voice of justice and of consanguinity. We must, therefore, acquiesce in the necessity, which denounces our Separation, and hold them, as we hold the rest of mankind, Enemies in War, in Peace Friends.

We, therefore, the Representatives of the united States of America, in General Congress, Assembled, appealing to the Supreme Judge of the world for the rectitude of our intentions, do, in the Name, and by the Authority of the good People of these Colonies, solemnly publish and declare, That these United Colonies are, and of Right ought to be Free and Independent States; that they are Absolved from all Allegiance to the British Crown, and that all political connection between them and the State of Great Britain, is and ought to be totally dissolved; and that as Free and Independent States, they have full Power to levy War, conclude Peace, contract Alliances, establish Commerce, and to do all other Acts and Things which Independent States may of right do. And for the support of this Declaration, with a firm reliance on the protection of divine Providence, we mutually pledge to each other our Lives, our Fortunes and our sacred Honor.

The signers of the Declaration represented the new states as follows:

Connecticut:

Roger Sherman, Samuel Huntington, William Williams, Oliver Wolcott

Delaware:

Caesar Rodney, George Read, Thomas McKean

Georgia:

Button Gwinnett, Lyman Hall, George Walton

Maryland:

Samuel Chase, William Paca, Thomas Stone, Charles Carroll of Carrollton

Massachusetts:

John Hancock, Samuel Adams, John Adams, Robert Treat Paine, Elbridge Gerry

New Hampshire:

Josiah Bartlett, William Whipple, Matthew Thornton

New Jersey:

Richard Stockton, John Witherspoon, Francis Hopkinson, John Hart, Abraham Clark

New York:

William Floyd, Philip Livingston, Francis Lewis, Lewis Morris

North Carolina:

William Hooper, Joseph Hewes, John Penn

Pennsylvania:

Robert Morris, Benjamin Rush, Benjamin Franklin, John Morton, George Clymer, James Smith, George Taylor, James Wilson, George Ross

Rhode Island:

Stephen Hopkins, William Ellery

South Carolina:

Edward Rutledge, Thomas Heyward, Jr., Thomas Lynch, Arthur Middleton, Jr.

Virginia:

George Wythe, Richard Henry Lee, Thomas Jefferson, Benjamin Harrison, Thomas Nelson, Jr., Francis Lightfoot Lee, Carter Braxton

* * * *

The United States Constitution

Preamble

We the People . . .

of the United States, in Order to form a more perfect Union, establish Justice, insure domestic Tranquility, provide for the common defense, promote the general Welfare, and secure the Blessings of Liberty to ourselves and our Posterity, do ordain and establish this Constitution for the United States of America.

Article. I. *Adopted 1787*

Section. 1.
All legislative Powers herein granted shall be vested in a Congress of the United States, which shall consist of a Senate and House of Representatives.

Section. 2.
The House of Representatives shall be composed of Members chosen every second Year by the People of the several States, and the Electors in each State shall have the Qualifications requisite for Electors of the most numerous Branch of the State Legislature.

No Person shall be a Representative who shall not have attained to the Age of twenty five Years, and been seven Years a Citizen of the United States, and who shall not, when elected, be an Inhabitant of that State in which he shall be chosen.

Representatives and direct Taxes shall be apportioned among the several States which may be included within this Union, according to their respective Numbers, which shall be determined by adding to the whole Number of free Persons, including those bound to Service for a Term of Years, and excluding Indians not taxed, three fifths of all other Persons.

The actual Enumeration shall be made within three Years after the first Meeting of the Congress of the United States, and within every subsequent Term of ten Years, in such Manner as they shall by Law direct. The Number of Representatives shall not exceed one for every thirty Thousand, but each State shall have at Least one Representative; and until such enumeration shall be made, the State of New Hampshire shall be entitled to choose three, Massachusetts eight, Rhode Island and

Providence Plantations one, Connecticut five, New York six, New Jersey four, Pennsylvania eight, Delaware one, Maryland six, Virginia ten, North Carolina five, South Carolina five and Georgia three.

When vacancies happen in the Representation from any State, the Executive Authority thereof shall issue Writs of Election to fill such Vacancies.

The House of Representatives shall chuse their Speaker and other Officers; and shall have the sole Power of Impeachment.

Section. 3.
The Senate of the United States shall be composed of two Senators from each State, chosen by the Legislature thereof, for six Years; and each Senator shall have one Vote.

Immediately after they shall be assembled in Consequence of the first Election, they shall be divided as equally as may be into three Classes. The Seats of the Senators of the first Class shall be vacated at the Expiration of the second Year, of the second Class at the Expiration of the fourth Year, and of the third Class at the Expiration of the sixth Year, so that one third may be chosen every second Year; and if Vacancies happen by Resignation, or otherwise, during the Recess of the Legislature of any State, the Executive thereof may make temporary Appointments until the next Meeting of the Legislature, which shall then fill such Vacancies.

No person shall be a Senator who shall not have attained to the Age of thirty Years, and been nine Years a Citizen of the United States, and who shall not, when elected, be an Inhabitant of that State for which he shall be chosen.

The Vice President of the United States shall be President of the Senate, but shall have no Vote, unless they be equally divided.

The Senate shall chuse their other Officers, and also a President pro tempore, in the absence of the Vice President, or when he shall exercise the Office of President of the United States.

The Senate shall have the sole Power to try all Impeachments. When sitting for that Purpose, they shall be on Oath or Affirmation. When the President of the United States is tried, the Chief Justice shall preside: And no Person shall be convicted without the Concurrence of two thirds of the Members present.

Judgment in Cases of Impeachment shall not extend further than to removal from Office, and disqualification to hold and enjoy any Office of honor, Trust or Profit under the United States: but the Party convicted shall nevertheless be liable and subject to Indictment, Trial, Judgment and Punishment, according to Law.

Section. 4.

The Times, Places and Manner of holding Elections for Senators and Representatives, shall be prescribed in each State by the Legislature thereof; but the Congress may at any time by Law make or alter such Regulations, except as to the Places of chusing Senators.

The Congress shall assemble at least once in every Year, and such Meeting shall be on the first Monday in December, unless they shall by Law appoint a different Day.

Section. 5.

Each House shall be the Judge of the Elections, Returns and Qualifications of its own Members, and a Majority of each shall constitute a Quorum to do Business; but a smaller number may adjourn from day to day, and may be authorized to compel the Attendance of absent Members, in such Manner, and under such Penalties as each House may provide.

Each House may determine the Rules of its Proceedings, punish its Members for disorderly Behaviour, and, with the Concurrence of two-thirds, expel a Member.

Each House shall keep a Journal of its Proceedings, and from time to time publish the same, excepting such Parts as may in their Judgment require Secrecy; and the Yeas and Nays of the Members of either House on any question shall, at the Desire of one fifth of those Present, be entered on the Journal.

Neither House, during the Session of Congress, shall, without the Consent of the other, adjourn for more than three days, nor to any other Place than that in which the two Houses shall be sitting.

Section. 6.

The Senators and Representatives shall receive a Compensation for their Services, to be ascertained by Law, and paid out of the Treasury of the United States. They shall in all Cases, except Treason, Felony and Breach of the Peace, be privileged from Arrest during their Attendance at the Session of their respective Houses, and in going to and returning from the same; and for any Speech or Debate in either House, they shall not be questioned in any other Place.

No Senator or Representative shall, during the Time for which he was elected, be appointed to any civil Office under the Authority of the United States which shall have been created, or the Emoluments whereof shall have been increased during such time; and no Person holding any Office under the United States, shall be a Member of either House during his Continuance in Office.

Section. 7.

All bills for raising Revenue shall originate in the House of Representatives; but the Senate may propose or concur with Amendments as on other Bills.

Every Bill which shall have passed the House of Representatives and the Senate, shall, before it become a Law, be presented to the President of the United States; If he approve he shall sign it, but if not he shall return it, with his Objections to that House in which it shall have originated, who shall enter the Objections at large on their Journal, and proceed to reconsider it. If after such Reconsideration two thirds of that House shall agree to pass the Bill, it shall be sent, together with the Objections, to the other House, by which it shall likewise be reconsidered, and if approved by two thirds of that House, it shall become a Law. But in all such Cases the Votes of both Houses shall be determined by Yeas and Nays, and the Names of the Persons voting for and against the Bill shall be entered on the Journal of each House respectively. If any Bill shall not be returned by the President within ten Days (Sundays excepted) after it shall have been presented to him, the Same shall be a Law, in like Manner as if he had signed it, unless the Congress by their Adjournment prevent its Return, in which Case it shall not be a Law.

Every Order, Resolution, or Vote to which the Concurrence of the Senate and House of Representatives may be necessary (except on a question of Adjournment) shall be presented to the President of the United States; and before the Same shall take Effect, shall be approved by him, or being disapproved by him, shall be repassed by two thirds of the Senate and House of Representatives, according to the Rules and Limitations prescribed in the Case of a Bill.

Section. 8.

The Congress shall have Power To lay and collect Taxes, Duties, Imposts and Excises, to pay the Debts and provide for the common Defence and general Welfare of the United States; but all Duties, Imposts and Excises shall be uniform throughout the United States;

To borrow Money on the credit of the United States;

To regulate Commerce with foreign Nations, and among the several States, and with the Indian Tribes;

To establish an uniform Rule of Naturalization, and uniform Laws on the subject of Bankruptcies throughout the United States;

To coin Money, regulate the Value thereof, and of foreign Coin, and fix the Standard of Weights and Measures;

To provide for the Punishment of counterfeiting the Securities and current Coin of the United States;

To establish Post Offices and post Roads;

To promote the Progress of Science and useful Arts, by securing for limited Times to Authors and Inventors the exclusive Right to their respective Writings and Discoveries;

To constitute Tribunals inferior to the supreme Court;

To define and punish Piracies and Felonies committed on the high Seas, and Offences against the Law of Nations;

To declare War, grant Letters of Marque and Reprisal, and make Rules concerning Captures on Land and Water;

To raise and support Armies, but no Appropriation of Money to that Use shall be for a longer Term than two Years;

To provide and maintain a Navy;

To make Rules for the Government and Regulation of the land and naval Forces;

To provide for calling forth the Militia to execute the Laws of the Union, suppress Insurrections and repel Invasions;

To provide for organizing, arming, and disciplining the Militia, and for governing such Part of them as may be employed in the Service of the United States, reserving to the States respectively, the Appointment of the Officers, and the Authority of training the Militia according to the discipline prescribed by Congress;

To exercise exclusive Legislation in all Cases whatsoever, over such District (not exceeding ten Miles square) as may, by Cession of particular States, and the acceptance of Congress, become the Seat of the Government of the United States, and to exercise like Authority over all Places purchased by the Consent of the Legislature of the State in which the Same shall be, for the Erection of Forts, Magazines, Arsenals, dock-Yards, and other needful Buildings; And

To make all Laws which shall be necessary and proper for carrying into Execution the foregoing Powers, and all other Powers vested by this Constitution in the Government of the United States, or in any Department or Officer thereof.

Section. 9.

The Migration or Importation of such Persons as any of the States now existing shall think proper to admit, shall not be prohibited by the Congress prior to the Year one thousand eight hundred and eight, but a Tax or duty may be imposed on such Importation, not exceeding ten dollars for each Person.

The privilege of the Writ of Habeas Corpus shall not be suspended, unless when in Cases of Rebellion or Invasion the public Safety may require it.

No Bill of Attainder or ex post facto Law shall be passed.

No capitation, or other direct, Tax shall be laid, unless in Proportion to the Census or Enumeration herein before directed to be taken.

No Tax or Duty shall be laid on Articles exported from any State.

No Preference shall be given by any Regulation of Commerce or Revenue to the Ports of one State over those of another: nor shall Vessels bound to, or from, one State, be obliged to enter, clear, or pay Duties in another.

No Money shall be drawn from the Treasury, but in Consequence of Appropriations made by Law; and a regular Statement and Account of the Receipts and Expenditures of all public Money shall be published from time to time.

No Title of Nobility shall be granted by the United States: And no Person holding any Office of Profit or Trust under them, shall, without the Consent of the Congress, accept of any present, Emolument, Office, or Title, of any kind whatever, from any King, Prince or foreign State.

Section. 10.

No State shall enter into any Treaty, Alliance, or Confederation; grant Letters of Marque and Reprisal; coin Money; emit Bills of Credit; make any Thing but gold and silver Coin a Tender in Payment of Debts; pass any Bill of Attainder, ex post facto Law, or Law impairing the Obligation of Contracts, or grant any Title of Nobility.

No State shall, without the Consent of the Congress, lay any Imposts or Duties on Imports or Exports, except what may be absolutely necessary for executing it's inspection Laws: and the net Produce of all Duties and Imposts, laid by any State on

Imports or Exports, shall be for the Use of the Treasury of the United States; and all such Laws shall be subject to the Revision and Controul of the Congress.

No State shall, without the Consent of Congress, lay any duty of Tonnage, keep Troops, or Ships of War in time of Peace, enter into any Agreement or Compact with another State, or with a foreign Power, or engage in War, unless actually invaded, or in such imminent Danger as will not admit of delay.

Article II. *Adopted 1787*

Section. 1.
The executive Power shall be vested in a President of the United States of America. He shall hold his Office during the Term of four Years, and, together with the Vice-President chosen for the same Term, be elected, as follows:

Each State shall appoint, in such Manner as the Legislature thereof may direct, a Number of Electors, equal to the whole Number of Senators and Representatives to which the State may be entitled in the Congress: but no Senator or Representative, or Person holding an Office of Trust or Profit under the United States, shall be appointed an Elector.

The Electors shall meet in their respective States, and vote by Ballot for two persons, of whom one at least shall not be an Inhabitant of the same State with themselves. And they shall make a List of all the Persons voted for, and of the Number of Votes for each; which List they shall sign and certify, and transmit sealed to the Seat of the Government of the United States, directed to the President of the Senate. The President of the Senate shall, in the Presence of the Senate and House of Representatives, open all the Certificates, and the Votes shall then be counted. The Person having the greatest Number of Votes shall be the President, if such Number be a Majority of the whole Number of Electors appointed; and if there be more than one who have such Majority, and have an equal Number of Votes, then the House of Representatives shall immediately chuse by Ballot one of them for President; and if no Person have a Majority, then from the five highest on the List the said House shall in like Manner chuse the President. But in chusing the President, the Votes shall be taken by States, the Representation from each State having one Vote; a quorum for this Purpose shall consist of a Member or Members from two-thirds of the States, and a Majority of all the States shall be necessary to a Choice. In every Case, after the Choice of the President, the Person having the greatest Number of Votes of the Electors shall be the Vice President. But if there should remain two or more who have equal Votes, the Senate shall chuse from them by Ballot the Vice President.

The Congress may determine the Time of chusing the Electors, and the Day on which they shall give their Votes; which Day shall be the same throughout the United States.

No person except a natural born Citizen, or a Citizen of the United States, at the time of the Adoption of this Constitution, shall be eligible to the Office of President; neither shall any Person be eligible to that Office who shall not have attained to the Age of thirty-five Years, and been fourteen Years a Resident within the United States.

In Case of the Removal of the President from Office, or of his Death, Resignation, or Inability to discharge the Powers and Duties of the said Office, the same shall devolve on the Vice President, and the Congress may by Law provide for the Case of Removal, Death, Resignation or Inability, both of the President and Vice President, declaring what Officer shall then act as President, and such Officer shall act accordingly, until the Disability be removed, or a President shall be elected.

The President shall, at stated Times, receive for his Services, a Compensation, which shall neither be increased nor diminished during the Period for which he shall have been elected, and he shall not receive within that Period any other Emolument from the United States, or any of them.

Before he enter on the Execution of his Office, he shall take the following Oath or Affirmation:-- "I do solemnly swear (or affirm) that I will faithfully execute the Office of President of the United States, and will to the best of my Ability, preserve, protect and defend the Constitution of the United States."

Section. 2.
The President shall be Commander in Chief of the Army and Navy of the United States, and of the Militia of the several States, when called into the actual Service of the United States; he may require the Opinion, in writing, of the principal Officer in each of the executive Departments, upon any Subject relating to the Duties of their respective Offices, and he shall have Power to Grant Reprieves and Pardons for Offenses against the United States, except in Cases of Impeachment.

He shall have Power, by and with the Advice and Consent of the Senate, to make Treaties, provided two thirds of the Senators present concur; and he shall nominate, and by and with the Advice and Consent of the Senate, shall appoint Ambassadors, other public Ministers and Consuls, Judges of the supreme Court, and all other Officers of the United States, whose Appointments are not herein otherwise provided for, and which shall be established by Law: but the Congress may by Law vest the Appointment of such inferior Officers, as they think proper, in the President alone, in the Courts of Law, or in the Heads of Departments.

The President shall have Power to fill up all Vacancies that may happen during the Recess of the Senate, by granting Commissions which shall expire at the End of their next Session.

Section. 3.
He shall from time to time give to the Congress Information of the State of the Union, and recommend to their Consideration such Measures as he shall judge necessary and expedient; he may, on extraordinary Occasions, convene both Houses, or either of them, and in Case of Disagreement between them, with Respect to the Time of Adjournment, he may adjourn them to such Time as he shall think proper; he shall receive Ambassadors and other public Ministers; he shall take Care that the Laws be faithfully executed, and shall Commission all the Officers of the United States.

Section. 4.
The President, Vice President and all civil Officers of the United States, shall be removed from Office on Impeachment for, and Conviction of, Treason, Bribery, or other high Crimes and Misdemeanors.

Article III. *Adopted 1787*

Section. 1.
The judicial Power of the United States, shall be vested in one supreme Court, and in such inferior Courts as the Congress may from time to time ordain and establish. The Judges, both of the supreme and inferior Courts, shall hold their Offices during good Behavior, and shall, at stated Times, receive for their Services a Compensation which shall not be diminished during their Continuance in Office.

Section. 2.
The judicial Power shall extend to all Cases, in Law and Equity, arising under this Constitution, the Laws of the United States, and Treaties made, or which shall be made, under their Authority;-- to all Cases affecting Ambassadors, other public Ministers and Consuls; to all Cases of admiralty and maritime Jurisdiction;-- to Controversies to which the United States shall be a Party;-- to Controversies between two or more States; between a State and Citizens of another State;-- between Citizens of different States; between Citizens of the same State claiming Lands under Grants of different States, and between a State, or the Citizens thereof, and foreign States, Citizens or Subjects.

In all Cases affecting Ambassadors, other public Ministers and Consuls, and those in which a State shall be Party, the Supreme Court shall have original Jurisdiction. In all the other Cases before mentioned, the supreme Court shall have appellate Jurisdiction, both as to Law and Fact, with such Exceptions, and under such

Regulations as the Congress shall make.

The Trial of all Crimes, except in Cases of Impeachment, shall be by Jury; and such Trial shall be held in the State where the said Crimes shall have been committed; but when not committed within any State, the Trial shall be at such Place or Places as the Congress may by Law have directed.

Section. 3.
Treason against the United States, shall consist only in levying War against them, or in adhering to their Enemies, giving them Aid and Comfort. No Person shall be convicted of Treason unless on the Testimony of two Witnesses to the same overt Act, or on Confession in open Court.

The Congress shall have Power to declare the Punishment of Treason, but no Attainder of Treason shall work Corruption of Blood, or Forfeiture except during the Life of the Person attainted.

Article. IV. *Adopted 1787*

Section. 1.
Full Faith and Credit shall be given in each State to the public Acts, Records, and judicial Proceedings of every other State. And the Congress may by general Laws prescribe the Manner in which such Acts, Records and Proceedings shall be proved, and the Effect thereof.

Section. 2.
The Citizens of each State shall be entitled to all Privileges and Immunities of Citizens in the several States.

A Person charged in any State with Treason, Felony, or other Crime, who shall flee from Justice, and be found in another State, shall on demand of the executive Authority of the State from which he fled, be delivered up, to be removed to the State having Jurisdiction of the Crime.

No Person held to Service or Labour in one State, under the Laws thereof, escaping into another, shall, in Consequence of any Law or Regulation therein, be discharged from such Service or Labour, but shall be delivered up on Claim of the Party to whom such Service or Labour may be due.

Section. 3.
New States may be admitted by the Congress into this Union; but no new State shall be formed or erected within the Jurisdiction of any other State; nor any State be

formed by the Junction of two or more States, or parts of States, without the Consent of the Legislatures of the States concerned as well as of the Congress.

The Congress shall have Power to dispose of and make all needful Rules and Regulations respecting the Territory or other Property belonging to the United States; and nothing in this Constitution shall be so construed as to Prejudice any Claims of the United States, or of any particular State.

Section. 4.
The United States shall guarantee to every State in this Union a Republican Form of Government, and shall protect each of them against Invasion; and on Application of the Legislature, or of the Executive (when the Legislature cannot be convened) against domestic Violence.

Article. V. *Adopted 1787*

The Congress, whenever two thirds of both Houses shall deem it necessary, shall propose Amendments to this Constitution, or, on the Application of the Legislatures of two thirds of the several States, shall call a Convention for proposing Amendments, which, in either Case, shall be valid to all Intents and Purposes, as part of this Constitution, when ratified by the Legislatures of three fourths of the several States, or by Conventions in three fourths thereof, as the one or the other Mode of Ratification may be proposed by the Congress; Provided that no Amendment which may be made prior to the Year One thousand eight hundred and eight shall in any Manner affect the first and fourth Clauses in the Ninth Section of the first Article; and that no State, without its Consent, shall be deprived of its equal Suffrage in the Senate.

Article. VI. *Adopted 1787*

All Debts contracted and Engagements entered into, before the Adoption of this Constitution, shall be as valid against the United States under this Constitution, as under the Confederation.

This Constitution, and the Laws of the United States which shall be made in Pursuance thereof; and all Treaties made, or which shall be made, under the Authority of the United States, shall be the supreme Law of the Land; and the Judges in every State shall be bound thereby, any Thing in the Constitution or Laws of any State to the Contrary notwithstanding.

The Senators and Representatives before mentioned, and the Members of the several State Legislatures, and all executive and judicial Officers, both of the United States and of the several States, shall be bound by Oath or Affirmation, to support this Constitution; but no religious Test shall ever be required as a Qualification to any Office or public Trust under the United States.

Article. VII. *Adopted 1787*

The Ratification of the Conventions of nine States, shall be sufficient for the Establishment of this Constitution between the States so ratifying the Same.

Done in Convention by the Unanimous Consent of the States present the Seventeenth Day of September in the Year of our Lord one thousand seven hundred and Eighty seven and of the Independence of the United States of America the Twelfth. In Witness whereof We have hereunto subscribed our Names.

George Washington - President and deputy from Virginia
New Hampshire - John Langdon, Nicholas Gilman
Massachusetts - Nathaniel Gorham, Rufus King
Connecticut – Wm. Saml. Johnson, Roger Sherman
New York - Alexander Hamilton

New Jersey – Wil: Livingston, David Brearley. Wm. Paterson. Jona: Dayton
Pennsylvania - Benjamin Franklin, Thomas Mifflin, Robt. Morris, Geo. Clymer, Thos. FitzSimons, Jared Ingersoll, James Wilson, Gouverneur Morris
Delaware - Geo. Read, Gunning Bedford Junior, John Dickinson, Richard Bassett, Jaco: Broom
Maryland - James McHenry, Dan of St Thos. Jenifer, Danl Carroll
Virginia - John Blair, James Madison, Junior
North Carolina – Wm. Blount, Richd. Dobbs Spaight, Hugh Williamson
South Carolina - J. Rutledge, Charles Cotesworth Pinckney, Charles Pinckney, Pierce Butler
Georgia - William Few, Abr Baldwin
Attest: William Jackson, Secretary

ARTICLES IN ADDITON TO, AND AMENDMENT OF THE CONSTITUTION OF THE UNITED STATES OF AMERICA, PROPOSED BY CONGRESS, AND RATIFIED BY THE LEGISLATURES OF THE SEVERAL STATES, PURSUANT TO THE FIFTH ARTICLE OF THE ORIGINAL CONSTITUTION.

Article. I.
(Adopted 1791; freedom of religion, speech, press, assembly, and petition.)

Congress shall make no law respecting an establishment of religion, or prohibiting the free exercise thereof; or abridging the freedom of speech, or of the press; or the right of the people peaceably to assemble, and to petition the Government for a redress of grievances.

Article. II.
(Adopted 1791; the right to keep and bear arms.)

A well regulated Militia, being necessary to the security of a free State, the right of the people to keep and bear Arms, shall not be infringed.

Article. III.
(Adopted 1791; prevention of compulsory billeting during peacetime.)

No Soldier shall, in time of peace be quartered in any house, without the consent of the Owner, nor in time of war, but in a manner to be prescribed by law.

Article. IV.
(Adopted 1791; security of person and possessions against search and seizure.)

The right of the people to be secure in their persons, houses, papers, and effects, against unreasonable searches and seizures, shall not be violated, and no Warrants shall issue, but upon probable cause, supported by Oath or affirmation, and particularly describing the place to be searched, and the persons or things to be seized.

Article. V.
(Adopted 1791; prevention of double-jeopardy; right against self-incrimination.)

No person shall be held to answer for a capital, or otherwise infamous crime, unless on a presentment or indictment of a Grand Jury, except in cases arising in the land or naval forces, or in the Militia, when in actual service in time of War or public danger;

nor shall any person be subject for the same offense to be twice put in jeopardy of life or limb; nor shall be compelled in any criminal case to be a witness against himself, nor be deprived of life, liberty, or property, without due process of law; nor shall private property be taken for public use, without just compensation.

Article. VI.
(Adopted 1791; Fair, speedy, and public trial by jury; right to counsel.)

In all criminal prosecutions, the accused shall enjoy the right to a speedy and public trial, by an impartial jury of the State and district wherein the crime shall have been committed, which district shall have been previously ascertained by law, and to be informed of the nature and cause of the accusation; to be confronted with the witnesses against him; to have compulsory process for obtaining witnesses in his favor, and to have the Assistance of Counsel for his defense.

Article. VII.
(Adopted 1791; right to trial by jury in common law suits.)

In Suits at common law, where the value in controversy shall exceed twenty dollars, the right of trial by jury shall be preserved, and no fact tried by a jury, shall be otherwise re-examined in any Court of the United States, than according to the rules of the common law.

Article. VIII.
(Adopted 1791; prevention of cruel or unusual punishments; and excessive fines)

Excessive bail shall not be required, nor excessive fines imposed, nor cruel and unusual punishments inflicted.

Article. IX.
(Adopted 1791; protection of rights not enumerated in the Constitution.)

The enumeration in the Constitution, of certain rights, shall not be construed to deny or disparage others retained by the people.

Article. X.
(Adopted 1791; reservation of the rights of individual states.)

The powers not delegated to the United States by the Constitution, nor prohibited by it to the States, are reserved to the States respectively, or to the people.

Article. XI.
(Adopted 1795; delineation of judicial powers of the US in certain cases.)

The Judicial power of the United States shall not be construed to extend to any suit in law or equity, commenced or prosecuted against one of the United States by Citizens of another State, or by Citizens or Subjects of any Foreign State.

Article. XII.
(Adopted 1804; procedures for electing President and Vice President.)

The Electors shall meet in their respective states, and vote by ballot for President and Vice-President, one of whom, at least, shall not be an inhabitant of the same state with themselves; they shall name in their ballots the person voted for as President, and in distinct ballots the person voted for as Vice-President, and they shall make distinct lists of all persons voted for as President, and of all persons voted for as Vice-President and of the number of votes for each, which lists they shall sign and certify, and transmit sealed to the seat of the government of the United States, directed to the President of the Senate;

The President of the Senate shall, in the presence of the Senate and House of Representatives, open all the certificates and the votes shall then be counted;

The person having the greatest Number of votes for President, shall be the President, if such number be a majority of the whole number of Electors appointed; and if no person have such majority, then from the persons having the highest numbers not exceeding three on the list of those voted for as President, the House of Representatives shall choose immediately, by ballot, the President. But in choosing the President, the votes shall be taken by states, the representation from each state having one vote; a quorum for this purpose shall consist of a member or members from two-thirds of the states, and a majority of all the states shall be necessary to a choice. And if the House of Representatives shall not choose a President whenever the right of choice shall devolve upon them, before the fourth day of March next following, then the Vice-President shall act as President, as in the case of the death or other constitutional disability of the President.

The person having the greatest number of votes as Vice-President, shall be the Vice-President, if such number be a majority of the whole number of Electors appointed, and if no person have a majority, then from the two highest numbers on the list, the Senate shall choose the Vice-President; a quorum for the purpose shall consist of two-thirds of the whole number of Senators, and a majority of the whole number shall be necessary to a choice. But no person constitutionally ineligible to the office of President shall be eligible to that of Vice-President of the United States.

Article. XIII.
(Adopted 1865, abolished slavery.)

Section 1. Neither slavery nor involuntary servitude, except as a punishment for crime whereof the party shall have been duly convicted, shall exist within the United States, or any place subject to their jurisdiction.

Section 2. Congress shall have power to enforce this article by appropriate legislation.

Article. XIV.
(Adopted 1868; guarantee of the rights of citizenship.)

Section 1. All persons born or naturalized in the United States, and subject to the jurisdiction thereof, are citizens of the United States and of the State wherein they reside. No State shall make or enforce any law which shall abridge the privileges or immunities of citizens of the United States; nor shall any State deprive any person of life, liberty, or property, without due process of law; nor deny to any person within its jurisdiction the equal protection of the laws.

Section 2. Representatives shall be apportioned among the several States according to their respective numbers, counting the whole number of persons in each State, excluding Indians not taxed. But when the right to vote at any election for the choice of electors for President and Vice-President of the United States, Representatives in Congress, the Executive and Judicial officers of a State, or the members of the Legislature thereof, is denied to any of the male inhabitants of such State, being twenty-one years of age, and citizens of the United States, or in any way abridged, except for participation in rebellion, or other crime, the basis of representation therein shall be reduced in the proportion which the number of such male citizens shall bear to the whole number of male citizens twenty-one years of age in such State.

Section 3. No person shall be a Senator or Representative in Congress, or elector of President and Vice-President, or hold any office, civil or military, under the United States, or under any State, who, having previously taken an oath, as a member of Congress, or as an officer of the United States, or as a member of any State legislature, or as an executive or judicial officer of any State, to support the Constitution of the United States, shall have engaged in insurrection or rebellion against the same, or given aid or comfort to the enemies thereof. But Congress may by a vote of two-thirds of each House, remove such disability.

Section 4. The validity of the public debt of the United States, authorized by law, including debts incurred for payment of pensions and bounties for services in suppressing insurrection or rebellion, shall not be questioned. But neither the United States nor any State shall assume or pay any debt or obligation incurred in aid of insurrection or rebellion against the United States, or any claim for the loss or emancipation of any slave; but all such debts, obligations and claims shall be held illegal and void.

Section 5. The Congress shall have power to enforce, by appropriate legislation, the provisions of this article.

Article. XV.

(Adopted 1870; right to vote regardless of race, color, or previous servitude.)

Section 1. The right of citizens of the United States to vote shall not be denied or abridged by the United States or by any State on account of race, color, or previous condition of servitude.

Section 2. The Congress shall have power to enforce this article by appropriate legislation.

Article. XVI.

(Adopted 1913; right of Congress to levy income taxes.)

The Congress shall have power to lay and collect taxes on incomes, from whatever source derived, without apportionment among the several States, and without regard to any census or enumeration.

Article. XVII.

(Adopted 1913; procedure for electing Senators.)

The Senate of the United States shall be composed of two Senators from each State, elected by the people thereof, for six years; and each Senator shall have one vote. The electors in each State shall have the qualifications requisite for electors of the most numerous branch of the State legislatures.

When vacancies happen in the representation of any State in the Senate, the executive authority of such State shall issue writs of election to fill such vacancies: *Provided*, That the legislature of any State may empower the executive thereof to make temporary appointments until the people fill the vacancies by election as the legislature may direct.

This amendment shall not be so construed as to affect the election or term of any Senator chosen before it becomes valid as part of the Constitution.

Article. XVIII.

(Adopted 1919; introduced prohibition.)

Section 1. After one year from the ratification of this article the manufacture, sale, or transportation of intoxicating liquors within, the importation thereof into, or the exportation thereof from the United States and all territory subject to the jurisdiction thereof for beverage purposes is hereby prohibited.

Section 2. The Congress and the several States shall have concurrent power to enforce this article by appropriate legislation.

Section 3. This article shall be inoperative unless it shall have been ratified as an amendment to the Constitution by the legislatures of the several States, as provided in the Constitution, within seven years from the date of the submission hereof to the States by the Congress.

Article. XIX.
(Adopted 1920; right of women to vote.)

The right of citizens of the United States to vote shall not be denied or abridged by the United States or by any State on account of sex.

Congress shall have power to enforce this article by appropriate legislation.

Article. XX.
(Adopted 1933; presidential term and succession; terms of other offices.)

Section 1. The terms of the President and Vice President shall end at noon on the 20th day of January, and the terms of Senators and Representatives at noon on the 3d day of January, of the years in which such terms would have ended if this article had not been ratified; and the terms of their successors shall then begin.

Section 2. The Congress shall assemble at least once in every year, and such meeting shall begin at noon on the 3d day of January, unless they shall by law appoint a different day.

Section 3. If, at the time fixed for the beginning of the term of the President, the President elect shall have died, the Vice President elect shall become President. If a President shall not have been chosen before the time fixed for the beginning of his term, or if the President elect shall have failed to qualify, then the Vice President elect shall act as President until a President shall have qualified; and the Congress may by law provide for the case wherein neither a President elect nor a Vice President elect shall have qualified, declaring who shall then act as President, or the manner in which one who is to act shall be selected, and such person shall act accordingly until a President or Vice President shall have qualified.

Section 4. The Congress may by law provide for the case of the death of any of the persons from whom the House of Representatives may choose a President whenever the right of choice shall have devolved upon them, and for the case of the death of any of the persons from whom the Senate may choose a Vice President whenever the right of choice shall have devolved upon them.

Section 5. Sections 1 and 2 shall take effect on the 15th day of October following the ratification of this article.

Section 6. This article shall be inoperative unless it shall have been ratified as an amendment to the Constitution by the legislatures of three-fourths of the several States within seven years from the date of its submission.

Article. XXI.
(Adopted 1933; repeal of prohibition, the 18th Amendment.)

Section 1. The eighteenth article of amendment to the Constitution of the United States is hereby repealed.

Section 2. The transportation or importation into any State, Territory, or possession of the United States for delivery or use therein of intoxicating liquors, in violation of the laws thereof, is hereby prohibited.

Section 3. The article shall be inoperative unless it shall have been ratified as an amendment to the Constitution by conventions in the several States, as provided in the Constitution, within seven years from the date of the submission hereof to the States by the Congress

Article. XXII.
(*Adopted 1951*; limiting Presidents to two terms.)

Section 1. No person shall be elected to the office of the President more than twice, and no person who has held the office of President, or acted as President, for more than two years of a term to which some other person was elected President shall be elected to the office of the President more than once. But this Article shall not apply to any person holding the office of President, when this Article was proposed by the Congress, and shall not prevent any person who may be holding the office of President, or acting as President, during the term within which this Article becomes operative from holding the office of President or acting as President during the remainder of such term.

Section 2. This article shall be inoperative unless it shall have been ratified as an amendment to the Constitution by the legislatures of three-fourths of the several States within seven years from the date of its submission to the States by the Congress.

Article. XXIII.
(Adopted 1961; grants Washington DC residents a vote in the Presidential elections.)

Section 1. The District constituting the seat of Government of the United States shall appoint in such manner as the Congress may direct: A number of electors of President and Vice President equal to the whole number of Senators and Representatives in Congress to which the District would be entitled if it were a State, but in no event more than the least populous State; they shall be in addition to those appointed by the States, but they shall be considered, for the purposes of the election of President and Vice President, to be electors appointed by a State; and they shall meet in the District and perform such duties as provided by the twelfth article of amendment.

Section 2. The Congress shall have power to enforce this article by appropriate legislation.

Article. XXIV.

(Adopted 1964; bars poll (voting) tax in federal elections.)

Section 1. The right of citizens of the United States to vote in any primary or other election for President or Vice President, for electors for President or Vice President, or for Senator or Representative in Congress, shall not be denied or abridged by the United States or any State by reason of failure to pay any poll tax or other tax.

Section 2. The Congress shall have power to enforce this article by appropriate legislation.

Article. XXV.

(Adopted 1967; order of Presidential succession.)

Section 1. In case of the removal of the President from office or of his death or resignation, the Vice President shall become President.

Section 2. Whenever there is a vacancy in the office of the Vice President, the President shall nominate a Vice President who shall take office upon confirmation by a majority vote of both Houses of Congress.

Section 3. Whenever the President transmits to the President pro tempore of the Senate and the Speaker of the House of Representatives his written declaration that he is unable to discharge the powers and duties of his office, and until he transmits to them a written declaration to the contrary, such powers and duties shall be discharged by the Vice President as Acting President.

Section 4. Whenever the Vice President and a majority of either the principal officers of the executive departments or of such other body as Congress may by law provide, transmit to the President pro tempore of the Senate and the Speaker of the House of Representatives their written declaration that the President is unable to discharge the powers and duties of his office, the Vice President shall immediately assume the powers and duties of the office as Acting President.

Thereafter, when the President transmits to the President pro tempore of the Senate and the Speaker of the House of Representatives his written declaration that no inability exists, he shall resume the powers and duties of his office unless the Vice President and a majority of either the principal officers of the executive department or of such other body as Congress may by law provide, transmit within four days to the President pro tempore of the Senate and the Speaker of the House of

Representatives their written declaration that the President is unable to discharge the powers and duties of his office. Thereupon Congress shall decide the issue, assembling within forty eight hours for that purpose if not in session. If the Congress, within twenty one days after receipt of the latter written declaration, or, if Congress is not in session, within twenty one days after Congress is required to assemble, determines by two thirds vote of both Houses that the President is unable to discharge the powers and duties of his office, the Vice President shall continue to discharge the same as Acting President; otherwise, the President shall resume the powers and duties of his office.

Article. XXVI.
(Adopted 1971; right to vote at age eighteen.)

Section 1. The right of citizens of the United States, who are eighteen years of age or older, to vote shall not be denied or abridged by the United States or by any State on account of age.
Section 2. The Congress shall have power to enforce this article by appropriate legislation.

Article. XXVII.
(Adopted 1992; postpones Representatives' pay raises until after new elections.)

No law, varying the compensation for the services of the Senators and Representatives, shall take effect, until an election of Representatives shall have intervened.

* * * * * * *

To find additional information about the Declaration of Independence, visit these sites on the Internet:

- National Archives and Records Administration http://www.pro.gov.uk/webarchive/
- Library of Congress: http://www.loc.gov/about/sitemap/

GEORGE WASHINGTON

1st *President.* George Washington was born in Westmoreland County, Virginia, on February 22, 1732. His father was a prosperous farmer and his grandparents emigrated from England to Virginia in the 1650s. Washington did not have formal schooling and started his first job as a surveyor when he was 16. He inherited Mount Vernon at the age of 21 when his Uncle Lawrence died. At this time, Governor Dinwiddie of Virginia sent Washington on a mission to keep the French out of the Ohio Valley. In 1755, when he was 23 he became commander in chief of the Virginia forces, and led them to victory in the French and Indian War. Washington married Martha Custis, a young widow. They did not have any children. In 1774 he served as a delegate to the first Continental Congress. He was unanimously elected commander in chief of the American forces. He helped write the peace treaty ending the American Revolution that was signed in 1783. George Washington was named president of the Federal Convention in 1787, and in 1789 was elected President of the United States. As the first president, he had to decide how to carry out the duties of the executive branch of the government. Washington felt this position should be devoid of politics and should represent all of the American people. The first Cabinet was made up of Thomas Jefferson, Secretary of State; Edmund Randolph, Attorney General; Alexander Hamilton, Secretary of the Treasury, and Henry Knox, Secretary of War. Washington stayed in office two terms and declined to serve a third term. He died at his home in Mount Vernon on December 14, 1799. **Washington was President from 1789-97 and the Vice President was John Adams.**

JOHN ADAMS

2nd President. John Adams was born October 19, 1735 in Quincy, Massachusetts. He attended school at a local academy. When he was 16 he enrolled in Harvard College and graduated in 1755. Adams began practicing law three years later. He married Abigail Smith in 1764. Together they had four children. In 1765 Adams wrote letters in the Boston Gazette protesting the Stamp Act imposed by the British Government that would tax public documents, newspapers, licenses, insurance policies and playing cards. Adams was elected to the Congress in 1774 and served in that capacity until 1777. He established his reputation as a champion of individual rights. He was one of the first to favor independence from Great Britain and in 1775 he recommended George Washington to be the commander in chief of the new Continental Army. He was appointed minister to the Netherlands in 1780. He successfully arranged loan and trade agreements between France, the Netherlands and America. Adams traveled to Paris in 1782 to negotiate a peace treaty with Great Britain, ending the Revolutionary War. He served as minister to Great Britain in 1785 until 1788 when he was elected vice president. Adams served as vice president two terms until he was elected President in 1797. John and Abigail Adams were the first to occupy the White House. It was undecorated and littered with building debris. Adams served only one term. He lost the election to Thomas Jefferson. He returned to Quincy, Massachusetts where he spent most of his time reading and writing essays. He died there on July 4, 1826. **Adams was President from 1797-1801 and the Vice President was Thomas Jefferson.**

THOMAS JEFFERSON

3rd President. Thomas Jefferson was born in Albermarle County, Virginia on April 13, 1743. He was born into a wealthy family and tutored at home. When he turned 17 he went to the College of William and Mary in Williamsburg; he learned to speak French, Italian and Spanish. Jefferson studied law for five years; he was admitted to the bar when he was 24 and practiced law until the American Revolution closed the courts. In 1772 he married Martha Wayles Shelton; they had three daughters. In 1776 when he was 33, he was chosen to write the Declaration of Independence. The same year he became a member of the United States House of Delegates. He introduced the Bill for Establishing Religious Freedom in 1779 and it which passed seven years later. Jefferson served two terms as Governor of Virginia. In 1782 his wife died; he never remarried. He was elected to the Continental Congress in 1783 where he presented the concept of the dollar and the decimal system of money. In 1785, George Washington appointed Jefferson, Ben Franklin's successor as minister to France; he lived there for five years. Jefferson left France at the start of the French Revolution in 1789. Jefferson ran against Adams for the presidency and lost by three electoral votes. He became the Vice President. In 1800 he ran against Adams again and won the office of president; he served two terms. As president, he bought the Louisiana Territory for $15 million doubling the size of the United States. In 1819 he chartered the University of Virginia. He died at this home, Monticello, on July 4, 1826. **Jefferson was President from 1801-09 and the Vice Presidents were Aaron Burr and George Clinton.**

JAMES MADISON

4th President. James Madison was born on March 16, 1751 in Port Conway, Virginia. He grew up on a family-owned tobacco farm called Montpellier, where his mother and grandmother taught him to read and write. When he turned 11 he went off to a boarding school in King and Queen County, Virginia. He loved to read. He attended Princeton University and graduated. In 1776 when Madison was 25, he was elected to the Virginia Revolutionary Convention. He served in the Virginia legislature for one year and was defeated when he ran for reelection. He was appointed a member of the Governor's Council, which managed the Revolutionary War efforts. In 1784 he was elected to the Virginia legislature. He married Dolley Payne Todd; they did not have any children. Madison attended the Annapolis Convention as a delegate and then the Constitutional Convention in Philadelphia, in 1787. His understanding of government made him a leader in creating the Constitution; he advocated a republican form of government in which the people are sovereign but rule through elected representatives. In 1789 he was elected to the United States House of Representatives, at the end of his term in 1801, he was appointed Secretary of State by Thomas Jefferson. In 1809 he was elected President. At this time, Britain and France were at war. After Britain attacked American ships, Madison reluctantly led our nation into war against Britain. During the War of 1812 the White House was burned by the British. Madison left Washington in 1817 after serving two terms as president. He lived on his farm developing new methods of agriculture until he died on June 28, 1836. **Madison was President from 1809-17 and the Vice Presidents were George Clinton and Elbridge Gerry.**

JAMES MONROE

5th President. James Monroe was born on April 28, 1758 in Westmoreland County, Virginia. He left home to attend William and Mary College when he was 16. He joined the Third Virginia Regiment two years later to fight in the American Revolution. In 1782 he was elected to the Virginia House of Delegates. In 1783 he became a member of the Congress of the Confederation and served until 1786. He married Elizabeth Kortright. In 1790 he was elected United States senator and served four years, until he was appointed minister to France by George Washington. He retired from that position and two years later he was elected Governor of Virginia. In 1803 Monroe went on a diplomatic mission to France to help negotiate the Louisiana Purchase. President Jefferson appointed Monroe minister to Great Britain and a year later minister to Spain. In 1810 he was elected to the Virginia legislature. He served one year and left when he was elected Governor again in 1811. The same year he was appointed Secretary of State and also served under James Madison as the Secretary of War between 1814–1815. In 1817 Monroe was elected President. During his administration he bought Florida from Spain and established the Canadian border. He authored the Monroe doctrine that warned nations in Europe not to take over or set up colonies in North and South America. His presidency was known as the Era of Good Feeling; a period of national optimism, expansion and growth. He left office in 1825 and moved to Loudon County, Virginia with his wife. After she died he lived with his daughter in New York City, where he died July 4, 1831. **Monroe was President from 1817-25 and the Vice President was Daniel D. Tompkins.**

JOHN Q. ADAMS

6th President. John Quincy Adams was born July 11, 1767 in Quincy, Massachusetts. He attended a village school until the headmaster went off to war. His parents were determined to see him elected President. He was tutored by his mother and he attended schools in Paris and the University of Leiden in the Netherlands. John Quincy Adams began a diary that he wrote in for over 60 years. He returned from the Netherlands in 1785 to attend Harvard College. He graduated in 1787 and started to practice law. At the age of 28 he was appointed minister to the Netherlands by George Washington. He married Louisa Johnson and they had three sons. He served as the minister of Prussia in 1797 until 1801. He often traveled with his father, President John Adams. Adams was elected to the United States Senate in 1803 where he served until 1808. In 1809 President Madison appointed Adams the first American minister to Russia. In 1812 war broke out between the United States and Great Britain. Adams served on the delegation that brought about the Peace Treaty of Ghent in 1814. He later became the minister of Great Britain until 1817 when President Monroe appointed Adams Secretary of State. He negotiated the Transcontinental Treaty of 1819 with Spain, which gave the United States access to the Pacific Ocean. Adams ran for President in 1825 and won; he served one term. Adams was frustrated by the politics of the presidency, because he believed he was the people's representative. He ran for Congress in 1831 where he was an active legislator. He died two days after he collapsed on the floor of the House of Representatives, on February 23, 1848. **Adams was President from 1825-29 and the Vice President was John C. Calhoun.**

ANDREW JACKSON

7th President. Andrew Jackson was born March 15, 1767 in a log cabin in the Waxhaw settlement of South Carolina, a few days after his father's death. He lived with his mother and two older brothers. He attended school enough to learn how to read. Jackson left home to fight in the Revolutionary War when he was 14; he was taken prisoner by a British officer. Shortly after he was released, his mother died. Jackson studied law in Salisbury, North Carolina and passed the bar exam. He traveled west with a group of pioneers and ended up in Tennessee where he married Rachel Donelson Robards in 1794. He was elected to the House of Representatives in 1796 and served one year in the United States Senate as the first representative from Tennessee. He had to return home when his farm failed. In 1802 he was elected major general of the Tennessee militia. In 1814 he defeated the Creek Indians during the war of 1812; he was promoted to major general of the United States Army. In 1815 he successfully fought off the British attack on New Orleans protecting the Louisiana territory. Jackson was elected Governor of the new territory of Florida in 1821. He served two years and returned to his home in Tennessee where he was elected to the Senate. Jackson ran for President in 1824 but lost to Martin Van Buren in a close race that was decided by the House of Representatives. In 1829 he was elected President and served two terms. He returned to his home in Nashville, Tennessee where he lived until he died on June 8, 1845. **Jackson was President from 1829-37 and the Vice Presidents were John C. Calhoun and Martin Van Buren**

MARTIN VAN BUREN

8th President. Martin Van Buren was born March 4, 1782, in Kinderhook, New York. He was the first President to be born under the American flag. His father owned a tavern that also served as a polling place so he heard politics discussed at an early age. He was a student at the Kinderhook Academy until he took a job in a local law office when he was 13. At the age of 18 he had a job in a law office in New York City. When he was 21 he returned home to practice law; he was a very successful lawyer. He married Hannah Hoes his childhood sweetheart in 1807; together they had four sons. Van Buren was elected to the Senate in 1812. He interrupted his second term in the Senate to run for governor. Van Buren served as governor of New York for ten weeks and then moved to Washington to perform the duties of Secretary of State in 1829. He was elected to the office of Vice President under Andrew Jackson in 1832. Van Buren was elected president in 1836, the first to be elected as a member of the Democratic Party. He became president in 1837—the same year that the United States entered a major depression. During his administration he instituted the Department of the Treasury to protect government funds due to the economic depression caused by investment in land speculation that was not backed by gold or silver. He was not reelected in 1840 because he did not believe the government should help people who had lost their jobs. Van Buren tried again for the presidency in 1848 as a Free Soil candidate and lost to Zachary Taylor. Martin Van Buren died in Kinderhook, New York, in 1862. **Van Buren was President from 1837-41 and the Vice President was Richard M. Johnson.**

WILLIAM H. HARRISON

9th President. William Henry Harrison was born on February 9, 1773 on his father's plantation in Charles City County, Virginia. His father was one of the signers of the Declaration of Independence and a friend of George Washington. Harrison was taught at home until he went to Hampden-Sidney College in Virginia. He joined the Army one year later— his father died the same year. In 1795 he married Anna Symmes; they had ten children. Harrison was appointed Secretary of the Northwest Territory by President John Adams and served as the delegate to Congress from that region. He was appointed Governor of the Indiana territory where he served for 12 years. He took millions of acres of land from Native Americans by battle or by treaty. And he recaptured Detroit from the British in the War of 1812. Harrison defeated the Shawnee Indians and destroyed their settlement. Harrison defeated the British and Indians at the Battle of the Thames during the War of 1812. He moved his family to North Bend, Ohio. In 1816 he was elected to Congress and in 1819 he was elected to the state senate. At the end of that term in 1825 he was elected to the United States Senate. President John Quincy Adams appointed Harrison minister to Columbia in 1828. He offended Columbian President Simon Bolivar and he was forced to leave the country in 1829. Harrison returned to his farm in North Bend. He was nominated by the Whig Party to run for President against Martin Van Buren, and lost. Harrison ran for President again four years later and won. He died of pneumonia only 30 days after taking office on April 4, 1841. **Harrison was President from March 4, 1841 to April 4, 1841. His Vice President was John Tyler.**

JOHN TYLER

10th President. John Tyler was born on March 29, 1790 in Charles City County, Virginia on Greenway Plantation. His mother died when he was seven; his father was a judge who was later elected Governor of Virginia. In 1802 Tyler entered the high school of the College of William and Mary in Williamsburg. He graduated from the college when he was only 17. He went on to study law and in 1811 he began to practice law in Richmond. That same year he was elected to the Virginia House of Delegates and served for five years. He married Letitia Christian in 1813; they had eight children. In 1816 Tyler was elected to the United States House of Representatives where he served for five years. He went on to serve in the Virginia House of Delegates from 1823 until 1825. He was elected Governor of Virginia in 1825 and served for two years until he was elected to the United States Senate. He served in the Senate for nine years. Tyler was elected Vice President when William Henry Harrison won the presidential election in 1841. A month later he took the oath of office as President. His wife died while he was President and he married Julia Gardiner; he was the first president to wed, while serving in office. Together they had seven children. President Tyler was instrumental in the annexation of Texas in 1845 and he established the United States Weather Bureau. As president he established trade between China and the United States. He left office in 1845 and moved to his plantation, Sherwood Forest in Virginia, where he lived with his family until he died on January 18, 1862. **Tyler was President from 1841-45 and since he was completing Harrison's term, there was no Vice President**

JAMES K. POLK

11th President. James Knox Polk was born on November 2, 1795 in Mecklenburg, Virginia. His family moved to Duck River, Tennessee when he was a child. Polk was too weak to work on the farm so his parents focused on developing his mind. He was taught by tutors and sent to private schools. In 1815 he enrolled in the University of North Carolina. He then went on to study law and he was admitted to the bar. Polk became a popular lawyer and he was elected to the state legislature. On January 1, 1824 he married Sarah Childress; they did not have any children. Polk was elected to the United States legislature in 1825. He served seven terms, until he was elected Governor of Tennessee. Polk served one term as Governor. Then he was nominated by the Democrats to run for President. He won and pledged to serve only one term. During Polk's administration, gold was discovered in 1848 in California and thousands of people migrated west. The Department of the Interior was established. Wisconsin, Iowa and Texas became states. Polk negotiated the Minnesota and Oregon boundaries with Britain, making them federal territories. He fought and won the Mexican War (1846-48) and annexed much of the southwest and California. Polk was known for working long hours, beginning his duties at daybreak and working at his desk until midnight. He was away from the White House only 37 days during his four-year term. He promised voters he would serve only one term as president. He died three months after he left office at his home in Nashville, Tennessee on June 15, 1849. **Polk was President from 1845-49 and the Vice President was George M. Dallas.**

ZACHARY TAYLOR

12th President. Zachary Taylor was born November 24, 1784 in Barbour, Virginia. His father was a Lieutenant Colonel in the Revolutionary War. His only formal education was at a small schoolhouse. His parents supplemented his education by teaching him at home. He worked on his family's plantation until he left home in 1808 to join the 7th Infantry Regiment. He married Margaret Mackall South in 1810; they had six children, two daughters died in childhood. Taylor fought in the War of 1812 rising in rank to major by 1816 in the 3rd Infantry Regiment. In 1832 Taylor fought in the Black Hawk War to drive the Sac Indians out of Illinois. In 1837 he received an honorary commission as brigadier general after he defeated the Seminole Indians in Florida during the Second Seminole War. He was not able to defeat the Indians in 1840; he was relieved of his command and assigned to fight the Indians in the Southwest. He fought on the Texas-Louisiana border; his army defeated the Mexicans at Palo Alto. His final victory was in the battle of Buena Vista where his troops were outnumbered 4-1. He was promoted to major general. The Whig Party nominated Zachary Taylor for President in 1848 and he won. Taylor was a soldier most of his life; he never held political office until he was elected president. He was inaugurated President, March 1849. As president he tried to use the army to prevent states from withdrawing from the United States. He was President for only one year. On July 4th, he participated in the ceremonies laying the cornerstone of the Washington monument. He had a stomach ailment after eating that day and he died July 9, 1850. **Taylor was President from 1849-50 and the Vice President was Millard Fillmore.**

MILLARD FILLMORE

13th *President.* Millard Fillmore was born January 7, 1800 in Locke, New York. He spent most of his early life on a farm and attended a local school. He was eager to improve his education and jumped at the chance to study law with County Judge Walter Wood as an apprentice when he was 19. In 1823 Fillmore passed the bar exam and began his law practice in East Aurora, New York. He married Abigail Powers and they had two children. The family moved to Buffalo where Fillmore was a very successful lawyer. In 1828 he was elected to the state legislature. In 1832 he was elected to the United States House of Representatives where he served for ten years. In 1842, he ran for Governor of New York and lost. Fillmore was elected Vice President in 1848 and was sworn in as President July 10, 1850 when Zachary Taylor died unexpectedly. During his administration the first practical fire engine was developed in Ohio. There was a lot of tension between the North and the South over slavery. Many slaves were fleeing to the north. He addressed the slavery issue through the Compromise of 1850 and signed the Fugitive Slave Law. Fillmore tried to prevent the Civil War by urging the northern and southern states to agree on a settlement. He was not nominated for a second term because neither expansionists nor slaveholders were pleased with his policies. He returned to Buffalo, New York; that same year his wife died. Fillmore ran again for President in 1856 but was unsuccessful. He remarried and lived in Buffalo participating in many civic activities in his community until his death on March 8, 1874. **Fillmore was President from 1850-53 and since he was completing Taylor's term, there was no Vice President.**

FRANKLIN PIERCE

14th *President.* Franklin Pierce was born in a log cabin on November 23,1804 in Hillsborough County, New Hampshire. He attended a local school. His father was a general in the Revolutionary War and served two terms as Governor. In 1820 Pierce enrolled in Bowdoin College in Brunswick, Maine, where he met a life long friend, Nathaniel Hawthorne. He went on to study law and was admitted to the bar in 1827. Pierce was elected to the state legislature where he served four years. He married Jane Means Appleton, they had three children, two died in childhood, only Benjamin survived. Pierce was elected to two terms in Congress and then to the United States Senate in 1832. He only served four years; he resigned his post because his wife did not like life in Washington D.C. They returned to Concord, New Hampshire where he practiced law. In 1847 Pierce enlisted as a private in the Concord Light Infantry in the Mexican War; he rose to brigadier general of volunteers commanding 2,500 men in Mexico. He could no longer fight after he suffered a leg wound from a fall from his horse. Pierce returned to practice law in Concord. He was nominated by the Democrats to run for President and won the election easily. On January 6, the Pierces were in a train wreck that killed their son Benjamin who was 11 years old. Mrs. Pierce who was brokenhearted lived in seclusion in the White House the entire term. During his presidency, Pierce bought parts of Arizona and New Mexico from Mexico. The Pierces toured Europe after he left office and returned to live in Concord, New Hampshire where he died on October 8, 1869. **Pierce was President from 1853-57 and the Vice President was William R. King, he died before their term even started.**

JAMES BUCHANAN

15th President. James Buchanan was born on April 23, 1791 in a log cabin in Cove Gap, Pennsylvania. He was the oldest of nine children. They went to school in Mercersburg, where his father owned a general store. He attended Dickinson College in Carlisle, Pennsylvania, and law school in Lancaster. He was a successful lawyer who interrupted his career to join a volunteer cavalry company to fight in the War of 1812. Buchanan returned home from the war and won a seat in the legislature, serving two terms. He returned to his law practice in Lancaster. He ran for Congress and served ten years. Andrew Jackson asked him to serve as minister to Russia; he negotiated the first trade agreement between Russia and the United States. Buchanan was elected to the Senate in 1835 where he served ten years. President Polk appointed Buchanan Secretary of State. He helped to arrange the Treaty of Guadalupe Hidalgo with Mexico in 1848, by which the United States purchased the region extending west from Texas to the Pacific Ocean. He wrote the Oregon Treaty of 1846 that settled the Northwestern boundary between Canada and the United States. Buchanan ran for President in 1852 and lost. He served as minister to Great Britain. Buchanan ran for President as a Democrat and won. He never married, so his niece Harriet Lane performed the duties of First Lady. Buchanan ran for a second term and was defeated by Lincoln. Before he left office, seven states seceded from the Union. He returned to his home in Lancaster, Pennsylvania, where he lived until he died June 1, 1868. **Buchanan was President from 1857-61 and the Vice President was John C. Breckenridge.**

ABRAHAM LINCOLN

16th President. Abraham Lincoln was born on February 12, 1809, in a log cabin in Hardin County, Kentucky. His parents were uneducated pioneers. His father, Thomas Lincoln, believed that there were better opportunities elsewhere, so the family migrated 100 miles to Indiana. Lincoln helped his father, who was a farmer. His mother died when he was nine and his father remarried. He attended school and loved to read. He was a master of the English language by reading every book he could hold. He practiced law in Springfield, Illinois and he was very popular in the court circuit. He married Mary Todd in 1842 and they had four boys. Edward died when he was four. Lincoln was elected to the United States House of Representatives. He ran for the Senate and lost. He was elected President in 1860 and guided the nation through the Civil War. It was his aim to preserve the union of the states. The Emancipation Proclamation was one of the most important features of his presidency. It did not free the slaves, but it gave all Americans the same rights under the Constitution. The Gettysburg Address was another important document. More than 60,000 soldiers died in a three-day battle in Gettysburg, Pennsylvania; he tried to bring the nation together in their grief. Their son William died in the White House when he was 11. Lincoln was elected to a second term in 1864 and the Civil War ended. The peace treaty was signed on April 11, 1865. Three days later, Lincoln was shot in the head by John Wilkes Booth of Virginia. He died from that gunshot wound April 15, 1865, in the White House. **Lincoln was President from 1861-65 and the Vice Presidents were Hannibal Hamlin and Andrew Johnson.**

ANDREW JOHNSON

17th President. Andrew Johnson was born in Raleigh, North Carolina on December 29, 1808. His father died when he was three, so his mother and brother worked to support themselves. He did not go to school and at 17 he was sent to apprentice with a tailor, where he learned to read. In 1826 when he was 18, his family moved to Greenville, Tennessee. He married Eliza McCardle a year later; she taught him how to write. He opened his own tailor shop, where young men in the town frequently stopped by to debate politics and public affairs. He was elected mayor of Greenville in 1830 and served three years. Johnson was elected to the Tennessee legislature for two terms: from 1835-37 and 1839-43. He served in the House of Representatives for ten years until he was elected Governor in 1853. After serving one term he was voted into office as a U.S. senator until he was elected the Military Governor of Tennessee during the Civil War. In 1864 he was nominated as the Vice President from the Republican Party and was inaugurated Vice President on March 4, 1865; he took office on April 15, as President after Lincoln was assassinated. In 1868 he was impeached by the House of Representatives, accused of breaking the law by dismissing the Secretary of War, Edwin Stanton. He was acquitted in the impeachment trial in 1869. His main accomplishment as president was the purchase of Alaska from Russia. He finished his term as President and returned to Tennessee in March, 1869. He was elected to the Senate in 1874 and died of a stroke a year later, July 31, 1875. **Johnson was President, from 1865-69 and since he was completing Lincoln's term, there was no Vice President.**

ULYSSES S. GRANT

18th President. Ulysses Simpson Grant was born on April 27, 1822 in Point Pleasant, Ohio. He was born to a pioneer family and he was agile with horses. He attended schools in Georgetown, Ohio. In 1839 he received an appointment to West Point Academy, he graduated as a first lieutenant, but he did not like the Army and wanted to join the cavalry but there were no openings. He fought in the Mexican-American War in 1846 until 1848 when the war ended. He married Julia Dent and they had two children. In 1854 he resigned from the Army. He tried farming and real estate and failed. He became a partner in a family leather business in Galena, Illinois. He returned to join the military when the Civil War broke out in 1861 as colonel of the 21st, Illinois Volunteer Regiment. He captured Vicksburg, Mississippi in 1863 giving the Union control of the Mississippi River. Lincoln promoted him to lieutenant general and commander of the Union armies. Grant accepted Lee's surrender April 9, 1865 in Appomattox, Virginia. The Congress appointed Grant full General– the first man to reach this rank since George Washington. He served briefly as secretary of war. Grant was a popular leader after the Civil War. He was elected President in 1869 and served two terms. During his administration the 15th Amendment was ratified giving Negroes the right to vote. The telephone was invented by Alexander Graham Bell. Grant went home to Mount McGregor, New York when he left the White House to write his memoirs. He died on July 23, 1885, one week after he completed his memoirs. **Grant was President, from 1869-77 and the Vice Presidents were Schuyler Colfax and Henry Wilson.**

RUTHERFORD B. HAYES

19th President. Rutherford Birchard Hayes was born in October 4, 1822 in Delaware, Ohio. His father died when he was born. He was educated at home and attended private schools. In 1838 he enrolled in Kenyon College in Gambier, Ohio. Hayes graduated in 1842 and went on to Harvard Law School. He graduated in 1845 and was admitted to the Ohio bar. He married Lucy Webb and they had eight children. In 1857 he was elected City Solicitor of Cincinnati. When the Civil War started he joined the Union Army. He was elected to Congress in 1864 but refused to leave the battlefield. Hayes advanced to the rank of brevetted major general. He was reelected to Congress in 1866 and served only one year because he was elected Governor of Ohio in 1867. He remained in office until 1871 when his term ended. He returned to Spiegel Grove where he practiced law and worked to develop public libraries. He ran again for Governor of Ohio in 1876 and won. He was nominated as the Republican candidate for President against Samuel J. Tilden. Hayes won in a close election that was disputed for four months and ultimately decided by an electoral commission. During his administration Hayes removed all federal troops from the South ending Reconstruction—the period after the Civil War during which the southern states were reorganized and made part of the Union once again. He tried to promote civil service reform. Hayes did not want to serve a second term and retired to his farm in Spiegel Grove, Ohio where he devoted his time to his philanthropic activities until he died January 17, 1893. **Hayes was President from 1877-81 and the Vice President was William A. Wheeler.**

JAMES A. GARFIELD

20th President. James Abram Garfield was born in Orange, Ohio on November 19, 1831. He was the last president to be born in a log cabin. His father died when he was two; he had three brothers and sisters. Since they had to work to help support their family, they could only attend school three months out of the year. He loved to read. At the age of 16 Garfield left home and joined the Navy in Cleveland. He had to return home when he became sick with malaria. He attended Williams College in Massachusetts. He graduated and became a teacher at Hiram College where he was once a student. A year later he became President of the College. He married Lucretia Rudolph in 1858; they had five children He was elected to the Ohio Senate in 1859 where he served for two years. Garfield volunteered for the Union Army in 1861. He fought bravely in the Civil War battles at Shiloh and Chickamauga. He resigned in 1863 with the rank of major general, when President Lincoln asked him to serve in Congress. In December 1863, at the age of 32, he entered the House of Representatives. He served as a legislator for 17 years and he became minority leader. He was nominated by the Republican Party and was elected President in 1881. Garfield was the first President to use a telephone. During his administration the great railroad expansion reached Texas. James Garfield was assassinated on July 2nd, while waiting for a train, only a few months after he took office. He died two months later in Elbron, New Jersey on September 19, 1881. **Garfield was President from March 4, 1881-September 17, 1881 and the Vice President was Chester A. Arthur.**

CHESTER A. ARTHUR

21st President. Chester Alan Arthur was born in Fairfield, Vermont on October 5, 1830. His father was a teacher and minister. Arthur entered Union College in Schenectady, New York when he was 15. He taught school to pay for his education and graduated with honors in 1848. He then went on to study law, while continuing to teach to support himself, until 1853 when he went to New York City to start his career as a lawyer. He took on two very controversial and groundbreaking cases. His law firm gained freedom for eight runaway slaves. He also represented Lizzie Jennings in a segregation case. She was told she was not allowed to ride on a New York City streetcar because she was black; she was awarded $500 in damages. The court decision stated that Negroes had the same right to ride on New York streetcars as anyone else. Chester Arthur married Ellen Lewis Herndon in 1859. During the Civil War, Arthur was appointed quartermaster general of New York State, supplying food, guns and tents to the soldiers. After that he became collector of customs for the port of New York. He held that job for seven years until he was elected Vice President. When James Garfield was assassinated only six months after he was sworn in as President, Chester Arthur became President on September 19, 1881. During his administration he passed the Pendleton Act, which helped to eliminate the corruption of political favors and he established the Civil Service Agency. After serving as president, he returned to New York and practice law. A year later Chester Arthur became ill and died at his home on November 18, 1886. **Arthur was President from 1881-85 and since he was completing the term of James Garfield, there was no Vice President.**

GROVER CLEVELAND

22nd and 24th President. (Stephen) Grover Cleveland was born in Caldwell, New Jersey on March 18, 1837. His family moved to Fayetteville, New York where he was educated in a one-room schoolhouse. In 1850 they moved to Clinton, New York where he attended the Liberal Institute. When he was 18 he visited an uncle in Buffalo, New York who asked for his help on his farm, in return for getting him a job in a law office. Four years later, Cleveland was admitted to the bar. In 1863 he became assistant district attorney of Erie County, New York for two years. His record of honesty and fairness led to his election as sheriff of Erie County; a job he held for two years. In 1882 Cleveland was elected mayor of Buffalo. He was in office one year, when the Democrats nominated him for governor of New York and he won. Two years later he was nominated for President, and won. Grover Cleveland married Frances Folsom, June 2, 1886 in a White House ceremony. He lost the 1888 presidential election to Benjamin Harrison. He returned to New York where he practiced law. He was nominated for President again by the Democrats and won. Cleveland was inaugurated as the 24th President in 1893. He served one term, he was not nominated by his party for a second term because he did not support the expansion of the United States into Hawaii and Cuba. His party turned to William Jennings Bryan who lost to William McKinley. Cleveland returned to Princeton, New Jersey where he lived until he died on June 24, 1908. **Cleveland was President from 1885-89 and the Vice President was Thomas A. Hendricks. He served as President again from 1893-97 and the Vice President was Adlai E. Stevenson.**

BENJAMIN HARRISON

23rd President. Benjamin Harrison was born August 20, 1833 in North Bend, Ohio, the grandson of President William Henry Harrison. He attended school in a log cabin and at the age of 14 he entered the Farmer's College near Cincinnati. Three years later he transferred to Miami University at Oxford, Ohio. Harrison went on to study law and he was admitted to the bar. In 1893 Harrison married Caroline Scott; they had two children. He started a law practice in Indianapolis, Indiana. Harrison joined the Union Army in 1862 during the Civil War and earned the position of general. When the war ended he returned to his law practice. In 1876 he ran unsuccessfully for Governor of Indiana. In 1881 Harrison was elected to the Senate, he served the six-year term. He was elected President in 1889. His wife died two weeks before the election. During his administration Montana, Idaho, Washington, North and South Dakota and Wyoming were admitted into the Union. The Sherman Anti-Trust Act was passed in 1890, which was designed to regulate monopolies and prices. American women were first allowed to vote in Wyoming in 1890. The first gasoline-powered automobile was built in 1893 by the Duryea brothers. Basketball was invented by James Naismith in 1891 in Springfield, Massachusetts. He fortified the Navy and established trade policies over seas. Harrison was defeated in the 1892 presidential election. After leaving office he returned to his home in Indianapolis. He married again in 1896 to Mary Scott Lord Dimmick; they had one child. He worked at his law practice in Indianapolis until he died March 13, 1901. **Harrison was President from 1889-93 and the Vice President was Levi P. Morton.**

WILLIAM MCKINLEY

25th President. William McKinley was born in Niles, Ohio on January 29, 1843. He attended school in Poland, Ohio and then went to Allegheny College in Meadville, Pennsylvania. At the age of 17 he left college before he graduated and started teaching in a rural school. The Civil War broke out in 1861. He enlisted as a private in the 23rd Ohio Volunteer Regiment. He was promoted to captain and served under Rutherford B. Hayes. When the Civil War ended in 1865 he returned to Ohio to study law. He opened a law office in 1867 and two years later he was elected county solicitor. He married Ida Saxton in 1871. They had two daughters who died at an early age. His wife suffered from epilepsy and needed special attention. In 1876 McKinley was elected to the House of Representatives where he served for 17 years. In 1892 he was elected Governor of Ohio. He served one term and was nominated by the Republican Party to run for President. He won and he was a popular president. He guided the country through the Spanish-American War that lasted 100 days. As a result of the war, Cuba gained its independence from Spain. He became known for the open door policy with China, which initiated the equality of trade with China and respect for its territorial integrity. McKinley was reelected to a second term but he was assassinated only six months after he was sworn in. He died on September 14, 1901, eight days after being shot at point blank range, when he was attending a reception in Buffalo, New York. He is buried in Canton, Ohio. **McKinley was President from 1897-1901 and the Vice Presidents were Garret A. Hobart and Theodore Roosevelt.**

THEODORE ROOSEVELT

26th *President.* Theodore Roosevelt was born on October 27, 1858 in New York City. He graduated from Harvard College. In 1880 he married Alice Hathaway Lee. He enrolled in Columbia University to study law. He was elected to the New York State Assembly and abandoned his studies. In his third term in 1884, Roosevelt's mother died, a few hours later his wife who had given birth to a baby girl also died. He served to the end of his term, and moved to the Dakota Territory to his ranch on the Little Missouri River. He married Edith Carow; they had five children. He accepted an appointment to the United States Civil Service Commission, serving for six years until he was appointed President of the New York City board of police commissioners. He resigned from the Police Board to become assistant Secretary of the Navy. A year later during the Spanish-American War, he organized and then commanded the first United States Volunteer Cavalry Regiment. In 1899 he was elected Governor of New York. Roosevelt was inaugurated Vice President of the United States, March 4, and sworn in as President upon the death of President William McKinley on September14, 1901. Roosevelt received the Nobel Peace Prize and he served two terms. He traveled to Africa after he left the White House, staying there a year and returned to New York City to organize the Bull Moose Party to run for President in 1912. He lost to Woodrow Wilson. He wrote his autobiography and then traveled to South America in 1913. Roosevelt died in his sleep at his home in Sycamore Hill, New York, on January 6, 1919. **Roosevelt was President from 1901-09 and the Vice President was Charles W. Fairbanks.**

WILLIAM HOWARD TAFT

27th *President.* William Howard Taft was born in Cincinnati, Ohio on September 15, 1857. His father was a judge and his son William was an excellent student. He entered Yale College in 1874 and graduated four years later. He returned to Cincinnati and became a successful lawyer. In 1881 he was appointed assistant prosecutor in a county court. He went on to become an official in the Bureau of the Internal Revenue Service. He did not like that job and resigned. He married Helen Herron in 1886; they had three children. In 1887 he was named a judge on the Ohio Superior Court. In 1890, President Benjamin Harrison offered Taft the post of Solicitor General of the United States, the second largest job of the Department of Justice. In 1892 he resigned to become a federal circuit judge. President William McKinley picked Taft to become head of a special commission to oversee the government of the Philippine Islands. In 1904 President Roosevelt appointed Taft Secretary of War. Taft was nominated for President by the Republican Party and won. He helped establish the 16th Amendment to the Constitution, which allows Congress to collect income tax. He served one term. In the next election he was opposed by Roosevelt and Woodrow Wilson. He retired to Yale University where he was a professor of constitutional law until he was appointed Chief Justice of the United States by President Harding in 1921. His administrative work as a Supreme Court justice is considered his greatest contribution to public service. He retired in 1930 due to poor health and died the same year on March 8 in Washington D.C. **Taft was President from 1909-13 and the Vice President was James S. Sherman.**

WOODROW WILSON

28th *President.* (Thomas) Woodrow Wilson was born in Staunton, Virginia, on December 28, 1856. He attended private school in Georgia and South Carolina. He enrolled in Davidson College in North Carolina and then Princeton University in New Jersey. He studied law at the University of Virginia and went on to Johns Hopkins University in Baltimore, where he graduated with a PhD in philosophy in 1886. He married Ellen Axson in 1885 and they had three daughters. He taught at Bryn Mawr College in Pennsylvania and at Wesleyan University in Connecticut. He later taught constitutional law at Princeton University and became the president of the University. In 1911 he was elected governor of New Jersey. He was elected President in 1913 during World War I. During his presidency 1,000,000 automobiles were produced. As President, Wilson reduced tariffs and created an income tax. The United States became involved in World War I when the Germans ambushed the *Lusitania*. Wilson was awarded the Nobel Peace Prize in 1919. He tried to introduce the concept of the League of Nations when he went to Paris to present the 14 Points of Peace and to help write the Versailles Treaty. The armistice was signed on November 11, 1918. Wilson had support for the League of Nations abroad, but when he proposed the idea at home, it was defeated in the Senate. Women were given the right to vote when the 19th Amendment was added to the Constitution in 1920. He left office in 1921. He died of a stroke while on a speaking tour in Washington, D.C. to promote the League of Nations, on February 3, 1924. **Wilson was President from 1913-21 and the Vice President was Thomas R. Marshall.**

WARREN G. HARDING

29th *President.* Warren Gamaliel Harding was born in Corsica, Ohio on November 2, 1865. He helped his father on the farm and attended school. He was the oldest of eight children; his father was a doctor. When he was 14, Harding enrolled in Ohio Central College in Iberia. He graduated and took a job as a schoolteacher but quit after one year. He moved to Marion, Ohio and studied law, but he didn't like it. He borrowed money from his father, and with two friends bought the newspaper, the Marion *Star*. His partners left the business and he ran it by himself. He married a widow, Florence Kling De Wolfe in 1891 who helped him make the newspaper successful. In 1899 he was nominated by the Republicans to the State Senate and won; he served four years. In 1903 he was elected Lieutenant Governor of Ohio and served for three years. He returned to Marion to work on the *Star*. In 1910 he ran for governor and lost. In 1914 he was elected to the United States Senate. He served a full term and he was sworn in as President in 1921. During his administration the first nonstop transcontinental airplane flight was made and the first commercial radio broadcast took place. During his presidency Harding established a budget system for the federal government but his administration was marred by the Tea Pot Dome scandal: federal oil reserves were illegally rented to private businesses. In June 1923, Harding and his wife took a trip across the country. He fell ill from food poisoning when he arrived in San Francisco after leaving Alaska. He developed pneumonia, and died suddenly on August 2, 1923. **Harding was President from 1921-23 and the Vice President was Calvin Coolidge.**

CALVIN COOLIDGE

30th President. Calvin Coolidge was born on July 4, 1872 in Plymouth Notch, Vermont. He went to school in a one-room schoolhouse. He was raised on a farm and enrolled in Black River Academy in Ludlow, Vermont when he was 13. He was admitted to Amherst College in Massachusetts and after graduation in 1895 he decided to become a lawyer. He took a job in a law office in Northampton, Massachusetts. While working, he studied for the bar exam and was admitted to the bar in 1897. He opened his own law office in the same town. In 1898 he was elected city councilman. He married Grace Goodhue in 1905; they had two sons. In 1907 Coolidge was elected to the Massachusetts legislature. Three years later he was voted mayor of Northampton. In 1912 he became a Massachusetts State senator. He was elected Lieutenant Governor of Massachusetts in 1916. Coolidge was elected governor in 1919. He served one year. He was elected Vice President of the United States in 1920. In 1923 he became President of the United States when President Warren Harding died on August 3. The next year he won the presidential election and served the full term. The 18th Amendment, Prohibition, was enforced during his presidency. During his term Coolidge worked to lower the national debt, limit government spending and lower taxes. Charles Lindbergh was the first man to fly alone nonstop across the Atlantic Ocean. Coolidge did not want to run for a second term. He retired to Northampton where he wrote his biography and lived a quiet life until he died of a heart attack on January 5, 1933. **Coolidge was President from 1923-29 and the Vice President was Charles G. Dawes.**

HERBERT C. HOOVER

31st President. Herbert Clark Hoover was born on August 10, 1874, in West Branch, Iowa. His father died of typhoid fever when he was six and his mother died of pneumonia when he was nine. He moved to Newberg, Oregon to live with his uncle, Dr. Henry Minthorn, where he attended a Quaker academy. When he was 16 he enrolled in Leland Stanford, a new university in Palo Alto, California. He graduated in 1895 and worked in the goldfields of Nevada. In 1927 a British company hired Hoover to manage its gold mines in Australia where he lived for two years. He returned home and married Lou Henry in 1899. During the next 15 years he worked on engineering projects in Europe, India, South Africa, and Egypt. He was rich enough to retire before he was 40 years old. Hoover was in London as Chairman of the American Relief Committee when World War I broke out. He also served on the Commission for Relief in Belgium. President Warren G. Harding appointed Hoover Secretary of Commerce at the end of World War I. Coolidge decided not to run again for President in 1928, so Hoover's opponent in the presidential election was New York Governor, Alfred E. Smith. Hoover won in a landslide victory. As president he worked to improve the standards of radio broadcasting, aviation and housing. During his administration the stock market crashed in 1929 leading to the Great Depression. He established a few programs to combat the crises, but he opposed aid through federal bureaucracy, so he did not provide enough relief and he was defeated in the 1932 election. He died on October 20, 1964 in New York City at the age of 90. **Hoover was President from 1929-33 and the Vice President was Charles Curtis.**

FRANKLIN D. ROOSEVELT

32nd President. Franklin Delano Roosevelt was born on January 30, 1882 in Hyde Park, New York. He was tutored by a governess and he took frequent trips to Europe. He graduated in 1904 from Harvard University and attended Columbia University Law School in New York. He married Eleanor Roosevelt, a distant cousin, in 1905; together they had six children. He passed the bar and joined a law firm on Wall Street. He ran for the New York Senate and won when he was 28 years old. President Woodrow Wilson appointed Roosevelt assistant Secretary of the Navy three years later in 1913. Roosevelt traveled to Canada where he contracted polio, leaving him paralyzed from the waist down for the rest of his life. In 1929 he ran for Governor of New York and won. He served one term; he was nominated by the Democratic Party to run for President. Roosevelt won in a landslide victory. He helped the nation through its economic depression by creating the Public Works Administration. In 1937 he was elected to a second term. By then the economy had reached the prosperity levels of the 1920s. Roosevelt was elected to a third term in 1941. On December 7 of that year, the United States Naval fleet was attacked by the Japanese Air Force. The United States declared war on Japan. On December 8, Roosevelt declared war with Italy and on December 11, war with Germany. The United States was embroiled in World War II when Roosevelt was elected to his fourth term as President. He died in office before the end of his first year on April 12, 1945. **Roosevelt was President from 1933-45 and the Vice Presidents were John N. Garner, Henry A. Wallace and Harry S Truman.**

HARRY S TRUMAN

33rd President. Harry S Truman was born in Lamar, Missouri on May 8, 1884. His father lost all of their money in the family business so Truman had to forgo college. He worked on the family farm for 10 years to help his family pay off their debts. During World War I, he joined the Missouri National Guard, commanding a field artillery battery. He married Bess Wallace in 1919; they had one daughter. Truman opened a men's clothing store in the post-war depression. He lost his life's savings and was left $20,000 in debt. He spent 15 years paying it off. He ran for County Judge and won. Truman had little knowledge about law so he attended the Kansas City School of Law in 1928. He served as a judge of the Jackson County Court until 1935 when he was elected to the United States Senate. He was elected to two terms as Senator of Missouri, but left the office to join Franklin D. Roosevelt's campaign for President. He was inaugurated Vice President of the United States, January 20, 1945 and became President four months later upon the death of Franklin D. Roosevelt, on April 12. Truman adopted Roosevelt's programs, renaming them the "Fair Deal." He worked hard for equal rights for all citizens of the United States. He was the only president to use the atomic bomb, when the United States attacked Hiroshima and Nagasaki, Japan on September 2, 1945, causing Japan to surrender and end World War II. Truman was elected to a second term but decided not to run for office in 1953. He returned to his home in Independence, Missouri where he devoted his time to the Truman Library. He died on December 26, 1972. **Truman was President from 1945-53 and the Vice President was Alben W. Barkley.**

DWIGHT D. EISENHOWER

34th President. Dwight David Eisenhower was born on October 14, 1890 in Denison, Texas. His family moved to Abilene, Kansas where his father bought a house on a 10-acre farm. He worked on the farm with his four brothers. Everyone called him "Ike". In high school he worked nights at the creamery where he studied his lessons. He was accepted into West Point and graduated in 1915, commissioned as a second lieutenant. A year later he married Mamie Geneva Doud. General Douglas MacArthur asked Eisenhower to assist him in planning the defense of the Philippines. In 1942 Eisenhower became the commanding General of the American Forces in the European theater and the Commander in Chief of the Allied forces in North Africa, commanding the British and American troops. A year later he was promoted to full (4-Star) general. On June 6, 1944, also known as D-Day, he was promoted to General of the Army. May 7, 1945 the German Army surrendered. For the next three years Eisenhower was Chief of Staff of the United States Army until he was named the President of Columbia University. In 1951, he was named Commander of NATO Forces in Europe. Eisenhower was nominated by the Republicans to run for President and won; he served two terms. While President, he suffered a severe heart attack in 1955 and another stroke. During his administration he was engaged in foreign policy, ending the Korean War, and negotiating with China, Cuba and Russia. He enforced the Supreme Court's decision to integrate schools. Eisenhower died on March 28, 1969 at Walter Reed Hospital in Washington D. C. **Eisenhower was President from 1953-61 and the Vice President was Richard M. Nixon.**

JOHN F. KENNEDY

35th President. John Fitzgerald Kennedy was born on May 29, 1917 in Brookline, Massachusetts; the second of nine children. He attended private schools and enrolled in Princeton University. The summer of his freshman year he traveled to London where he fell ill with jaundice. It took two years for him to fully recover. He resumed his studies at Harvard and graduated with honors in 1940. When the Japanese attacked Pearl Harbor, Kennedy signed up with the Navy and became a Naval Commander. He had to leave the Navy in 1944 after contracting malaria. He was elected to Congress in 1946 in a Boston district; and re-elected twice. In 1953 he was elected Senator of Massachusetts defeating Henry Cabot Lodge. That same year he married Jacqueline Bouvier; they had three children, Patrick Bouvier died a few days after he was born. Kennedy was sworn in as President in 1960. His administration was marked by the invasion of the Bay of Pigs in Cuba, in which anti-Castro Cubans invaded Cuba but failed to complete their mission, causing a stand off. Kennedy succeeded in forcing the Soviet Union to remove missiles from Cuba and in keeping troops in Berlin to prevent the spread of Communism. Kennedy supported the Civil Rights movement led by Martin Luther King. He advocated improved medical care for the elderly and space exploration. Kennedy initiated the Peace Corps, an organization of volunteers who carry education and skills to underdeveloped countries. Kennedy was assassinated while traveling in a motorcade through Dallas, Texas on November 22, 1963. **Kennedy was President from 1961-63 and the Vice President was Lyndon B. Johnson.**

LYNDON B. JOHNSON

36th President. Lyndon Baines Johnson was born in Stonewall, Texas on August 27, 1908. He attended public schools and after realizing how hard it was to find a good job without a college education, he enrolled in Southwest Texas State Teachers College. He graduated in 1930. He taught for one year in the Texas schools and left to enter politics. He worked for a Congressman and was then appointed Texas state director of the National Youth Administration. He married Claudia Alta Taylor better known as Lady Bird; they had two daughters. Johnson was elected to the United States House of Representatives in 1937. He joined the Navy when the Japanese attacked Pearl Harbor on December 7, 1941, and fought in the Pacific Theater. He returned from the war in July 1942 when President Roosevelt ordered all members of Congress, in military service to return to their duties. In 1948 he was elected to the Senate and became the Democratic majority leader in 1954. He was elected Vice President when John F. Kennedy won the presidential election in 1959. Johnson took the oath of office 99 minutes after Kennedy was assassinated in Dallas, Texas in 1963. He passed Kennedy's bills on civil rights, poverty and conservation. Johnson ran for President in 1964 and won. During his administration he was known for creating the "Great Society" that promoted equal rights for all and provided health insurance for the elderly. Johnson decided not to run for re-election in 1968 because of the national turmoil of the Vietnam War, which escalated during his administration. He returned to his ranch in Texas where he died on January 22, 1973. **Johnson was President from 1963-69 and the Vice President was Hubert H. Humphrey.**

RICHARD M. NIXON

37th President. Richard Milhous Nixon was born on January 9, 1913 in Yorba Linda, California. His family moved to Whittier, California, where he and his five brothers attended public school and worked at his father's store. He enrolled in Whittier College when he was 17. In 1934, Nixon received a scholarship to Duke University Law School in North Carolina. After graduation, he tried to get a job with a law firm in New York, but he was turned down every time and returned to Whittier, California, where he met and married Patricia Ryan. They had two daughters. Nixon joined the Navy in 1942 rising to lieutenant commander by the end of the war. He was elected to Congress in 1947 and served there until he was elected to the United States Senate in 1951. In 1959 he was elected Vice President under Eisenhower. He served eight years as an active leader when Eisenhower was sick. Nixon ran for President in 1960 and lost against Kennedy. Two years later, he ran for Governor of California and lost. He moved to New York to practice law. He was nominated for President by the Republican Party in 1968, and won. Nixon began bringing troops home from Vietnam; in his second term the war ended with the victory of North Vietnam. During his administration the Apollo 11 space flight landed a man on the moon. Nixon rebuilt ties with China, after 21 years of silence. The *Washington Post* began investigating the illegal activities of Nixon and his aides during the 1972 election campaign. In 1974 he faced impeachment and resigned as president. He retired to his home in San Clemente, California where he died on April 22, 1994. **Nixon was President from 1969-74 and the Vice Presidents were Spiro T. Agnew and Gerald R. Ford.**

GERALD R. FORD

38th President. Gerald Rudolph Ford was born on July 14, 1913 in Omaha, Nebraska. His family moved to Grand Rapids, Michigan where he attended school. He was a popular student and athlete– he played football in high school and college at the University of Michigan. Ford went on to law school and he was admitted to the bar. He joined the Navy in 1942 to fight in World War II; he was promoted to the rank of lieutenant commander and remained on active duty until 1946. After the war he ran for a seat in Congress and won. He spent 25 years in the United States House of Representatives. Ford married Elizabeth Bloomer Warren in 1948; they had four children. In 1973, after the resignation of Vice President Spiro T. Agnew, President Nixon asked Ford to be the Vice President. A year later, Ford replaced Nixon as President, when Nixon resigned his presidency because of the Watergate scandal. Ford is the only President and Vice President who was not elected to the office. As President, Ford gave Nixon a pardon, "for all offenses against the United States." The act made him unpopular with many people. Ford, who was a Republican, had many differences with the Democrat-controlled Congress. He vetoed over 50 bills. He successfully controlled inflation during his administration, but there was high unemployment during his term in office. He was successful with his foreign policies– he championed détente, and peace among nations. Ford tried to bring the nation together after Nixon left the White House, but he lost his bid for a second term in 1976. Ford and his wife Betty are active citizens and live in Rancho Mirage, California. **Ford was President from 1974-77 and the Vice President was Nelson A. Rockefeller.**

JAMES E. CARTER, JR.

39th President. James Earl Carter, Jr. was born on October 1, 1924 in Plains, Georgia. Carter grew up on a peanut farm. He left home to attend the United States Naval Academy at Annapolis, Maryland where he studied nuclear physics, and graduated in 1946. He served in the Navy for seven years and returned home to run the family business. He married Rosalynn Smith, they had four children. Carter served on the school board and educational advisory boards. He was a strong advocate for Civil Rights. In 1962 he was elected to the United States Senate; he served one term. He ran for Governor of Georgia and won. Carter was nominated by the Democratic Party to run for President in 1976, and won. As President he established diplomatic ties with China and helped to write the Camp David Accords; a peace treaty between Egypt and Israel. He helped negotiate the Salt II treaty with the Soviet Union. During his administration Carter was criticized for the poor economy. In 1979, a group of Iranian students took the staff of the United States Embassy, hostage. Efforts to negotiate the release of the 52 hostages in Iran consumed the later half of his term. He was not successful and this was perceived as weakness and made him unpopular. He was defeated in the 1980 election. After his term as President, Carter started the Carter Center to introduce democracy throughout the world. He won the Nobel Peace prize. Carter and his wife Rosalynn live in Plains, Georgia. They have actively promoted humanitarian programs and peace throughout the world since he left the office of President. **Carter was President from 1977-81 and the Vice President was Walter F. Mondale.**

RONALD W. REAGAN

40th President. Ronald Wilson Reagan was born on February 6, 1911 in Tampico, Illinois. He attended public schools and enrolled in Eureka College in Illinois, where he studied economics and sociology. After graduating in 1932 he worked as a radio announcer in Iowa. In 1937 he started his career as an actor, appearing in more than 50 movies during his 27 year career. In 1940 he married Jane Wyman, they had a daughter and adopted a son; their marriage ended in divorce in 1948. Reagan joined the United States Air Force in 1942; he served three years. In 1952 he married actress Nancy Davis; they have a daughter and son. Reagan ran for Governor of California as a Republican in 1967 and won. He was elected President in a landslide victory over Jimmy Carter in 1980. Moments after he was sworn into office the hostages were released in Iran. During his administration he set out to strengthen military defense. He tried to stop inflation and improve business growth by instituting supply-side economics, which called for cutting taxes and government programs, but resulted in high budget deficits. He also promoted Social Security and tax reforms. In 1981 Reagan was wounded in an assassination attempt. During his second term the Berlin Wall came down, uniting East and West Germany who had been divided since World War II. His administration was criticized for the sale of weapons to Iran in exchange for hostages and the use of money to aid rebels in Nicaragua. Ten years before his death, Reagan suffered from Alzheimer's disease. He died in Bel Air, California on June 6, 2004. **Reagan was President from 1981-89 and the Vice President was George H. W. Bush.**

GEORGE H. W. BUSH

41st President. George Herbert Walker Bush was born on June 12, 1924 in Milton, Massachusetts. He attended private schools in Connecticut and Phillips Academy in Andover, Massachusetts. In 1942 he served in the Navy as a pilot during World War II and received the Distinguished Flying Cross after being shot down over the Pacific in 1944. He married Barbara Pierce in 1945. They had six children; one died of leukemia when she was four. In 1948 he graduated from Yale University. He started a development company, which dealt in Texas oil and gas properties. Bush tried running for the Senate and lost. He was elected to Congress in 1966. He tried again to run for Senator of Texas, and lost. He served as the ambassador to the United Nations and head of the Central Intelligence Agency. In 1974, he was appointed by President Ford as the U.S. liaison officer to the People's Republic of China. He served eight years as Vice President when Reagan was elected President in 1980. Bush was elected President in 1988. In 1989, he ordered an invasion of Panama to remove the country's leader, General Manuel Noriega. In 1990, Iraq invaded Kuwait and President Bush led the worldwide forces against Iraq in the Persian Gulf War. During his administration Bush supported democratization in the Soviet Union, which dissolved in 1991, and in Eastern Europe. He raised taxes to combat budget deficits. His popularity diminished in late 1991 due to the economic recession. He lost the election to Bill Clinton. George Bush and his wife Barbara live in Texas. **Bush was President from 1989-93 and the Vice President was J. Danforth Quayle.**

WILLIAM J. CLINTON

42nd President. William Jefferson Clinton was born William Jefferson Blythe IV on August 19, 1946, in Hope, Arkansas. His father died in a traffic accident a few months later. His mother wed Roger Clinton; he took the family name in high school. He attended public schools in Arkansas. He moved to Washington D.C. to attend Georgetown University. In 1968, he won a Rhodes Scholarship to Oxford University in London. He received a law degree from Yale University in 1973. Clinton returned to Arkansas and ran for Congress and lost. He married Hillary Rodham; they have one daughter. He taught at the University of Arkansas. In 1976, Clinton was elected Arkansas Attorney General. He ran for Governor of Arkansas in 1978 and won. He lost his bid for a second term, but he regained the office four years later, and served until he won the presidential nomination against incumbent George Bush in 1992. As President, Clinton successfully reduced the national debt. He promoted international trade and persuaded Congress to pass the North American Free Trade Agreement (NAFTA) between the United States, Canada and Mexico. Clinton was a global proponent for an expanded NATO. He was reelected in 1996. He supported use of NATO forces to remove Serbian troops from Kosovo in 1999. He improved relations with Russia and China. During his second term, in 1998 Clinton was the second president impeached by the House of Representatives. He was accused of giving false court testimony. He was tried in the Senate and found not guilty of the charges. Clinton and his wife Hillary live in New York, where she is a Senator. **Clinton was President from 1993-2001 and the Vice President was Albert Gore, Jr.**

GEORGE W. BUSH

43rd President. George Walker Bush was born on July 6, 1946 in New Haven, Connecticut the son of the 41st President, George H.W. Bush. He grew up in Houston, Texas where he attended private schools. He received a bachelor's degree from Yale University in 1968, and then served as an F-102 fighter pilot in the Texas Air National Guard. He enrolled in Harvard Business School in 1975 and received a Master of Business Administration. He married Laura Welch; they have twin daughters. He bought the Texas Rangers baseball franchise in 1989 by forming a partnership with a group of investors. He served as general manager and partner of the team, until he was elected Governor of Texas in 1994. Bush was the first Governor of Texas to be elected to consecutive four-year terms when he was re-elected in 1998. Since he was sworn in as president in 2001, President Bush has signed into law initiatives to improve public schools by raising standards, requiring accountability, and strengthening local control. He has signed tax relief that lowered the income tax rate. He has increased pay and benefits for America's military. After the attacks on the World Trade Center, Bush created the Office of Homeland Security to protect Americans against terrorism by making citizens better informed of suspected attacks and more vigilant. American troops have liberated Afghanistan from the Taliban regime. An American-led coalition of a few allied countries invaded Iraq in March 2003 to overthrow Saddam Hussein. The United States is helping Iraq establish a democratic form of government under new leadership. **Bush became President in 2001 and the Vice President is Richard B. Cheney.**

The Fifty United States

Britain's American colonies broke with the mother country in 1776 and were recognized as the new nation of the United States of America following the Treaty of Paris in 1783. During the 19th and 20th centuries, 37 new states were added to the original 13 as the nation expanded across the North American continent and acquired a number of overseas possessions. The two most traumatic experiences in the nation's history were the Civil War (1861-65) and the great Depression of the 1930s. Buoyed by victories in World Wars I and II and the end of the Cold War in 1991, the US remains the world's most powerful nation-state. The economy is marked by steady growth, low unemployment and inflation, and rapid advances in technology.

From the CIA Factbook, 2003

Size: 3,716,829 square miles   **Population:** 290,342,554

Motto "In God We Trust", adopted July 30, 1956. The original motto adopted September 9, 1776, was:
                    "E Pluribus Unum" (Out of many, one).

The Great Seal The Great Seal of the United States shows an American bald eagle with a ribbon in its mouth bearing the Latin words 'e pluribus unum". In its talons are the arrows of war and an olive branch of peace. On the back of the Great Seal is an unfinished pyramid with an eye (the eye of Providence) above it. The seal was approved by Congress on June 20, 1782.

Bird Eagle, adopted June 20, 1782

Song "The Star Spangled Banner," written by Francis Scott Key in 1814, was adopted by Congress as the National anthem in 1931.

Flower Rose, chosen by President Ronald Reagan when he was in office.

Washington, DC   *The Nation's Capital*

Washington, DC is the capital of the United States. It became the capital of the United States in 1800, when the federal government moved from Philadelphia. The city was named after George Washington. It was built to be the legislative, administrative, and judicial center of the United States.

There are 715 libraries—only 118 fewer than New York City. The Residence Bill of July 16, 1790, established the federal district which was first called the Territory of Columbia and the federal city, the City of Washington. The name changed to the District of Columbia in 1793. The first act passed by the District's first legislative assembly on August 3, 1871, created the corporate seal of the District of Columbia.

The present system of government (since 1975) provides for an elected mayor and city council, but reserves for Congress veto power over the budget and legislation and direct control over most of the federal buildings and monuments. The Twenty-Third Amendment (1961) to the Constitution gave inhabitants the right to vote in presidential elections.

Size 61 square miles
Population 572,059
Great Seal Justice, who is a woman, is shown on the seal hanging a wreath
         on a statue of George Washington.
Motto Justia Omnibus, Justice for all
Flower American beauty rose
Bird Woodthrush
Tree   Scarlet Oak

420

Alabama (AL)

The Yellowhammer State

Alabama comes from an Indian word for "tribal town."

Capital	Montgomery
Size	52,419 square miles
Population	4,486,508
Statehood	December 14, 1819; 22nd State
Motto	"We Dare Defend Our Rights"
Tree	Southern Pine
Bird	Yellowhammer
Flower	Camellia
Song	"Alabama," by Julia S. Tutwiler Music by Edna Goeckel Gussen

Alaska (AK)

The Last Frontier

Alaska comes from alakshak, the Eskimo word meaning "peninsula" or "land that is an island."

Capital	Juneau
Size	663,267 square miles
Population	643,786
Statehood	January 3, 1959, 49th State
Motto	"North to the Future"
Tree	Sitka Spruce
Bird	Willow Ptarmigan
Flower	Forget-Me-Not
Song	"Alaska's Flag" by Marie Drake Music by Elinor Dusenbury

NOTE: The abbreviations following the state name are U.S. Postal abbreviations.

Arizona (AZ)

The Grand Canyon State

Arizona comes from a Pima Indian word "Arizonac", meaning "little spring place."

Capital	Phoenix
Size	114,006 square miles
Population	5,456,453
Statehood	February 14, 1912 48th State
Motto	"Ditant Deus (God Enriches)"
Tree	Palo Verde
Bird	Cactus Wren
Flower	Saguaro Cactus Blossom
Song	"Arizona" by Margaret Rowe Clifford Music by Maurice Blumenthal

Arkansas (AR)

The Land of Opportunity

Arkansas is a variation of Quapaw, the name of an Indian tribe, which means "south wind."

Capital	Little Rock
Size	53,182 square miles
Population	2,710,079
Statehood	June 15, 1836 25th State
Motto	"Regnat Populus" (The People Rule)
Tree	Southern Pine
Bird	Mockingbird
Flower	Apple Blossom
Songs	"Arkansas," Music and Words by Wayland Holyfield "Oh, Arkansas," Music and Words by Terry Rose & Gary Klaff

California (CA) *The Golden State*

California was named by Spanish explorers after an imaginary island in the story, *Las Sergas de Esplandian*, by Garcia Ordonez de Montalvo, circa 1500.

Capital	Sacramento
Size	163,696 square miles
Population	35,116,033
Statehood	September 9, 1850, 31st State
Motto	"Eureka" (I have found it)
Tree	California Redwood
Bird	California Quail
Flower	California Poppy
Song	"I Love You, California"

Colorado (CO) *The Centennial State*

Colorado comes from an Algonquin Indian word meaning "long river place."

Capital	Denver
Size	104,100 square miles
Population	4,506,542
Statehood	August 1, 1876 38th State
Motto	"Nil Sine Numine" (Nothing without providence)"
Tree	Colorado Blue Spruce
Bird	Lark Bunting
Flower	Columbine
Song	"Where the Columbines Grow" by A.J. Flynn

Connecticut (CT)   *The Constitution State*

Connecticut comes from an Algonquin Indian word Quinnehtukqut, meaning "beside the long tidal river."

Capital	Hartford
Size	5,544 square miles
Population	3,460,503
Statehood	January 9, 1788 5th State
Motto	"Quitranstulit Sustined" (He who, transplanted, sustains)
Tree	White Oak
Bird	American Robin
Flower	Mountain Laurel
Song	"Yankee Doodle"

Delaware (DE)   *The First State*

Delaware is named after Sir Thomas West, Baron La Warr, the English governor of Virginia in colonial times.

Capital	Dover
Size	2,489 square miles
Population	807,385
Statehood	December 7, 1787, 1st State
Motto	"Liberty and Independence"
Tree	American Holly
Bird	Blue Hen Chicken
Flower	Peach Blossom
Song	"Our Delaware"

Florida (FL)

The Sunshine State

Florida, means "feast of flowers or flowery" in Spanish. The territory was named by the explorer Ponce de Leon, who discovered it on Easter.

Capital	Tallahassee
Size	65,755 square miles
Population	16,713,149
Statehood	March 3, 1845, 27th State
Motto	"In God We Trust"
Tree	Sabal Palmetto
Bird	Mockingbird
Flower	Orange Blossom
Song	"Swanee River" ("Old Folks at Home") by Stephen Foster

Georgia (GA)

The Peach State

Georgia was named after King George II of England, who granted the settlers the right to create a colony there in 1732.

Capital	Atlanta
Size	59,425 square miles
Population	8,560,310
Statehood	January 2, 1788, 4th State
Motto	"Wisdom, Justice, and Moderation"
Tree	Live Oak
Bird	Brown Thrasher
Flower	Cherokee Rose
Song	"Georgia" by Robert Loveman; music by Lollie Belle Wylie

Hawaii (HI)   *The Aloha State*

Hawaii comes from Hawaiki, or Owhyhee, the native Polynesians' word for "homeland."

Capital	Honolulu
Size	10,931 square miles
Population	1,244,898
Statehood	August 21, 1959, 50th State
Motto	"The Life of the Land is Preserved by Righteousness"
Tree	Kukui
Bird	Hawaiian Goose
Flower	Yellow Hibiscus
Song	"Our Own Hawaii" by King Kalakaua Music by Henry Berger

Idaho (ID)   *The Gem State*

Idaho is an invented name whose meaning is unknown. It may have come from a Kiowa Apache name for the Comanche Indians.

Capital	Boise
Size	83,574 square miles
Population	1,341,131
Statehood	July 3, 1890, 43rd State
Motto	"It Endures Forever" (Esto Perpetua)
Tree	Western White Pine
Bird	Mountain Bluebird
Flower	Syringa
Song	"Here We Have Idaho"

Illinois (IL)

Illinois comes from Illini, an Algonquin Indian word meaning "tribe of superior men" or "warriors."

Capital	Springfield
Size	57,918 square miles
Population	12,600,620
Statehood	December 3, 1818, 21st State
Motto	"State Sovereignty, National Union"
Tree	White Oak
Bird	Cardinal
Flower	Meadow Violet
Song	"Illinois"

Indiana (IN)

the Hoosier State

Indiana means "land of the Indians."

Capital	Indianapolis
Size	36,420 square miles
Population	6,159,068
Statehood	December 11, 1816, 19th State
Motto	"The Crossroads of America"
Tree	Tulip Tree
Bird	Cardinal
Flower	Peony
Song	"On the Banks of the Wabash Far Away" by Paul Dresser

Iowa (IA)

Iowa comes from an Indian word meaning "this is the place" or "the Beautiful land."

Capital	Des Moines
Size	56,276 square miles
Population	2,936,760
Statehood	December 28, 1846, 29th State
Motto	"Our Liberties We Prize and Our Rights We Will Maintain"
Tree	Oak
Bird	American Goldfinch
Flower	Wild Rose
Song	"The Song of Iowa" by Major S.H.M. Byers To the tune of "O Tannenbaum"

Kansas (KS)

Kansas comes from the name of a Sioux Indian word meaning, "people of the south wind."

Capital	Topeka
Size	82,282 square miles
Population	2,715,884
Statehood	January 29, 1861 , 34th State
Motto	"To the Stars Through Difficulty"
Tree	Cottonwood
Bird	Western Meadowlark
Flower	Sunflower
Song	"Home on the Range" by Dr. Brewster M. Higley; music by Daniel Kelly

Kentucky (KY)

From an Iroquoian word "Ken-tah-ten",
meaning "land of tomorrow."

Capital	Frankfort
Size	40,411 square miles
Population	4,092,891
Statehood	June 1, 1792, 15th State
Motto	"United We Stand, Divided We Fall"
Tree	Tulip Poplar
Bird	Cardinal
Flower	Goldenrod
Song	"My Old Kentucky Home" by Stephen Foster

Louisiana (LA)

Louisiana is named in honor of
Louis XIV of France.

Capital	Baton Rouge
Size	51,840 square miles
Population	4,482,646
Statehood	April 30, 1812, 18th State
Motto	"Union, Justice, and Confidence"
Tree	Bald Cypress
Bird	Brown Pelican
Flower	Magnolia
Song	"Song of Louisiana" by Vashti R. Stopher

Maine (ME)

The Pine Tree State

Maine was first used to distinguish the mainland from the offshore islands. It has been considered a complement to Henrietta Maria, queen of Charles I of England. She was said to have owned the province of Mayne in France.

Capital	Augusta
Size	35,385 square miles
Population	1,294,464
Statehood	March 15, 1820, 23rd State
Motto	"Dirigo (I Lead)"
Tree	White Pine
Bird	Black Capped Chickadee
Flower	White Pine Cone and Tassel
Song	"State of Maine Song" by Roger Vinton Snow

Maryland (MD)

The Old Line State

Maryland was named in honor of Queen Henrietta Maria, who granted the right to establish an English colony there.

Capital	Annapolis
Size	12,407 square miles
Population	5,458,137
Statehood	April 28, 1788, 7th State
Motto	"Strong Deeds, Gentle Words"
Tree	White Oak
Bird	Baltimore Oriole
Flower	Black-Eyed Susan
Song	"Maryland, My Maryland" by James R. Randall

Massachusetts (MA)

The Bay State

Massachusetts is named after the Massachusett tribe of Native Americans, meaning "at or about the great hill."

Capital	Boston
Size	10,555 square miles
Population	6,427,801
Statehood	February 6, 1788, 6th State
Motto	"By The Sword We Seek Peace, But Peace Only Under Liberty"
Tree	American Elm
Bird	Black Capped Chickadee
Flower	Mayflower
Song	"All Hail to Massachusetts"

Michigan (MI)

The Wolverine State

Michigan comes from the Chippewa Indian words "mica gama" meaning "great or large lake," referring to Lake Michigan.

Capital	Lansing
Size	96,705 square miles
Population	10,050,446
Statehood	January 26, 1837, 26th State
Motto	"If You Seek a Pleasant Peninsula, Look About You"
Tree	White Pine
Bird	Robin
Flower	Apple Blossom
Song	"Michigan, My Michigan"

Minnesota (MN)

The Land of 10, 000 Lakes

Minnesota comes from a Dakota Sioux Indian word meaning "sky-tinted water."

Capital	Saint Paul
Size	86,943 square miles
Population	5,019,720
Statehood	May 11, 1858, 32nd State
Motto	"Star of the North"
Tree	Red (Norway) Pine
Bird	Common Loon
Flower	Lady Slipper
Song	"Hail Minnesota"

Mississippi (MS)

The Magnolia State

Mississippi comes from an Indian word meaning "Father of Waters."

Capital	Jackson
Size	48,430 square miles
Population	2,871,782
Statehood	December 10, 1817, 20th State
Motto	"By Valor and Arms"
Tree	Magnolia
Bird	Mockingbird
Flower	Magnolia Blossom
Song	"Go Mississippi"

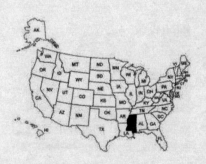

Missouri (MO)

Missouri is named after the Algonquin Indian tribe; it means "town of the large canoes."

Capital	Jefferson City
Size	69,709 square miles
Population	5,672,579
Statehood	August 10, 1821, 24th State
Motto	"Let the Welfare of the People be the Supreme Law"
Tree	Flowering Dogwood
Bird	Bluebird
Flower	White Hawthorn Blossom
Song	"Missouri Waltz" by J.R. Shannon

Montana (MT)

Montana was chosen from a Latin dictionary by J.M. Ashley. It is a Latinized Spanish word meaning "mountainous."

Capital	Helena
Size	147,046 square miles
Population	909,453
Statehood	November 8, 1889, 41st State
Motto	"Gold and Silver"
Tree	Ponderosa Pine
Bird	Western Meadowlark
Flower	Bitterroot
Song	"Montana" by Charles C. Cohen Music by Joseph E. Howard

Nebraska (NE)

The Cornhusker State

Nebraska comes from an Oto Indian word meaning "flat river" or "broad water," their name for the Platte River.

Capital	Lincoln
Size	77,358 square miles
Population	1,729,180
Statehood	March 1, 1867, 37th State
Motto	"Equality Before the Law"
Tree	Cottonwood
Bird	Western Meadowlark
Flower	Goldenrod
Song	"Beautiful Nebraska"

Nevada (NV)

The Silver State

Nevada was given the name by Spanish explorers. It is the Spanish word meaning "snowcapped," referring to the Sierra Nevada Mountains.

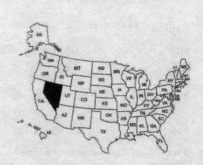

Capital	Carson City
Size	110,567 square miles
Population	2,173,491
Statehood	October 31, 1864, 36th State
Motto	"All for our Country"
Tree	Single-Leaf Pinon
Bird	Mountain Bluebird
Flower	Sagebrush
Song	"Home Means Nevada"

New Hampshire (NH)

The Granite State

New Hampshire comes from the English county of Hampshire.

Capital	Concord
Size	9,283 square miles
Population	1,275,056
Statehood	June 21, 1788, 9th State
Motto	"Live Free or Die"
Tree	White Birch
Bird	Purple finch
Flower	Purple Lilac
Song	"Old New Hampshire" by John F. Holmes; music by Maurice Hoffman

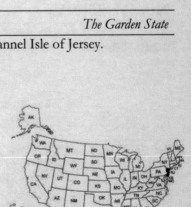

New Jersey (NJ)

The Garden State

New Jersey was named for the English Channel Isle of Jersey.

Capital	Trenton
Size	8,721 square miles
Population	8,590,300
Statehood	December 18, 1787, 3rd State
Motto	"Liberty and Prosperity"
Tree	Red Oak
Bird	American Goldfinch
Flower	Violet
Song	*None Listed*

New Mexico (NM)

Land of Enchantment

New Mexico was given its name by 16th-century Spaniards in Mexico.

Capital	Santa Fe
Size	121,598 square miles
Population	1,855,059
Statehood	January 6, 1912, 47th State
Motto	"It Grows as it Goes"
Tree	Nut Pine
Bird	Roadrunner
Flower	Yucca
Song	"O Fair New Mexico" by Elizabeth Garrett

New York (NY)

The Empire State

New York was named in honor of the Duke of York and Albany after the English took it from Dutch settlers.

Capital	Albany
Size	54,556 square miles
Population	19,157,532
Statehood	July 26, 1788, 11th State
Motto	"Excelsior"
Tree	Sugar Maple
Bird	Bluebird
Flower	Rose
Song	"I Love New York"

North Carolina (NC) *The Tar Heel State*

North Carolina, the northern part of the English colony of Carolana, was named in honor of Charles I of England.

Capital	Raleigh
Size	53,819 square miles
Population	8,320,146
Statehood	November 21, 1789, 12th State
Motto	"To Be Rather Than to Seem"
Tree	Southern Pine
Bird	Cardinal
Flower	Flowering Dogwood
Song	"The Old North State" by William Gaston

North Dakota (ND) *The Peace Garden State*

North Dakota is named after the Sioux tribe, meaning "allies."

Capital	Bismark
Size	70,704 square miles
Population	634,110
Statehood	November 2, 1889, 39th State
Motto	"Liberty and Union, Now and Forever, One and Inseparable"
Tree	American Elm
Bird	Western Meadowlark
Flower	Wild Prairie Rose
Song	"North Dakota Hymn" by James W. Foley Music by C.S. Putnam

Ohio (OH)

The Buckeye State

Ohio comes from an Iroquoian word meaning "great river."

Capital	Columbus
Size	44,828 square miles
Population	11,421,267
Statehood	March 1, 1803, 17th State
Motto	"With God All Things Are Possible"
Tree	Ohio Buckeye
Bird	Cardinal
Flower	Red Scarlet Carnation
Song	"Beautiful Ohio"

Oklahoma (OK)

The Sooners State

Oklahoma comes from two Choctaw Indian words meaning "red people."

Capital	Oklahoma City
Size	69,903 square miles
Population	3,493,714
Statehood	November 16, 1907, 46th State
Motto	"Labor Conquers all Things"
Tree	Redbud
Bird	Scissor-Tailed Fly Catcher
Flower	Mistletoe
Song	"Oklahoma" by R. Rogers and O. Hammerstein

Oregon (OR)

The Beaver State

Oregon's name origin is unknown, however, it is generally accepted that the name, first used by Jonathan Carver in 1778, was taken from the writings of Major Robert Rogers, an English army officer.

Capital	Salem
Size	98,381 square miles
Population	3,521,515
Statehood	February 14, 1859, 33rd State
Motto	"She Flies With Her Own Wings"
Tree	Douglas Fir
Bird	Western Meadowlark
Flower	Oregon Grape
Song	"Oregon, My Oregon," by J.A. Buchman Music By Henry B. Murtagh

Pennsylvania (PA)

The Keystone State

Pennsylvania means "Penn's Woodland," the name given to the colony founded by Admiral Sir William Penn, father of William Penn.

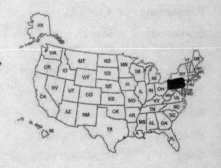

Capital	Harrisburg
Size	46,058 square miles
Population	12,335,091
Statehood	December, 12, 1787, 2nd State
Motto	"Virtue, Liberty and Independence"
Tree	Hemlock
Bird	Ruffed Grouse
Flower	Mountain Laurel
Song	"Pennsylvania"

Rhode Island (RI)

The Ocean State

Rhode Island's name comes from the Greek Island of Rhodes.

Capital	Providence
Size	1,545 square miles
Population	1,069,725
Statehood	May 29, 1790, 13th State
Motto	"Hope"
Tree	Red Maple
Bird	Rhode Island Red
Flower	Wood Violet
Song	"Rhode Island" by T. Clark Brown

South Carolina (SC)

The Palmetto State

South Carolina, the southern part of the English colony of Carolana, was named in honor of Charles I of England.

Capital	Columbia
Size	32,020 square miles
Population	4,107,183
Statehood	May 23, 1788, 8th State
Motto	"While I Breathe, I Hope"
Tree	Palmetto
Bird	Carolina Wren
Flower	Yellow Jasmine
Song	"Carolina" by Henry Timrod Music by Anna Custis Burgess

South Dakota (SD)

The Mount Rushmore State

South Dakota comes from a Sioux Indian word meaning "ally" or "friend."

Capital	Pierre
Size	77,121 square miles
Population	761,063
Statehood	November 2, 1889; 40th State
Motto	"Under God the People Rule"
Tree	Black Hills Spruce
Bird	Ring-necked Pheasant
Flower	Pasque flower
Song	"Hail, South Dakota"

Tennessee (TN)

The Volunteer State

Tennessee comes from "Tanasi," the name of Cherokee Indian villages on the Little Tennessee River.

Capital	Nashville
Size	42,146 square miles
Population	5,797,280
Statehood	June 1, 1796, 16th State
Motto	"Agriculture and Commerce"
Tree	Tulip Tree
Bird	Mockingbird
Flower	Iris
Songs	"Tennessee Waltz" (1965); "My Homeland, Tennessee" (1925); "When It's Iris Time in Tennessee" (1935); "My Tennessee" (1955); "Rocky Top" (1982); "Tennessee" (1992)

Texas (TX)

Texas comes from a word meaning "friends" or "allies," used by the Spanish to describe some of the American Indians living there.

Capital	Austin
Size	268,581 square miles
Population	21,779,893
Statehood	December 29, 1845; 28th State
Motto	"Friendship"
Tree	Pecan
Bird	Mockingbird
Flower	Bluebonnet
Song	"Texas, Our Texas," by William Marsh and Gladys Y. Wright

Utah (UT)

Utah comes from the Ute tribe, meaning "people of the mountains."

Capital	Salt Lake City
Size	84,904 square miles
Population	2,316,256
Statehood	January 4, 1896, 45th State
Motto	"Industry"
Tree	Blue Spruce
Bird	California Gull
Flower	Sego Lily
Song	"Utah, We Love Thee" by Evan Stephens

Vermont (VT)

The Green Mountain State

Vermont comes from the French "vert mont," meaning "green mountain."

Capital	Montpelier
Size	9,615 square miles
Population	616,592
Statehood	March 4, 1791; 14th State
Motto	"Freedom and Unity"
Tree	Sugar Maple Tree
Bird	Hermit Thrush
Flower	Red Clover
Song	"These Green Mountains" by Diane Martin

Virginia (VA)

Old Dominion State

Virginia is named in honor of Elizabeth, "Virgin Queen" of England.

Capital	Richmond
Size	42,774 square miles
Population	7,293,542
Statehood	June 25, 1788, 10th State
Motto	"Thus Always to Tyrants"
Tree	Dogwood
Bird	Cardinal
Flower	Flowering Dogwood
Song	"Carry Me Back to Old Virginia" by James A. Bland

Washington (WA)

Evergreen State

Washington is named in honor of George Washington, the first president of the United States.

Capital	Olympia
Size	71,300 square miles
Population	6,068,996
Statehood	November 11, 1889; 42nd State
Motto	"By and By"
Tree	Western Hemlock
Bird	Willow Goldfinch
Flower	Rhododendron
Song	"Washington My Home" by Mrs. Helen Davis

West Virginia (WV)

Mountain State

West Virginia was named by the people of western Virginia, who formed their own government during the Civil War.

Capital	Charleston
Size	24,231 square miles
Population	1,801,873
Statehood	June 20, 1863, 35th State
Motto	"Mountaineers Are Always Free"
Tree	Sugar Maple
Bird	Cardinal
Flower	Rhododendron
Song	"West Virginia Hills"

Wisconsin (WI)

The Badger State

Wisconsin comes from a Chippewa name that is believed to mean "grassy place."

Capital	Madison
Size	65,499 square miles
Population	5,441,196
Statehood	May 29, 1848; 30th State
Motto	"Forward"
Tree	Sugar Maple
Bird	Robin
Flower	Violet
Song	"On Wisconsin" by Dr. Phillip A. Forsbeck Music by William T. Purdy

Wyoming (WY)

The Equality State

Wyoming is named after the Delaware Indian word, meaning "mountains and valleys alternating"; the same as the Wyoming Valley in Pennsylvania.

Capital	Cheyenne
Size	97,818 square miles
Population	498,703
Statehood	July 10, 1890, 44th State
Motto	"Equal Rights"
Tree	Cottonwood
Bird	Western Meadowlark
Flower	Indian Paintbrush
Song	"Wyoming State Song" by Charles E. Winter Music by George E. Knapp

MAPS OF THE WORLD

The World

Total area: 510.072 million sq km (316.96 million sq mi).

Land area: 148.94 million sq km (92.55 million sq mi).

Water area: 361.132 million sq km (224.41 million sq mi).
Note: 70.8% of the world is water, 29.2% is land.

The Four Oceans:
Pacific Ocean: 64,186,300 square miles; 12,925 feet deep
Atlantic Ocean: 33,420,000 square miles; 11,730 feet deep
Indian Ocean: 28,350,500 square miles; 12,598 feet deep
Arctic Ocean: 5,105,700 square miles; 3,407 feet deep

- **Terrain:** Highest elevation is Mt. Everest at 8,850 m (29,035 ft) and lowest land depression is the Dead Sea at –411 m (–1,349 ft) below sea level.

- **Land use:** Arable land: 10%. Permanent crops: 1%. Meadows and pastures: 26%. Forests and woodlands: 32%. Other: 31% (1993 est.). Irrigated land: 2,481,250 sq km (1,541,849 sq mi).

* **Note**: *The maps of the world and the seven continents are from The World Factbook 2004.*

ASIA

Asia measures 16,990,000 square miles and holds 61% of the world's population. Asia is comprised of many countries including Russia, China, Mongolia, Iran and India.

AUSTRALIA

Australia, the only continent that is also a country, measures 2,978,000 square miles. It is in the Southern Hemisphere and holds .5% of the world's population.

AFRICA

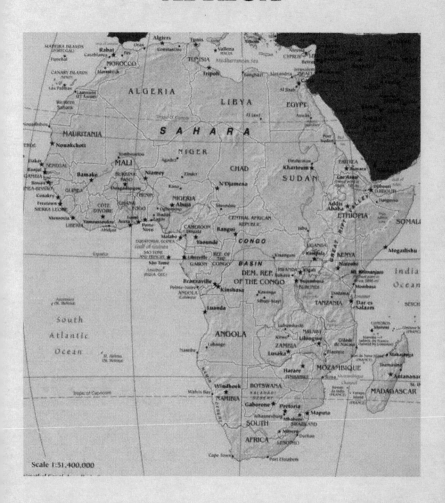

Africa measures 11,657,000 square miles and holds 12.7% of the world's population. The equator goes right through the middle of Africa.

NORTH AMERICA

North America measures 9,348,000 square miles. The United States, Canada and Mexico are the major countries in North America which holds 7.9% of the world's population.

SOUTH AMERICA

South America measures 6,885,000 square miles and is in the Southern Hemisphere. South America contains 5.9% of the world's population.

ANTARCTICA

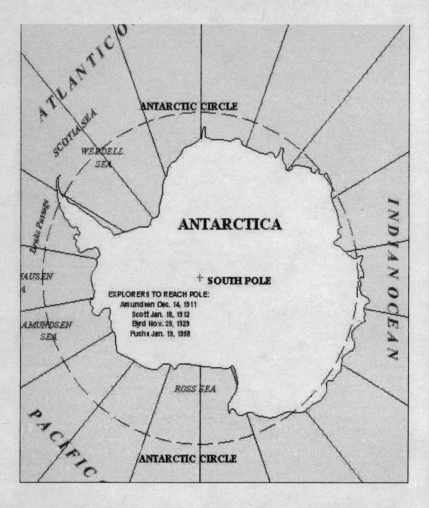

Located south of the Antarctic Circle, Antarctica is 5,404,000 square miles. Antarctica the coldest and windiest place on earth, there are no indigenous inhabitants, but there are seasonally staffed research stations. The Antarctic Treaty governs the continent of Antarctica.

EUROPE

Europe is 4,051,000 square miles and holds 12% of the world's population. Many countries make up the continent of Europe including France, Italy and Germany.

FACTS ABOUT COUNTRIES

The information gathered here is derived from <u>The World Factbook 2004 </u>compiled by the Central Intelligence Agency.

AFGHANISTAN
Location: Southern Asia, north and west of Pakistan, east of Iran
Area: 250,000 sq mi (647,500 sq km)
Area – comparative: slightly smaller than Texas
Population: 28,513,677
Population growth rate: 4.92%
Nationality: noun: Afghan(s), adjective: Afghan
Ethnic groups: Pashtun 42%, Tajik 27%, Hazara 9%, Uzbek 9%, others 13%
Religions: Sunni Muslim 80%, Shi`a Muslim 19%, other 1%
Languages: Pashtu 35%, Afghan Persian (Dari) 50%, Turkic languages 11%, 30 minor languages 4%
Government type: transitional
Capital: Kabul
Budget: revenues: $200 million, expenditures: $550 million   **Currency:** afghani (AFA)

ALBANIA
Location: Southeastern Europe, bordering the Adriatic Sea and Ionian Sea, between Greece and Serbia and Montenegro
Area: 11,099 sq mi (28,748 sq km)
Area – comparative: slightly smaller than Maryland
Population: 3,544,808
Population growth rate: 0 .51%
Nationality: noun: Albanian(s), adjective: Albanian
Ethnic groups: Albanian 95%, Greeks 3%, other 2%
Languages: Albanian (Tosk is the official dialect), Greek
Religions: Muslim 70%, Albanian Orthodox 20%, Roman Catholic 10%
Government type: emerging democracy
Capital: Tirana
Budget: revenues: $1.65 billion, expenditures: $ 2 billion , including capital of $406 million
Currency: lek (ALL)

ALGERIA
Location: Northern Africa, bordering the Mediterranean Sea, between Morocco and Tunisia
Area: 919,590 sq mi (2,381,740 sq km)
Area- comparative: slightly less than 3.5 times the size of Texas
Population: 32,129,324
Population growth rate: 1.28%
Nationality: noun: Algerian(s), adjective: Algerian
Ethnic groups: Arab-Berber 99%, European less than 1%
Languages: Arabic (official), French, Berber dialects
Religions: Sunni Muslim (state religion) 99%, Christian and Jewish 1%
Government type: republic
Capital: Algiers

Budget: revenues: $18.5 billion, expenditures: $22.7billion, including capital expenditures of $5.8 billion
Currency: Algerian dinar (DZD)

AMERICAN SAMOA
Location: Oceania, group of islands in the South Pacific Ocean approximately half way between Hawaii and New Zealand
Area: 77 sq mi (199 sq km)
Area-comparative: slightly larger than Washington, DC
Population: 57,902
Population growth rate: 0.04%
Nationality: noun: American Samoan(s), adjective: American Samoan
Ethnic groups: Samoan (Polynesian) 89%, Caucasian 2%, Tongan 4%, and other 5%
Language: Samoan (closely related to Hawaiian and other Polynesian languages), English
Religion: Christian Congressionalist 50%, Roman Catholic 20%, Protestant and other 30%
Government: NA
Capital: Pago Pago
Budget: revenues $121 million (37%-local, 63% US Grants)expenditures:$127 million, capital $NA
Currency: US dollar (USD)

ANDORRA
Location: Southwestern Europe, between France and Spain
Area: 180 sq mi (468 sq km)
Area –comparative: 2.5 times the size of Washington, DC
Population: 69,865
Population growth rate: 1%
Nationality: noun: Andorran(s), adjective: Andorran
Ethnic groups: Spanish 43%, Andorran 33%, Portuguese 11%, French 7%, other 6%
Languages: Catalan (official), French, Castilian, Portuguese
Religion: Roman Catholic (predominant)
Government type: parliamentary democracy
Capital: Andorra la Vella
Budget: revenues: $385 million, expenditures: $342 million   **Currency:** euro (EUR)

ANGOLA
Location: Southern Africa, bordering the South Atlantic Ocean, between Namibia and Democratic Republic of the Congo
Area: 481,351 sq mi (1,246,700 sq km)
Area-comparative: slightly less than twice the size of Texas
Population: 10,978,552
Population growth rate: 1.93%
Nationality: noun: Angolan(s), adjective: Angolan
Ethnic groups: Ovimbundu 37%, Kimbundu 25%, Bakongo 13%, Mestico 2%, European 1%, other 22%
Languages: Portuguese (official), Bantu and other African languages
Religions: indigenous beliefs 47%, Roman Catholic 38%, Protestant 15%
Government: Republic, nominally a multiparty democracy with a strong presidential system

Capital: Huanda
Budget: revenues: $928 million;; expenditures: $25 billion, including capital of $963 million
Currency: kwanza(AOA)

ANGUILLA (overseas territory of the UK)
Location: islands between the Caribbean Sea and the North Atlantic Ocean, east of Puerto Rico
Area: 39 sq mi (102 sq km)

Area-comparative: about one half the size of Washington, DC
Population: 13,008
Population growth rate: 1.98%
Nationality: noun: Anguillan(s), adjective: Anguillan
Ethnic groups: black (predominant), mulatto, white
Religions: Anglican 40%, Methodist 33%, other 12%, Seventh-Day Adventist 7%, Baptist 5% Catholic 3%
Government: NA
Capital: The Valley
Budget: revenues: $22.8 million, expenditures: $22.5 million   **Currency:** East Caribbean dollar (XCD)

ANTIGUA AND BARBUDA
Location: islands between the Caribbean Sea and North Atlantic Ocean, east-southeast of Puerto-Rico
Area: 171 sq mi (total 443 sq km)
Area-comparative: 2.5 times the size of Washington, DC
Population: 68,320
Population growth rate: 0.6%
Nationality: noun: Antiguan(s), Barbudan(s), adjective: Antiguan, Barbudan
Ethnic groups: black, British, Portuguese, Lebanese, Syrian
Languages: English (official), local dialects
Religions: Christian, (predominantly Anglican with other Protestant, and some Roman Catholic)
Government type: constitutional monarchy with UK-style parliament
Capital: Saint John`s
Budget: revenues: $123.7 million, expenditures: $145.9 million   **Currency:** East Caribbean dollar (XCD)

ARGENTINA
Location: Southern South America, bordering the South Atlantic Ocean, between Chile and Uruguay
Area: 1,068,297 sq mi (2,766,890 sq km)
Area-comparative: slightly less than three tenths the size of the US
Population: 39,144,753
Population growth rate: 1.02%
Nationality: noun: Argentine(s), adjective: Argentine
Ethnic groups: white (mostly Spanish and Italian) 97%, mestizo, or other nonwhite groups 3%
Languages: Spanish (official), English, Italian, German, French
Religions: Roman Catholic 92%, Protestant 2%, Jewish 2%, other 4%
Government type: republic
Capital: Buenos Aires
Budget: revenues: $44 billion, expenditures $48 billion, including capital of $NA
Currency: Argentine peso (ARS)

ARMENIA
Location: Southwestern Asia, east of Turkey
Area: 11,505 sq mi (29,800 sq km)
Area-comparative: slightly smaller than Maryland
Population: 2,991,360
Population growth rate: -0.32%
Nationality: noun: Armenian(s), adjective: Armenian
Ethnic groups: Armenian 93%, Azeri 1%, Russian 2%, other (including Kurd) 4%
Languages: Armenian 96%, Russian 2%, other 2%
Religions: Armenian Apostolic 94%, other Christian 4%, Yezidi 2%
Government type: republic
Capital: Yerevan
Budget: revenues: $402 million, expenditures: $482 million   **Currency:** dram (AMD)

ARUBA (part of the Kingdom of the Netherlands)
Location: Caribbean, island in the Caribbean Sea, north of Venezuela
Area: 74.5 sq mi (193 sq km)
Area-comparative: slightly larger than Washington, DC
Population: 71,218
Population growth rate: 0.51%
Nationality: noun: Aruban(s), adjective: Aruban: Dutch
Ethnic groups: mixed white/ Caribbean Amerindian 80%
Languages: Dutch (official), Papiamento (a dialect) English, Spanish
Religions: Roman Catholic 82%, Protestant 8%, Hindu, Muslim, Confucian, Jewish
Government type: parliamentary democracy
Capital: Oranjestad
Budget: revenues: $135.81 million, expenditures: $ 147 million   **Currency:** Aruban guilder/florin (AWG)

AUSTRALIA
Location: Oceania, continent between the Indian Ocean and the South Pacific Ocean
Area: 2,967,895 sq mi (7,686,850 sq km)
Area-comparative: slightly smaller than the US contiguous 48 states
Population: 19,913,144
Population growth rate: 0.9%
Nationality: noun: Australian(s), adjective: Australian
Ethnic groups: Caucasian 92%, Asian 7%, other (including Aboriginal) 1%
Languages: English (official), native languages
Religions: Roman Catholic, Anglican, and other Protestant 74%, other or nonreligious 26%
Government type: democratic,federal-state system, recognizing British monarch as sovereign
Capital: Canberra
Budget: revenues: $86.8 billion, expenditures: $ 84.1 billion   **Currency:** Australian dollar (AUD)

AUSTRIA
Location: Central Europe, north of Italy and Slovenia
Area: 32,377 sq mi (83,858 sq km)
Area-comparative: slightly smaller than Maine
Population: 8,174,762
Population growth rate: 0.14%
Nationality: noun: Austrian(s), adjective: Austrian
Ethnic groups: German 88%, other 12%
Language: German (official)
Religions: Roman Catholic 74%, Protestant 5%, Muslim 4%, other 17%
Government type: federal republic
Capital: Vienna
Budget: revenues: $67 billion, expenditures: $70 billion, including capital of $NA   **Currency:** euro (EUR)

AZERBAIJAN
Location: Southwestern Asia, bordering the Caspian Sea, between Iran and Russia
Area: 33,436 sq mi (86,600 sq km)
Area-comparative: slightly smaller than Maine
Population: 7,868,385
Population growth rate: 0.52%
Nationality: noun: Azerbaijani(s), adjective: Azerbaijani
Ethnic groups: Azeri 90%, Dagestani 3.2%, Russian 2.5%, Armenian 2% , other 2.3%
Languages: Azerbaijani (Azeri) 89%, Russian 3%, Armenian 2%, other 6%

Religions: Muslim 93%, other (including Armenian Orthodox) 4%, Russian Orthodox 3%
Government type: republic
Capital: Baku (Baki)
Budget: revenues: $786 million, expenditures: $ 807 million   **Currency:** Azerbaijani manta (AZM)

THE BAHAMAS
Location: Caribbean, chain of islands in the North Atlantic Ocean, northeast of Cuba
Area: 5,382 sq mi (13,940 sq km)
Area-comparative: slightly smaller than Connecticut
Population: 299,697
Population growth rate: 0.72%
Nationality: noun: Bahamian(s), adjective: Bahamian
Ethnic groups: black 85%, white 12%, Asian and Hispanic 3%
Languages: English (official), Creole
Religions: Baptist 32%, Anglican 20%, Roman Catholic 19%, Methodist 6%, Church of God 6%, Protestant 12%, other 5%
Government type: constitutional parliamentary democracy
Capital: Nassau
Budget: revenues: $918.5 million, expenditures: $956.5 million and capital expenditures of $106.7 million
Currency: Bahamian dollar (BSD)

BAHRAIN
Location: Middle East, east of Saudi Arabia
Area: 257 sq mi (665 sq km)
Area-comparative: 3.5 times the size of Washington, DC
Population: 677,886
Population growth rate: 1.56%
Nationality: noun: Bahraini(s), adjective: Bahraini
Ethnic groups: Bahraini 63%, Asian 19%, other Arab 10%, Iranian 8%
Languages: Arabic (official), English, Farsi, Urdu
Religions: Shi'a Muslim 70%, Sunni Muslim 30%
Government type: constitutional hereditary monarchy
Capital: Manama
Budget: revenues: $1.8 billion, expenditures: $2.2 billion, including capital expenditures of $700 million
Currency: Bahraini dinar (BHD)

BANGLADESH
Location: Southern Asia, bordering the Bay of Bengal, between Burma and India
Area: 55,598 sq mi (144,000 sq km)
Area-comparative: slightly smaller than Iowa
Population: 141,340,476
Population growth rate: 2.08%
Nationality: noun: Bangladeshi(s), adjective: Bangladeshi
Ethnic groups: Bengali 98%, other 2%
Languages: Bangla (official), English
Religions: Muslim 83%, Hindu 16%, other 1%
Government type: parliamentary democracy
Capital: Dhaka
Budget: revenues: $4.9 billion, expenditures: $6.8 billion   **Currency:** taka (BDT)

BARBADOS
Location: Caribbean, island in the North Atlantic Ocean, northeast of Venezuela
Area: 166 sq mi (430 sq km)

Area-comparative: 2.5 times the size of Washington, DC
Population: 278,289
Population growth rate: 0.36%
Nationality: noun: Barbadian(s) or Bajan, adjective: Barbadian or Bajan
Ethnic groups: black 90%, white 4%, Asian and mixed 6%
Languages: English (official)
Religions: Protestant 67%, Roman Catholic 4%, none 17%, other 12%
Government type: parliamentary democracy
Capital: Bridgetown
Budget: revenues: $847 million, expenditures: $886 million   **Currency:** Barbadian dollar (BBD)

BELARUS

Location: Eastern Europe, east of Poland
Area: 80,154 sq mi (207,600 sq km)
Area-comparative: slightly smaller than Kansas
Population: 10,310,520
Population growth rate: -0.11%
Nationality: noun: Belarusian(s); adjective: Belarusian
Ethnic groups: Belarusian 81%, Russian, 11%, Polish, Ukrainian, and other 8%
Languages: Belarusian (official), Russian (official); other
Religions: Eastern Orthodox 80%, other (including Roman Catholic, Protestant, Jewish and Muslim) 20%
Government type: republic
Capital: Minsk
Budget: revenues: $4 billion, expenditures: $ 4.1 billion   **Currency:** Belarusian ruble (BYB/BYR)

BELGIUM

Location: Western Europe, bordering the North Sea, between France and Netherlands
Area: 11,779 sq mi (30,510 sq km)
Area-comparative: about the size of Maryland
Population: 10,348,276
Population growth rate: 0.16%
Nationality: noun: Belgian(s), adjective: Belgian
Ethnic groups: Fleming 58%, Walloon 31%, mixed and other 11%
Languages: Dutch 60%, French 40%
Religions: Roman Catholic 75%, other (including Protestant) 25%
Government type: federal parliamentary democracy under a constitutional monarch
Capital: Brussels
Budget: revenues: $113.4 billion, expenditures: $106 billion   **Currency:** euro (EUR)

BELIZE

Location: Middle America, bordering the Caribbean Sea, between Guatemala and Mexico
Area: 8,867 sq mi (22,966 sq km)
Area-comparative: slightly smaller than Massachusetts
Population: 272,945
Population growth rate: 2.39%
Nationality: noun: Belizean(s), adjective: Belizean
Ethnic groups: Mestizo 49%, Creole 25%, Maya 11%, Garifuna 6%, other 9%
Languages: English (official), Spanish, Mayan, Garifuna, Creole
Religions: Roman Catholic 50%, Protestant 27%, other/nonreligious 23%
Government type: parliamentary democracy
Capital: Belmopan
Budget: revenues: $224 million, expenditures: $ 209 million   **Currency:** Belizean dollar (BZD)

BENIN

Location: Western Africa, bordering the Bight of Benin, between Nigeria and Togo
Area: 43,482 sq mi (112,620 sq km)
Area-comparative: slightly smaller than Pennsylvania
Population: 7,250,033
Population growth rate: 2.89%
Nationality: noun: Beninese (singular and plural), adjective: Beninese
Ethnic groups: African 99% (42 ethnic groups, including Fon, Adja, Yoruba, and Bariba), Europeans 1%
Languages: French (official), Fon and Yoruba, tribal languages (at least six major ones in north)
Religions: indigenous beliefs 50%, Christian 30%, Muslim 20%
Government type: republic under multiparty democratic rule
Capital: Porto-Novo(official capital) ; Cotonou is the seat of government
Budget: revenues: $ 377.4 million, expenditures: $ 561.8 million
Currency: Communaute Financiere Africaine franc (XOF)

BERMUDA *(overseas territory of the UK)*

Location: North America, group of islands in the North Atlantic Ocean, east of North Carolina (US)
Area: 20.5 sq mi (53.3 sq km)
Area-comparative: about one third the size of Washington, DC
Population: 64,935
Population growth rate: 0.68%
Nationality: *noun*, Bermudian(s); *adjective*, Bermudian
Ethnic groups: black 58%, white 36%, other 6%
Languages: English (official), Portuguese
Religions: non-Anglican Protestant 39%, Anglican 27%, Roman Catholic 15%, other 19%
Government type: parliamentary British overseas territory with internal self-government
Capital: Hamilton
Budget: revenues: $671.1 million, expenditures: $594.6 million, including capital of $55 million
Currency: Bermudian dollar (BMD)

BHUTAN

Location: Southern Asia, between China and India
Area: 18,146 sq mi (47,000 sq km)
Area-comparative: about half the size of Indiana
Population: 2,185,569
Population growth rate: 2.12%
Nationality: noun: Bhutanese (singular and plural), adjective: Bhutanese
Ethnic groups: Bhote 50%, ethnic Nepalese 35%, indigenous or migrant tribes 15%
Languages: Dzongkha (official), Tibetan, Nepalese dialects
Religions: Lamaistic Buddhism 75%, Indian- and Nepalese-influenced Hinduism 25%
Government type: monarchy, special treaty relationship with India
Capital: Thimphu
Budget: revenues: $146 million, expenditures: $ 152 million **Currency:** ngultrum; Indian rupee (BTN,INR)

BOLIVIA

Location: Central South America, southwest of Brazil
Area: 424,162 sq mi (1,098,580 sq km)
Area-comparative: slightly less than three times the size of Montana
Population: 8,724,156
Population growth rate: 1.56%
Nationality: noun: Bolivian(s), adjective: Bolivian

Ethnic groups: Ouechua 30%, Mestizo 30%, Aymara 25%, white 15%
Languages: Spanish (official), Ouechua (official), Aymara (official)
Religions: Roman Catholic 95%, other 5%
Government type: republic
Capital: La Paz (seat of government); Sucre (legal capital and seat of judiciary)
Budget: revenues: $4 billion, expenditures: $4 billion   **Currency:** boliviano (BOB)

BOSNIA AND HERZEGOVINA
Location: Southeastern Europe, bordering the Adriatic Sea and Croatia
Area: 19,740 sq mi (51,129 sq km)
Area-comparative: slightly smaller than West Virginia
Population: 4,007,608
Population growth rate: 0.45%
Nationality: noun: Bosnian(s); **adjective:** Bosnian
Ethnic groups: Serb 37%, Bosniak 48%, Croat 14%, other 1%
Languages: Croatian, Serbian, Bosnian
Religions: Muslim 40%, Orthodox 31%, Roman Catholic 15%, Protestant 4%, other 10%
Government type: emerging federal democratic republic
Capital: Sarajevo
Budget: revenues: $1.9 billion; expenditures: $2.2 billion   **Currency:** marka (BAM)

BOTSWANA
Location: Southern Africa, north of South Africa
Area: 231,803 sq mi (600,370 sq km)
Area-comparative: slightly smaller than Texas
Population: 1,561,973
Population growth rate: -0.89%
Nationality: noun: Motswana (singular); Batswana (plural); **adjective:** Motswana; Batswana
Ethnic groups: Tswana 79%, Kalanga 11%, Basarwa 3% and other 7%
Languages: English (official), Setswana
Religions: indigenous beliefs 85%, Christian 15%
Government type: parliamentary republic
Capital: Gaborone
Budget: revenues: $2.3 billion; expenditures: $2.4 billion   **Currency:** pula (BWP)

BRAZIL
Location: Eastern South America, bordering the Atlantic Ocean
Area: 3,285,618 sq mi (8,511,965 sq km)
Area-comparative: slightly smaller than the US
Population: 184,101,109
Population rate growth: 1.11%
Nationality: noun: Brazilian(s), **adjective:** Brazilian
Ethnic groups: white (including Portuguese, German, and Italian) 55%, mixed 38%, black 6%, other 1%
Languages: Portuguese (official), Spanish, English, French
Religions: Roman Catholic 80%, other 20%
Government type: federative republic
Capital: Brasilia
Budget: revenues: $100.6 billion; expenditures: $91.6 billion   **Currency:** real (BRL)

BRITISH VIRGIN ISLANDS (overseas territory of the UK)
Location: Caribbean, between the Caribbean Sea and the North Atlantic Ocean, east of Puerto Rico
Area: 59 sq mi (153 sq km)

461

Area-comparative: about 0.9 times the size of Washington, DC
Population: 22,187
Population growth rate: 2.06%
Nationality: noun: British Virgin Islander(s); adjective: British Virgin Islander
Ethnic groups: black 83%, white, Indian, Asian and mixed
Language: English (official)
Religions: Protestant 86%, Catholic10%, other 4%
Government type: NA (overseas territory of the UK)
Capital: Road Town
Budget: revenues: $121.5 million; expenditures: $115.5 million   **Currency:** US dollar (USD)

BRUNEI
Location: Southeastern Asia, bordering the South China Sea and Malaysia
Area: 2,227 sq mi (5,770 sq km)
Area-comparative: slightly smaller than Delaware
Population: 365,251
Population growth rate: 1.95%
Nationality: noun: Bruneian(s); adjective: Bruneian
Ethnic groups: Malay 67%, Chinese 15%, indigenous 6%, other 12%
Languages: Malay (official), English, Chinese
Religions: Muslim (official) 67%, Buddhist 13%, Christian 10%, indigenous beliefs and other 10%
Government type: constitutional sultanate
Capital: Bandar Seri Begawan
Budget: revenues: $2.5 billion; expenditures: $ 2.6 billion   **Currency:** Bruneian dollar (BND)

BULGARIA
Location: Southeastern Europe, bordering the Black Sea, between Romania and Turkey
Area: 42,811 sq mi (110,910 sq km)
Area-comparative: slightly larger than Tennessee
Population: 7,517,973
Population growth rate: -0.92%
Nationality: noun: Bulgarian(s); adjective: Bulgarian
Ethnic groups: Bulgarian 84%, Turk 9%, Roma 5%, other 2%
Languages: Bulgarian (official), secondary languages closely correspond to ethic breakdown
Religions: Bulgarian Orthodox 83%, Muslim 12%, other (including Jewish and Roman Catholic) 4%
Government type: parliamentary democracy
Capital: Sofia
Budget: revenues: $6.68 billion; expenditures: $6.69 billion, including capital of $NA
Currency: lev (BGN)

BURKINA FASO
Location: Western Africa, north of Ghana
Area: 105,841 sq mi (274,200 sq km)
Area-comparative: slightly larger than Colorado
Population: 13,574,820
Population growth rate: 2.57%
Nationality: noun: Burkinabe (singular and plural); adjective: Burkinabe
Ethnic groups: Mossi over 40%, Gurunsi, Senufo, Lobi, Bobo, Mande, Fulani
Languages: French (official), Sudanic tribal languages spoken by 90%
Religions: Muslim 50%, indigenous beliefs 40%, Christian (mainly Roman Catholic) 10%
Government type: parliamentary republic
Capital: Ouagadougou

462

Budget: revenues: $316 million; expenditures: $NA
Currency: Communaute Financiere Africaine franc (XOF)

BURMA
Location: Southeastern Asia, bordering the Andaman Sea and the Bay of Bengal
Area: 261,901 sq mi (678,500 sq km)
Area-comparative: slightly smaller than Texas
Population: 42,720,196
Population growth rate: 0.47%
Nationality: noun: Burmese (singular and plural); adjective: Burmese
Ethnic groups: Burman 68%, Shan 9%, Karen 7%, Rakhine 4%, Chinese 3%, Indian 2%, other 7%
Languages: Burmese, minority ethnic groups have their own languages
Religions: Buddhist 89%, Christian 4%, Muslim 4%, other 3%
Government type: military regime
Capital: Rangoon
Budget: revenues: $7.9 billion; expenditures: $ 12.2 billion   **Currency:** kyat (MMK)

BURUNDI
Location: Central Africa, east of Democratic Republic of the Congo
Area: 10,742 sq mi (27,830 sq km)
Area-comparative: slightly smaller than Maryland
Population: 6,231,221
Population growth rate: 2.2%
Nationality: noun: Burundian(s); adjective: Burundian
Ethnic groups: Hutu (Bantu) 84%, Tutsi 14%, Twa 1%, other 1%
Languages: Kirundi (official), French (official), Swahili
Religions: Roman Catholic 62%, indigenous beliefs 23%, Protestant 5%, Muslim 10%
Government type: republic
Capital: Bujumbura
Budget: revenues: $125 million; expenditures: $176 million   **Currency:** Burundi franc (BIF)

CAMBODIA
Location: Southeastern Asia, bordering the Gulf of Thailand, between Thailand, Vietnam, and Laos
Area: 69,881 sq mi (181,040 sq km)
Area-comparative: slightly smaller than Oklahoma
Population: 13,363,421
Population growth rate: 1.8%
Nationality: noun: Cambodian(s); adjective: Cambodian
Ethnic groups: Khmer 90%, Vietnamese 5%, Chinese 1%, other 4%
Languages: Khmer (official) 95%, French, English
Religions: Theravada Buddhism 95%, other 5%
Government type: multiparty democracy under a constitutional monarchy
Capital: Phnom Penh
Budget: revenues: $438 million; expenditures: $630 million, including capital expenditures of $291 million
Currency: riel (KHR)

CAMEROON
Location: Western Africa, bordering the Bight of Biafra, between Equatorial Guinea and Nigeria
Area: 183,519 sq mi (475,440 sq km)
Area-comparative: slightly larger than California
Population: 16,063,678
Population growth rate: 1.97%

463

Nationality: noun: Cameroonian(s); adjective: Cameroonian
Ethnic groups: Cameroon Highlanders 31%, Equatorial Bantu 19%, Kirdi 11%, Fulani 10%, Northwestern Bantu 8%, Eastern Nigritic 7%, other African 13%, non-African less than 1%
Languages: 24 major African language groups, English (official), French (official)
Religions: indigenous beliefs 40%, Christian 40%, Muslim 20%
Government type: unitary republic; multiparty presidential regime
Capital: Yaounde
Budget: revenues: $2.2 billion; expenditures: $ 2.1 billion
Currency: Communaute Financiere Africaine franc (XAF)

CANADA

Location: Northern North America, bordering the North Atlantic Ocean, Arctic Ocean and North Pacific Ocean, north of the 48 mainland states of the U.S.
Area: 3,850,790 sq mi (9,984,670 sq km)
Area-comparative: somewhat larger than the US
Population: 32,507,874
Population growth rate: 0.92%
Nationality: noun: Canadian(s), adjective: Canadian
Ethnic groups: British Isles origin 28%, French origin 23%, other European 15%, other (primarily Asian) 6%, Amerindian 2%,mixed background 26%
Languages: English (official), French (official)
Religions: Roman Catholic 46%, Protestant 36%, other 18%
Government type: confederation with parliamentary democracy
Capital: Ottawa
Budget: revenues: $178.6 billion; expenditures: $161.4 billion   **Currency:** Canadian dollar (CAD)

CAPE VERDE

Location: Western Africa, group of islands in the North Atlantic Ocean, west of Senegal
Area: 1,556 sq mi (4,033 sq km)
Area-comparative: slightly larger than Rhode Island
Population: 415,294
Population growth rate: 0.73%
Nationality: noun: Cape Verdean(s); adjective: Cape Verdean
Ethnic groups: Creole 71%, African 28%, European 1%
Language: Portuguese (official), Crioulo (a blend of Portuguese and West African words)
Religion: Roman Catholic (infused with indigenous beliefs), Protestants
Government type: republic
Capital: Praia
Budget: revenues: $112 million; expenditures: $198 million   **Currency:** Cape Verdean escudo

CAYMAN ISLANDS (overseas territory of the UK)

Location: Caribbean, island group in Caribbean Sea, nearly one half of the way from Cuba to Honduras
Area: 101 sq mi (262 sq km)
Area-comparative: 1.5 times the size of Washington, DC
Population: 43,103
Population growth rate: 2.71%
Nationality: noun: Caymanian(s); adjective: Caymanian
Ethnic groups: mixed 40%, white 20%, black 20%, various other 20%
Language: English
Religions: United Church (Presbyterian and Congregational), Anglican, Baptist, Church of God, Protestant, Roman Catholic
Government type: British crown colony

Capital: George Town
Budget: revenues: $ 265.2 million, expenditures: $248.9 million
Currency: Caymanian dollar (KYD)

CENTRAL AFRICAN REPUBLIC
Location: Central Africa, north of Democratic Republic of the Congo
Area: 240,471 sq mi (622,984 sq km)
Area-comparative: slightly smaller than Texas
Population: 3,742,482
Population growth rate: 1.56%
Nationality: noun: Central African(s); adjective: Central African
Ethnic groups: Baya 33%, Banda 27%, Mandjia 13%, Sara 10%, Mboum 7%, M'Baka 4%,Yakoma 4%, other 2%
Language: French (official), Sangho (lingua franca and national language), tribal languages
Religion: indigenous beliefs 35%, Protestant 25%, Catholic 25%, Muslim 15%
Government type: republic
Capital: Bangui
Budget: revenues: $NA; expenditures: $NA
Currency: Communaute Financiere Africaine franc (XAF)

CHAD
Location: Central Africa, south of Libya
Area: 495,624 sq mi (1,284,000 sq km)
Area-comparative: slightly more than 3 times the size of California
Population: 9,538,544
Population growth rate: 3%
Nationality: noun: Chadian(s); adjective: Chadian
Ethnic groups: Sara , Sudanic Arab , Mayo-Kebbi peoples , Kanem-Bornu peoples , Ouaddaï peoples , Hadjeray , Tangale peoples , Gorane peoples , Fitri-Batha peoples , Fulani , 200 distinct groups, other
Languages: French (official), Arabic (official), Sara, Sango, over 120 different languages and dialects
Religions: Muslim 51%, Christian 35%, animist 7%, other 7%
Government type: republic
Capital: N`Djamena
Budget: revenues: $198 million; expenditures: $218 million, including capital expenditures of $146 million
Currency: Communaute Financiere Africaine franc (XAF)

CHILE
Location: Southern South America, bordering the South Atlantic Ocean and South Pacific Ocean, between Argentina and Peru
Area: 292,182 sq mi (756,950 sq km)
Area-comparative: slightly smaller than twice the size of Montana
Population: 15,823,957
Population growth rate: 1.01%
Nationality: noun: Chilean(s); adjective: Chilean
Ethnic groups: white and white-Amerindian 95%, Amerindian 3%, other 2%
Language: Spanish (official)
Religions: Roman Catholic 89%, Protestant 11%
Government type: republic
Capital: Santiago
Budget: revenues: $14.7 billion; expenditures: $15.6 billion, including capital of $NA
Currency: Chilean peso (CLP)

CHINA

Location: Eastern Asia, bordering the East China Sea, Korea Bay, Yellow Sea, and South China Sea, between North Korea and Vietnam

Area: 3,704,426 sq mi (9,596,960 sq km)

Area-comparative: slightly smaller than the US

Population: 1,298,847,624

Population growth rate: 0.57%

Nationality: noun: Chinese (singular and plural); adjective: Chinese

Ethnic groups: Han Chinese 92%, Zhuang, Uyger, Hui, Yi, Tibetan, Miao, Manchu, Mongol, Buyi, Korean, and other nationalities 8%

Languages: Standard Chinese or Mandarin (official), Yue, Wu, Minbei, Minnan, Xiang, Gan, Hakka dialects, minority languages

Religions: officially Atheist, Daoist (Taoist), Buddhist, Muslim 1%-2%, Christian 3%- 4%

Government type: Communist state

Capital: Beijing

Budget: revenues: $224.8 billion; expenditures: $267.1 billion   **Currency:** yuan (CNY)

COLOMBIA

Location: Northern South America, bordering the Caribbean Sea, between Panama and Venezuela, bordering the North Pacific Ocean, between Ecuador and Panama

Area: 439,619 sq mi (1,138,910 sq km)

Area-comparative: slightly less than 3 times the size of Montana

Population: 42,310,775

Population growth rate: 1.53%

Nationality: noun: Colombian(s); adjective: Colombian

Ethnic groups: mestizo 58%, white 20%, mulatto 14%, black 4%, mixed black- Amerindian 4%

Language: Spanish (official)

Religions: Roman Catholic 90%, other 10%

Government type: republic, executive branch dominates government structure

Capital: Bogota

Budget: revenues: $24 billion; expenditures: $ 25.6 billion   **Currency:** Colombian peso (COP)

COMOROS

Location: Southern Africa, group of islands in the Mozambique Channel, about two-thirds of the way between northern Madagascar and northern Mozambique

Area: 838 sq mi (2,170 sq km)

Area-comparative: slightly more than 12 times the size of Washington, DC

Population: 651,901

Population growth rate:2.94%

Nationality: noun: Comoran(s); adjective: Comoran

Ethnic groups: Antalote, Cafre, Makoa, Oimatsaha, Sakalava

Languages: Arabic (official), French (official), Shikomoro (a blend of Swahili and Arabic)

Religions: Sunni Muslim 98%, Roman Catholic 2%

Government type: independent republic

Capital: Moroni

Budget: revenues: $27.6 million; expenditures: $NA   **Currency:** Comoran franc (KMF)

CONGO, DEMOCRATIC REPUBLIC OF THE (ZAIRE)

Location: Central Africa, northeast of Angola

Area: 905,328 sq mi (2,345,410 sq km)

Area-comparative: slightly less than one-fourth the size of the US

Population: 56,625,039

Population growth rate: 2.9%
Nationality: noun: Congolese (singular and plural); adjective: Congolese or Congo
Ethnic groups: Over 200 African ethnic groups the majority are Bantu, the four largest tribes-Mongo,
Luba, Kongo, and the Mangbetu-Azande- make up about 45% of the population
Languages: French (official), Lingala, Kingwana (a dialect of Kiswahili or Swahili), Kikongo, Tshiluba
Religions: Roman Catholic 50%, Protestant 20%, Kimbanguist 10%, Muslin 10%, other 10%
Government type: dictatorship
Capital: Kinshasa
Budget: revenues: $ 269 million; expenditures: $ 244 million, including capital expenditures of $24 million
Currency: Congolese franc (KMF)

CONGO, REPUBLIC OF THE

Location: Western Africa, bordering the South Atlantic Ocean, between Angola and Gabon
Area: 132,012 sq mi (342,000 sq km)
Area-comparative: slightly smaller than Montana
Population: 2,998,040
Population growth rate: 1.42%
Nationality: noun: Congolese (singular or plural); adjective: Congolese or Congo
Ethnic groups: Kongo 48%, Sangha 20%, Teke 17%, M'Bochi 12%, other 3%
Languages: French (official), Lingala and Monokutuba, many local languages and dialects
Religions: Christian 50%, animist 48%, Muslim 2%
Government type: republic
Capital: Brazzaville
Budget: revenues: $870 million; expenditures: $970 million
Currency: Communaute Financiere Africaine franc (XAF)

COSTA RICA

Location: Central America, bordering both the Caribbean Sea and the North Pacific Ocean, between
Nicaragua and Panama
Area: 19,724 sq mi (51,100 sq km)
Area-comparative: slightly smaller than West Virginia
Population: 3,956,507
Population growth rate: 1.52%
Nationality: noun: Costa Rican(s); adjective: Costa Rican
Ethnic groups: white and mestizo 94%, Black 3%, Amerindian 1%, Chinese 1%, other 1%
Language: Spanish (official) , English
Religions: Roman Catholic 76%, Evangelical 13%, other 11%
Government type: democratic republic
Capital: San Jose
Budget: revenues: $ 2.17 billion; expenditures $ 2.84 billion, including capital $NA
Currency: Costa Rican colon (CRC)

COTE d'IVOIRE

Location: Western Africa, bordering the North Atlantic Ocean, between Ghana and Liberia
Area: 124,469 sq mi (322,460 sq km)
Area-comparative: slightly larger than New Mexico
Population: 17,327,724
Population growth rate: 2.11%
Nationality: noun: Ivorian(s); adjective: Ivorian
Ethnic groups: Akan 42%, Voltaiques or Gur 17%, Northern Mandes 16%, Krous 11%, Southern Mandes
10%, other 4%
Languages: French (official), 60 native dialects with Dioula the most widely spoken

467

Religions: Muslim 35-40%, Christian 20-30%, indigenous 25-40%
Government type: republic
Capital: Yamoussoukro(official) ;Abidjan remains the commercial and administrative center
Budget: revenues: $1.72 billion; expenditures: $2.4 billion, including capital expenditures of $420 million
Currency: Communaute Financiere Africaine franc (XOF)

CROATIA
Location: Southeastern Europe, bordering the Adriatic Sea, between Bosnia, Herzegovina, and Slovenia
Area: 21,824 sq mi (56,542 sq km)
Area-comparative: slightly smaller than West Virginia
Population: 4,496,869   **Population growth rate:** -0.02%
Nationality: noun: Croat(s), Croatian(s); adjective: Croatian
Ethnic groups: Croat 89%, Serb 5%, other 6%
Languages: Croatian 96%, other 4%
Religions: Catholic 88%, Orthodox 4%, Muslim 1%, Protestant 0.3%, other 1%, unknown 5.97%
Government type: presidential/parliamentary democracy
Capital: Zagreb
Budget: revenues: $8.6 billion; expenditures: $9 billion   **Currency:** Croatian kuna (HRK)

CUBA
Location: Caribbean, island between the Caribbean Sea and the North Atlantic Ocean, south of Florida
Area: 42,791 sq mi (110,860 sq km)
Area-comparative: slightly smaller than Pennsylvania
Population: 11,308,764
Population growth rate: 0.34%
Nationality: noun: Cuban(s); adjective: Cuban
Ethnic groups: mulatto 51%, white 37%, black 11%, Chinese 1%
Language: Spanish (official)
Religions: nominally 85% Roman Catholic prior to Castro assuming power; Protestants, Jehovah`s
          Witnesses, Jews and Santeria are represented
Government type: Communist state
Capital: Havana
Budget: revenues: $14.9 billion; expenditures: $15.6 billion   **Currency:** Cuban peso (CUP)

CYPRUS
Location: Middle East, island in the Mediterranean Sea, south of Turkey
Area: 3,571 sq mi (9,250 sq km)
Area-comparative: about 0.6 times the size of Connecticut
Population: 775,927
Population growth rate: 0.55%
Nationality: noun: Cypriot(s); adjective: Cypriot
Ethnic groups: Greek 77%, Turkish 18%, other 5%
Languages: Greek (official), Turkish (official), English
Religions: Greek Orthodox 78%, Muslim 18%, other (including Maronite and Armenian Apostolic) 4%
Government type: republic
Capital: Nicosia
Budget: revenues: Greek Cypriot area-$4.4 billion, Turkish Cypriot area $231.3 million
          expenditures: $3.7 billion
Currency: Greek Cypriot area: Cypriot pound, Turkish Cypriot area: Turkish lira (CYP:TRL)

CZECH REPUBLIC
Location: Central Europe, southeast of Germany
Area: 30,403 sq mi (78,866 sq km)

Area-comparative: slightly smaller than South Carolina
Population: 10,246,178
Population growth rate: -0.05%
Nationality: noun: Czech(s); adjective: Czech
Ethnic groups: Czech 81%, Moravian 13%, Slovak 3%, Polish , German and other 3%
Language: Czech (official)
Religions: Roman Catholic 39.2%, Protestant 4.6%, Orthodox 3%, other 13.4%, atheist 39.8%
Government type: parliamentary democracy
Capital: Prague
Budget: revenues: $16.7 billion; expenditures: $18 billion   **Currency:** Czech koruna (CZK)

DENMARK
Location: Northern Europe, bordering the Baltic Sea and the North Sea, on a peninsula north of Germany
Area: 16,634 sq mi (43,094 sq km)
Area-comparative: slightly less than twice the size of Massachusetts
Population: 5,413,392
Population growth rate: 0.35%
Nationality: noun: Dane(s); adjective: Danish
Ethnic groups: Scandinavian, Inuit, Faroese, German, Turkish, Iranian, Somali
Languages: Danish (official), Faroese, Greenlandic, German
Religions: Evangelical Lutheran 95%, other Protestant and Roman Catholic 3%, Muslim 2%
Government type: constitutional monarchy
Capital: Copenhagen
Budget: revenues: $52.9 billion; expenditures: $51.3 billion, including capital expenditures of $500 million
Currency: Danish krone (DKK)

DJIBOUTI
Location: Eastern Africa, bordering the Gulf of Aden and the Red Sea, between Eritrea and Somalia
Area: 8,878 sq mi (23,000 sq km)
Area-comparative: slightly smaller than Massachusetts
Population: 466,900
Population growth rate: 2.1%
Nationality: noun: Djiboutian(s); adjective: Djiboutian
Ethnic groups: Somali 60%, Afar 35%, French, Arab, and Ethiopian and Italian 5%
Languages: French (official), Arabic (official), Somali, Afar
Religions: Muslim 94%, Christian 6%
Government type: republic
Capital: Djibouti
Budget: revenues: $135 million; expenditures: $182 million   **Currency:** Djiboutian franc (DJF)

DOMINICA
Location: Caribbean, island between the Caribbean Sea and the North Atlantic Ocean, about one-half of the way from Puerto Rico to Trinidad and Tobago
Area: 290 sq mi (754 sq km)
Area-comparative: slightly more than 4 times the size of Washington, DC
Population: 69,278
Population growth rate: -0.45%
Nationality: noun: Dominican(s); adjective: Dominican
Ethnic groups: black, mixed black and European, European, Syrian, Carib Amerindian
Languages: English (official), French patois
Religions: Roman Catholic 77%, Protestant 15%, other/nonreligious 8%
Government type: parliamentary democracy

Capital: Roseau
Budget: revenues: $73.9 million; expenditures: $84.4 million
Currency: East Caribbean dollar (XCD)

DOMINICAN REPUBLIC
Location: Caribbean, eastern two-thirds of the island of Hispaniola, between the Caribbean Sea and the North Atlantic Ocean, east of Haiti
Area: 18,810 sq mi (48,730 sq km)
Area-comparative: slightly more than twice the size of New Hampshire
Population: 8,833,634
Population growth rate: 1.33%
Nationality: noun: Dominican(s); adjective: Dominican
Ethnic groups: Multiracial (black and white) 73%, white16%, black 11%
Language: Spanish (official)
Religions: Roman Catholic 95%, other 5%
Government type: representative democracy
Capital: Santo Domingo
Budget: revenues: $2.9 billion; expenditures: $1.1 billion   **Currency:** Dominican peso (DOP)

ECUADOR
Location: Western South America, bordering the Pacific Ocean at the Equator
Area: 109,454 sq mi (283,560 sq km)
Area-comparative: slightly smaller than Nevada
Population: 13,212,742
Population growth rate: 1.03%
Nationality: noun: Ecuadorian(s): adjective: Ecuadorian
Ethnic groups: mestizo 65%, Amerindian 25%, Spanish and others 7%, black 3%
Languages: Spanish (official), Amerindian languages (particularly Quechua)
Religions: Roman Catholic 95%, other 5%
Government type: republic
Capital: Quito
Budget: revenues: $6.3 billion; expenditures: planned $5.3 billion   **Currency:** US dollar (USD)

EGYPT
Location: Northern Africa, bordering the Mediterranean Sea, between Libya and the Gaza Strip
Area: 386,560 sq mi (1,001,450 sq km)
Area-comparative: slightly more than three times the size of New Mexico
Population: 76,117,421
Population growth rate: 1.83%
Nationality: noun: Egyptian(s); adjective: Egyptian
Ethnic groups: Egyptians, Bedouins, and Berbers 99%, other including Greek, Nubian, and Armenian 1%
Languages: Arabic (official), English, French
Religions: Muslim (primarily Sunni) 94%, other (including Coptic Christian) 6%
Government type: republic
Capital: Cairo
Budget: revenues: $14 billion; expenditures: $18.1billion, including capital expenditures of $2.7 billion
Currency: Egyptian pound (EGP)

EL SALVADOR
Location: Central America, bordering the North Pacific Ocean, between Guatemala and Honduras
Area: 8,121 sq mi (21,040 sq km)
Area-comparative: slightly smaller than Massachusetts

470

Population: 6,587,541
Population growth rate: 1.78%
Nationality: noun: Salvadoran(s); adjective: Salvadoran
Ethnic groups: mestizo 90%, Amerindian 1%, white 9%
Languages: Spanish (official), Nahua
Religions: Roman Catholic 83%, Protestant and other 17%
Government type: republic
Capital: San Salvador
Budget: revenues: $2.1 billion; expenditures: $2.5 billion
Currency: US dollar (USD)

EQUATORIAL GUINEA
Location: Western Africa, bordering the Bight of Biafra, between Cameroon and Gabon
Area: 10,827 sq mi (28,051 sq km)
Area-comparative: slightly smaller than Maryland
Population: 523,051
Population growth rate: 2.43%
Nationality: noun: Equatorial Guinean(s) or Equatoguinean(s)
Ethnic groups: Bioko (primarily Bubi, some Fernandinos), Rio Muni (primarily Fang)
Languages: Spanish (official), French (official), pidgin English, Fang, Bubi , Ibo
Religions: nominally Christian and predominantly Roman Catholic, pagan practices
Government type: republic
Capital: Malabo
Budget: revenues: $200 million; expenditures: $158 million
Currency: Communaute Financiere Africaine franc (XAF)

ERITREA
Location: Eastern Africa, bordering the Red Sea, between Djibouti and Sudan
Area: 46,830 sq mi (121,320 sq km)
Area-comparative: slightly larger than Pennsylvania
Population: 4,447,307
Population growth rate: 2.57%
Nationality: noun: Eritrean(s); adjective: Eritrean
Ethnic groups: ethnic Tigrinya 50%, Tigre and Kunama 40%, Afar 4%, Saho 3%, other 3%
Languages: Afar, Arabic, Tigre and Kunama, Tigrinya, other Cushitic languages
Religions: Muslim, Coptic Christian, Roman Catholic, Protestant
Government type: transitional government
Capital: Asmara
Budget: revenues: $206.4 million; expenditures: $615.7 million
Currency: nakfa (ERN)

ESTONIA
Location: Eastern Europe, bordering the Baltic Sea and Gulf of Finland, between Latvia and Russia
Area: 17,457 sq mi (45,226 sq km)
Area-comparative: slightly smaller than New Hampshire and Vermont combined
Population: 1,341,664
Population growth rate: -0.66%
Nationality: noun:
Ethnic groups: Estonian 65%, Russian 28%, Ukrainian 3%, Belarusian 2%, other 2%
Languages: Estonian (official), Russian, Ukrainian, Finnish, other
Religions: Evangelical Lutheran, Russian Orthodox, Estonian Orthodox, Baptist, Methodist, Seventh-
Day Adventist, Roman Catholic, Pentecostal, Word of Life, Jewish
Government type: parliamentary republic

Capital: Tallinn
Budget: revenues: $3.13 billion; expenditures: $3 billion, including capital of $NA
Currency: Estonian kroon (EEK)

ETHIOPIA
Location: Eastern Africa, west of Somalia
Area: 435,071 sq mi (1,127,127 sq km)
Area-comparative: slightly less than twice the size of Texas
Population: 67,851,281
Population growth rate: 1.89%
Nationality: noun: Ethiopian(s); adjective: Ethiopian
Ethnic groups: Oromo 40%, Amhara and Tigre 32%, Sidamo 9%, Shankella 6%, Somali 6%, Afar 4%, Gurage 2%, other 1%
Languages: Amharic (official), Tigrinya, Orominga, Guaragigna, Somali, Arabic, other local languages
Religions: Muslim 45%-50%, Ethiopian Orthodox 35%-40%, animist 12%, other 3%-8%
Government type: federal republic
Capital: Addis Ababa
Budget: revenues: $1.4 billion; expenditures: $2.4 billion, including capital expenditures of $788 million
Currency: birr (ETB)

FIJI
Location: Oceania, island group in the South Pacific Ocean, 2/3's of the way from Hawaii to New Zealand
Area: 7,052 sq mi (18,270 sq km)
Area-comparative: slightly smaller than New Jersey
Population: 880,874
Population growth rate: 1.41%
Nationality: noun: Fijian(s); adjective: Fijian
Ethnic groups: Fijian 51%, Indian 44%, Europeans, Pacific Islanders, and overseas Chinese 5%
Language: English (official), Fijian, Hindustani
Religion: Christian 52%, Hindu 38%, Muslim 8%, other 2%
Government type: republic
Capital: Suva
Budget: revenues: $427.9 million; expenditures: $531.4 million   **Currency:** Fijian dollar (FJD)

FINLAND
Location: Northern Europe, bordering the Baltic Sea, Gulf of Bothnia, and Gulf of Finland
Area: 130,094 sq mi (337,030 sq km)
Area-comparative: slightly smaller than Montana
Population: 5,214,512
Population growth rate: 0.18%
Nationality: noun: Finn(s); adjective: Finnish
Ethnic groups: Finn 93%, Swede 6%, other 1%
Languages: Finnish (official), Swedish (official)
Religions: Evangelical Lutheran 89%, Russian Orthodox 1%, none 9%, other 1%
Government type: republic
Capital: Helsinki
Budget: revenues: $36.1 billion; expenditures: $ 31 billion   **Currency:** euro (EUR)

FRANCE
Location: Western Europe, bordering the Bay of Biscay and English Channel, between Belgium and Spain, southeast of the UK; bordering the Mediterranean Sea, between Italy and Spain
Area: 211,153 sq mi (547,030 sq km)
Area-comparative: slightly less than twice the size of Colorado

Population: 60,424,213
Population growth rate: 0.39%
Nationality: noun: Frenchman(men), Frenchwoman(women; adjective: French
Ethnic groups: Celtic and Latin with Teutonic Slavic, North African, Indochinese, Basque
Language: French (official)
Religions: Roman Catholic 83%-88%, Muslim 5%-10%, Protestant 2%, Jewish 1%, unaffiliated 4%
Government type: republic
Capital: Paris
Budget: revenues: $217billion; expenditures: $ 274 billion, including capital expenditures of $ 23 billion
Currency: euro (EUR)

GABON

Location: Western Africa, bordering the Atlantic Ocean at the Equator, between Republic of the Congo and Equatorial Guinea
Area: 103,319 sq mi (267,667 sq km)
Area-comparative: slightly smaller than Colorado
Population: 1,355,246
Population growth rate: 2.5%
Nationality: noun: Gabonese (singular and plural); adjective: Gabonese
Ethnic groups: Bantu tribes including four major tribal groups; Fang, Bapounou, Nzebi and Obamba
Languages: French (official), Fang, Myene, Nzebi, Bapounou/Eschira, Bandjabi
Religions: Christian 55%-75%, animist, Muslim 1%
Government type: republic, multiparty presidential regime
Capital: Libreville
Budget: revenues: $1.8 billion; expenditures: $1.8 billion, including capital expenditures of $310 million
Currency: Communaute Financiere Africaine franc (XAF)

GAMBIA

Location: Western Africa, bordering the North Atlantic Ocean and Senegal
Area: 4,362 sq mi (11,300 sq km)
Area-comparative: slightly less than twice the size of Delaware
Population: 1,546,848
Population growth rate: 2.98%
Nationality: noun: Gambian(s); adjective: Gambian
Ethnic groups: African (including Mandinka, Fula, and Wolof) 99%, non-African 1%
Languages: English (official), Mandinka, Wolof, Fula
Religions: Muslim 90%, Christian 9%, indigenous beliefs 1%
Government type: republic under multiparty democratic rule
Capital: Banjul
Budget: revenues: $ 90.5 million; expenditures: $80.9 million   **Currency:** dalasi (GMD)

GAZA STRIP

Location: Middle East, bordering the Mediterranean Sea, between Egypt and Israel
Area: 360 sq km
Area-comparative: slightly more than twice the size of Washington, DC
Population: 1,324,991
Population growth rate: 3.83%
Nationality: noun: NA; adjective: NA
Ethnic groups: Palestinian Arab and other 99.4%, Jewish 0.6%
Languages: Arabic, Hebrew, English
Religions: Muslim (predominantly Sunni) 98.7%, Christian 0.7%, Jewish 0.6%
Government type: NA

Capital: NA
Budget: revenues: $676.6 million; expenditures: $1.155 billion, including capital of $NA
Currency: new Israeli shekel (ILS)

GEORGIA
Location: Southwestern Asia, bordering the Black Sea, between Turkey and Russia
Area: 26,904 sq mi (69,700 sq km)
Area-comparative: slightly smaller than South Carolina
Population: 4,693,892
Population growth rate: -0.36%
Nationality: noun: Georgian(s); adjective: Georgian
Ethnic groups: Georgian 70%, Armenian 8%, Russian 6%, Azeri 6%, Ossetian 3%, Abkhaz 2%, other 5%
Languages: Georgian (official), Russian, Armenian, Azeri, other
Religions: Georgian Orthodox 65%, Russian Orthodox 10%, Muslim 11%, Armenian Apostolic 8%, other
Government type: republic
Capital: T`bilisi
Budget: revenues: $499 million; expenditures: $554 million   **Currency:** lari (GEL)

GERMANY
Location: Central Europe, bordering the Baltic Sea and North Sea, between the Netherlands and Poland
Area: 137,810 sq mi (357,021 sq km)
Area-comparative: slightly smaller than Montana
Population: 82,424,609
Population growth rate: 0.02%
Nationality: noun: German(s); adjective: German
Ethnic groups: German 92%,Turkish 2%, other 6%
Language: German (official)
Religion: Protestant 34%, Roman Catholic 34%, Muslim 4%, unaffiliated or other 28%
Government type: federal republic
Capital: Berlin
Budget: revenues: $ 802 billion; expenditures: $825 billion   **Currency:** euro (EUR)

GHANA
Location: Western Africa, bordering the Gulf of Guinea, between Cote d'Ivoire and Togo
Area: 92,431 sq mi (239,460 sq km)
Area-comparative: slightly smaller than Oregon
Population: 20,757,032
Population growth rate: 1.36%
Nationality: noun: Ghanaian9s); adjective: Ghanaian
Ethnic groups: black African 98%, (major tribes- Akan 44%, Moshi-Dagomba 16%, Ewe 13%, Ga 8%)
Language: English (official), African languages (including Akan, Moshi-Dagomba, Ewe and Ga)
Religion: indigenous beliefs 21%, Muslim 16%, Christian 63%
Government type: constitutional democracy
Capital: Accra
Budget: revenues: $ 1.6 billion; expenditures: $1.9 billion   **Currency:** cedi (GHC)

GREECE
Location: Southern Europe, bordering the Aegean Sea, Ionian Sea, and the Mediterranean Sea
Area: 50,929 sq mi (131,940 sq km)
Area-comparative: slightly smaller than Alabama
Population: 10,647,529
Population growth rate: 0.2%

Nationality: noun: Greek(s); adjective: Greek
Ethnic groups: Greek 98%, other 2%
Languages: Greek (official), English, French
Religions: Greek Orthodox 98%, Muslim 1%, other 1%
Government type: parliamentary republic
Capital: Athens
Budget: revenues: $45 billion; expenditures: $47.6 billion   **Currency:** euro (EUR)

GRENADA

Location: Caribbean island between the Caribbean Sea and Atlantic Ocean, north of Trinidad and Tobago
Area: 133 sq mi (344 sq km)
Area-comparative: twice the size of Washington, DC
Population: 89,357
Population growth rate: 0.14%
Nationality: noun: Grenadian(s); adjective: Grenadian
Ethnic groups: black 82%, Multiracial (white and black)13%, other 4%
Language: English (official), French patois
Religion: Roman Catholic 53%, Anglican 13.8%, other Protestant 33.2%
Government type: constitutional monarchy with Westminster-style parliament
Capital: Saint George`s
Budget: revenues: $85.8 million; expenditures: $102.1 million, also, capital expenditures of $28 million
Currency: East Caribbean dollar (XCD)

GUATEMALA

Location: Central America, bordering the North Pacific Ocean, between El Salvador and Mexico, and
          bordering the Gulf of Honduras (Caribbean Sea) between Honduras and Belize
Area: 42,031 sq mi (108,890 sq km)
Area-comparative: slightly smaller than Tennessee
Population: 14,280,596
Population growth rate: 2.61%
Nationality: noun: Guatemalan(s); adjective: Guatemalan
Ethnic groups: Mestizo 55%, Amerindian or predominantly Amerindian 43%, whites and others 2%
Languages: Spanish, Amerindian languages
Religions: Roman Catholic, Protestant, traditional Mayan beliefs
Government type: constitutional democratic republic
Capital: Guatemala
Budget: revenues: $2.3 billion; expenditures: $2.7 billion, including capital expenditures of $ 750 million
Currency: quetzal (GTQ) US dollar (USD), others allowed

GUINEA

Location: Western Africa, bordering the North Atlantic Ocean, between Guinea-Bissau and Sierra Leone
Area: 94,901 sq mi (245,857 sq km)
Area-comparative: slightly smaller than Oregon
Population: 9,246,462
Population growth rate: 2.37%
Nationality: noun: Guinean(s); adjective; Guinean
Ethnic groups: Peuhl 40%, Malinke 30%, Soussou 20%, other tribes 10%
Language: French (official), tribal languages
Religion: Muslim 85%, Christian 8%, indigenous beliefs 7%
Government type: republic
Capital: Conakry
Budget: revenues: $395.7 million; expenditures: $472.4million   **Currency:** Guinean franc (GNF)

GUINEA-BISSAU
Location: Western Africa, bordering the North Atlantic Ocean, between Guinea and Senegal
Area: 13,942 sq mi (36,120 sq km)
Area-comparative: slightly less than three times the size of Connecticut
Population: 1,388,363
Population growth rate: 1.99%
Nationality: noun: Guinean(sh); adjective Guinean(s)
Ethnic groups: Balanta 30%, Fula 20%, other 17%, Majaca 12%, Mandinga 13%, Papel 7%
Languages: Portuguese (official), Crioulo, African languages
Religion: indigenous beliefs 50%, Muslim 45%, Christian 5%
Government type: republic, multiparty since mid-1991
Capital: Bissau
Budget: revenues: $NA; expenditures: $NA
Currency: Communaute Financiere Africaine franc (XOF);(GWP)

GUYANA
Location: Northern South America, bordering North Atlantic Ocean, between Suriname and Venezuela
Area: 82,978 sq mi (214,970 sq km)
Area-comparative: slightly smaller than Idaho
Population: 705,803
Population growth rate: 0.61%
Nationality: noun:Guyanese (singular and plural); adjective: Guyanese
Ethnic groups: East Indian 50%, black 36%, Amerindian 7%, white, Chinese, and mixed 7%
Languages: English (official), Amerindian dialects, Creole, Hindi, Urdu
Religions: Christian 50%, Hindu 35%, Muslim 10%, other 5%
Government type: republic within the Commonwealth
Capital: Georgetown
Budget: revenues $227 million; expenditures: $235.2 million,
        including capital expenditures of $93.4 million
Currency: Guyanese dollar (GYD)

HAITI
Location: Caribbean, western one-third of the island of Hispaniola, west of the Dominican Republic
Area: 10,711 sq mi (27,750 sq km)
Area-comparative: slightly smaller than Maryland
Population: 7,656,166
Population growth rate: 1.71%
Nationality: noun: Haitian(s); adjective: Haitian
Ethnic groups: black 95%, mulatto and white 5%
Languages: French (official), Creole (official)
Religions: Roman Catholic 80%, Protestant 16%, 10% Baptist, Pentecostal 4%, and Adventist 1%, other
        1%, other/nonreligious 4%
Government type: elected government
Capital: Port-au-Prince
Budget: revenues: 273 million; expenditures: $361 million
Currency: gourde (HTG)

HONDURAS
Location: Central America, bordering the Caribbean Sea, between El Salvador and Nicaragua
Area: 43,267 sq mi (112,090 sq km)
Area-comparative: slightly larger than Tennessee

Population: 6,823,568
Population growth rate: 2.24%
Nationality: noun: Honduran(s); adjective: Honduran
Ethnic groups: mestizo (mixed Amerindian and European) 90%, Amerindian 7%, black 2%, white 1%
Languages: Spanish (official), Amerindian dialects
Religions: Roman Catholic 97%, other 3%
Government type: democratic constitutional republic
Capital: Tegucigalpa
Budget: revenues: $607 million; expenditures: $411.9 million, including capital expenditures of $106 million   **Currency:** lempira (HNL)

HUNGARY

Location: Central Europe, northwest of Romania
Area: 35,910 sq mi (93,030 sq km)
Area-comparative: slightly smaller than Indiana
Population: 10,032,375
Population growth rate: - 0.25%
Nationality: noun: Hungarian(s); adjective: Hungarian
Ethnic groups: Hungarian 89.9%, Roma 4%, German 2.6%, Serb 2%, Slovak 0.8%, Romanian 0.7%
Languages: Hungarian 98.2%, other 1.8%
Religion: Roman Catholic 67.5%, Calvinist 20%, Lutheran 5%, atheist and other 7.5%
Government type: parliamentary democracy
Capital: Budapest
Budget: revenues $13 billion; expenditures: $14.4 billion, including capital of $NA
Currency: forint (HUF)

ICELAND

Location: Northern Europe, island between the Greenland Sea and the North Atlantic Ocean
Area: 39,758 sq mi (103,000 sq km)
Area-comparative: slightly smaller than Kentucky
Population: 293,966
Population growth rate: 0.97%
Nationality: noun, Icelanders(s); adjective, Icelandic
Ethnic groups: a mixture of descendents of Norse and Celts 94%, population of foreign origin 6%
Languages: Icelandic, English, Nordic languages, German widely spoken
Religions: Evangelical Lutheran 87%, other Protestant and Roman Catholic 6%, other 7%
Government type: constitutional republic
Capital: Reykjavik
Budget: revenues: $3.5 billion , expenditures: $3.3 billion, including capital expenditures of $467 million
Currency: Icelandic krona (ISK)

INDIA

Location: Southern Asia, bordering the Arabian Sea and the Bay of Bengal, between Burma and Pakistan
Area: 1,269,009 sq mi (3,287,590 sq km)
Area-comparative: slightly more than one-third the size of the United States
Population: 1,065,070,607
Population growth-rate: 1.44%
Nationality: noun, Indian(s); adjective: Indian
Ethnic groups: Indo-Aryan 72%, Dravidian 25%, Mongoloid and other 3%
Language: Hindi (official), English and 14 other official regional languages
Religions: Hindu 81.3%, Muslim 12%, Christian 2.3%, Sikh 1.9%, Buddhist, Jain, Parsi 2.5%
Government type: federal republic

Capital: New Delhi
Budget: revenues: $48.3 billion; expenditures: $78.2 billion, including capital expenditures of $13.5 billion
Currency: Indian rupee (INR)

INDONESIA
Location: Southeastern Asia, group of islands between the Indian Ocean and the Pacific Ocean
Area: 740,904 sq mi (1,919,440 sq km)
Area-comparative: slightly less than three times the size of Texas
Population: 238,452,952
Population growth rate: 1.49%
Nationality: noun: Indonesian(s); adjective: Indonesian
Ethnic groups: Javanese 45%, Sundanese 14%, Madurese 7.5%, other 26%
Languages: Bahasa Indonesia (official modified form of Malay), English, Dutch, Javanese, local dialects
Religions: Muslim 88%, Protestant 5%, Roman Catholic 3%, Hindu 2%, Buddhist 1%, other 1%
Government: republic
Capital: Jakarta
Budget: revenues: $39 billion; expenditures: $43 billion, including capital of $NA
Currency: Indonesian rupiah (IDR)

IRAN
Location: Middle East, bordering the Gulf of Oman, the Persian Gulf, and the Caspian Sea,
                between Iraq and Pakistan
Area: 636,128 sq mi (1,648,000 sq km)
Area-comparative: slightly larger than Alaska
Population: 69,018,924
Population growth rate: 1.07%
Nationality: noun: Iranian(s), adjective: Iranian
Ethnic groups: Persian 51%, Azeri 24%, Gilaki and Mazandarani 8%, Kurd 7%, Arab 3%, Lur 2%, Baloch
                2%, Turkmen 2%, other 1%
Languages: Persian and Persian dialects 58%, Turkic and Turkic dialects 26%, Kurdish 9%, Luri 2%,
                Balochi 1%, Arabic 1%, Turkish 1%, other 2%
Religions: Shi'a Muslim 89%, Sunni Muslim 9%, Zoroastrian, Jewish, Christian, and Baha'i 2%
Government type: theocratic republic
Capital: Tehran
Budget: revenues: $41.2 billion; expenditures: $43.4 billion, including capital of $7.6
Currency: Iranian rial (IRR)

IRAQ
Location: Middle East, bordering the Persian Gulf, between Iran and Kuwait
Area: 168,710 sq mi (437,072 sq km)
Area-comparative: slightly more than twice the size of Idaho
Population: 25,374,691
Population growth rate: 2.74%
Nationality: noun, Iraqi(s); adjective: Iraqi
Ethnic groups: Arab 75%-80%, Kurdish 15%-20%,Turkoman, Assyrian or other 5%
Languages: Arabic, Kurdish (official in Kurdish regions), Assyrian, Armenian
Religions: Muslim 97% (Shi'a 60% - 65%, Sunni 32% - 37%), Christian or other 3%
Government type: in transition following April 2003 defeat of Saddam Husayn regime by US-led coalition
Capital: Baghdad
Budget: revenues: $12.8 billion $NA; expenditures: $13.4 billion $NA, including capital of $NA
Currency: Iraqi dinar (NID)

IRELAND

Location: Western Europe, island in the North Atlantic Ocean, west of Great Britain
Area: 27,128 sq mi (70,280 sq km)
Area-comparative: slightly larger than West Virginia
Population: 3,969,558
Population growth rate: 1.16%
Nationality: noun: Irishman(men), Irishwoman(women), Irish (collective plural); adjective: Irish
Ethnic groups: Celtic, English
Languages: English (official), Irish (official)
Religions: Roman Catholic 92%, Church of Ireland 3%, other 5%
Government type: republic
Capital: Dublin
Budget: revenues: $30.7 billion; expenditures: $30.5 billion, including capital expenditures of $5.5 billion
Currency: euro(EUR)

ISRAEL

Location: Middle East, bordering the Mediterranean Sea, between Egypt and Lebanon
Area: 8,017 sq mi (20,770 sq km)
Area-comparative: slightly smaller than New Jersey
Population: 6,199,008
Population growth rate: 1.29%
Nationality: noun: Israeli(s); adjective: Israeli
Ethnic groups: Jewish (including European/American-born, Israeli-born, and African-born) 80%, non-Jewish 20% (primarily Arab)
Languages: Hebrew (official), Arabic, English
Religions: Jewish 80%, Muslim 15%, other (including Christian) 5%
Government type: parliamentary democracy
Capital: Jerusalem
Budget: revenues: $38.5 billion: expenditures: $45.1 billion
Currency: new Israeli shekel (ILS)

ITALY

Location: Southern Europe, peninsula extending into the central Mediterranean Sea, northeast of Tunisia
Area: 116,275 sq mi (301,230 sq km)
Area-comparative: slightly larger than Arizona
Population: 58,057,477
Population growth rate: 0.09%
Nationality: noun: Italian(s); adjective: Italian
Ethnic groups: Italian, small clusters of German, French, Slovene, and Albanian
Languages: Italian (official), German, French, Slovene
Religions: Roman Catholic 98%, other 2%
Government type: republic
Capital: Rome
Budget: revenues: $504 billion; expenditures: $517 billion   **Currency:** euro (EUR)

JAMAICA

Location: Caribbean, island in the Caribbean Sea, south of Cuba
Area: 4,242 sq mi (10,991 sq km)
Area-comparative: slightly smaller than Connecticut
Population: 2,713,130
Population growth rate: 0.66%
Nationality: noun: Jamaican(s); adjective: Jamaican

Ethnic groups: black 90%, Multiracial (black and white) 7%, other 2%, East Indian 1%
Language: English (official), Creole
Religion: Protestant 61%, other (including indigenous beliefs) 35%, Roman Catholic 4%
Government type: constitutional parliamentary democracy
Capital: Kingston
Budget: revenues: $ 2.23 billion: expenditures: $2.56 billion, including capital expenditures of $232 million
Currency: Jamaican dollar (JMD)

JAPAN
Location: Eastern Asia, island chain between the North Pacific Ocean and the Sea of Japan, east of the Korean Peninsula
Area: 145,844 sq mi (377,835 sq km)
Area-comparative: slightly smaller than California
Population: 127,333,002
Population growth rate: 0.08%
Nationality: noun: Japanese (singular and plural); adjective: Japanese
Ethnic groups: Japanese 99%, other 1%
Language: Japanese (official)
Religion: observe both Shinto and Buddhist 84%, other 16% (including Christian 0.7%)
Government type: constitutional monarchy with a parliamentary government
Capital: Tokyo
Budget: revenues: $380 billion; expenditures: $746 billion , including capital of about $71 billion
Currency: yen (JPY)

JORDAN
Location: Southwestern Asia, northwest of Saudi Arabia
Area: 35,628 sq mi (92,300 sq km)
Area-comparative: slightly smaller than Indiana
Population: 5,611,202
Population growth rate: 2.67%
Nationality: noun: Jordanian(s); adjective: Jordanian
Ethnic groups: Arab 98%, Circassian 1%, Armenian 1%
Languages: Arabic (official), English
Religions: Sunni Muslim 92%, Christian 6%, other 2% (several small Shi'a Muslim and Druze populations)
Government type: constitutional monarchy
Capital: `Amman
Budget: revenues: $3.3 billion; expenditures: $3.7 billion, including capital expenditures of $582 million
Currency: Jordanian dinar (JOD)

KAZAKHSTAN
Location: Central Asia, northwest of China
Area: 1,048,877 sq mi (2,717,300 sq km)
Area-comparative: slightly less than four times the size of Texas
Population: 15,143,704
Population growth rate: 0.26%
Nationality: noun: Kazakhstani(s); adjective: Kazakhstani
Ethnic groups: Kazakh (Qazaq) 53%, Russian 30%, Ukrainian 4%, German 3%, Uzbek 2%, other 8%
Language: Russian (official), Kazakh (official)
Religion: Muslim 47%, Russian Orthodox 44%, other (including Protestant) 9%
Government type: republic
Capital: Astana
Budget: revenues: $4.2 billion; expenditures: $5.1 billion
Currency: tenge (KZT)

480

KENYA
Location: Eastern Africa, bordering the Indian Ocean, between Somalia and Tanzania
Area: 224,903 sq mi (582,650 sq km)
Area-comparative: slightly more than twice the size of Nevada
Population: 32,021,856
Population growth rate: 1.14%
Nationality: Kenyan(s); adjective: Kenyan
Ethnic groups: Kikuyu 22%, Luhya/ Luo 27%, Kalenjin 12%, Kamba 11%, Kisii 6%, Meru 6%, other 16%
Language: English (official), Kiswahili (official), numerous indigenous languages
Religion: Protestant 45%, Roman Catholic 33%, indigenous beliefs 10%, Muslim 10%, other 2%
Government type: republic
Capital: Nairobi
Budget: revenues: $2.91 billion; expenditures: $2.97 billion   **Currency:** Kenyan shilling (KES)

KIRIBATI
Location: a group of islands in the Pacific Ocean, about one-half of the way from Hawaii to Australia
Area: 313 sq mi (811 sq km)
Area-comparative: four times the size of Washington, DC
Population: 100,798
Population growth rate: 2.25%
Nationality: noun: I-Kiribati (singular and plural); adjective: I-Kiribati
Ethnic groups: predominantly Micronesian with some Polynesian
Language: I-Kiribati, English (official)
Religion: Roman Catholic 52%, Protestant 40%, Seventh-Day Adventist, Baha'i, and Church of God 6%
Government type: republic
Capital: Tarawa
Budget: revenues: $28.4 million; expenditures: $ 37.2 million   **Currency:** Australian dollar (AUD)

KOREA, NORTH
Location: Eastern Asia, northern half of the Korean Peninsula bordering the Korea Bay and the Sea of
          Japan, between China and South Korea
Area: 46,541 sq mi (120,540 sq km)
Area-comparative: slightly smaller than Mississippi
Population: 22,697,553
Population growth rate: 0.98%
Nationality: noun: Korean(s); adjective: Korean
Ethnic groups: Korean
Language: Korean (official)
Religion: traditional Buddhist and Confucianist
Government type: Communist state one-man dictatorship
Capital: Pyongyang
Budget: revenues: $NA; expenditures: $NA   **Currency:** North Korean won (KPW)

KOREA, SOUTH
Location: Eastern Asia, southern half of the Korean Peninsula bordering Yellow Sea and Sea of Japan
Area: 38,013 sq mi (98,480 sq km)
Area-comparative: slightly larger than Indiana
Population: 48,598,175
Population growth rate: 0.62%
Nationality: noun: Korean; adjective: Korean
Ethnic groups: homogeneous (except for about 20,000 Chinese)

Languages: Korean, English widely taught in junior high and high school
Religions: no affiliation 46%, Christian 26%, Buddhist 26%, Confucianist 1%, other 1%
Government type: republic
Capital: Seoul
Budget: revenues: $127 billion; expenditures: $108 billion,
including capital expenditures of $23.5 billion
Currency: South Korean won (KRW)

KUWAIT

Location: Southwestern Asia, the Middle East, bordering the Persian Gulf, between Iraq and Saudi Arabia
Area: 6,878 sq mi (17,820 sq km)
Area-comparative: slightly smaller than New Jersey
Population: 2,257,549
Population growth rate: 3.36%
Nationality: noun: Kuwaiti(s); adjective: Kuwaiti
Ethnic groups: Kuwaiti 45%, other Arab 35%, South Asian 9%, Iranian 4%, other 7%
Languages: Arabic (official), English widely spoken
Religions: Muslim 85% (Shi'a 30%, Sunni 70%), Christian, Hindu, Parsi, and other 15%
Government type: nominal constitutional monarchy
Capital: Kuwait
Budget: revenues: $11 billion; expenditures: $19.5 billion
Currency: Kuwaiti dinar

KYRGYZSTAN

Location: Central Asia, west of China
Area: 76,621 sq mi (198,500 sq km)
Area-comparative: slightly smaller than South Dakota
Population: 5,081,429
Population growth rate: 1.25%
Nationality: noun: Kyrgyzstani(s); adjective: Kyrgyzstani
Ethnic groups: Kyrgyz 52.4%, Russian 18%, Uzbek 12.9%, other 11.8%, Ukrainian 2.5%, German 2.4%
Languages: Kyrgyz (official), Russian (official)
Religions: Muslim 75%, Russian Orthodox 20%, other 5%
Government type: republic
Capital: Bishkek
Budget: revenues: $207.4 million; expenditures: $238.7 million
Currency: Kyrgyzstani som (KGS)

LAOS

Location: Southeastern Asia, northeast of Thailand, west of Vietnam
Area: 91,405 sq mi (236,800 sq km)
Area-comparative: slightly larger than Utah
Population: 6,068,117
Population growth rate: 2.44%
Nationality: noun: Lao(s) or Laotian(s); adjective: Lao or Laotian
Ethnic groups: Lao Loum 68%, Lao Theung 22%, Lao Soung 9%, ethnic Vietnamese/Chinese 1%
Languages: Lao (official), French, English
Religions: Buddhist 60%, other (including indigenous beliefs) 40%
Government type: Communist state
Capital: Vientiane
Budget: revenues: $211 million; expenditures: $462 million
Currency: kip (LAK)

LATVIA

Location: Eastern Europe, bordering the Baltic Sea, between Estonia and Lithuania
Area: 24,931 sq mi (64,589 sq km)
Area-comparative: slightly larger than West Virginia
Population: 2,306,306
Population growth rate: -0.71%
Nationality: noun: Latvian(s); adjective: Latvian
Ethnic groups: Latvian 57.7%, Russian 29.6%, Belarusian 4.1%, Ukrainian 2.7%, Polish 2.5%, Lithuanian 1.4%, other 2%
Languages: Latvian (official), Lithuanian, Russian
Religions: Lutheran, Roman Catholic, Russian Orthodox
Government type: parliamentary democracy
Capital: Riga
Budget: revenues: $2.4 billion; expenditures: $2.6 billion   **Currency:** Latvian lat (LVL)

LEBANON

Location: Middle East, bordering the Mediterranean Sea, between Israel and Syria
Area: 4,015 sq mi (10,400 sq km)
Area-comparative: about 0.7 times the size of Connecticut
Population: 3,777,218
Population growth rate: 1.3%
Nationality: noun: Lebanese (singular and plural); adjective: Lebanese
Ethnic groups: Arab 95%, Armenian 4%, other 1%
Languages: Arabic (official), French, English, Armenian
Religions: Muslim 70%, Christian 30%
Government type: republic
Capital: Beirut
Budget: revenues: $4.5 billion; expenditures: $7.1 billion, including capital of $NA
Currency: Lebanese pound (LBP)

LESOTHO

Location: Southern Africa, an enclave of South Africa
Area: 11,717 sq mi (30,355 sq km)
Area-comparative: slightly smaller than Maryland
Population: 1,865,040
Population growth rate: 0.14%
Nationality: noun: Mosotho (singular), Basotho (plural); adjective: Basotho
Ethnic groups: Sotho 99.7%, other 0.3%
Languages: Sesotho (official), English (official), Zulu, Xhosa
Religions: Christian 80%, indigenous beliefs 20%
Government type: parliamentary constitutional monarchy
Capital: Maseru
Budget: revenues: $76 million; expenditures: $80 million including capital expenditures of $15 million
Currency: loti; South African rand ((LSL;ZAR)

LIBERIA

Location: Western Africa, bordering the North Atlantic Ocean, between Cote d'Ivorie and Sierra Leone
Area: 42,989 sq mi (111,370 sq km)
Area-comparative: slightly larger than Tennessee
Population: 3,390,635
Population growth rate: 2.7%
Nationality: noun: Liberian(s); adjective: Liberian

Ethnic groups: indigenous African tribes 95%, Americo-Liberians 2.5%, Congo people 2.5%
Languages: English 20% (official),some 20 tribal languages
Religions: indigenous beliefs 40%, Christian 40%, Muslim 20%
Government type: republic
Capital: Monrovia
Budget: revenues: $85.4 million; expenditures: $90.5 million
Currency: Liberian dollar (LRD)

LIBYA
Location: Northern Africa, bordering the Mediterranean Sea, between Egypt and Tunisia
Area: 679,182 sq mi (1,759,540 sq km)
Area-comparative: slightly larger than Alaska
Population: 5,631,585
Population growth rate: 2.37%
Nationality: noun: Libyan(s); adjective: Libyan
Ethnic groups: Arab-Berber 97%, other 3%
Languages: Arabic, Italian, English
Religions: Sunni Muslim 97%, other 3%
Government type: Jamahiriya (a state of the masses)
Capital: Tripoli
Budget: revenues: $10.2 billion; expenditures: $7.8 billion
Currency: Libyan dinar (LYD)

LIECHTENSTEIN
Location: Central Europe, between Austria and Switzerland
Area: 62 sq mi (160 sq km)
Area-comparative: about 0.9 times the size of Washington D.C.
Population: 33,436
Population growth rate: 0.86%
Nationality: noun: Liechtensteiner(s); adjective: Liechtenstein
Ethnic groups: Alemannic 86%, Italian, Turkish and other 14%
Languages: German (official), Alemannic dialect
Religions: Roman Catholic 76.2%, Protestant 7%, unknown 10.6%, other 6.2%
Government type: hereditary constitutional monarchy
Capital: Vaduz
Budget: revenues:$424.2 million; expenditures:$ 414.1 million
Currency: Swiss franc(CHF)

LITHUANIA
Location: Eastern Europe, bordering the Baltic Sea, between Latvia and Russia
Area: 25,167 sq mi (65,200 sq km)
Area-comparative: slightly larger than West Virginia
Population: 3,607,899
Population growth rate: -0.33%
Nationality: noun :Lithuanian(s); adjective: Lithuanian
Ethnic groups: Lithuanian 80.6%, Russian 8.7%, Polish 7%, other 2.1%, Belarusian 1.6%
Languages: Lithuanian (official), Polish, Russian
Religions: Roman Catholic, Russian Orthodox, Lutheran, Protestant, Christian, Baptist, Muslim, Jewish
Government: parliamentary democracy
Capital: Vilnius
Budget: revenues: $1.59 billion; expenditures: $1.77 billion
Currency: litas (LTL)

LUXEMBOURG

Location: Western Europe, between France and Germany
Area: 998 sq mi (2,586 sq km)
Area-comparative: slightly smaller than Rhode Island
Population: 462,690
Population growth rate: 1.28%
Nationality: noun: Luxembourger(s); adjective: Luxembourg
Ethnic groups: Celtic base (with French and German blend), Portuguese, Italian, and European
Languages: Luxembourgish, German, French
Religions: Roman Catholic 87%, Protestant 13%, Jews and Muslims
Government type: constitutional monarchy
Capital: Luxemborg
Budget: revenues: $5.5 billion; expenditures: $5.5 billion,
    including capital expenditures of $760 million   **Currency:** euro (EUR)

MACEDONIA

Location: Southeastern Europe, north of Greece
Area: 9,778 sq mi (25,333 sq km)
Area-comparative: slightly larger than Vermont
Population: 2,071,210
Population growth rate: 0.39%
Nationality: noun: Macedonian(s); adjective: Macedonian
Ethnic groups: Macedonian 64.2%, Albanian 25.2%, Turkish 3.8%, Roma 2.7%, Serb 1.8%, other 2.3%
Languages: Macedonian 68%, Albanian 25%, Turkish 3%, Serb- Croatian 2%, other 2%
Religions: Macedonian Orthodox 70%, Muslim 29%, other 1%
Government type: parliamentary democracy
Capital: Skopje
Budget: revenues: $1.03 billion; expenditures: $1.05 billion, including capital of $80 million $NA
Currency: Macedonian denar (MKD)

MADAGASCAR

Location: Southern Africa, island in the Indian Ocean, east of Mozambique
Area: 226,597 sq mi (587,040 sq km)
Area-comparative: slightly less than twice the size of Arizona
Population: 17,501,871
Population growth rate: 3.03%
Nationality: noun: Malagasy (singular and plural); adjective: Malagasy
Ethnic groups: Malayo-Indonesian, Cotiers, French, Indian, Creole, Comoran
Languages: French (official), Malagasy (official)
Religions: indigenous beliefs 52%, Christian 41%, Muslim 7%
Government type: republic
Capital: Antananarivo
Budget: revenues: $459.8 million; expenditures: $798.9 million, including capital of $331 million
Currency: Malagasy franc (MGF)

MALAWI

Location: Southern Africa, east of Zambia
Area: 45,733 sq mi (118,480 sq km)
Area-comparative: slightly smaller than Pennsylvania
Population: 11,906,855

Population growth rate: 2.14%
Nationality: noun: Malawian(s); adjective: Malawian
Ethnic groups: Chewa, Nyanja, Tumbuko, Yao, Lomwe, Sena, Tonga, Ngoni, Ngonde, Asian, European
Languages: English (official), Chichewa (official)
Religions: Protestant 55%, Roman Catholic 20%, Muslim 20%, other 5%
Government type: multiparty democracy
Capital: Lilongwe
Budget: revenues: $490 million; expenditures: $523 million   **Currency:** Malawian kwacha (MWK)

MALAYSIA
Location: Southeastern Asia, peninsula and northern one-third of the island of Borneo, south of Vietnam
Area: 127,283 sq mi (329,750 sq km)
Area-comparative: slightly larger than New Mexico
Population: 23,522,482
Population growth rate: 1.83%
Nationality: noun: Malaysian(s); adjective: Malaysian
Ethnic groups: Malay and other indigenous 58%, Chinese 24%, other 10%, Indian 8%,
Languages: Bahasa Melayu (official), English, Chinese dialects, Indian languages
Religions: Muslim, Buddhist, Hindu, Daoist, Sikh, and Christian
Government type: constitutional monarchy
Capital: Kuala Lumpur
Budget: revenues: $25.2 billion; expenditures: $28.9 billion, including capital of $9.4 billion
Currency: ringgit (MYR)

MALDIVES
Location: Southern Asia, group of atolls in the Indian Ocean, south-southwest of India
Area: 116 sq mi (300 sq km)
Area-comparative: about 1.7 times the size of Washington D.C.
Population: 339,330
Population growth rate: 2.86%
Nationality: noun: Maldivian(s); adjective: Maldivian
Ethnic groups: South Indians, Sinhalese, Arabs
Languages: Maldivian Divehi (official), English
Religion: Sunni Muslim 100%
Government: republic
Capital: Male
Budget: revenues: $224 million; expenditures: $282 million, including capital expenditures of $80 million
Currency: rufiyaa (MVR)

MALI
Location: Western Africa, southwest of Algeria
Area: 478,640 sq mi (1,240,000 sq km)
Area-comparative: slightly less than twice the size of Texas
Population: 11,956,788
Population growth rate: 2.78%
Nationality: noun: Malian(s); adjective: Malian
Ethnic groups: Mande (including Bambara, Malinke, and Sarakole) 50%, Peul 17%, Voltaic 12%, Tuareg and Moor 10%, Songhai 6%, other 5%
Languages: French (official), Bambara 80%, numerous African languages
Religions: Muslim 90%, indigenous beliefs 9%, Christian 1%
Government type: republic
Capital: Bamako

Budget: revenues: $764 million; expenditures: $828 million
Currency: Communaute Financiere Africaine franc (XOF)

MALTA
Location: Southern Europe, islands in the Mediterranean Sea, south of Italy
Area: 122 sq mi (316 sq km)
Area-comparative: slightly less than twice the size of Washington, DC
Population: 396,851
Population growth rate: 0.42%
Nationality: noun: Maltese (singular and plural); adjective: Maltese
Ethnic groups: Maltese
Languages: Maltese (official), English (official)
Religions: Roman Catholic 98%, other 2%
Government type: republic
Capital: Valletta
Budget: revenues: $1.5 billion; expenditures: $1.6 billion   **Currency:** Maltese lira (MTL)

MARSHALL ISLANDS
Location: Oceania, group of atolls and reefs in the North Pacific Ocean, about one-half of the way from
          Hawaii to Australia
Area: 70 sq mi (181.3 sq km)
Area-comparative: about the size of Washington D.C.
Population: 57,738
Population growth rate: 2.29%
Nationality: noun: Marshallese (singular and plural); adjective: Marshallese
Ethnic groups: Micronesian
Languages: English and Marshallese are both official languages, 2 major Marshallese dialects, Japanese
Religions: Christian (mostly Protestant)
Government type: constitutional government in free association with the United States
Capital: Majuro
Budget: revenues: $42 million; expenditures: $40 million   **Currency:** United States dollar (USD)

MAURITANIA
Location: Northern Africa, bordering the North Atlantic Ocean, between Senegal and Western Sahara
Area: 397,850 sq mi (1,030,700 sq km)
Area-comparative: slightly larger than three times the size of New Mexico
Population: 2,998,563
Population growth rate: 2.91%
Nationality: noun, Mauritanian(s); adjective: Mauritanian
Ethnic groups: multiracial Maur and Black 40%, Maur 30%, Black 30%
Languages: Hassaniya Arabic (official), Pulaar, Soninke, Wolof (official), French
Religion: Muslim 100%
Government type: republic
Capital: Nouakchott
Budget: revenues: $421 million; expenditures: $378 million including capital expenditures of $154 million
Currency: ouguiya (MRO)

MAURITIUS
Location: Southern Africa, island in the Indian Ocean, east of Madagascar
Area: 787 sq mi (2,040 sq km)
Area-comparative: almost 11 times the size of Washington D.C.

Population: 1,220,481
Population growth rate: 0.81%
Nationality: noun: Mauritian(s); adjective: Mauritian
Ethnic groups: Indo-Mauritian 68%, Creole 27%, Sino-Mauritian 3%, Franco-Mauritian 2%
Languages: English (official), Creole, French (official), Hindi, Urdu, Hakka, Bhojpuri
Religions: Hindu 52%, Roman Catholic 26%, Muslim 16.6%, other 3.1%, Protestant 2.3%
Government type: parliamentary democracy
Capital: Port Louis
Budget: revenues:$1.1 billion; expenditures: $1.2 billion   **Currency:** Mauritian rupee (MUR)

MEXICO

Location: located in the southern part of North America, bordering the Caribbean Sea and the Gulf of
          Mexico, between Guatemala and the United States
Area: 761,404 sq mi (1,972,550 sq km)
Area-comparative: slightly less than three times the size of Texas
Population: 104,959,594
Population growth rate: 1.18%
Nationality: noun: Mexican(s); adjective: Mexican
Ethnic groups: Mestizo 60%, Amerindian or predominantly Amerindian 30%, Caucasian 9%, other 1%
Languages: Spanish (official), Mayan and Nahuatl dialects
Religions: Roman Catholic 89%, Protestant 6%, other 5%
Government type: federal republic
Capital: Mexico
Budget: revenues: $174 billion; expenditures: $176 billion, including capital of $NA
Currency: Mexican peso (MXN)

MICRONESIA

Location: Federated state of Oceania, island group in the North Pacific Ocean, about three-quarters of
          the way from Hawaii to Indonesia
Area: 271 sq mi (702 sq km)
Area-comparative: four times the size of Washington D.C.
Population: 108,155
Population growth rate: -0.02%
Nationality: noun:Micronesian(s); adjective: Micronesian; Chuukese, Kosraen(s), Pohnpein(s), Yapese
Ethnic groups: nine ethnic Micronesian and Polynesian groups
Languages: English (official), Trukese, Pohnpeian, Yapese, Kosrean, Ulithian, Woleaian, Nukuoro,
          Kapingamarangi
Religions: Roman Catholic 50%, Protestant 47%, other 3%
Government type: constitutional government in the association with the United States
Capital: Palikir
Budget: revenues: $161 million ($69 million less grants); expenditures: $160 million
Currency: United States dollar (USD)

MOLDOVA

Location: Eastern Europe, northeast of Romania
Area: 13,060 sq mi (33,843 sq km)
Area-comparative: slightly larger than Maryland
Population: 4,446,455
Population growth rate: 0.18%
Nationality: noun: Moldovan(s); adjective: Moldovan
Ethnic groups: Moldavian and Romanian 64.5%, Ukrainian 13.8%, Russian 13%, Jewish 1.5%, Gagauz
          Bulgarian 2% and other 5.2%

Languages: Moldovan (official), Russian, Gagauz
Religions: Eastern Orthodox 98%, Jewish 1.5%, Baptist and other 0.5%
Government type: republic
Capital: Chisinau
Budget: revenues: $536 million; expenditures: $594 million
Currency: Moldovan leu (MDL)

MONACO
Location: Western Europe, bordering the Mediterranean Sea, on the southern coast of France, near the border with Italy
Area: .75 sq mi (1.95 sq km)
Area-comparative: about 3 times the size of The Mall in Washington, DC
Population: 32,270
Population growth rate: 0.44%
Nationality: noun: Monegasque(s); adjective: Monegasque or Monacan
Ethnic groups: French 47%, Monegasque 16%, Italian 16%, other 21%
Languages: French (official), English, Italian, Monegasque
Religions: Roman Catholic 90%
Government type: constitutional monarchy
Capital: Monaco
Budget: revenues: $518 million; expenditures: $531 million   **Currency:** euro (EUR)

MONGOLIA
Location: Northern Asia, between China and Russia
Area: 604,090 sq mi (1,565,000 sq km)
Area-comparative: slightly smaller than Alaska
Population: 2,751,314
Population growth rate: 1.43%
Nationality: noun: Mongolian(s); adjective: Mongolian
Ethnic groups: Mongol 85%, Turkic 7%, Tungusic 4.6%, other 3.4%
Languages: Khalkha Mongol 90%, Turkic, Russian
Religions: Tibetan Buddhist Lamaisian 96%, Muslim, Shamanism, and Christian 4%
Government type: parliamentary
Capital: Ulaanbaatar
Budget: revenues: $387 million; expenditures: $428 million, including capital of $NA
Currency: togrog / tughrik (MNT)

MOROCCO
Location: Northern Africa, bordering the North Atlantic Ocean and the Mediterranean Sea, between Algeria and Western Sahara
Area: 172,368 sq mi (446,550 sq km)
Area-comparative: slightly larger than California
Population: 32,209,101
Population growth rate: 1.61%
Nationality: noun: Moroccan(s); adjective: Moroccan
Ethnic groups: Arab-Berber 99%, other 1%
Languages: Arabic (official), Berber dialects, French
Religions: Muslim 98.7%, Christian 1.1%, Jewish 0.2%
Government: constitutional monarchy
Capital: Rabat
Budget: revenues: $13.8 billion; expenditures: $14 billion, including capital expenditures of $2.1 billion
Currency: Moroccan dirham (MAD)

MOZAMBIQUE
Location: South-eastern Africa, bordering the Mozambique Channel, between South Africa and Tanzania
Area: 309,414 sq mi (801,590 sq km)
Area-comparative: slightly less than twice the size of California
Population: 18,811,731
Population growth rate: 1.22%
Nationality: noun: Mozambican(s); adjective: Mozambican
Ethnic groups: indigenous tribal groups 99.66%, Europeans 0.06%, Euro-Africans 0.2%, Indians 0.08%
Languages: Portuguese (official), indigenous dialects
Religions: indigenous beliefs 50%, Christian 30%, Muslim 20%
Government type: republic
Capital: Maputo
Budget: revenues: $393.1 million; expenditures: $1.025 billion,
including capital expenditures of $479.4 million   **Currency:** metical(MZM)

NAMIBIA
Location: Southern Africa, bordering the South Atlantic Ocean, between Angola and South Africa
Area: 318,611 sq mi (825,418 sq km)
Area-comparative: slightly more than half the size of Alaska
Population: 1,954,033
Population growth rate: 1.25%
Nationality: noun: Namibian(s); adjective: Namibian
Ethnic groups: black 87.5%, white 6%, mixed 6.5%,
Languages: German, English 7% (official), Afrikaans 60%, German 32%, indigenous languages:
Oshivambo, Herero, Nama
Religions: Christian (at least 50% Lutheran) 80%, indigenous beliefs 20%
Government type: republic
Capital: Windhoek
Budget: revenues: $883 million; expenditures: $950 million
Currency: Namibian dollar; South African rand (NAD;ZAR)

NAURU
Location: Oceania, island in the South Pacific Ocean, south of the Marshall Islands
Area: 8 sq mi (21 sq km)
Area-comparative: about 0.1 times the size of Washington D.C.
Population: 12,809
Population growth rate: 1.87%
Nationality: noun: Nauruan(s); adjective: Nauruan
Ethnic groups: Nauruan 58%, other Pacific Islander 26%, Chinese 8%, European 8%
Languages: Nauruan (official), English
Religions: Protestant 67%, Roman Catholic 33%
Government type: republic
Capital: government offices in Yaren District
Budget: revenues: $23.4 million; expenditures: $64.8 million
Currency: Australian dollar (AUD)

NEPAL
Location: Southern Asia, between China and India
Area: 54,349 sq mi (140,800 sq km)
Area-comparative: slightly larger than Arkansas
Population: 27,070,666
Population growth rate: 2.23%

490

Nationality: noun: Nepalese (singular and plural); adjective: Nepalese
Ethnic groups: Brahman, Chetri, Newar, Gurung, Magar, Tamang, Rai, Limbu, Sherpa, Tharu, and others
Languages: Nepali (official, spoken by 90% of the population), 12 other languages and 30 dialects
Religions: Hindu 86.2%, Buddhism 7.8%, Islam 3.8%, other 2.2%
Government type: parliamentary democracy and constitutional monarchy
Capital: Kathmandu
Budget: revenues: $665 million; expenditures: $1.1 billion   **Currency:** Nepalese rupee (NPR)

NETHERLANDS

Location: Western Europe, bordering the North Sea, between Belgium and Germany
Area: 16,029 sq mi (41,526 sq km)
Area-comparative: slightly less than twice the size of New Jersey
Population: 16,318,199
Population growth rate: 0.57%
Nationality: noun: Dutchman (men), Dutchwoman (women); adjective: Dutch
Ethnic groups: Dutch 83%, other 17%
Languages: Dutch (official), Frisian (official)
Religions: Roman Catholic 31%, Protestant 21%, Muslim 4.4%, other 3.6%, unaffilitated 40%
Government type: constitutional monarchy
Capital: Amsterdam, The Hague is the seat of government
Budget: revenues: $134 billion; expenditures: $134 billion   **Currency:** euro (EUR)

NEW ZEALAND

Location: Oceania, islands in the South Pacific Ocean, southeast of Australia
Area: 103,710 sq mi (268,680 sq km)
Area-comparative: about the size of Colorado
Population: 3,993,817
Population growth rate: 1.05%
Nationality: noun: New Zealander(s); adjective: New Zealand
Ethnic groups: New Zealand European 74.5%, Maori 9.7%, Asian 7.4%, European 4.6%, other 3.8%
Languages: English (official), Maori (official)
Religions: Anglican 24%, Presbyterian 18%, Roman Catholic 15%, Methodist 5%, Baptist 2%, Protestant 3%, unspecified or none 33%
Government type: parliamentary democracy
Capital: Wellington
Budget: revenues: $29.2 billion; expenditures: $31.2 billion   **Currency:** New Zealand dollar (NZD)

NICARAGUA

Location: Central America, bordering both the Caribbean Sea and the North Pacific Ocean, between Costa Rica and Honduras
Area: 49,985 sq mi (129,494 sq km)
Area-comparative: slightly smaller than the state of New York
Population: 5,359,759
Population growth rate: 1.97%
Nationality: noun: Nicaraguan(s); adjective: Nicaraguan
Ethnic groups: mestizo 69%, white 17%, black 9%, Amerindian 5%
Language: Spanish (official)
Religion: Roman Catholic 85%, Protestant 15%
Government type: republic
Capital: Managua
Budget: revenues: $726 million; expenditures: $908 million
Currency: gold cordoba (NIO)

491

NIGER

Location: Western Africa, southeast of Algeria
Area: 489,062 sq mi (1,267,000 sq km)
Area-comparative: slightly less than twice the size of Texas
Population: 11,360,538
Population growth rate: 2.67%
Nationality: noun: Nigerien(s); adjective: Nigerien
Ethnic groups: Hausa 56%, Djerma 22%, Fula 8.5%, Tuareg 8%, Beri Beri 4.3%, Arab, Toubou, and Gourmantche 1.2%
Languages: French (official), Hausa, Djerma
Religions: Muslim 80%, remainder indigenous beliefs and Christian
Government type: republic
Capital: Niamey
Budget: revenues: $320 million; expenditures: $320 million, including capital expenditures of $178 million
Currency: Communaute Financiere Africaine franc (XOF)

NIGERIA

Location: Western Africa, bordering the Gulf of Guinea, between Benin and Cameroon
Area: 356,574 sq mi (923,768 sq km)
Area-comparative: slightly more than twice the size of California
Population: 137,253,133
Population growth rate: 2.45%
Nationality: noun: Nigerian; adjective: Nigerian
Ethnic groups: Hausa, Fulani 29%, Yoruba 21%, Igbo 18%, Ijaw 10%, Kanuri 4%, Ibibio 3.5%, Tiv 2.5%
Languages: English (official), Hausa, Yoruba, Igbo, Fulani
Religions: Muslim 50%, Christian 40%, indigenous beliefs 10%
Government type: republic transitioning from military to civilian rule
Capital: Abuja
Budget: revenues: $3.4 billion; expenditures: $3.6 billion   **Currency:** naira (NGN)

NORWAY

Location: Northern Europe, bordering the North Sea and the North Atlantic Ocean, west of Sweden
Area: 125,149 sq mi (324,220 sq km)
Area-comparative: slightly larger than New Mexico
Population: 4,574,560
Population growth rate: 0.41%
Nationality: noun: Norwegian(s); adjective: Norwegian
Ethnic groups: Norwegian, Sami 20,000
Language: Norwegian (official)
Religions: Evangelical Lutheran 86%, other Protestant and Roman Catholic 3%, other/nonreligious 11%
Government type: constitutional monarchy
Capital: Oslo
Budget: revenues: $71.7 billion; expenditures: $57.6 billion   **Currency:** Norwegian krone (NOK)

OMAN

Location: Middle East, bordering the Arabian Sea, Gulf of Oman, and the Persian Gulf, between Yemen and United Arab Emirates
Area: 82,010 sq mi (212,460 sq km)
Area-comparative: slightly smaller than Kansas
Population: 2,903,165
Population growth rate: 3.35%
Nationality: noun: Omani(s); adjective: Omani
Ethnic groups: Arab, Baluchi, Indian, Pakistani, Sri Lankan, Bangladeshi, African

Languages: Arabic (official), English, Baluchi, Urdu, Indian dialects
Religions: Ibadhi Muslim 75%, other including Sunni Muslim, Shi'a Muslim, Hindu 25%
Government type: monarchy
Capital: Muscat
Budget: revenues: $9.2 billion; expenditures: $6.9 billion
Currency: Omani rial (OMR)

PAKISTAN
Location: Southern Asia, bordering the Arabian Sea, between India on the east and Iran and Afghanistan on the west and China in the north
Area: 310, 321 sq mi (803,940 sq km)
Area-comparative: slightly less than twice the size of California
Population: 159,196,336
Population growth rate: 1.98%
Nationality: noun: Pakistani(s); adjective: Pakistani
Ethnic groups: Punjabi, Sindhi, Pashtun, Baloch, Muhajir
Languages: Punjabi 48%, Sindhi 12%, Siraiki 10%, Pashtu 8%, Urdu (official) 8%, Balochi 3%, Hindko 2%, Brahui 1%, English
Religions: Sunni Muslim 77%, Shi'a Muslim 20%, other (including Christian and Hindu) 3%
Government type: federal republic
Capital: Islamabad
Budget: revenues: $12.2 billion; expenditures: $12.3 billion , including capital of $NA
Currency: Pakistani rupee(PKR)

PALAU
Location: Oceania, group of islands in the North Pacific Ocean, southeast of the Philippines
Area: 177 sq mi (458 sq km)
Area-comparative: slightly more than 2.5 times the size of Washington, DC
Population: 20,016
Population growth rate: 1.46%
Nationality: noun: Palauan(s); adjective: Palauan
Ethnic groups: Palauan 70%, Asian 28%, white 2%
Languages: English, Palauan, Sonsorolese, Angaur, Japanese, and Tobi (all official within certain states)
Religions: Roman Catholic 49%, Protestant religions 18%, Modekngei,(religion indigenous to Palau) 33%
Government type: constitutional government in free association with the U.S.
Capital: Koror
Budget: revenues: $57.7 million; expenditures: $80.8 million and capital expenditures of $17.1 million
Currency: United States dollar (USD)

PANAMA
Location: Central America, bordering both the Caribbean Sea and the North Pacific Ocean, between Colombia and Costa Rica
Area: 30,185 sq mi (78,200 sq km)
Area-comparative: slightly smaller than South Carolina
Population: 3,000,463
Population growth rate: 1.31%
Nationality: noun: Panamanian(s); adjective: Panamanian
Ethnic groups: mestizo 70%, Amerindian and Multiracial (West Indian) 14%, white 10%, Amerindian 6%
Languages: Spanish (official), English 14%
Religions: Roman Catholic 85%, Protestant 15%
Government type: constitutional democracy
Capital: Panama
Budget: revenues: $1.9 billion: expenditures: $2 billion, including capital expenditures of $471 million
Currency: balboa; United States dollar (PAB;USD)

PAPUA NEW GUINEA

Location: Oceania, group of islands including the eastern half of the island of New Guinea, east of Indonesia

Area: 178,656 sq mi (462,840 sq km)

Area-comparative: slightly larger than California

Population: 5,420,280

Population growth rate: 2.3%

Nationality: noun: Papua New Guinean(s); adjective: Papua New Guinean

Ethnic groups: Melanesian, Papuan, Negrito, Micronesian, Polynesian

Languages: English spoken by 1-2% , pidgin English widespread, Motu

Religions: indigenous beliefs 34%, Roman Catholic 22%, Lutheran 16%, other Protestant 10%, Anglican 5%, Evangelical Alliance 4%, Seventh Day Adventist 1%

Government type: constitutional monarchy with parliamentary democracy

Capital: Port Moresby

Budget: revenues: $894 million; expenditures: $1.1 billion, including capital expenditures of $344 million

Currency: kina (PGK)

PARAGUAY

Location: Central South America, northeast of Argentina

Area: 157,005 sq mi (406,750 sq km)

Area-comparative: slightly smaller than California

Population: 6,191,368

Population growth rate: 2.51%

Nationality: noun: Paraguayan(s); adjective: Paraguayan

Ethnic groups: mestizo (Mixed Spanish and Amerindian) 95%

Languages: Spanish (official), Guarani

Religions: Roman Catholic 90%, Mennonite and other Protestant

Government type: constitutional republic

Capital: Asuncion

Budget: revenues: $1.3 billion; expenditures: $2 billion, including capital expenditures of $700 million

Currency: guarani (PYG)

PERU

Location: Western South America, bordering the South Pacific Ocean, between Chile and Ecuador

Area: 496,094 sq mi (1,285,220 sq km)

Area-comparative: slightly smaller than Alaska

Population: 27,544,305

Population growth rate: 1.39%

Nationality: noun: Peruvian(s); adjective: Peruvian

Ethnic groups: Amerindian 45%, mestizo 37%, white 15%, black, Japanese, Chinese, and other 3%

Languages: Spanish (official), Quechua (official), Aymara

Religions: Roman Catholic 90%, other 10%

Government type: constitutional republic

Capital: Lima

Budget: revenues: $10.4 billion; expenditures: $10.4 billion

Currency: nuevo sol (PEN)

PHILIPPINES

Location: Southeastern Asia, group of islands between the Philippine Sea and the South China Sea, east of Vietnam

Area: 115,800 sq mi (300,000 sq km)

Area-comparative: slightly larger than Arizona
Population: 86,241,697
Population growth rate: 1.88%
Nationality: noun: Filipino(s); adjective: Philippine
Ethnic groups: Christian Malay 91.5%, Muslim Malay 4%, other 3%, Chinese 1.5%
Languages: Filipino(based on Tagalog) and English are official languages; eight major dialects
Religions: Roman Catholic 83%, Protestant 9%, Muslim 5%, Buddhist and other 3%
Government type: republic
Capital: Manila
Budget: revenues: $10.9 billion: expenditures: $15 billion
Currency: Philippine peso (PHP)

POLAND
Location: Central Europe, east of Germany
Area: 120,696 sq mi (312,685 sq km)
Area-comparative: slightly smaller than New Mexico
Population: 38,626,349
Population growth rate: 0.02%
Nationality: noun: Pole(s); adjective: Polish
Ethnic groups: Polish 96.7%, German 0.4%, Ukrainian 0.1%, Belarusian 0.1% , other 2.7%
Language: Polish (official)
Religions: Roman Catholic 95%, other (including Eastern Orthodox and Protestant) 5%
Government type: republic
Capital: Warsaw
Budget: revenues: $49.6 billion; expenditures: $52.3 billion
Currency: zloty (PLN)

PORTUGAL
Location: Southwestern Europe, bordering the North Atlantic Ocean, west of Spain
Area: 35,663 sq mi (92,391 sq km)
Area-comparative: slightly smaller than Indiana
Population: 10,524,145
Population growth rate: 0.41%
Nationality: noun: Portuguese (singular and plural); adjective: Portuguese
Ethnic groups: homogeneous Mediterranean stock, African
Language: Portuguese (official) , Mirandese (official-but locally used)
Religions: Roman Catholic 94%, Protestant
Government: parliamentary democracy
Capital: Lisbon
Budget: revenues: $45 billion; expenditures: $48 billion
Currency: euro (EUR)

PUERTO RICO (commonwealth associated with the US)
Location: Caribbean, island between the Caribbean Sea and the North Atlantic Ocean, east of the
　　　　　Dominican Republic
Area: 3514 sq mi (9,104 sq km)
Area-comparative: slightly less than three times the size of Rhode Island
Population: 3,897,960
Population growth rate: 0.49%
Nationality: noun: Puerto Rican(s) (United States citizens); adjective: Puerto Rican
Ethnic groups: white 80%, black 8%, mixed and other 10.9%, Amerindian 0.4%, Asian 0.2%
Languages: Spanish, English

Religions: Roman Catholic 85%, Protestant and other 15%
Government type: commonwealth
Capital: San Juan
Budget: revenues: $6.7 billion; expenditures: $9.6 billion
Currency: United States dollar (USD)

QATAR
Location: Middle East, peninsula bordering the Persian Gulf and Saudi Arabia
Area: 4,415 sq mi (11,437 sq km)
Area-comparative: slightly smaller than Connecticut
Population: 840,290
Population growth rate: 2.74%
Nationality: noun: Qatari(s); adjective: Qatari
Ethnic groups: Arab 40%, Pakistani 18%, Indian 18%, other 14%, Iranian 10%
Languages: Arabic (official), English
Religions: Muslim 95%
Government type: traditional monarchy
Capital: Doha
Budget: revenues: $5 billion; expenditures: $5.5billion, including capital of $2.2 billion
Currency: Qatari riyal (QAR)

ROMANIA
Location: Southeastern Europe, bordering the Black Sea, between Bulgaria and Ukraine
Area: 91,675 sq mi (237,500 sq km)
Area-comparative: slightly smaller than Oregon
Population: 22,355,551
Population growth rate: -0.11%
Nationality: noun: Romanian(s); adjective: Romanian
Ethnic groups: Romanian 89.5%, Hungarian 6.6%, Roma 2.5%, Ukrainian 0.3%, German 0.3%, Russian 0.2%, Turkish 0.2%, other 0.4%
Languages: Romanian (official), Hungarian, German
Religions: Eastern Orthodox 87%, Protestant 6.8%, Catholic 5.6%, other, mostly Muslim 0.4%
Government type: republic
Capital: Bucharest
Budget: revenues:$ 8.2 billion; expenditures: $9.6 billion, including capital of $NA
Currency: leu (ROL)

RUSSIA
Location: Northern Asia, bordering the Arctic Ocean, between Europe and the North Pacific Ocean
Area: 6,591,027sq mi (17,075,200 sq km)
Area-comparative: approximately 1.8 times the size of the United States
Population: 143,782,338
Population growth rate: -0.45%
Nationality: noun: Russian(s); adjective: Russian
Ethnic groups: Russian 81.5%, Tatar 3.8%, Ukrainian 3%, Chuvash 1.2%, Bashkir 0.9%, Belarusian 0.8%, Moldavian 0.7% , other 8.1%
Languages: Russian (official), many others
Religions: Russian Orthodox, Muslim, other
Government type: federation
Capital: Moscow
Budget: revenues: $70 billion; expenditures: $62 billion
Currency: Russian ruble (RUR)

RWANDA
Location: Central Africa, east of Democratic Republic of the Congo
Area: 10,166 sq mi (26,338 sq km)
Area-comparative: slightly smaller than Maryland
Population: 7,954,013
Population growth rate: 1.82%
Nationality: noun: Rwandan(s); adjective: Rwandan
Ethnic groups: Hutu 84%, Tutsi 15%, Twa 1%
Languages: Kinyarwanda (official), French (official), English (official), Kiswhahilli
Religions: Roman Catholic 56.5%, Protestant 26%, Adventist 11.1%, Muslim 4.6%, indigenous beliefs 0.1%, none 1.7%
Government type: republic, presidential, multiparty system
Capital: Kigali
Budget: revenues: $199.3 million; expenditures: $445 million   **Currency:** Rwandan franc (RWF)

SAINT KITTS AND NEVIS
Location: Caribbean, islands in the Caribbean Sea, about one-third of the way from Puerto Rico to Trinidad and Tobago
Area: 101 sq mi (261 sq km)
Area-comparative: 1.5 times the size of Washington, D C
Population: 38,836
Population growth rate: 0.25%
Nationality: noun: Kittitian(s), Nevisian(s); adjective: Kittitian, Nevisian
Ethnic groups: predominately black, some British, Portuguese, and Lebanese
Language: English (official)
Religions: Anglican, other Protestant, Roman Catholic
Government type: constitutional monarchy with Westminister-style parliament
Capital: Basseterre
Budget: revenues: 89.7 million; expenditures: $128.2 million and capital expenditures of $19.5 million
Currency: East Caribbean dollar (XCD)

SAINT LUCIA
Location: island between the Caribbean Sea and North Atlantic Ocean, north of Trinidad and Tobago
Area: 238 sq mi (616 sq km)
Area-comparative: 3.5 times the size of Washington, D C
Population: 164,213
Population growth rate: 1.27%
Nationality: noun: Saint Lucian(s); adjective: Saint Lucian
Ethnic groups: black 90%, mixed 6%, East Indian 3%, white 1%
Languages: English (official), French patois
Religions: Roman Catholic 90%, Anglican 3%, other Protestant 7%
Government type: Westminster-style parliamentary democracy
Capital: Castries
Budget: revenues: $141.2 million; expenditures: $146.7 million, including capital expenditures of $25.1 million   **Currency:** East Caribbean dollar (XCD)

SAINT VINCENT AND THE GRENADINES
Location: Caribbean, islands in the Caribbean Sea, north of Trinidad and Tobago
Area: 150 sq mi (389 sq km)
Area-comparative: twice the size of Washington D.C.
Population: 117,193
Population growth rate: 0.31%

497

Nationality: noun: Saint Vincentian(s) or Vincentians(s); adjective: Saint Vincentian or Vincentian
Ethnic groups: black 66%, mixed 19%, East Indian 6%, Carib Amerindian 2% , other 7%
Languages: English, French patois
Religions: Anglican 47%, Methodist 28%, Roman Catholic 13%, Hindu, Seventh-Day Adventist, other, Protestant 12%
Government type: parliamentary democracy; independent sovereign state within the Commonwealth
Capital: Kingstown
Budget: revenues: $94.6 million; expenditures: $85.8 million   **Currency:** East Caribbean dollar (XCD)

SAMOA
Location: Oceania, group of islands in the South Pacific Ocean, about one-half of the way from Hawaii to New Zealand
Area: 1,136 sq mi (2,944 sq km)
Area-comparative: slightly smaller than Rhode Island
Population: 177,714
Population growth rate: -0.25%
Nationality: noun: Samoan(s); adjective: Samoan
Ethnic groups: Samoan 92.6%, Euronesians 7%, Europeans 0.4%
Languages: Samoan (official), English
Religions: Christian 99.7%, other 0.3%
Government type: constitutional monarchy under native chief
Capital: Apia
Budget: revenues: $105 million; expenditures: $119 million   **Currency:** tala (SAT)

SAN MARINO
Location: Southern Europe, an enclave in central Italy
Area: 24 sq mi (61 sq km)
Area-comparative: about 0.3 times the size of Washington, DC
Population: 28,503
Population growth rate: 1.33%
Nationality: noun: Sammarinese (singular and plural); adjective: Sammarinese
Ethnic groups: Sammarinese, Italian
Language: Italian (official)
Religion: Roman Catholic
Government type: Independent republic
Capital: San Marino
Budget: revenues: $400 million; expenditures: $400 million   **Currency:** euro (EUR)

SAO TOME AND PRINCIPE
Location: Western Africa, islands straddling the Equator in the Gulf of Guinea, west of Gabon
Area: 386 sq mi (1,001 sq km)
Area-comparative: more than 5 times the size of Washington, DC
Population: 181,565
Population growth rate: 3.18%
Nationality: Sao Tomean(s); adjective: Sao Tomean
Ethnic groups: mestico, angolares, forros, servicais, tongas, Europeans
Language: Portuguese (official)
Religions: Christian 80%, including Roman Catholic, Evangelical Protestant, Seventh Day Adventist
Government type: republic
Capital: Sao Tome
Budget: revenues: $58 million; expenditures: $114 million, including capital expenditures of $54 million
Currency: dobra (STD)

SAUDI ARABIA

Location: Middle East, bordering the Persian Gulf and the Red Sea, north of Yemen
Area: 756,785 sq mi (1,960,582 sq km)
Area-comparative: slightly more than one-fifth the size of the US
Population: 25,795,938
Population growth rate: 2.44%
Nationality: noun: Saudi(s); adjective: Saudi or Saudi Arabian
Ethnic groups: Arab 90%, Afro-Asian 10%
Language: Arabic (official)
Religion: Muslim 100%
Government type: monarchy
Capital: Riyadh
Budget: revenues: $78.7 billion; expenditures: $66.7 billion , including capital of $NA
Currency: Saudi riyal (SAR)

SENEGAL

Location: Western Africa, bordering the North Atlantic Ocean, between Guinea-Bissau and Mauritania
Area: 75,729 sq mi (196,190 sq km)
Area-comparative: slightly smaller than South Dakota
Population: 10,852,147
Population rate growth: 2.52%
Nationality: noun: Senegalese (singular and plural); adjective: Senegalese
Ethnic groups: Wolof 43.3%, Pular 23.8%, Serer 14.7%, Jola 3.7%, Mandink 3%,
Soninke 1.1%, European and Lebanese 1% , other 9.4%
Languages: French (official), Wolof, Pulaar, Jola, Mandinka
Religions: Muslim 94%, indigenous beliefs 1%, Christian (primarily Roman Catholic) 5%
Government type: republic under multiparty democratic rule
Capital: Dakar
Budget: revenues: $1.373 billion; expenditures: $1.373 billion,
including capital expenditures of $357million
Currency: Communaute Financiere Africaine franc (XOF)

SEYCHELLES

Location: Eastern Africa, group of islands in the Indian Ocean, northeast of Madagascar
Area: 176 sq mi (455 sq km)
Area-comparative: 2.5 times the size of Washington, DC
Population: 80,832
Population growth rate: 0.45%
Nationality: noun: Seychellois (singular and plural); adjective: Seychellois
Ethnic groups: mixed French, African, Indian, Chinese, and Arab
Languages: English (official), French (official), Creole
Religions: Roman Catholic 86.6%, Anglican 6.8%, other 4.1%, Christian 2.5%
Government type: republic
Capital: Victoria
Budget: revenues: $249 million; expenditures: $262 million   **Currency:** Seychelles rupee (SCR)

SIERRA LEONE

Location: Western Africa, bordering the North Atlantic Ocean, between Guinea and Liberia
Area: 27,692 sq mi (71,740 sq km)
Area-comparative: slightly smaller than South Carolina
Population: 5,883,889
Population growth rate: 2.27%

Nationality: noun: Sierra Leonean(s); adjective: Sierra Leonean
Ethnic groups: 20 native African tribes 90%, Creole 10%
Languages: English (official), Mende, Temne, Krio
Religions: Muslim 60%, indigenous beliefs 30%, Christian 10%
Government type: constitutional democracy
Capital: Freetown
Budget: revenues: $96 million; expenditures: $351 million   **Currency:** leone (SLL)

SINGAPORE
Location: Southeastern Asia, islands between Malaysia and Indonesia
Area: 267 sq mi (692.7 sq km)
Area-comparative: slightly more than 3.5 times the size of Washington, DC
Population: 4,353,893
Population growth rate: 1.71%
Nationality: noun: Singaporean(s); adjective: Singapore
Ethnic groups: Chinese 76.7%, Malay 14%, Indian 7.9%, other 1.4%
Languages: Chinese (official), Malay, Tamil, English
Religions: Buddhist, Muslim, Christian, Hindu, Sikh, Taoist, Confucianist
Government type: parliamentary republic
Capital: Singapore
Budget: revenues: $17.4 billion; expenditures: $17.6 billion, including capital of $5.6 billion
Currency: Singapore dollar (SGD)

SLOVAKIA
Location: Central Europe, south of Poland
Area: 18,854 sq mi (48,845 sq km)
Area-comparative: about twice the size of New Hampshire
Population: 5,423,567
Population growth rate: 0.14%
Nationality: noun: Slovak(s); adjective: Slovak
Ethnic groups: Slovak 85.7%, Hungarian 10.6%, Roma 1.6%, Czech, Moravian, other 2.1%
Languages: Slovak (official), Hungarian
Religions: Roman Catholic 60.3%, atheist 9.7%, Protestant 8.4%, Orthodox 4.1% , other 17.5%
Government type: parliamentary democracy
Capital: Bratislava
Budget: revenues: $5.2 billion; expenditures: $5.6 billion
Currency: Slovak koruna (SKK)

SLOVENIA
Location: Central Europe, eastern Alps bordering the Adriatic Sea, between Austria and Croatia
Area: 7,825 sq mi (20,273 sq km)
Area-comparative: slightly smaller than New Jersey
Population: 2,011,473
Population growth rate: -0.01%
Nationality: noun: Slovene(s); adjective: Slovenian
Ethnic groups: Slovene 92%, Croat 1%, Serb 0.5%, Bosniak 0.3% , Hungarian 0.4%, other 5.8%
Languages: Slovenian 91%, Serbo-Croatian 6%, other 3%
Religions: Roman Catholic 71%, other (including Atheist, Lutheran, and Muslim) 29%
Government type: parliamentary democratic republic
Capital: Ljubljana
Budget: revenues: $9.9 billion; expenditures: $10.5 billion, including capital of $NA
Currency: tolar (SIT)

500

SOLOMON ISLANDS

Location: Oceania, group of islands in the South Pacific Ocean, east of Papua New Guinea
Area: 10,981 sq mi (28,450 sq km)
Area-comparative: slightly smaller than Maryland
Population: 523,617
Population growth rate: 2.76%
Nationality: noun: Solomon Islander(s); adjective: Solomon Islander
Ethnic groups: Melanesian 93%, Polynesian 4%, Micronesian 2%, European 1%
Languages: Melanesian, Polynesian, 120 indigenous languages , English is official but spoken by 1-2% of the population
Religions: Anglican 45%, Roman Catholic 18%, United (Methodist/Presbyterian) 12%, Baptist 9%, Seventh-Day Adventist 7%, other Protestant 5%, indigenous beliefs 4%
Government type: parliamentary democracy tending toward anarchy
Capital: Honiara
Budget: revenues: $38 million; expenditures: $NA   **Currency:** Solomon Islands dollar (SBD)

SOMALIA

Location: Eastern Africa, bordering the Gulf of Aden and the Indian Ocean, east of Ethiopia
Area: 246,135 sq mi (637,657 sq km)
Area-comparative: slightly smaller than Texas
Population: 8,304,601
Population growth rate: 3.41%
Nationality: noun: Somali(s); adjective: Somali
Ethnic groups: Somali 85%, Bantu and other non-Somali 15%
Languages: Somali (official), Arabic, Italian, English
Religion: Sunni Muslim 100%
Government type: no permanent national government; transitional, parliamentary national government
Capital: Mogadishu
Budget: revenues: $NA; expenditures: $NA   **Currency:** Somali shilling(SOS)

SOUTH AFRICA

Location: Southern Africa, at the southern tip of the continent of Africa
Area: 470,886 sq mi (1,219,912 sq km)
Area-comparative: slightly less than twice the size of Texas
Population: 42,718,530
Population growth rate: -0.25%
Nationality: noun: South African(s); adjective: South African
Ethnic groups: black 75%, white 14%, Multiracial (black and white) 9%, Indian 2%
Languages: 11 official languages, including Afrikaans, English, Ndebele, Pedi, Sotho, Swazi, Tsonga, Tswana, Venda, Xhosa, Zulu
Religions: Christian 68%, indigenous beliefs and animist 29%, Muslim 2%, other 1%
Government type: republic
Capital: Pretoria (administrative), Cape Town (legislative center), Bloemfontein (judicial center)
Budget: revenues: $22.6 billion; expenditures: $24.7 billion
Currency: rand (ZAR)

SPAIN

Location: Southwestern Europe, bordering the Bay of Biscay, Mediterranean Sea, North Atlantic Ocean, and Pyrenees Mountains, southwest of France
Area: 194,846 sq mi (504,782 sq km)
Area-comparative: slightly more than twice the size of Oregon
Population: 40,280,780
Population growth rate: 0.16%

Nationality: noun: Spaniard(s); adjective: Spanish
Ethnic groups: composite of Mediterranean and Nordic stock
Languages: Castilian Spanish (official), Catalan, Galician, Basque
Religions: Roman Catholic 94%, other 6%
Government type: parliamentary democracy
Capital: Madrid
Budget: revenues: $105 billion; expenditures: $109 billion, including capital expenditures of $12.8 billion
Currency: euro (EUR)

SRI LANKA
Location: Southern Asia, island in the Indian Ocean, south of India
Area: 25,325 sq mi (65,610 sq km)
Area-comparative: slightly larger than West Virginia
Population: 19,905,165
Population growth rate: 0.81%
Nationality: noun: Sri Lankan(s); adjective: Sri Lankan
Ethnic groups: Sinhalese 74%, Tamil 18%, Moor 7%, Burgher, Malay, and Vedda 1%
Languages: Sinhala 74%, Tamil 18%, other 8%
Religions: Buddhist 70%, Hindu 15%, Christian 8%, Muslim 7%
Government type: republic
Capital: Colombo; note- Sri Jayewardenepura Kotte is the legislative capital
Budget: revenues: $2.8 billion; expenditures: $4.1 billion
Currency: Sri Lankan rupee (LKR)

SUDAN
Location: Northern Africa, bordering the Red Sea, between Egypt and Eritrea
Area: 967,242 sq mi (2,505,810 sq km)
Area-comparative: slightly more than one-quarter the size of the US
Population: 39,148,162
Population growth rate: 2.64%
Nationality: noun: Sudanese (singular and plural); adjective: Sudanese
Ethnic groups: black 52%, Arab 39%, Beja 6%, foreigners 2%, other 1%
Languages: Arabic (official), Nubian, Ta Bedawie, diverse dialects of Nilotic, English
Religions: Sunni Muslim 70%, indigenous beliefs 25%, Christian 5%
Government type: authoritarian regime-ruling military junta took power in 1989, government is run by an alliance of the military and the National Congress Party (NCP)
Capital: Khartoum
Budget: revenues: $1.4 billion; expenditures: $1.6 billion, capital of $304 million
Currency: Sudanese dinar (SDD)

SURINAME
Location: Northern South America, bordering the North Atlantic Ocean, between French Guiana and Guyana
Area: 63,022 sq mi (163,270 sq km)
Area-comparative: slightly larger than Georgia
Population: 436,935
Population growth rate: 0.31%
Nationality: noun: Surinamer(s); adjective: Surinamese
Ethnic groups: Hindustani 37%, Creole 31%, Javanese 15%, Maroons 10%, Amerindian 2%, Chinese 2%, other 2%, white 1%
Languages: Dutch (official), English, Sranang Tongo, Hindustani, Javanese
Religions: Hindu 27%, Protestant 25%, Roman Catholic 22.8%, Muslim 19.6%, indigenous beliefs 5%

Government type: constitutional democracy
Capital: Paramaribo
Budget: revenues: $393 million; expenditures: $403 million, including capital expenditures of $34 million
Currency: Surinamese guilder (SRG)

SWAZILAND
Location: Southern Africa, between Mozambique and South Africa
Area: 6,702 sq mi (17,363 sq km)
Area-comparative: slightly smaller than New Jersey
Population: 1,169,241
Population growth rate: 0.55%
Nationality: noun: Swazi(s); adjective: Swazi
Ethnic groups: African 97%, European 3%
Languages: English, siSwati
Religions: Zionist 40%, Roman Catholic 20%, Muslim 10%, Anglican, Bahai, Methodist, Mormon, Jewish and other 30%
Government type: monarchy, independent member of Commonwealth
Capital: Mbabane; Lobamba is the royal and legislative capital
Budget: revenues: $448 million: expenditures: $506.9 miliion, including capital expenditures of $147 million
Currency: lilangeni (SZL)

SWEDEN
Location: Northern Europe, bordering the Baltic Sea, Gulf of Bothnia, Kattegat, and Skagerrak, between Finland and Norway
Area: 173,686 sq mi (449,964 sq km)
Area-comparative: slightly larger than California
Population: 8,986,400
Population growth rate: 0.18%
Nationality: noun: Swede(s); adjective: Swedish
Ethnic groups: Swedes and Finnish and Sami minorities; Finns, Yugoslavs, Danes, Norwegians, Greeks
Languages: Swedish (official); small Sami-and Finnish-speaking minorities
Religions: Lutheran 87%, Roman Catholic. Orthodox, Baptist, Muslim, Jewish, Buddhist
Government: constitutional monarchy
Capital: Stockholm
Budget: revenues: $119 billion; expenditures: $110 billion
Currency: Swedish krona (SEK)

SWITZERLAND
Location: Central Europe, east of France, north of Italy
Area: 15,938 sq mi (41,290 sq km)
Area-comparative: slightly less than twice the size of New Jersey
Population: 7,450,867
Population growth rate: 0.54%
Nationality: noun: Swiss (singular and plural); adjective: Swiss
Ethnic groups: German 65%, French 18%, Italian 10%, other %, Romansch 1%
Languages: German (official) 63.7%, French 19.2%, other 8.9%, Italian 7.6%, Romansch 0.6%
Religions: Roman Catholic 46.1%, Protestant 40%, none 8.9%, other 5%
Government type: federal republic
Capital: Bern
Budget: revenues: $30 billion; expenditures: $30 billion
Currency: Swiss franc (CHF)

503

SYRIA

Location: Middle East, bordering the Mediterranean Sea, between Lebanon and Turkey
Area: 71,479 sq mi (185,180 sq km)
Area comparative: slightly larger than North Dakota
Population: 18,056,874
Population growth rate: 2.4%
Nationality: noun: Syrian(s); adjective: Syrian
Ethnic groups: Arab 90.3%, Kurds, Armenians, and other 9.7%
Languages: Arabic (official), Kurdish, Armenian, Aramaic, Circassian, French, English
Religions: Sunni Muslim 74%, Alawite, Druze, and other Muslim sects 16%, Christian (various sects) 10%
Government type: republic under military regime since March 1963
Capital: Damascus
Budget: revenues: $4.3 billion: expenditures: $8.6 billion, including capital expenditures of $3.6 billion
Currency: Syrian pound (SYP)

TAIWAN

Location: Eastern Asia, islands bordering the East China Sea, Philippine Sea, South China Sea, and Taiwan Strait, north of the Philippines, off the southeastern coast of China
Area: 13,888 sq mi (35,980 sq km)
Area-comparative: slightly smaller than Maryland and Delaware combined
Population: 22,749,838
Population growth rate: 0.64%
Nationality: noun: Chinese/Taiwanese (singular and plural); adjective: Chinese/Taiwanese
Ethnic groups: Taiwanese (including Hakka) 84%, mainland Chinese 14%, aborigine 2%
Languages: Mandarin Chinese (official), Taiwanese (Min), Hakka dialects
Religions: mixture of Buddhist, Confucian, and Taoist 93%, Christian 4.5%, other 2.5%
Government type: multiparty democratic regime headed by popularly-elected president and unicameral legislature
Capital: Taipei
Budget: revenues: $52.5 billion; expenditures: $ 63 billion, including capital of $14.4
Currency: new Taiwan dollar (TWD)

TAJIKISTAN

Location: Central Asia, west of China
Area: 55,237 sq mi (143,100 sq km)
Area-comparative: slightly smaller than Wisconsin
Population: 7,011,556
Population growth rate: 2.14%
Nationality: noun: Tajikstani(s); adjective: Tajikstani
Ethnic groups: Tajik 64.9%, Uzbek 25%, other 6.6%, Russian 3.5%
Languages: Tajik (official), Russian
Religions: Sunni Muslim 85%, Shi'a Muslim 5% other (including Russian Orthodox) 10%,
Government type: republic
Capital: Dushanbe
Budget: revenues: $502 million; expenditures: $520 million, including capital expenditures of $86 million
Currency: somoni (TJS)

TANZANIA

Location: Eastern Africa, bordering the Indian Ocean, between Kenya and Mozambique
Area: 364,804 sq mi (945,087 sq km)
Area-comparative: slightly larger than twice the size of California

Population: 36,588,225
Population growth rate: 1.95%
Nationality: noun: Tanzanian(s); adjective: Tanzanian
Ethnic groups: African 99%, other 1%
Languages: Kiswahili or Swahili (official), English, Arabic, many local languages
Religions: Muslim 35%, indigenous beliefs 35%, Christian 30%
Government type: republic
Capital: Dar es Salaam (administrative); Dodoma (legislative)
Budget: revenues: $1.2 billion; expenditures: $2 billion , including capital of $NA
Currency: Tanzanian shilling(TZS)

THAILAND
Location: Southeastern Asia, bordering the Andaman Sea and the Gulf of Thailand, southeast of
          Myanmar
Area: 198,404 sq mi (514,000 sq km)
Area-comparative: slightly more than twice the size of Wyoming
Population: 64,865,523
Population growth rate: 0.91%
Nationality: noun: Thai (singular and plural); adjective: Thai
Ethnic groups: Thai 75%, Chinese 14%, other 11%
Languages: Thai (official), English, regional dialects
Religions: Buddhism 95%, Muslim 3.8%, other 0.6%, Christianity 0.5%, Hinduism 0.1%,
Government type: constitutional monarchy
Capital: Bangkok
Budget: revenues: $21 billion; expenditures: $22 billion , including capital of $5 billion
Currency: baht (THB)

TOGO
Location: Western Africa, bordering the Bight of Benin, between Benin and Ghana
Area: 21,919 sq mi (56,785 sq km)
Area-comparative: slightly smaller than West Virginia
Population: 5,556,812
Population growth rate: 2.27%
Nationality: noun: Togolise; adjective: Togolese
Ethnic groups: native African (37 tribes, largest are Ewe, Mina, and Kabre) 99%, European and Syrian-
          Lebanese less than 1%
Languages: French (official), Ewe, Mina, Kabye, Dagomba
Religions: indigenous beliefs 51%, Christian 29%, Muslim 20%
Government type: republic under transition to multiparty democratic rule
Capital: Lome
Budget: revenues: $232 million; expenditures: $252 million
Currency: Communaute Financiere Africaine franc(XOF)

TONGA
Location: Oceania, group of islands in the South Pacific Ocean, about two-thirds of the way from Hawaii
          to New Zealand
Area: 289 sq mi (748 sq km)
Area-comparative: four times the size of Washington, DC
Population: 110,237
Population growth rate: 1.94%
Nationality: noun: Tongan(s); adjective: Tongan
Ethnic groups: Polynesian, Europeans (about 300)

Languages: Tongan (official), English
Religions: Christian (Free Wesleyan church claims 30,000 adherents)
Government type: hereditary constitutional monarchy
Capital: Nuku'alofa
Budget: revenues: $39.9 million; expenditures: $52.4 million, including capital expenditures of $1.9 million
Currency: pa'anga(TOP)

TRINIDAD AND TOBAGO
Location: Caribbean, islands between the Caribbean Sea and the North Atlantic Ocean, northeast of Venezuela
Area: 1,979 sq mi (5,128 sq km)
Area-comparative: slightly smaller than Delaware
Population: 1,096,585
Population growth rate: -0.71%
Nationality: noun: Trinidadian(s), Tobagonian(s); adjective: Trinidadian, Tobagonian
Ethnic groups: East Indian 40.3%, black 39.5%, mixed 18.4%, Chinese and other 1.2%, white 0.6%
Languages: English (official), Hindi, French, Spanish, Chinese
Religions: Roman Catholic 29.4%, other 26.7%, Hindu 23.8%, Anglican 10.9%, Muslim 5.8%, Presbyterian 3.4%
Government type: parliamentary democracy
Capital: Port-of-Spain
Budget: revenues: $1.54 billion; expenditures: $1.6 billion, including capital expenditures of $117.3 million
Currency: Trinidad and Tobago dollar (TTD)

TUNISIA
Location: Northern Africa, bordering the Mediterranean Sea, between Algeria and Libya
Area: 63,153 sq mi (163,610 sq km)
Area-comparative: slightly larger than Georgia
Population: 9,974,722
Population growth rate: 1.01%
Nationality: noun: Tunisian(s); adjective: Tunisian
Ethnic groups: Arab 98%, European 1%, Jewish and other 1%
Languages: Arabic (official), French
Religions: Muslim 98%, Christian 1%, Jewish and other 1%
Government type: republic
Capital: Tunis
Budget: revenues: $5.2 billion; expenditures: $5.7 billion, including capital expenditures of $1.6 billion
Currency: Tunisian dinar (TND)

TURKEY
Location: Southeastern Europe, Southwestern Asia, bordering the Black Sea, between Bulgaria and Georgia, between Greece and Syria
Area: 301,304 sq mi (780,580 sq km)
Area-comparative: slightly larger than Texas
Population: 68,893,918
Population growth rate: 1.13%
Nationality: noun: Turk(s); adjective: Turkish
Ethnic groups: Turkish 80%, Kurdish 20%
Languages: Turkish (official), Kurdish, Arabic, Armenian, Greek
Religions: Muslim 99.8% (primarily Sunni); Christians and Jews 0.2%
Government type: republican parliamentary democracy
Capital: Ankara
Budget: revenues: $42.4 billion; expenditures: $69.1 billion   **Currency:** Turkish lira (TRL)

TURKMENISTAN
Location: Central Asia, bordering the Caspian Sea, between Iran and Kazakhstan
Area: 188,406 sq mi (488,100 sq km)
Area-comparative: slightly larger than California
Population: 4,863,169
Population growth rate: 1.81%
Nationality: noun: Turmen(s); adjective: Turkmen
Ethnic groups: Turkmen 85%, Uzbek 5%, Russian 4%, other 6%
Languages: Turkmen 72%(official), Russian 12%, Uzbek 9%, other 7%
Religions: Muslim 89%, Eastern Orthodox 9%, unknown 2%
Government type: republic
Capital: Ashgabat
Budget: revenues: $588.6 million; expenditures: $658.2 million
Currency: Turkmen manta (TMM)

UGANDA
Location: Eastern Africa, west of Kenya
Area: 91,111 sq mi (236,040 sq km)
Area-comparative: slightly smaller than Oregon
Population: 26,404,543
Population growth rate: 2.97%
Nationality: noun: Ugandan(s); adjective: Ugandan
Ethnic groups: Baganda 17%, Ankole 8%, Basogo 8%, Iteso 8%,Bakiga 7%, Langi 6%, Rwanda 6%, Bagisu 5%, Acholi 4%, Lugbara 4%, Bunyoro 3%, Batoro 3%, Alur 2%, Bagwere 2%, Bakonjo 2%, Jopodhola 2%, Karamojong 2%, Rundi 2%, non-African (European, Asian, Arab) 1% , other 8%
Languages: English (official), Ganda, Luganda, Swahili, Nilo-Saharan languages, Arabic
Religions: Roman Catholic 33%, Protestant 33%, indigenous beliefs 18%, Muslim 16%
Government type: republic
Capital: Kampala
Budget: revenues: $959 million; expenditures: $1.04 billion   **Currency:** Ugandan shilling (UGX)

UKRAINE
Location: Eastern Europe, bordering the Black Sea, between Poland and Russia
Area: 233,028 sq mi (603,700 sq km)
Area-comparative: slightly smaller than Texas
Population: 47,732,079
Population growth rate: -0.66%
Nationality: noun: Ukrainian(s); adjective: Ukrainian
Ethnic groups: Ukrainian 77.8%, Russian 17.3%, other 1.8 %, Belarusian 0.6%, Moldovan 0.5%, Crimean Tatar 0.5%, Bulgarian 0.4%, Hungarian 0.3%, Romanian 0.3%, Polish 0.3%, Jewish 0.2%
Languages: Ukrainian, Russian, Romanian, Polish, Hungarian
Religions: Ukrainian Orthodox, Protestant, Roman Catholic, Ukrainian Catholic, and Jewish
Government type: republic
Capital: Kiev (Kyyiv)
Budget: revenues: $10.2 billion; expenditures: $11.1 billion
Currency: hryvnia (UAH)

UNITED ARAB EMIRATES
Location: Middle East, bordering the Gulf of Oman and the Persian Gulf, between Oman and Saudi Arabia

Area: 31,992 sq mi (82,880 sq km)
Area-comparative: slightly smaller than Maine
Population: 2,523,915
Population growth rate: 1.57%
Nationality: noun: Emirati(s); adjective: Emirati
Ethnic groups: South Asian 50%, Arab and Iranian 23%, Emirati 19%, other (including Westerners and East Asians) 8%
Languages: Arabic (official), Persian, English, Hindi, Urdu
Religions: Muslim 96%, other (including Christian and Hindu) 4%
Government type: federation with specified powers delegated to the UAE federal government and other powers reserved to member emirates
Capital: Abu Dhabi
Budget: revenues: $15.6 billion; expenditures: $23.5 billion, including capital of $3.4 billion
Currency: Emiriati dirham(AED)

UNITED KINGDOM

Location: Western Europe, between the North Atlantic Ocean and the North Sea, northwest of France
Area: 94,500 sq mi (244,820 sq km)
Area-comparative: slightly smaller than Oregon
Population: 60,270,708
Population growth rate: 0.29%
Nationality: noun: Briton(s), British (collective plural); adjective: British
Ethnic groups: English 81.5%, Scottish 9.6%, West Indian, Indian, Pakistani, and other 2.8%, Irish 2.4%, Welsh 1.9%, Ulster 1.8%
Religions: Anglican and Roman Catholic, Muslim, Presbyterian, Methodist, Sikh, Hindu, Jewish
Languages: English, Welsh, Scottish form of Gaelic
Government type: constitutional monarchy
Capital: London
Budget: revenues: $565 billion; expenditures: $540 billion   **Currency:** British pound(GBP)

UNITED STATES OF AMERICA

Location: North America, bordering both the North Atlantic Ocean and the North Pacific Ocean, between Canada and Mexico
Area: 3,716,829 sq mi (9,631,418 sq km)
Area-comparative: about half the size of Russia; about three-tenths the size of Africa; about half the size of South America(or slightly larger than Brazil); slightly larger than China; about two and half times the size of western Europe
Population: 293,027,571
Population growth rate: 0.92%
Nationality: noun: American(s); adjective: American
Ethnic groups: white 77.1%, black 12.9%, Asian 4.2%, other 4%, Amerindian and Alaska native 1.5%, native Hawaiian and other Pacific Islander 0.3%
Languages: English (official), Spanish
Religions: Protestant 56%, Roman Catholic 28%, nonreligious 10%, other 4%, Jewish 2%
Government type: Constitution-based federal republic; strong democratic tradition
Capital: Washington, DC
Budget: revenues: $1.946 trillion; expenditures: $2.052 trillion   **Currency:** United States dollar (USD)

URUGUAY

Location: Southern South America, bordering the South Atlantic Ocean, between Argentina and Brazil
Area: 68,021 sq mi (176,220 sq km)
Area-comparative: slightly smaller than the state of Washington

Population: 3,399,237
Population growth rate: 0.51%
Nationality: noun: Uruguayan(s); adjective: Uruguayan
Ethnic groups: white 88%, mestizo 8%, black 4%
Languages: Spanish, Portunol, or Brazilero
Religions: Roman Catholic 66%, other or not religious 31%, Protestant 2%, Jewish 1%
Government type: constitutional republic
Capital: Montevideo
Budget: revenues: $2.74 billion; expenditures: $3.4 billion, including capital expenditures of $193 million
Currency: Uruguayan peso(UYU)

UZBEKISTAN

Location: Central Asia, north of Afghanistan
Area: 172,696 sq mi (447,400 sq km)
Area-comparative: slightly larger than California
Population: 26,410,416
Population growth rate: 1.65%
Nationality: noun: Uzbekistani; adjective: Uzbekistani
Ethnic groups: Uzbek 80%, Russian 5.5%, Tajik 5%, Kazakh 3%, Karakalpak 2.5%, other 4%
Languages: Uzbek 74.3%, Russian 14.2%, other 7.1%, Tajik 4.4%
Religions: Muslim 88% (primarily Sunni), Eastern Orthodox 9%, other 3%
Government type: republic; authoritarian presidential rule, with little power outside the executive branch
Capital: Tashkent
Budget: revenues: $2.42 billion; expenditures: $2.45 billion, including capital of $NA
Currency: Uzbekistani sum(UZS)

VANUATU

Location: Oceania, group of islands in the South Pacific Ocean, about three-quarters of the way from Hawaii to Australia
Area: 4,709 sq mi (12,200 sq km)
Area-comparative: slightly larger than Connecticut
Population: 202,609
Population growth rate: 1.57%
Nationality: noun: Ni-Vanuatu (singular and plural); adjective: Ni-Vanuatu
Ethnic groups: indigenous Melanesian 98%, French, Vietnamese, Chinese, other Pacific Islanders
Languages: English, French, Bislama, and more than 100 local languages
Religions: Presbyterian 36.7%, other 15.7%, Anglican 15%, Catholic 15%, indigenous beliefs 7.6%, Seventh-Day Adventist 6.2%, Church of Christ 3.8%
Government type: parliamentary republic
Capital: Port-Vila (Etate)
Budget: revenues: $94.4 million; expenditures: $99.8 million, including capital expenditures of $30.4 million   **Currency:** vatu (VUV)

VENEZUELA

Location: Northern South America, bordering the Caribbean Sea and the North Atlantic Ocean, between Colombia and Guyana
Area: 352,051 sq mi (912,050 sq km)
Area-comparative: slightly more than twice the size of California
Population: 25,017,387
Population growth rate: 1.44%
Nationality: noun: Venezuelan(s); adjective: Venezuelan
Ethnic groups: Spanish, Italian, Portuguese, Arab, German, African, indigenous people

Languages: Spanish (official), numerous indigenous dialects
Religions: Roman Catholic 96%, Protestant 2%, other 2%
Government type: federal republic
Capital: Caracas
Budget: revenues: $20.1 billion; expenditures: $23.3billion, including capital of $2.6 billion
Currency: bolivar (VEB)

VIETNAM
Location: Southeastern Asia, bordering the Gulf of Thailand, Gulf of Tonkin, and South China Sea,
        alongside China, Laos, and Cambodia
Area: 127,210 sq mi (329,560 sq km)
Area-comparative: slightly larger than New Mexico
Population: 82,689,518
Population growth rate: 1.3%
Nationality: noun: Vietnamese (singular and plural); adjective: Vietnamese
Ethnic groups: Vietnamese 85-90%, Chinese, Hmong, Thai, Khmer, Cham, mountain groups
Languages: Vietnamese (official), Chinese, English, (increasingly favored as 2nd language), French,
        Khmer, tribal languages
Religions: Buddhist, Hoa Hao, Cao Dai, Taoist, Roman Catholic, indigenous beliefs, Muslim, Protestant,
Government type: Communist state
Capital: Hanoi
Budget: revenues: $5.3 billion; expenditures: $5.6 billion, including capital expenditures of $1.8 billion
Currency: dong (VND)

VIRGIN ISLANDS
Location: islands between the Caribbean Sea and the North Atlantic Ocean, east of Puerto Rico
Area: 136 sq mi (352 sq km)
Area-comparative: twice the size of Washington, DC
Population: 108,775
Population growth rate: -0.05%
Nationality: noun: Virgin Islander(s); adjective; Virgin Islander
Ethnic groups: West Indian 81%, US mainland 13%, Puerto Rican 4%, other 2%
Languages: English (official), Spanish, Creole
Religions: Baptist 42%, Roman Catholic 34%, Episcopalian 17%, other 7%
Government type: NA
Capital: Charlotte Amalie
Budget: revenues: $364.4 million; expenditures: $364.4 million   **Currency:** U S dollar (USD)

WEST BANK
Location: Middle East, west of Jordan
Area: 2,262 sq mi (5,860 sq km)`
Area-comparative: slightly smaller than Delaware
Population: 2,311,204
Population growth rate: 3.21%
Nationality: noun: NA; adjective: NA
Ethnic groups: Palestinian Arab and other 83%, Jewish 17%
Languages: Arabic, Hebrew, English
Religions: Muslim 75%, Jewish, 17%, Christian and other 8%
Government type: NA
Capital: NA
Budget: revenues: $676.6 million; expenditures: $1.155 billion, including capital expenditures of $NA
Currency: new Israeli shekel; Jordanian dinar (ILS/JOD)

YEMEN
Location: Middle East, bordering the Arabian Sea, Gulf of Aden, and Red Sea, between Oman and Saudi Arabia
Area: 203,796 sq mi (527,970 sq km)
Area-comparative: slightly larger than twice the size of Wyoming
Population: 20,024,867
Population growth rate: 3.44%
Nationality: noun: Yemeni(s); adjective: Yemeni
Ethnic groups: predominantly Arab, some Afro-Arab, South Asians, Europeans
Language: Arabic
Religions: Muslim including Shaf'i and Shi'a, small numbers of Jewish, Christian, and Hindu
Government type: republic
Capital: Sanaa
Budget: revenues: $3.78 billion; expenditures: $4.13 billion, including capital of $NA
Currency: Yemeni rial (YER)

ZAMBIA
Location: Southern Africa, east of Angola
Area: 290,509 sq mi (752,614 sq km)
Area-comparative: slightly larger than Texas
Population: 10,462,436
Population growth rate: 1.47%
Nationality: noun: Zambian(s); adjective: Zambian
Ethnic groups: African 98.7%, European 1.1%, other 0.2%
Languages: English(official), Bemba, Kaonda, Lozi, Lunda, Luvale, Nyanja, 70 indigenous languages
Religions: Christian 50%-75%, Muslim and Hindu 24%-49%, indigenous beliefs 1%
Government type: republic
Capital: Lusaka
Budget: revenues: $1.2 billion; expenditures: $1.25 billion
Currency: Zambian kwacha(ZMK)

ZIMBABWE
Location: Southern Africa, between South Africia and Zambia
Area: 150,764 sq mi (390,580 sq km)
Area-comparative: slightly larger than Montana
Population: 12,671,860
Population growth rate: 0.68%
Nationality: noun: Zimbabwean(s); adjective: Zimbabwean
Ethnic groups: Shona 82%, Ndebele 14%, other 2%, mixed and Asian 1%, white less than 1%
Languages: English (official), Shona, Sindebele, and numerous other tribal dialects
Religions: syncretic 50%,, Christian 25%, indigenous beliefs 24%, Muslim and other 1%
Government type: parliamentary democracy
Capital: Harare
Budget: revenues: $1.4 billion; expenditures: $2 billion, including capital of $NA
Currency: Zimbabwean dollar (ZWD)

The World

Area: 510.072 million sq km;
land: 148.94 million sq km;
water: 361.132 million sq km.

Note: 70.8% of the world's surface is *water*,
29.2% is *land*.

Area-comparative:
land area about 16 times the size of the US

Population: 6,379,157,361

Population growth rate: 1.14%

Languages: Chinese Mandarin 14.37%; Hindi 6.02%; English 5.61%; Spanish 5.59%, Bengali 3.4%, Portuguese 2.63%, Russian 2.75%, Japanese 2.06%, German 1.64%, Korean 1.28%, French 1.27%, other 53.38%

Religions: Christians 32.79%; (of which Roman Catholics 17.33%, Protestants 5.62%, Orthodox 3.51%, Anglicans 1.31%, other 5.02%) Muslims 19.6%, Hindus 13.31%, Buddhists 5.88%, Sikhs o.38%, Jews 0.24%, other religions 12.83%, non-religious 12.53%, atheists 2.44%.

2006 Calendar with Islamic dates in North America
ISLAMIC DATES BEGIN AT SUNSET THE PRECEDING EVENING

NY= NewYear Aa= Aashurah MN= Miladun-Nabi IM= Isra/Me'raj NS= NisfuSha'ban RM= Ramadan LQ= Lailatul-Qadr EF= Eidul-Fitr YA= Yaum-i-Arafah EA= Eidul-Adha

Solar Eclipses
T=Total A=Annular H=Hybrid(Tot./Ann.) P=Partial

Lunar Eclipses
t=Total(Umbral) p=Partial(Umbral) n=Penumbral

January — ZHJ 1426/MHM 1427

S	M	T	W	T	F	S
1	2	3	4	5	6	7
ZH	2	3	4	5	6	7
8	9	10	11	12	13	14
8	YA	EA	11	12	13	14
15	16	17	18	19	20	21
15	16	17	18	19	20	21
22	23	24	25	26	27	28
22	23	24	25	26	27	28
29	30	31				
29	30	NY				

February — Muharram 1427

S	M	T	W	T	F	S
			1	2	3	4
			2	3	4	5
5	6	7	8	9	10	11
6	7	8	9	Aa	11	12
12	13	14	15	16	17	18
13	14	15	16	17	18	19
19	20	21	22	23	24	25
20	21	22	23	24	25	26
26	27	28				
27	28	29				

March — Safar/Rabi-I

S	M	T	W	T	F	S
			1	2	3	4
			SF	2	3	4
5	6	7	8	9	10	11
5	6	7	8	9	10	11
12	13	14	15	16	17	18
12	13	n	15	16	17	18
19	20	21	22	23	24	25
19	20	21	22	23	24	25
26	27	28	29	30	31	
26	27	28	T	??	RA	

April — Rabi-I/Rabi-II

S	M	T	W	T	F	S
						1
						2
2	3	4	5	6	7	8
3	4	5	6	7	8	9
9	10	11	12	13	14	15
10	11	MN	13	14	15	16
16	17	18	19	20	21	22
17	18	19	20	21	22	23
23	24	25	26	27	28	29
24	25	26	27	28	29	R T
30						
2						

May — Rabi-II/Jumada-I

S	M	T	W	T	F	S
	1	2	3	4	5	6
	3	4	5	6	7	8
7	8	9	10	11	12	13
9	10	11	12	13	14	15
14	15	16	17	18	19	20
16	17	18	19	20	21	22
21	22	23	24	25	26	27
23	24	25	26	27	28	29
28	29	30	31			
JA	2	3	4			

June — Jumada-I/Jumada-II

S	M	T	W	T	F	S
				1	2	3
				5	6	7
4	5	6	7	8	9	10
8	9	10	11	12	13	14
11	12	13	14	15	16	17
15	16	17	18	19	20	21
18	19	20	21	22	23	24
22	23	24	25	26	27	28
25	26	27	28	29	30	
29	30	JT	2	3	4	

July — Jumada-II/Rajab

S	M	T	W	T	F	S
						1
						5
2	3	4	5	6	7	8
6	7	8	9	10	11	12
9	10	11	12	13	14	15
13	14	15	16	17	18	19
16	17	18	19	20	21	22
20	21	22	23	24	25	26
23	24	25	26	27	28	29
27	28	29	??	RJ	2	3
30	31					
4	5					

August — Rajab/Sha'ban

S	M	T	W	T	F	S
		1	2	3	4	5
		6	7	8	9	10
6	7	8	9	10	11	12
11	12	13	14	15	16	17
13	14	15	16	17	18	19
18	19	20	21	22	23	24
20	21	22	23	24	25	26
25	IM	28	29	SB	2	
27	28	29	30	31		
3	4	5	6	7		

September — Sha'ban/Ramadan

S	M	T	W	T	F	S
					1	2
					8	9
3	4	5	6	7	8	9
10	11	12	13	p	NS	16
10	11	12	13	14	15	16
17	18	19	20	21	22	23
17	18	19	20	21	22	23
24	25	26	27	28	29	30
24	25	26	27	28	29	30
RM	2	3	4	5	6	7

October — Ramadan/Shawaal

S	M	T	W	T	F	S
1	2	3	4	5	6	7
8	9	10	11	12	13	14
8	9	10	11	12	13	14
15	16	17	18	19	20	21
15	16	17	18	19	20	21
22	23	24	25	LQ	27	28
22	23	24	25	26	27	28
29	30	EF	2	3	4	5
30	31					
6	7					

November — Shawwal/Zul-Qa'dah

S	M	T	W	T	F	S
		1	2	3	4	
		9	10	11	12	
5	6	7	8	9	10	11
13	14	15	16	17	18	19
12	13	14	15	16	17	18
19	20	21	22	23	24	25
20	21	22	23	24	25	26
27	28	29	30	ZQ	2	3
26	27	28	29	30		
4	5	6	7	8		

December — Zul-Qa'dah/Zul-Hijjah

S	M	T	W	T	F	S
					1	2
					9	10
3	4	5	6	7	8	9
11	12	13	14	15	16	17
10	11	12	13	14	15	16
18	19	20	21	22	23	24
17	18	19	20	21	22	23
25	26	27	28	29	ZH	2
24	25	26	27	28	29	30 WA
3	4	5	6	7	8	
JI						
EA						

? Slight chance, month may start on this day.   ?? Very slight chance, month may start on this day.

Calendar for year 2006 (United States)

January 2006
Su	Mo	Tu	We	Th	Fr	Sa
1	2	3	4	5	6	7
8	9	10	11	12	13	14
15	16	17	18	19	20	21
22	23	24	25	26	27	28
29	30	31				

February 2006
Su	Mo	Tu	We	Th	Fr	Sa
			1	2	3	4
5	6	7	8	9	10	11
12	13	14	15	16	17	18
19	20	21	22	23	24	25
26	27	28				

March 2006
Su	Mo	Tu	We	Th	Fr	Sa
			1	2	3	4
5	6	7	8	9	10	11
12	13	14	15	16	17	18
19	20	21	22	23	24	25
26	27	28	29	30	31	

April 2006
Su	Mo	Tu	We	Th	Fr	Sa
						1
2	3	4	5	6	7	8
9	10	11	12	13	14	15
16	17	18	19	20	21	22
23	24	25	26	27	28	29
30						

May 2006
Su	Mo	Tu	We	Th	Fr	Sa
	1	2	3	4	5	6
7	8	9	10	11	12	13
14	15	16	17	18	19	20
21	22	23	24	25	26	27
28	29	30	31			

June 2006
Su	Mo	Tu	We	Th	Fr	Sa
				1	2	3
4	5	6	7	8	9	10
11	12	13	14	15	16	17
18	19	20	21	22	23	24
25	26	27	28	29	30	

July 2006
Su	Mo	Tu	We	Th	Fr	Sa
						1
2	3	4	5	6	7	8
9	10	11	12	13	14	15
16	17	18	19	20	21	22
23	24	25	26	27	28	29
30	31					

August 2006
Su	Mo	Tu	We	Th	Fr	Sa
		1	2	3	4	5
6	7	8	9	10	11	12
13	14	15	16	17	18	19
20	21	22	23	24	25	26
27	28	29	30	31		

September 2006
Su	Mo	Tu	We	Th	Fr	Sa
					1	2
3	4	5	6	7	8	9
10	11	12	13	14	15	16
17	18	19	20	21	22	23
24	25	26	27	28	29	30

October 2006
Su	Mo	Tu	We	Th	Fr	Sa
1	2	3	4	5	6	7
8	9	10	11	12	13	14
15	16	17	18	19	20	21
22	23	24	25	26	27	28
29	30	31				

November 2006
Su	Mo	Tu	We	Th	Fr	Sa
			1	2	3	4
5	6	7	8	9	10	11
12	13	14	15	16	17	18
19	20	21	22	23	24	25
26	27	28	29	30		

December 2006
Su	Mo	Tu	We	Th	Fr	Sa
					1	2
3	4	5	6	7	8	9
10	11	12	13	14	15	16
17	18	19	20	21	22	23
24	25	26	27	28	29	30
31						

Holidays and Observances

Date	Observance
Jan 1	New Year's Day
Jan 2	Last day of Chanukah (Jewish)
Jan 2	'New Year's Day' observed
Jan 10	Eid-al-Adha (Islamic)
Jan 16	Martin Luther King Day
Jan 31	Muharramn/New Year (Islamic)
Feb 14	Valentine's Day
Feb 20	President's Day
Apr 11	Prophet's Birthday (Islamic)
Apr 13	First day of Passover (Jewish)
Apr 14	Good Friday (Christian)
Apr 16	Easter Sunday (Christian)
Apr 17	Easter Monday (Christian)
Apr 20	Last day of Passover (Jewish)
May 14	Mother's Day
May 29	Memorial Day
Jun 18	Father's Day
Jul 4	Independence Day
Sep 4	Labor Day
Sep 23	Rosh Hashana (Jewish)
Sep 24	Ramadan begins (Islamic)
Oct 2	Yom Kippur (Jewish)
Oct 9	Columbus Day
Oct 13	Last day of Sukkot (Jewish)
Oct 19	Lailat Ul Qadr (Islamic)
Oct 24	Eid-al-Fitr (Islamic)
Oct 31	Halloween
Nov 10	'Veterans Day' observed
Nov 11	Veterans Day
Nov 23	Thanksgiving Day
Dec 16	First day of Chanukah (Jewish)
Dec 23	Last day of Chanukah (Jewish)
Dec 25	Christmas Day (Christian)
Dec 31	Eid-al-Adha (Islamic)

Calendar for year 2006 (Israel)

January 2006

Su	Mo	Tu	We	Th	Fr	Sa
1	2	3	4	5	6	7
8	9	10	11	12	13	14
15	16	17	18	19	20	21
22	23	24	25	26	27	28
29	30	31				

February 2006

Su	Mo	Tu	We	Th	Fr	Sa
			1	2	3	4
5	6	7	8	9	10	11
12	13	14	15	16	17	18
19	20	21	22	23	24	25
26	27	28				

March 2006

Su	Mo	Tu	We	Th	Fr	Sa
			1	2	3	4
5	6	7	8	9	10	11
12	13	14	15	16	17	18
19	20	21	22	23	24	25
26	27	28	29	30	31	

April 2006

Su	Mo	Tu	We	Th	Fr	Sa
						1
2	3	4	5	6	7	8
9	10	11	12	13	14	15
16	17	18	19	20	21	22
23	24	25	26	27	28	29
30						

May 2006

Su	Mo	Tu	We	Th	Fr	Sa
	1	2	3	4	5	6
7	8	9	10	11	12	13
14	15	16	17	18	19	20
21	22	23	24	25	26	27
28	29	30	31			

June 2006

Su	Mo	Tu	We	Th	Fr	Sa
				1	2	3
4	5	6	7	8	9	10
11	12	13	14	15	16	17
18	19	20	21	22	23	24
25	26	27	28	29	30	

July 2006

Su	Mo	Tu	We	Th	Fr	Sa
						1
2	3	4	5	6	7	8
9	10	11	12	13	14	15
16	17	18	19	20	21	22
23	24	25	26	27	28	29
30	31					

August 2006

Su	Mo	Tu	We	Th	Fr	Sa
		1	2	3	4	5
6	7	8	9	10	11	12
13	14	15	16	17	18	19
20	21	22	23	24	25	26
27	28	29	30	31		

September 2006

Su	Mo	Tu	We	Th	Fr	Sa
					1	2
3	4	5	6	7	8	9
10	11	12	13	14	15	16
17	18	19	20	21	22	23
24	25	26	27	28	29	30

October 2006

Su	Mo	Tu	We	Th	Fr	Sa
1	2	3	4	5	6	7
8	9	10	11	12	13	14
15	16	17	18	19	20	21
22	23	24	25	26	27	28
29	30	31				

November 2006

Su	Mo	Tu	We	Th	Fr	Sa
			1	2	3	4
5	6	7	8	9	10	11
12	13	14	15	16	17	18
19	20	21	22	23	24	25
26	27	28	29	30		

December 2006

Su	Mo	Tu	We	Th	Fr	Sa
					1	2
3	4	5	6	7	8	9
10	11	12	13	14	15	16
17	18	19	20	21	22	23
24	25	26	27	28	29	30
31						

Holidays and Observances

Jan 1	Tevet 1	Chanukah VII
Jan 2	Tevet 2	Chanukah VIII
Jan 10	Tevet 10	Asarah B'Tevet
Feb 13	Shevat 15	Tu B'Shevat (Arbor Day)
Mar 13	Adar 13	Ta'anit Esther (Fast of Esther)
Mar 14	Adar 14	Purim (Deliverance of the Jews)
Mar 15	Adar 15	Shushan Purim
Apr 13	Nisan 15	Pesach I (First day of Passover)
Apr 14	Nisan 16	Pesach II (Passover)
Apr 15	Nisan 17	Pesach III (Passover)
Apr 16	Nisan 18	Pesach IV (Passover)
Apr 17	Nisan 19	Pesach V (Passover)
Apr 18	Nisan 20	Pesach VI (Passover)
Apr 19	Nisan 21	Pesach VII (Last day of Passover)
Apr 25	Nisan 27	Yom HaShoah (Holocaust memorial)
May 2	Iyar 4	Yom HaZikaron (Memorial Day)
May 3	Iyar 5	Yom HaAtzmaut (Independence Day)
May 16	Iyar 18	Lag B'Omer
May 26	Iyar 28	Yom Yerushalayim (Jerusalem Day)
Jun 2	Sivan 6	Shavuot (Pentecost)
Jul 13	Tammuz 17	Fast of Shiva Asar B'Tammuz
Aug 3	Av 9	Tisha B'Av
Sep 23	Tishri 1	Rosh Hashana (New Year)
Sep 24	Tishri 2	Rosh Hashana II (New Year day 2)
Sep 25	Tishri 3	Tzom Gedaliah
Oct 2	Tishri 10	Yom Kippur
Oct 7	Tishri 15	Sukkot I
Oct 8	Tishri 16	Sukkot II
Oct 9	Tishri 17	Sukkot III
Oct 10	Tishri 18	Sukkot IV
Oct 11	Tishri 19	Sukkot V
Oct 12	Tishri 20	Sukkot VI
Oct 13	Tishri 21	Sukkot VII/Hoshanah Rabah
Oct 14	Tishri 22	Shmini Atzeret/Simchat Torah
Dec 16	Kislev 25	Chanukah I (Holiday of lights)
Dec 17	Kislev 26	Chanukah II
Dec 18	Kislev 27	Chanukah III
Dec 19	Kislev 28	Chanukah IV
Dec 20	Kislev 29	Chanukah V
Dec 21	Kislev 30	Chanukah VI/Rosh Chodesh Tevet
Dec 22	Tevet 1	Chanukah VII
Dec 23	Tevet 2	Chanukah VIII
Dec 31	Tevet 10	Asarah B'Tevet

515

THE PLANETS

THE PLANETS - SUMMARY

	Distance From Sun (000 km)	Radius (km)	Planet Mass
Mercury	57,910	24393.30	e23
Venus	108,200	6052 4.87	e24
Earth	149,600	6378 5.98	e24
Mars	227,940	3397 6.42	e23
Jupiter	778,330	71492 1.90	e27
Saturn	1,426,940	60268 5.69	e26
Uranus	2,870,990	25559 8.69	e25
Neptune	4,497,070	24764 1.02	e26
Pluto	5,913,520	1160 1.31	e22

Energy is to Mass times the speed of light squared.

$E = mc^2 =$ 0.111 x 300,000,000 x 300,000,000= **10,000,000,000,000,000 Joules**

There are nine planets that revolve around the sun in our solar system. A planet is a body of matter that shines by reflected light. Like Earth, the other planets do not give off light of their own. When you see a planet shining brightly in the night sky, it is giving off the sun's light. These planets travel around the sun in fixed paths called orbits. A planet's gravitational pull depends on two things: size and density. The term *density* refers to the amount of matter contained in a given space. A block of iron and a block of ice that are equal in size are not equal in weight; the block of iron weighs more than the block of ice. A scientist would say that the iron is more *dense*; the block of iron has more matter in it.

The sun is by far the largest object in the solar system. It is about 4.5 billion years old. Since its birth it has used up about half of the hydrogen in its core. It will continue to radiate for another 5 billion years, but eventually it will run out of hydrogen fuel. The sun contains more than 99.8% of the total mass of the solar system (Jupiter contains most of the rest). The sun is composed of 75% hydrogen and 25% helium. The diameter of the sun is 864,000 miles— more than 100 time's Earth's diameter.

Every star in the night sky is the sun of another solar system. Therefore, our sun would look like a small twinkle if you saw it from the planet of another solar system.

MERCURY is the planet closest to the sun and the eighth largest. It was named for Mercury, who in Greek myths was the speedy messenger of the gods. Mercury revolves around the sun in 88 days, faster than any other planet. Its orbit has the shape of an ellipse—a sort of flattened circle. Because of its irregular shape, sometimes it is only about 28,550,000 miles away from the sun, while at other times it is about 43,350,000 miles away. Its diameter is approximately the same as the width of the Atlantic Ocean. Mercury has a surface temperature of about – 9.6 degrees Fahrenheit on its dark side. The planet is surrounded by a thin atmosphere of carbon dioxide. Mercury's pull of gravity is much weaker than Earth's. A man weighing 140 pounds on Earth would weigh about 50 pounds on Mercury. He would be able to jump over an elephant.

VENUS is the second planet from the sun and the sixth largest. It is named the Evening Star because it shines brighter than any other planet or star. Venus travels around the sun in a circular orbit. It revolves around the sun in 225 days. It is almost the same size as the Earth, its diameter is 7,600 miles. Venus' pull of gravity is also nearly the same as the Earth's. A man who weighed 150 pounds on Earth would weigh about 135 pounds on Venus. Venus is the planet closest to the Earth, surrounded by a layer of white clouds composed of sulfuric acid. These clouds completely obscure our view of the surface. A glowing halo of hydrogen surrounds Venus, about 1,800 miles above its surface. The atmosphere of Venus is made up chiefly of carbon dioxide gas. The surface temperature of the planet is about 536 degrees (hot enough to melt lead).

EARTH is the third planet from the sun, about 93 million miles from the sun. It is the fifth largest planet and the densest major body in the solar system. Simply put, the Earth is a ball of rock and metal that is about 4.5 to 4.6 billion years old. The estimated weight of our planet is 6,580,000,000,000,000,000,000 tons.

Three-fourths of Earth's surface is water. One-fifth of the land surface is mountains. Earth travels around the sun in a circular orbit. It takes about 365 days to make one orbit of the sun. It takes almost 24 hours to rotate once on its axis. It is surrounded by a thin atmosphere of nitrogen and oxygen. The oxygen in Earth's atmosphere is produced and maintained by biological processes. Without life there would be no free oxygen. Earth's diameter is 7,913 miles. Everything falls to the ground, since there is a strong pull of gravity. The hottest recorded temperature on Earth is 58 degrees Celsius in Lybia which is 136.4 degrees Fahrenheit, and the coldest was -89 degrees Celsius (-128.2 degrees Fahrenheit) in Vostok, Antarctica where nobody lives. The lowest recorded temperature in an inhabited region is -72 degrees Celsius (-97.6 degrees Fahrenheit) in the Siberian Village of Oymyakon, Russia. The temperature of the Earth's core is believed to be 4,000 degrees Celsius.

MARS

MARS is the fourth planet from the sun, about 142 million miles away. It is the seventh largest planet; it takes 687 days to revolve around the sun. Its orbit is an ellipse. Mars is called the Red Planet because it shines in the sky with a reddish color. At its closest, it is 128,500,000 miles from the sun, at its farthest it is 154,500,000 miles. Mars is a small, rocky world that measures 4,200 miles in diameter. It contains the largest volcano in the solar system, Olympus Mons, which is 15 miles high, almost three times higher than Mount Everest, the highest mountain on Earth. Spirit and Opportunity are Rovers sent to Mars to see if life did exist on the planet. The Martian atmosphere is made up mainly of carbon dioxide, there is little if any oxygen. A man weighing 150 pounds on Earth would weigh 57 pounds on Mars. The hottest temperature on Mars is about 70 degrees Fahrenheit on a hot day at the equator, at the poles which are covered with ice in the winter, the temperatures drop to 150 degrees below zero.

JUPITER is the fifth planet from the sun, about 484 million miles away. It is the largest planet; its diameter is almost 11 times the size of the earth. Jupiter takes 11.9 years to revolve around the sun. The planet spins very quickly, it takes less than 10 hours to rotate once. Jupiter radiates more energy into space than it receives from the Sun. It is surrounded by clouds that are thousands of miles deep, composed of 90% hydrogen and 10% helium. The temperature at the top of the cloud layers is about 202 degrees below zero. The gases surrounding the planet are moving at different speeds, creating bands of colors in the cloud layers. The pull of gravity is about two and a half times stronger than Earth's. A man weighing 150 pounds on Earth would weigh about 397 pounds on Jupiter.

SATURN is the sixth planet from the sun, about 888 million miles away. It was the first known planet. Galileo was the first to observe it with a telescope in 1610, since it is a very large planet. It takes nearly 29 years to make one trip around the sun. Saturn is not quite as large as Jupiter. Its diameter is about nine times that of Earth. Saturn rotates very quickly. The planet takes only 10 hours and two minutes to turn once on its axis. Saturn's atmosphere is surrounded by 75% hydrogen and 25% helium. The winds on Saturn blow at about 900 miles an hour—about six times stronger than hurricane-force winds on Earth. Its gravitational pull is a little less than the pull of Earth. Saturn can be distinguished by the flat rings that circle its equator that seem to be composed primarily of water and ice, but they may also include rocky particles with icy coatings.

URANUS is the seventh planet from the sun, about 1.8 billion miles away. It is the third largest, about four times larger than Earth. A year on Uranus, or the time it takes to revolve around the sun, is 84 years on Earth. It spins on its side; its axis is tilted nearly 98 degrees. Observed through a telescope, Uranus appears to be slightly blue, the result of absorption of red light by methane in the upper atmosphere, surrounded by 11 rings that appear as a broad band of a silvery color around its equator. It is also surrounded by an atmosphere composed of 83% hydrogen, 15% helium and 2% methane. Uranus is composed primarily of rock and various ices. The gravitational pull is a little more than Earth's. A man weighing 150 pounds on Earth would weigh 162 pounds on Uranus. It is believed to be very hot beneath its atmosphere, but its exact surface temperature is not known.

NEPTUNE is the eighth planet from the sun, about 2.8 billion miles away from it. Pluto's orbit is so eccentric, it sometimes crosses the orbit of Neptune making Neptune the most distant planet from the sun for a few years. Its diameter is 30,603 miles, and it takes 165 years to revolve around the sun. Neptune has rapid winds confined to bands of latitude; Neptune's winds are the fastest in the solar system, reaching 2000 km/hour. Like Jupiter and Saturn, Neptune has an internal heat source— it radiates more than twice as much energy as it receives from the sun. Through the telescope, it looks like a twin of Uranus. It has almost the same diameter and it also appears as a dim blue-green object with a few bands of color around it. Few details can be seen with a telescope. Scientists believe that the surface of the planet is very hot. A man weighing 150 pounds on Earth would weigh about 165 pounds on Neptune.

PLUTO is the planet furthest away from the sun, about 3.6 billion miles away. It is the only planet that has not been visited by a spacecraft. Its diameter is half the width of the earth. Any atmosphere surrounding the planet must be frozen. The temperatures probably never rise beyond 348 degrees below zero. Pluto's orbit is a long ellipse. It rotates in the opposite direction from most of the other planets. Sometimes Pluto is as far as 4,534,000,000 miles away from the sun and it comes as close 2,758,000,000 miles to the sun. This means that at times Pluto is about 35,000,000 miles closer to the sun than Neptune. Pluto's journey around the sun takes about 248.4 earth years.

Sign Language

The American Manual Alphabet

Drawings show a side view. In actual practice the letters should face the persons with whom you are communicating.

522

BRAILLE

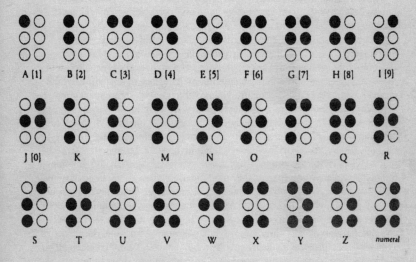

A [1] B [2] C [3] D [4] E [5] F [6] G [7] H [8] I [9]

J [0] K L M N O P Q R

S T U V W X Y Z numeral

The numeral symbol indicates that the letters A–J are being used as digits.

"There is a wonder in reading Braille that the sighted will never know: to touch words and have them touch you back."
Jack Fiebig

The Longest Word
in the English Language

Methionylglutaminylarginyltyrosylglutamylserylleucylphenylalanylalanyl-
glutaminylleucyllysylglutamylarginyllysylglutamylglysylalanylphenylalanylvalyl-
Prolylphenylalanylyalylthreonylleucylglcycylaspartylprolylglicylisoleucy-
glutamylglutaminlserylleucyllysylisoleucylaspartylthreonylleucylisoleu-
cylglutamylalanylglyclyalanylaspartylalanylleucyglutamylleucylgluycyliso-
leucylproluylphenylalanyserylaspartyprolylleucelalanylaspartylglycylprolylthreo-
Nylisolleucyglutaminylasparaginylalanythreonylleucylarginylalanylphenylalany-
Lalanylalanylglycylvalylthreonylprolylalanylglutaminylcysteinylphenylalanyl-
Glglutamylmethionylleucyalanylleucylisoleucylarginylglutaminyllysylhistidyl-
Prolyuthreonylisoleucylprolylisoleucylglycylleucylleucylmethionyltyrosylalany-
Lasbaraginylleucylvalylphenylalanylsparaginyyllysylglycylisoleucylaspartyl-
Glutamylphenylalanylyltyrosylalanylglutaminylcysteinylglutamyllysylvalyl-
Glycylvalylspartylserylvalylleucylvallalanylaspartylvalylprolylvalvlglutaminyl-
Glutamylserylalanylprolylphenylalalrginylglutaminylalanylalanylleucylarginyl-
Histidylasparaginylvalylalalprolylisoleucylphenylalanylisoleucylcysteinylproly-
Prolylaspartylalanylaspartylaspartysspartyleucylleucylarginylglutaminylisoleucy-
Lalanylseryltyroslglycylarginylglycyltyrosylthreonyltyrosylleucylleucylsery-
larginlalanylglycylvalylthreonylglycylalanylglutamylasparaginylarginyla-
nylalanylleucylprolylleucylaspaaginylhistidylleucylvalylalanyllysylleucylly-
sylglutamyltyrosylasparagimylalanylalanyprolylprolylleucylglutaminylglycyl-
phenlalanylglycylisoleyucylserylalanylprolylaspartylglutaminylvalyllysylalany-
lalanylisoleucylalspartylalanylglycylalanylalanylglycylalanylasoleucylse-
rylglycylserylalanylisoleucylbalyllysylisoleucylisoleucylglutamylglutaminyl-
histidylasparaginylisoleucylglutamylpronylglu-0-tamyllysylmethionylluecy-
lalanylalanyoeucyllysylvalylphenylalanylvalylglutamilylprolylmethionylly-
sylalanylalanylthreonylarginylserine.

This word consists of 1,909 letters. It is the term for the formula
$C_{1289}H_{2051}N_{343}O_{375}S_8$. A Tryptophan synthetase A protein,
an enzyme that has 267 amino acids.